Essentials of Practical Microtechnique

Essentials of Practical Microtechnique

by The Late ALBERT E. GALIGHER

and

EUGENE N. KOZLOFF, Ph.D.

Professor of Zoology and
Resident Associate Director
Friday Harbor Laboratories
University of Washington

SECOND EDITION

75 Illustrations

LEA & FEBIGER Philadelphia · 1971

ISBN: 0-8121-0356-4

Library of Congress Catalog Card Number: 70-152023

Published in Great Britain by Henry Kimpton Publishers, London

PRINTED IN THE UNITED STATES OF AMERICA

Preface

In the seven years since the First Edition of this book was published, I have received suggestions, queries, and criticisms from many persons. Their help and encouragement, given freely and with a view to improving the book and extending its usefulness, are sincerely appreciated. Dr. Donald J. Zinn, of the University of Rhode Island, deserves special recognition for his efforts along this line.

Of the numerous techniques proposed for inclusion in the Second Edition, I have added a few that are of particular importance. It seemed best to omit methods that I was unable to try, or that I tried and found to be no more satisfactory than some that have served the same purposes faithfully for many years. As in the First Edition, priority has been given to widely accepted procedures that teach the basic principles of microtechnique. I have, in general, avoided methods whose utility is limited to demonstrating specific chemical constituents of cells and tissues, because these should be learned with the help of books that treat histochemistry thoroughly. I have also given minimal attention to the use of fugitive dyes and "short methods" which, although they may quickly give pathologists, parasitologists, and biologists in some other fields the information they need, often do not yield preparations of really high quality or of permanence. Some more or less obsolete techniques have been withdrawn.

I hope that users of this book will find that the new chapter on embedding in Epon and sectioning with glass knives compensates for some of the deficiencies they may discover in other portions. Although it has been written with the preparation of material for light microscopy in mind, it will enable students to gain some experience with the procedures required for electron microscopy. Two of my associates in the University of Washington gave valuable assistance with this chapter: Mr. John Boykin helped me to learn the Epon method, and Dr. Richard Cloney provided much information that is needed to solve the logistic problems connected with it. I must also give credit to the firm of Ivan Sorvall, Inc. for permitting me to draw freely from literature it has published about the use of its ultramicrotomes.

I am indebted to a number of persons for help in the preparation of new manuscript material, but especially to my daughter, Rae. My wife, Anne, again managed the enterprise of proofreading. Mrs. Susan Congleton made most of the new line drawings, but preparatory work on some of them was done by Mrs. Janet Mackas.

The editorial staff of Lea & Febiger made every effort to improve the text of both old and new portions, and was patient in dealing with an author who changed his mind so often.

Eugene N. Kozloff

Friday Harbor, Washington

Preface to First Edition

When Albert E. Galigher published his *Essentials of Practical Microtechnique* in 1934, the reputation of his commercial laboratory in Berkeley was already well established. As a biologist, he knew what each slide preparation should show, and he did not compromise for less than the best. With a view to improving and standardizing important technical procedures, Mr. Galigher devoted considerable time to experimentation and to comparing related methods. The usefulness of his book was due in large part to the fact that he had thoroughly tested every method he described.

Essentials of Practical Microtechnique fell out of print within a rather short time, but a few stain-spattered copies continue to serve the original owners or their beneficiaries. A somewhat oblique but nevertheless warm testimonial to its value came from a fellow zoologist who told me it was the only book which had been stolen from him twice.

Before his passing in 1960, Mr. Galigher had revised and enlarged over half of the text of his book. The unfinished manuscript was placed in my hands for further revision, and I was also charged with the responsibility for preparing a new version of remaining portions. I have preserved the general format originally adopted by Mr. Galigher. The various methods are introduced by a discussion of the principles involved, and the actual steps in preparing material are usually consecutively numbered or lettered. Explanations are inserted wherever necessary. The results which should be obtained by each method are described, and in many cases illustrated by a photomicrograph.

In the Preface to *Essentials of Practical Microtechnique*, Mr. Galigher acknowledged the valuable suggestions and criticisms offered by Dr. James E. Lynch of the University of Washington. I also offer sincere thanks to Dr. Lynch. His own skill and wide experience with procedures of microtechnique made him the logical person to read the entire manuscript. A number of additions which he proposed will prove useful to many workers. Dr. Lynch's thoroughness in weeding out errors and inconsistencies has undoubtedly spared me considerable embarrassment. However, I must assume responsibility for any mistakes which remain, and I shall be grateful if even minor errors are brought to my attention.

It is a pleasure to acknowledge the helpful courtesies extended to me by Mrs. Galigher and by her daughter, Mrs. Mary Groesbeck. Dr. Charles E. Moritz, of the editorial staff of Lea & Febiger, prodded only very gently, and he was quick to offer counsel whenever I asked for it. Many colleagues and industrial firms helped me with specific little problems. The line drawings were prepared by Mrs. Kay Bittick of the Department of Medical Illustration, University of Oregon Medical School. A number of manufacturers generously supplied photographs or other illustrations of equipment, and the sources are acknowledged in connection with the figures.

Over a period of nearly two years, five nimble typists worked at the task of converting hundreds of pages of ugly handwriting into legible copy. I thank each one of them: Miss Gail Burton, Miss Holly Heid, Miss Jill Retzloff, Miss Martha Robinson, and Mrs. Sue Kingsland. My wife, Anne, worked at correcting the proofs and made this aspect of getting a book before the public slightly more attractive.

EUGENE N. KOZLOFF

Portland, Oregon

Contents

CHAPTER 1

Introduction

Although speculations on the subject of light and optics date back to the ancient Greeks, the use of lenses for practical purposes began in the thirteenth century, after Roger Bacon called attention to the fact that a curved piece of glass "makes things appear larger." The introduction of spectacles led to the development of more powerful lenses, and eventually to the rude instruments of the early microscopists. Through generations of painstaking work, the light microscope has been brought to its present high standard of perfection. Its lenses are practically flawless and its mechanical parts are finished with great precision, so that the optical system functions with an efficiency which approaches the limits theoretically possible with visible light. As the microscope has been improved for study of the physical aspects of life, methods of preparing organisms or their parts for microscopic examination have likewise been refined.

Microtechnique, or *microscopical technique*, in the broadest sense of the term, embraces all methods of studying specimens with the microscope. It therefore includes manipulation of the instrument itself, as well as procedures for staining and mounting specimens, cutting them into thin sections, or preparing them in a variety of other ways for observation with the microscope.

The purpose of this book is to describe and to explain, insofar as it is possible to do this in limited space, a selection of well-proved methods for preparing zoological materials. Chapter 5 deals briefly with illuminating and examining specimens with the microscope.

ORIGIN OF MODERN MICROTECHNIQUE

In the last quarter of the eighteenth century, English microscopists began to take an interest in making preparations of specimens for study, and most of the progress in microtechnique, until about the middle of the nineteenth century, can be credited to them. Many of the English microscopists of those times were curiosity-seekers, rather than serious scientific investigators. Their methods had little influence upon the growth of scientific biology until the latter half of the nineteenth century, when they were adopted—in some cases rather reluctantly—by European biologists.

By the middle of the nineteenth century, the compound microscope had been developed to a form roughly resembling that of modern instruments, and the foundations of modern biology had been strengthened by the enunciation of the Cell Theory by Schleiden and Schwann. Up to this time, the biologists of Germany, who were by far the most active investigators in the field, had employed only rather primitive methods of making microscopical preparations. They had studied the anatomy of small organisms from intact specimens, or by means of dissection, and had succeeded in observing many of the microscopic structures of the tissues of larger organisms by tearing off small shreds of tissue, or by dissecting out small portions and then teasing them apart or crushing them in a "compressorium." In the years following 1850, it gradually became apparent to German investigators that further progress would require more refined methods of microtechnique. They began to draw upon the methods already in use in England, or to arrive at them independently, and soon improved these techniques and invented new ones.

BASIC PROBLEMS AND METHODS OF MICROTECHNIQUE

In preparing biological specimens for detailed study with the microscope, three primary problems present themselves. These problems, and the methods which have been devised to overcome them, are outlined below.

Subdivision

The compound microscope is ordinarily used to form an image from light which has been transmitted through the specimen. Therefore it is essential that the object to be examined be thin and translucent. This may be the case with intact protozoans, rotifers, and other small organisms. However, the tissues of larger organisms, being organized in dense and opaque masses, must be subdivided into thin portions through which light will pass. Subdivision is necessary also because microscope lenses have little depth of focus, and the depth of focus decreases as the magnification increases. If the specimen is more than a few microns (1 micron = 0.001 millimeter) thick, the out-of-focus images of overlying and underlying parts may make it impossible to obtain a clear view of individual structures. For this reason, it is sometimes necessary to subdivide even relatively small and transparent organisms in order to study finer details. Methods of subdivision fall into three major categories.

Dissection. For study under low or moderate magnifications, the organs of small animals may be removed and mounted, either intact or after further dissection. For example, the nerve cord or one of the branching tracheal systems of an insect may be treated in this way.

Dissections which are too opaque for study by transmitted light are often studied to greater advantage with the aid of reflected light.

Spreading and Dissociation. STRETCHING. The time-honored method of preparing natural membranes, such as mesenteries and subcutaneous connective tissue, by stretching them into a thin layer, continues to be a valuable one for this special purpose. It is not, however, applicable to most other types of tissue.

SMEARING. This method is used principally for spreading out the cells of blood and other liquid or semi-liquid materials. It has many applications for the detection of parasitic microorganisms in blood, fecal material, and exudates. Smears may also be made from soft tissues such as the testis or bone marrow, and are commonly used in making counts of chromosomes and in studying various kinds of cell granules. Smearing is sometimes combined with dissociation.

DISSOCIATION. Dissociation includes several methods that originated early, and these techniques are designed to accomplish the separation of cells and intercellular substances from one another.

Teasing, or mechanical separation by means of fine instruments, was the principal method of dissociation in use to about 1865, and has been employed to the present day for fibrous tissues such as tendons and nerves.

Dissociation by chemical agents (sometimes called "maceration") provides a method of separating tissues composed of delicate cells held together by intercellular cements. Soaking small pieces of tissue in weak alcohol or certain other reagents was the best means for preparing some cells for study until the time when efficient sectioning methods came into use. It still continues to serve some useful purposes. Dissociation by means of digestive enzymes was introduced soon after the action of these substances was demonstrated, about 1880.

Corrosion, or the isolation of hard parts (such as scales or spicules) by destruction of the surrounding soft parts, has been employed in one way or another for a very long time.

Sectioning. Since about 1880, methods of cutting sections have gradually replaced smearing techniques and dissociation for the study of most types of material. The great advantage of sections is that they show the microscopic structures in their natural relations. Sections were employed in the study of animal histology as early as 1840, and it is known that, at least 70 years earlier, botanists commonly sectioned their material. The softness of most animal tissues offers a serious obstacle to cutting sections, which can be overcome only by hardening and supporting them in some way. Until about 1875, the method of hardening in general use was that of placing tissues in alcohol ("spirits of wine"), and only crude techniques for supporting them were employed. The sections made in early days were therefore too thick and uneven to demonstrate clearly fine details of structure. As more efficient methods of hardening and supporting tissues were introduced, improvements in methods for cutting sections were made also.

Microtomes, or machines for cutting thin sections, were used by English microscopists as early as 1770. These instruments, however, were used chiefly for sectioning plant stems, and were not adopted by German and other European scientists. The first microtome used in Germany was a simple, hand-held instrument invented by Oschatz about 1843. But until about 1865, German biologists continued to favor freehand sections. The advent of better staining methods emphasized the need for thinner and more uniform sections, and improved microtomes began to appear in Germany and in other European countries.

The practice of embedding tissues in nitrocellulose or paraffin developed side by side with improvements in prototypes of the modern sliding microtome. The invention of the rotary microtome shortly before 1890 greatly increased the usefulness of the paraffin method. Cutting sections with a rotary microtome is rapid, and the slight heat produced when the stationary knife cuts through the block of paraffin containing the tissue welds successive sections into a continuous ribbon, thus making it possible to mount sections in serial order.

The development of electron microscopy has required the application of embedding materials that are sufficiently hard so that sections a fraction of a micron in thickness can be cut with special microtomes. Several kinds of transparent plastics are used for this purpose. Because most of these materials also can be sectioned at a thickness of about 1 or 2μ, and because they infiltrate tissues without causing much distortion, they have achieved considerable importance in light microscopy.

Fixation and Hardening

Before tissues can be embedded in a medium that can be sectioned with precision, they must be prepared to withstand passage through a series of reagents. The term *fixation* is applied to the initial step in the process of preserving and hardening tissues. A fixing reagent, or *fixative*, must kill the cells effectively and rapidly, preserving the nucleus and other structures as faithfully as possible, and making these structures more receptive to stains. If certain chemical constituents are to be demonstrated, the fixative must preserve these also.

When it was finally realized that the most essential function of killing and hardening tissues is that of "fixing" the cells so that their morphology is modified as little as possible, progress was made in developing reagents which would serve this purpose. The old method of preserving specimens in alcohol was replaced by fixation in potassium dichromate followed by hardening in alcohol. This method was popular between 1860 and 1880.

By 1875 several fairly good selective stains had come into use, and the study of cell structure and cell division had become a most promising subject. Progress in cytology, however, could only be made through the development of good methods for preserving delicate cell structures. The concerted attack on this problem by cytologists soon brought revolutionary developments. Picric acid, chromic acid, acetic acid, formaldehyde, osmium tetroxide, and a number of other substances and mixtures were tried successfully. In the years following 1880, the term "fixation" was adopted for the killing and initial hardening of tissues, and the use of the term "hardening" was restricted to the subsequent hardening involved in preparing them for sectioning or other treatment.

Most of the fixative mixtures in common use today were formulated before 1920. Many of them seem to have survived simply because they are less inferior than others. Surprisingly little enthusiasm has been shown in the last several decades for working out better methods of fixation for histological and cytological materials. The innovations in techniques of fixation have been concerned largely with preservation and identification of specific substances within cells and tissues, and with preparing material for electron microscopy.

Embedding biological materials in relatively hard epoxy resins, methacrylates, and similar media preparatory to cutting ultra-thin sections has presented electron microscopists with various new problems of fixation. Electron microscopists have strived for excellent preservation of such structures as cell and nuclear membranes, mitochondria, endoplasmic reticulum, ribosomes, centrioles, and cilia. However, because electron microscopy is so completely different from light microscopy, methods of fixation developed for the former often are not applicable to the latter, at least as far as embedding tissues in paraffin or nitrocellulose is concerned. Nevertheless, some of the reagents rather recently introduced for electron microscopy (glutaraldehyde) deserve careful study in case any of their virtues can be applied to good advantage in procedures used for light microscopy. Electron microscopists have experimented extensively with buffering agents to control the pH of fixatives, and much of their success is related to this aspect. Techniques for light microscopy could undoubtedly be greatly improved by similar experimentation.

Optical Differentiation

Staining. With some exceptions, the microscopic structures of organisms possess insufficient color or other optical differentiation to render them plainly visible. In order to supplement what can be learned about these structures in living material, it is necessary to increase their visibility by artificial means. Apparently it was the work of Gerlach, published in 1858, which effectively called the attention of biologists to the fact that differentiation could be achieved by staining tissues with a dye which would color some of their cellular and intercellular structures more deeply than others. Gerlach used carmine, extracted from cochineal insects, for the differential staining of nuclei and other structures in human tissues. A number of investigators had previously used cochineal extracts for essentially the same purpose, but the publications of Gerlach first brought the practice of staining into general use. The rapid development of staining methods which took place in subsequent years will be mentioned later.

For several years, carmine was the only dye in regular use for biological staining. Aniline dyes first appeared on the market in 1860, and 2 years later Beneke applied one of them to histological staining. As more and more of these beautiful and brilliant dyes were manufactured, many of them were utilized by biologists.

Hematoxylin, extracted from logwood, was used more or less experimentally by several investigators soon after 1860, but its application in biological staining was not really successful until Böhmer discovered that the combination of this substance with an alum yielded a fine deep blue nuclear stain. Subsequent developments in staining with hematoxylin have largely been based upon this fact.

In 1867, Schwartz conceived the idea of using picric acid (a yellow dye) and carmine (red) to stain different structures of one specimen in contrasting colors. This was the beginning of "double staining," which has been given wide application, as in the use of hematoxylin (blue) and eosin (pink). In 1891, Flemming introduced his famous "triple stain," a combination of safranin (a red nuclear stain), crystal violet (a purple stain for mitotic spindles), and orange G (a yellow cytoplasm stain).

The use of a metallic compound for staining was first described by von Recklinghausen in 1862. He developed a method for outlining cells by treating them with silver nitrate in the presence of strong light. A few years later, gold chloride was applied in a technique for demonstrating nerve endings. Elaborate methods of impregnating nerve cells and their processes with salts of silver and mercury were eventually devised, especially by Golgi. The methods of Golgi, supplemented by many more recent ones, have contributed a great deal to our understanding of the structure of nerves, and of the morphological relationships of nerve endings to effector organs.

Mounting. Mounting specimens in media which increase their transparency, or which, because of their refractive index, make certain details more clearly visible, aids greatly in studying the material with transmitted light. In the case of small organisms, the use of an appropriate mounting medium may obviate the necessity for dissecting or sectioning them, except for working out certain fine details.

Until 1832, when Prichard suggested the use of a mixture of a gum and isinglass, specimens which could withstand desiccation reasonably well were mounted dry. Canada balsam was introduced in 1835, and for a time was used principally for mounting sections of stems and other specimens which first could be dried. It was not until 1851, when Clarke worked out a method for dehydrating specimens in alcohol and then clearing them in turpentine, that balsam became a practical mounting medium for soft animal tissues. Balsam and various other natural and synthetic resins are the most generally useful mounting media today. However, various mixtures containing glycerol, gelatin, and water-soluble gums, or combinations of these, are commonly employed for mounting materials which, for one reason or another, cannot be mounted in resinous media.

SOME GENERAL REMARKS ON THE PREPARATION AND STUDY OF MICROSCOPICAL MATERIAL

Although modern methods of microtechnique enable us to differentiate structures which may otherwise not be clearly visible, most preparations must be studied with caution and a certain amount of suspicion. Fixation, dehydration, staining, clearing, and mounting all require treatment of cells and tissues with reagents which will alter various cellular or intercellular substances. The amount and the kind of alteration which particular structures may undergo obviously depend upon the ways in which the reagents used in the course of making preparations affect their physical and chemical natures. For example, lipid inclusions will be removed completely by solvents such as toluene. Shrinkage and other forms of distortion affecting whole cells or structures within cells are also among the more common causes of alteration. Some reagents may induce the formation of precipitates or coagulates, and these *artifacts* may be misconstrued as definite structures.

An important matter which should be kept in mind is that a choice of methods should be made with reference to what is known or suspected concerning the nature of the structures to be demonstrated, and that alterations likely to occur during technical procedures should be taken into consideration when conclusions are to be drawn from a study of the preparations. If good judgment is exercised in selecting and carrying out methods of fixation and other techniques, alteration may be limited and a majority of structures may be demonstrated with remarkable fidelity. Carelessness in selection and execution of methods may result in failure to bring out important structural details, or may encourage distortion and the formation of artifacts, and will therefore inevitably lead to false conclusions. The reliability of observations based on fixed and stained material is therefore to a very large extent dependent upon the judgment and skill applied in making the preparations.

CHAPTER 2

Laboratory Equipment and Supplies

At least certain portions of this chapter should prove useful to biologists responsible for organization of laboratories and for selection of equipment and expendable supplies. It may also be of help to investigators or students learning more or less independently to prepare materials by a wide variety of methods. It gives suggestions not only concerning major items of equipment, but also concerning glassware, reagents, other supplies, and reference books which nearly every laboratory is likely to need.

LOCATION AND FURNISHING OF THE LABORATORY

The room used for microtechnical work must be well lighted and ventilated. A room having windows on two or more sides, one of which faces north, is perhaps ideal for the purpose. It is important that the laboratory be located where it will remain as cool as possible during the summer, and also that adequate facilities be provided for heating it during the winter. Extremes of temperature may influence certain procedures unfavorably.

The laboratory should contain at least one large sink, to which is attached a long draining board with a gentle slope. It is helpful to have a water pipe (about a foot above the draining board and parallel to it) provided with small outlet taps at intervals of about 6 inches. This arrangement makes it possible to conduct simultaneously several operations requiring running water.

The provisions for a supply of distilled water will vary in different laboratories. If distilled water is not piped into the laboratory, it may be convenient to operate a small still, preferably elevated above or to one side of the sink, with a glass or polyethylene carboy beneath it for accumulating and dispensing the water.

Supplies of filtered tap water or spring water, stored in glass or polyethylene jugs, will be useful for washing filamentous organisms or microscopic organisms. The air bubbles in water freshly drawn from a tap are a nuisance when they accumulate in such material or on the surface of dishes used for examination.

The laboratory should be furnished with several stout tables to hold microtomes and other apparatus for general use, and shelves or cabinets for reagent bottles, apparatus, and books. For the use of each student there should be a large table, at least 30 by 60 inches and of convenient height (about 30 inches). If the tables do not have drawers provided with locks, a locker should be available for each person. The work tables should be arranged in such a way that they receive adequate but not excessive illumination.

On or near each table there should be an outlet for electricity and, if possible, an outlet for gas. Durable polyethylene wastebaskets or pails should be provided, and the room should contain at least one fire extinguisher. A large wall clock with a sweep second hand is a decided asset.

REFERENCES

Only a few particularly important reference works will be mentioned. At least some of these should be available for consultation in the laboratory.

Baker, J. R., 1958. *Principles of Biological Microtechnique.* New York: Barnes and Noble, Inc. A thorough treatment of principles involved in fixation, embedding, staining, and mounting; not intended to be a formulary or manual, however.

Baker, J. R., 1966. *Cytological Technique*, 5th ed. New York: Barnes and Noble, Inc Similar to the above, but more brief.

Conn, H. J. (Editor), 1969. *Biological Stains*, 8th ed., revised by R. D. Lillie. Baltimore: The Williams & Wilkins Co. Gives detailed information concerning properties of dyes used for biological staining.

Gurr, E., 1960. *Encyclopaedia of Microscopic Stains.* Baltimore: The Williams & Wilkins Co. Gives much useful information concerning properties and applications of dyes.

Stecher, P. G., Windholz, M., and D. S. Leahy (Editors), 1968. *The Merck Index and Encyclopedia of Chemicals and Drugs*, 8th ed. Rahway, New Jersey: Merck & Co. Perhaps the best single book for information concerning physical constants, physiological effects, and other properties of many chemicals, drugs, anesthetics, and pharmaceuticals which a technician may have occasion to use.

Stain Technology (periodical, 1926–). Baltimore: The Williams & Wilkins Co. A journal devoted to recording improvements and new or modified methods in microtechnique.

APPARATUS AND SUPPLIES

The following lists of apparatus, chemicals, and other supplies have been compiled to meet the requirements of students intending to perform most of the basic techniques described in this book. They do not include reagents needed for some special methods. It will be a simple matter to modify these lists if certain methods are to be omitted, or if other methods are to be added. In this chapter, and occasionally elsewhere in the book, suggestions will be given concerning the selection of equipment and supplies.

Equipment for Individual Use

COMPOUND MICROSCOPE. A moderate-priced instrument of any dependable manufacturer is suitable for microtechnical work. It should be equipped with a substage condenser and at least a 10× ("low power"; equivalent focus 16 mm.) and a 40× ("high power"; equivalent focus 4 mm.) objective. A very low-power objective (about 3.5×) of long working distance is also desirable, and an oil immersion objective (about 90×; equivalent focus 2 mm.) will be necessary if the quality of certain cytological preparations is to be assessed in the laboratory as they are completed. Eyepieces of 5× and 10× (or 8×) are satisfactory. The practice of relegating worn-out or obsolete microscopes to the microtechnique laboratory is rather unwise, because staining processes cannot be controlled successfully without the aid of fairly good optical and mechanical equipment. If the stage of the microscope is protected by a thin glass plate of proper size, and precautions are taken to avoid touching the objectives and other parts with reagents, no damage is likely to be incurred.

MICROSCOPE LAMP. An artificial source of light is not only convenient, because of its dependability, but it is almost indispensable in localities subject to much cloudy weather. Some suggestions which may prove helpful in choosing a microscope lamp are given in Chapter 5.

MICROTOME KNIFE. If possible, each student should use, sharpen, and be responsible for a separate knife. Knives used promiscuously soon become damaged, and are a detriment to the work of everyone. There are microtome knives of many sizes, shapes, and grades of steel, as well as adapters for safety razor blades. The very important matter of selecting knives and caring for them is discussed in Chapter 13.

GLASSWARE. The following items of glassware should be available in any laboratory where a variety of microtechnical procedures are to be carried out. The number each student is likely to need is suggested.

6 to 8 Stender dishes; assorted sizes ranging from about 35 to 60 mm. in diameter.

18 to 24 Staining jars. Coplin jars (Coplin dishes) are probably the most suitable containers for handling small numbers of slides, despite the fact that they are rather difficult to clean. Other types, including some inexpensive substitutes are discussed on page 247.

8 to 12 Syracuse dishes (Syracuse "watch-glasses"). These are convenient for observing and selecting material under a dissecting microscope, and also for embedding. If they are to be used for embedding, the inside diameter should be a trifle larger at the top than at the bottom; otherwise it will be difficult to remove paraffin blocks from them. Avoid very shallow dishes of this sort.

3 Graduated cylinders; at least one of 10 ml. capacity, one of 100 ml. capacity, one of 1000 ml. capacity; other sizes, especially 25 ml. and 50 ml., will also prove useful.

2 Beakers, low form with lip; one of about 150 ml. capacity, one of 400 or 500 ml. capacity.

2 Glass funnels; one about 2½ inches, one about 5 inches in diameter.

5 Bottles, narrow-mouthed, preferably with glass stoppers, capacity about 1,000 ml. These are for storage of stocks of distilled water, absolute alcohol, 95% alcohol, 70% alcohol, and toluene or xylene.

15 Bottles, narrow-mouthed, preferably with glass stoppers, capacity 100 to 250 ml. These are convenient for alcohols, clearing agents, fixatives, and staining solutions used in moderate quantities.

3 Bottles, wide-mouthed, with glass stoppers, capacity about 100 or 125 ml. These are useful for solutions of nitrocellulose.

3 Bottles, narrow-mouthed, with glass stoppers (or pipettes), capacity about 50 or 60 ml. These are for liquids such as tincture of iodine, concentrated hydrochloric acid, glycerol, or other reagents likely to be needed in very small amounts.

2 Balsam bottles, with glass rods.

12 Shell vials, with corks to fit; 15 × 50 mm. and 30 × 90 mm. are convenient sizes. Shell vials have straight sides and, for the majority of purposes, are preferable to "homeopathic" vials with constricted necks.

12 Specimen jars, with screw caps; 1 and 2 ounce capacity. Such jars are preferable to corked bottles, particularly for handling material which is to be dehydrated and infiltrated with paraffin or nitrocellulose.

1 Tissue-washing apparatus, as illustrated on page 135.

Ribbon Trays (for storing paraffin sections). Cardboard boxes (with covers), about 8 × 10 inches, and not more than about ¾-inch deep, will serve very well, provided their bottom surface is smooth. Ribbon trays may also be made from sheets of stiff cardboard and narrow strips of wood.

Slide Boxes. Several boxes with a capacity of 25 or 100 slides should be available. It is true economy to purchase well-made boxes. Cheap boxes are likely to warp and fall to pieces, or they may be made so inaccurately that slides either will not fit into the grooves or will fall out of them. If boxes with a capacity of 25 slides are used, those with slip-over covers are preferable to those with fit-in covers. Slides can be removed from them more easily, and the boxes (especially the covers) are generally more durable. However, beware of boxes in which the top and bottom are of equal depth, and therefore indistinguishable until the box is opened! The storage and arrangement of slide collections are discussed in Chapter 28.

Dissecting Instruments. These must be of good quality steel. The following list of essentials may be supplemented as necessary.

1 Scalpel, preferably with a steel handle. A Bard-Parker scalpel, or a similar type which accommodates removable blades manufactured in various shapes, is desirable but more expensive.

- 1 pair Scissors, with heavy blades, for general work and cutting labels.
- 1 pair Scissors, with fine-pointed blades, for delicate dissections.
- 1 pair Forceps, with heavy straight tips.
- 1 pair Forceps, with fine curved tips.
- 2 Needles, mounted in wooden handles.
- 1 Section-lifter, medium size.

MISCELLANEOUS SMALL ITEMS.

- 1 Thermometer, 0° to 100° C.
- 1 Bunsen burner, micro burner, or alcohol lamp. A gas micro burner is apt to be the most generally useful, but an alcohol lamp is essential for some techniques requiring a flame of relatively low temperature.
- 1 Wooden block, about 8 × 4 × 2 inches, into one side of which are drilled 2 or 3 rows of holes about ¾ inch deep. The holes, which are to serve as receptacles for shell vials, should be of several diameters, ranging from about ⅝ inch to 1¼ inches.
- 1 Strip of sheet cork (or thin, smooth, soft wood), size about 3 × 6 inches. This is to be cut up, as needed, and used as a substrate upon which pieces of tissue may be pinned out for fixation.
- 1 Bottle brush, medium size.
- 6 Medicine droppers.
- 1 Pipette, made by inserting a 4-inch length of glass tubing into a rubber bulb of the size used in small syringes, such as an ear syringe. This is very useful for withdrawing liquids quickly from containers of material.
- 2 Small camel's hair brushes (No. 4 or 5), for manipulating sections at the microtome.
- 1 Celluloid ruler (with metric and inch scales), about 6 inches long.
- 1 Wax pencil, for marking on glass.
- 100 Gummed labels, 1-inch square, for labeling slides.
- 50 Gummed labels, about 1 × 2 inches, for labeling bottles and dishes.
- 100 Cards (3 × 5 inches or 4 × 6 inches), or a notebook, for records.
- 5 Pieces of cloth for wiping slides and coverglasses. These must be soft and free of lint. New cloths must be washed thoroughly in order to remove the dressing from them.
- 1 package Filter paper, qualitative grade, diameter about 15 or 20 cm.
- 1 package Disposable laboratory wiping tissues.
- 1 booklet Lens paper.
- 2 gross Microscope slides, 1 × 3 inches.
- ½ ounce Coverglasses, square, 22 mm., No. 1 thickness.
- 1 ounce Coverglasses, circular, 22 mm., No. 1 thickness.
- ½ ounce Coverglasses, circular, 18 mm., No. 1 thickness.
- 1 ounce Coverglasses, rectangular, 22 × 50 mm., No. 1 thickness.

Equipment for General Use

ROTARY AND SLIDING MICROTOMES. One or more instruments of each type should be available, depending upon the number of workers in the laboratory. The selection and care of these instruments, which are matters of much importance, will be discussed at length in Chapter 13. A number of circular metal object holders (Fig. 34, p. 231) of assorted sizes, or small blocks or pegs of hardwood, should be provided for use with the rotary microtome. Blocks of compressed fiber, slotted on one side (Fig. 40, p. 259), are the best object holders for tissues that are to be embedded in nitrocellulose and sectioned on a sliding microtome. Several sizes, from about $\frac{1}{2} \times \frac{1}{2} \times \frac{3}{4}$ inch to $1 \times 1 \times \frac{3}{4}$ inch should be available.

FREEZING APPARATUS. A very practical type is one which utilizes liquid carbon dioxide and which can be clamped into a standard sliding microtome or into a so-called clinical microtome.

DISSECTING MICROSCOPES. If possible, wide-field binocular microscopes should be provided for use in finding small organisms, dissecting, inspecting sections, and examining material to check on the progress of staining and destaining. If the strictest economy is necessary, simple magnifying lenses can be made to serve in some situations.

SLIDE TURNTABLE. If mounts are to be made in non-resinous media, a turntable for ringing the mounts with gold-size or other cements will be desirable.

ELECTRIC HOT PLATE. A small hot plate (with enclosed heating element) is desirable, though not indispensable.

CENTRIFUGE. This is an optional but very useful piece of equipment for concentrating small specimens, such as protozoa. A hand centrifuge will probably be adequate for most purposes.

INCUBATORS. A bacteriological incubator is most convenient for drying slides to which paraffin sections have been affixed, and also for use in certain procedures of fixation and staining.

PARAFFIN OVEN. Various types of apparatus can be used for keeping melted paraffin at a constant temperature. Any electric oven which has an accurate thermostatic control will serve the purpose, provided it quickly re-establishes the temperature for which it is set after the door is opened and closed.

RECEPTACLES FOR MELTED PARAFFIN. There should be at least two or three metal receptacles for each grade of paraffin. Some of the more elaborate ovens have special paraffin reservoirs. Cheap coffee percolators, from which the inner fittings have been removed, are excellent. Metal food cans of small or medium size are perfectly acceptable; they can be provided with spouts by pinching their rims. Glass receptacles are too fragile to be recommended for stocks of melted paraffin.

PARAFFIN FILTER. A copper funnel with a hot-water jacket is most convenient, but any metal funnel can be used if the paraffin is well heated before being poured into it.

EQUIPMENT FOR SHARPENING MICROTOME KNIVES. Individual preferences in this matter differ greatly. A discussion of some methods and apparatus for knife-sharpening will be found in Chapter 13. The least expensive outfit, and a very satisfactory one, consists of a fine-grained yellow Belgian hone, a smooth slate hone, and a strop of the finest calf skin mounted smoothly upon a wooden block. The selection of these items is considered in the chapter to which reference has been made.

BALANCES AND WEIGHTS. An inexpensive balance, such as a trip balance accurate to about 0.1 gram, should be kept in the laboratory, unless all reagents are prepared in a central storeroom. It is desirable that there be access to an analytical balance when preparation of certain solutions requires precision.

WARMING TABLE FOR STRETCHING PARAFFIN SECTIONS. An electric table with thermostatic control is desirable, but an inexpensive substitute can be improvised from a sheet of metal and an electric lamp (see p. 244).

DISSECTING BOARD OR TABLE. A very convenient form which can be made at little cost is illustrated on page 121. More elaborate types may be purchased.

DISSECTING AND INJECTING INSTRUMENTS. These include bone-cutting forceps, bone saw, cartilage shears, long forceps (8 to 10 inches) for reaching into specimen jars, a small (10 ml.) glass or hard-rubber injecting syringe, and a large (100 ml.) glass or hard-rubber injecting syringe. Metal syringes are satisfactory for injecting gelatinous media, but not for injecting fixatives.

STOCK BOTTLES FOR REAGENTS. There should be several dozen of these, ranging in capacity from about 100 ml. to about 2 liters or more. Those used to contain strong acids and other corrosive reagents should be of glass and must have glass stoppers. However, polyethylene bottles with screw caps serve very well for almost all other reagents. Distilled water and large quantities of physiological saline solutions are commonly stored in 5-gallon glass or polyethylene carboys fitted with siphons or faucets.

GRADUATED MEASURING OR SEROLOGICAL PIPETTES. These are needed for measuring the standard solutions used to control the hydrogen ion concentration of stains and in other procedures requiring a higher standard of accuracy than it is possible to achieve with a graduated cylinder. A few of 1 ml., 10 ml., and 25 ml. capacity should be available.

MISCELLANEOUS GLASSWARE and PORCELAIN. The following should be on hand:

glass stirring rods

glass tubing (small and medium sizes)

culture dishes

glass plates (various sizes, from about 3 × 4 inches to 10 × 10 inches)

The need for other glassware depends upon the nature of the work to be undertaken, and the size of the group doing it. Items generally required are:

beakers
Erlenmeyer flasks
test tubes
centrifuge tubes
Petri dishes (100 mm., 150 mm.)
large staining dishes (see p. 246)
crystallizing dishes
aquaria
bottles for specimens

Receptacles which are to be heated should be of resistant glass, such as "Pyrex." It is well to avoid glassware which is either unusually thin or very thick and clumsy.

A porcelain mortar and pestle, and one or two porcelain evaporating dishes having a diameter of about 5 inches, are likely to be necessary.

General Supplies. These include:

gum rubber tubing (or polyethylene tubing) of several sizes
cork and rubber stoppers
cork borer
copper wire gauze, coarse and fine
hacksaw blades, fine-toothed
small triangular files
diamond pencil for writing on glass slides
enamelware or polyethylene pans about 2½ inches deep
ring stands
tripod stands
water bath

Chemicals, Dyes, and Mounting Media

The following chemicals, dyes, and mounting media should be available in a typical laboratory, in which whole mounts and histological and cytological materials are to be prepared. Omissions or additional requirements will vary, as also will the quantities needed. As a general indication of the relative quantities of different substances required, those likely to be used in large quantities (several pounds or several gallons for a class of about 20 students) are marked with an asterisk (*), and those which are likely to be needed only in small quantities are marked with a dagger (†). Items not marked in either way probably will be required in moderate amounts (perhaps a pound or two for a class of about 20 students).

Any of the well known brands of laboratory chemicals may be used with confidence. Chemicals of the "reagent" or "c.p." grade should be employed for preparation of staining or fixing solutions. For dehydrating and clearing agents, the "U.S.P." grade is satisfactory. Cheaper "technical" grade chemicals are satisfactory for some purposes, as in the preparation of the potassium dichromate-sulfuric acid mixture for cleaning glassware.

Dehydrating Agents. ETHYL ALCOHOL (ETHANOL)*. This is generally used for dehydration. It may be purchased tax-free by educational institutions or hospitals, but the heavy tax levied on other consumers will probably make it necessary for them to use a substitute.

Commercial ethyl alcohol (approximately 95%) is employed in preparing all weaker solutions and most laboratories will find it economical to purchase it in large cans or barrels. Absolute ethyl alcohol is considerably more expensive, and absolute methyl alcohol may be substituted for it. Isopropyl alcohol may be substituted in certain procedures.

METHYL ALCOHOL (METHANOL)*. Methyl alcohol is inexpensive and tax-free. It is quite as satisfactory as ethyl alcohol for making up staining solutions, for dehydrating objects which are to be embedded by the paraffin or nitrocellulose method, and also for dehydrating specimens or sections before clearing and mounting in resinous media.

Absolute methyl alcohol is generally used for making all lower strengths. The fumes of methyl alcohol are toxic if inhaled in considerable quantities, but this substance may be used with safety if the room is well ventilated and if reasonable care is exercised to avoid inhaling the vapors.

ISOPROPYL ALCOHOL. This is inexpensive and tax-free, and may be used for dehydration procedures prior to embedding specimens in paraffin, or before clearing and mounting entire specimens or sections in resinous media. However, isopropyl alcohol cannot be used in the nitrocellulose method of embedding, and the low solubility of many standard dyes in isopropyl alcohol makes it unfit for preparation of staining solutions. In general, it is probably best not to use isopropyl alcohol in staining or destaining procedures.

When sold in bulk, isopropyl alcohol may contain a small amount of water. If so, a supply which is known to be absolute should be used for final dehydration, unless it is to be followed by a clearing agent such as terpineol-toluene which will absorb traces of water as it clears.

Clearing Agents. TOLUENE*; XYLENE. Toluene is the most useful single solvent for clearing specimens to be mounted in resinous media, for de-alcoholizing material to be embedded in paraffin, and for removing paraffin from sections mounted on slides. Xylene also clears well and is quite suitable for de-paraffining. However, as an intermediate between alcohol and paraffin, it is inferior to toluene, because it tends to make some tissues very hard and therefore more difficult to section.

TERPINEOL*; PHENOL; BEECHWOOD CREOSOTE. To clear whole specimens in which minute traces of water may remain, even after treatment with absolute alcohol, or for clearing nitrocellulose sections which cannot be passed through absolute alcohol because it will dissolve the supporting nitrocellulose, the addition of terpineol, phenol (crystals), or beechwood creosote is indicated. All of these are miscible with 95% alcohol, so if they are added to toluene or xylene, the mixtures will complete dehydration and de-alcoholization at the same time as they clear. Terpineol is recommended, for it is less dangerous to the skin and mucous membranes and its odor is more pleasing than that of either phenol or beechwood creosote.

GLYCEROL (GLYCERIN). Glycerol is valuable for dehydrating, clearing, and mounting certain types of materials and is an ingredient of certain mounting media and staining solutions. Glycerol is also applied as a thin coating on glass or porcelain dishes used for embedding specimens in paraffin. Most commercial preparations of glycerol have slight impurities and therefore are sold as glycerin. They are perfectly acceptable for microtechnique.

Embedding Media. PARAFFIN. Only paraffins known to be suitable for embedding should be used. For most purposes, a paraffin having a melting point of about 56° C., serves well. However, for certain purposes, a paraffin with a much lower melting point, about 45° C., is useful. For additional information concerning the selection of paraffins for embedding, see page 216.

NITROCELLULOSE. Parlodion (Mallinckrodt Chemical Works) and low-viscosity nitrocellulose (Hercules Powder Co.) are particularly well adapted for embedding purposes. For coating sections on slides, Parlodion is recommended.

Mounting Media. CANADA BALSAM. The resin of the balsam fir has long been employed for mounting sections or entire specimens. Being a natural substance and much affected by the way it is prepared for commerce, balsam varies greatly as a market product. Selection of samples suitable for microtechnical work is discussed on page 453.

SYNTHETIC RESINS. A number of good products are available. Because they preserve stains well and have little tendency to discolor, synthetic resins are now more widely used than natural resins. A list of trade names and manufacturers or suppliers is found on page 452.

Acids, Hydroxides, Salts, and Other Chemicals

ACIDS

acetic acid, glacial*
chromic acid
hydrochloric acid*
nitric acid
oxalic acid
phosphomolybdic acid
picric acid
sulfuric acid*
trichloroacetic acid

HYDROXIDES

ammonium hydroxide
potassium hydroxide
sodium hydroxide

Salts

aluminum ammonium sulfate
aluminum potassium sulfate
calcium carbonate
calcium chloride, anhydrous†
cobalt nitrate†
cupric sulfate†
ferric ammonium sulfate*
mercuric chloride*
mercuric oxide†
magnesium chloride
magnesium sulfate ("Epsom salts")
potassium phosphate, monobasic†
potassium chloride†
potassium dichromate*
potassium iodide†
silver nitrate†
sodium bicarbonate
sodium bisulfite, anhydrous†
sodium carbonate
sodium chloride*
sodium sulfite, anhydrous†
sodium tetraborate

Miscellaneous Chemicals

acetone
aniline oil
chloral hydrate†
chlorobutanol†
ether
formalin*
iodine (crystals)†
menthol (recrystallized)†

Dyes. It is a matter of utmost importance to obtain dyes which can be relied upon to stain tissue and cellular elements in a precise and reproducible manner. This fact was impressed upon American biologists during World War I, when German sources were cut off and many domestic substitutes proved unsuitable for microtechnical work. The manufacture of dependable biological dyes in America has been helped along by the *Commission on the Standardization of Biological Stains.* Samples of dyes approved by the Commission are almost always dependable when used for the purposes for which they are certified.

Recommended Dyes. Throughout the chapters on staining, suggestions will be given concerning the choice of dyes for certain purposes. Following is a list of dyes likely to be used in a course in microtechnique or in connection with more or less standard procedures employed in research laboratories.

acid fuchsin†
aniline blue, water soluble†
basic fuchsin†
carmine†
crystal violet†
eosin Y (eosin yellowish, soluble in water and alcohol)†
fast green FCF†
Giemsa's stain†
hematoxylin, light crystals†
Janus green B†
methyl green†
methylene blue (methylene blue chloride)†
neutral red†
orcein†
orange G†
safranin O†
Sudan IV (or Sudan black B)†

CHAPTER 3

Preparation for Laboratory Work

The importance of keeping equipment clean, protecting it from damage, and arranging it in a manner for convenient use cannot be overemphasized. A disorderly laboratory table reduces efficiency. Dirty slides and glassware, inaccurately labeled bottles, and mechanical devices which do not operate properly lead to further frustration.

CLEANING GLASSWARE

Foreign matter of most kinds can be removed by washing glassware with hot water and soap, or with one of a number of good cleaning agents recommended for laboratory glassware, and then rinsing and drying it.

Receptacles to which resins or paraffin have adhered must be soaked in toluene or xylene or some other solvent (waste toluene or xylene can be saved for this purpose) and then washed with hot water and soap.

Insoluble organic residues and precipitates of stains or metallic salts should be removed by soaking glassware in the potassium dichromate-sulfuric acid cleaning fluid described below. The time required for thorough cleaning varies from a few minutes to several days, after which the glassware must be rinsed thoroughly with water in order to free it from all traces of acid.

Potassium Dichromate-Sulfuric Acid Cleaning Mixture for Glassware

Potassium dichromate	200 gm.
Water	1 liter
Sulfuric acid, concentrated (cheapest grade)	750 ml.

Make the mixture in a vessel of heat-resistant glass. First dissolve the potassium dichromate in the water, warming this if you wish to accelerate the process. When the solution has cooled, slowly pour in the sulfuric acid, while stirring the mixture with a glass rod. This will generate some heat. When the solution has cooled again, store it in a glass-stoppered bottle. The mixture may be used repeatedly, until it becomes dark green in color.

CLEANING SLIDES AND COVERGLASSES

NEW SLIDES AND COVERGLASSES. Most microscope slides and coverglasses now on the market have been cleaned and then packaged to prevent them from becoming dusty or greasy. However, once the box has been opened and some slides or coverglasses removed, dust and lint are likely to settle on those remaining in the box. Accidentally touching them with one's fingers may leave traces of oil which also must be removed before they are used. As a general rule, it is a good idea to clean all slides and coverglasses again, except those which are carefully removed when the box is first opened.

To clean unused slides or coverglasses, allow them to stand for a time in 90 or 95% alcohol. Remove one slide or coverglass at a time and wipe it carefully on both sides with a soft, lint-free cloth. A clean, old handkerchief, a piece of old cotton shirt or bedsheet, or washed cheesecloth will be suitable. Public laundries can often supply bundles of small cloths torn from old sheets or toweling, and among these will usually be found some pieces which are ideal for wiping slides or coverglasses.

To wipe a coverglass, hold its edges between the thumb and first finger of the left hand. Cover the thumb and first finger of the right hand with the cloth and wipe both sides of the coverglass simultaneously, keeping the fingertips opposite each other and exerting only the gentlest pressure between them. At first, some breakage is certain to occur, but a careful worker will soon develop the necessary skill.

Slides and coverglasses which have not been thoroughly cleaned before packaging should be treated in the same way.

The effectiveness of cleaning in alcohol may be tested by placing a drop of water on one of the cleaned pieces and tilting it in various directions. Should the water spread evenly, the absence of grease is proved, and the remainder of the lot may be cleaned with confidence in the same way. If, on the contrary, the water forms droplets and ridges, grease is still present, and other means must be employed to remove it. Soaking them in a mixture of equal quantities of ether and absolute alcohol for 15 minutes is likely to be effective if alcohol alone is not.

Many workers prefer to clean slides and coverglasses as they are needed. Others clean a larger supply and repack them carefully so that they do not become dusty or greasy, or load them into special dispensers manufactured for this purpose. Dispensers certainly do offer some advantages, but somehow slides withdrawn from them often appear to be in need of a brief dip in alcohol followed by wiping with a lint-free cloth.

Used Slides and Coverglasses. If they have no resinous material adhering to them, dirty slides may be cleaned by abrasion with moistened scouring powder (such as Bon Ami), washing them thoroughly in water, and then allowing them to stand awhile in 90 or 95% alcohol. After being wiped dry in the conventional manner, they should be ready for use. If, for any reason, the slides are not properly cleaned by this method, they may be placed in the potassium dichromate-sulfuric acid cleaning solution for at least a day or two, washed in water, and then allowed to stand for an hour or more in water alkalized with a little ammonium hydroxide.

After this, they should be washed thoroughly in water, and then dipped in alcohol before being wiped dry.

Old preparations in resinous media may be soaked in waste toluene or xylene until the coverglasses separate from the slides. After treatment in one or more additional changes of solvent to remove as much of the resin as possible, the slides and coverglasses may be cleaned with the aid of scouring powder or the potassium dichromate-sulfuric acid cleaning solution, as described in the preceding paragraph.

In some laboratories staffed by expert personnel, cleaning old slides and coverglasses may prove costly by comparison with purchase of new ones. Moreover, slides and coverglasses from old permanent preparations are often pitted, dulled, or chipped to the extent that they simply are not worth the effort and expense of reclaiming them.

PREPARATION OF REAGENTS

The names and formulas of fixatives, stains, and all other reagents required for the various procedures described in this book are given in connection with the methods themselves. In the majority of college laboratories, certain reagents are mixed in large quantities and kept in stock bottles from which students and investigators fill their bottles and dishes as the need arises. Mixtures which do not keep well are usually prepared just before use, from dry chemicals or stock solutions of ingredients. A list of chemicals required for the preparation of most reagents necessary for the standard procedures described in this manual has been given in Chapter 2.

Rules for Measurement. It is necessary to measure the ingredients of solutions with sufficient accuracy to insure consistently good results. In preparation of most reagents used in microtechnical work, however, the degree of accuracy required in quantitative chemistry is unnecessary. Ordinarily, liquids may be measured in graduated cylinders, and solids may be weighed on trip balances accurate to 0.1 gram, or on simple balances equipped with a variety of brass and aluminum weights.

In the majority of formulas given in this book, the quantity of each substance is plainly stated in grams or milliliters. In certain cases, however, it is more convenient to refer to solutions in terms of percent concentration, either as grams of a solid per 100 ml. of solvent, or as milliliters of liquid mixed with another liquid to make a total of 100 ml. Thus, to make a 4% aqueous solution of sodium hydroxide, 4 gm. of sodium hydroxide are dissolved in 100 ml. of water. To make a 2% aqueous solution of hydrochloric acid, 2 ml. of concentrated hydrochloric acid are added to 98 ml. of water.

A quick method for diluting a solution to any lower percentage is to measure, in a graduated cylinder, the number of milliliters of the stock solution which is equal to the percentage desired, and then to add the solvent until the total volume equals the number representing the percentage of the stock solution. For example, to prepare 50% alcohol from 95% alcohol, pour 50 ml. of the latter into a graduated cylinder and add water to bring the level of the liquid to 95 ml. To make a 4% solution of sodium hydroxide from a 20% solution, use 4 ml. of the stock solution and add water to bring the total volume up to 20 ml.

LABELING BOTTLES

Gummed labels of paper or plastic are suitable. A size of about 1 × 2 inches is good for most purposes. It is preferable to use waterproof ink, India ink, or a typewriter for writing upon the labels. After paper labels have been applied to the bottles, it is a good idea to coat them with a thin coat of a waterproof varnish.

ARRANGING EQUIPMENT

For the sake of efficiency, the various pieces of glassware and apparatus used in microtechnique must be organized in an orderly manner. The arrangement illustrated in Figure 1 will prove quite suitable for most work of a rather general nature, and can be adapted to tables of various shapes and sizes and to those with or without sinks. Pieces of heavy wrapping paper or shelving paper may be ruled into squares so that bottles of alcohol and other reagents used in routine procedures have spaces in front of them for vials or dishes. This system allows the laboratory worker to keep track of material being processed. It minimizes the danger that a step will be accidentally overlooked, or that the worker will forget which reagent each dish or vial contains.

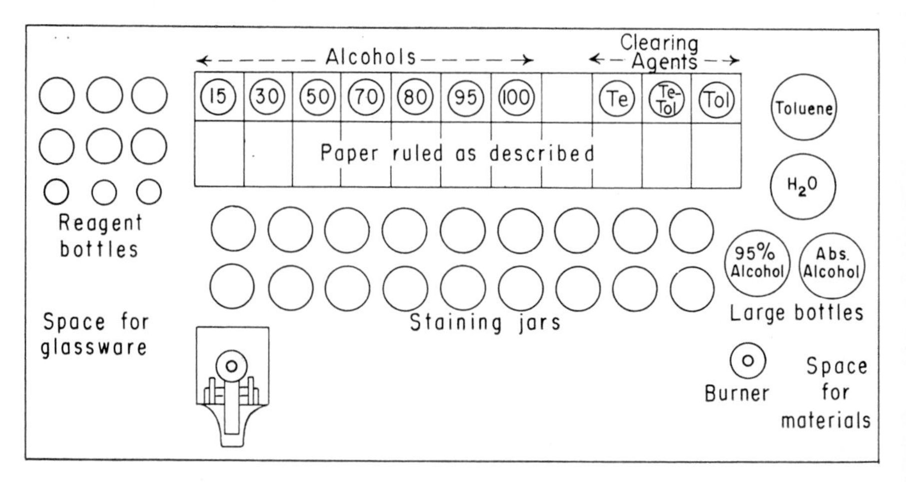

FIG. 1. A convenient arrangement of equipment on the work table.

CHAPTER 4

Important General Rules

Work Intelligently!

1. Before using a technique which is new to you, read the general explanation of the method and make certain that you understand the purposes for which the method is designed. This will enable you, later on, to select appropriate methods for various types of material.

2. Before performing each step, be sure that you understand how it is to be carried out.

3. Familiarize yourself with the histological or cytological structures which should be demonstrated by each preparation. This knowledge will help you to carry out the various operations with good judgment, and to evaluate the reliability of your preparations.

4. Keep an accurate record of each lot of material processed. These records can be kept conveniently on 3 × 5 inch or 4 × 6 inch cards, or in a small notebook, one card or page being devoted to each lot of material. A comprehensive record form, suitable for many types of preparations, may be worked out to meet the needs of a particular laboratory, technician, or investigator. At first, it is a good idea to record nearly every detail, such as the length of time the material remained in each grade of alcohol. After one becomes familiar with the routine methods, it is sufficient to record only the more important facts, such as the names of the fixatives and stains used, the time of exposure to each, the thickness of sections, and similar information. Note carefully any deviation of your finished preparations from the results expected.

Sometimes it is convenient to write down information concerning the various reagents employed and time of exposure on labels affixed to the dishes in which the specimens are handled, and later to copy the data into the permanent record.

5. Plan your work systematically. Make memoranda—on a calendar pad or in a notebook—of things to be done during future laboratory periods.

6. Label your finished preparations carefully. Upon each label note the number of the corresponding record card, the date, and other pertinent information suggested on page 466.

7. Do not be discouraged if your first efforts with any method are unsuccessful. Mastery of some procedures can be attained only through

careful study and experience. If you should fail with a particular method, make certain that you understand the principles upon which the method is based. Compare your record with the printed directions and try to locate the causes of failure. Then try again, modifying your procedures accordingly.

Work Carefully and Neatly!

8. Keep your equipment in order. Have a definite and convenient place for every article, and keep it there except when it is actually in use elsewhere.

9. Clean the lenses of your microscope frequently, using nothing but lens paper, aided if necessary by a few drops of distilled water or clean xylene or toluene. When examining wet slides, cover the stage of the microscope with a glass plate. When your microscope is not in use, keep it in its case or cover it so that it is protected from dust and moisture.

10. Keep all apparatus in good condition, so that you can use it to the best advantage. Keep the microtome clean and properly oiled. After using a microtome knife, carefully wipe it and replace it in its box. Keep your dissecting instruments clean and sharp. If an item of apparatus appears to be out of order, have it repaired before its malfunction leads to trouble and inefficiency.

11. When weighing chemicals, protect the pans of the balances with pieces of paper, counterbalancing these before adding chemicals or weights.

12. Do not use metal instruments for handling corrosive chemicals such as mercuric chloride and chromic acid. Use spatulas of porcelain, glass, plastic, horn, wood, or stiff paper.

13. Use only clean glassware. Clean all dirty glassware as promptly as possible. Never replace glassware in its customary place until it has been thoroughly cleaned.

14. Employ only reagents which are clean and pure enough for the purpose for which they are to be used. To prevent them from becoming contaminated, the bottles must always be tightly stoppered except when they are in use.

If a reagent becomes cloudy or forms a precipitate, consult the text or an instructor in order to determine whether it is fit for use. If it is still usable, filter it; otherwise, discard it at once.

15. Label reagent bottles fully and clearly. In the case of fixatives and stains, especially those which tend to deteriorate, write the date of mixing directly on the label.

16. If pipettes are used for handling reagents such as alcohol or toluene, label these with a narrow band of gummed paper or plastic tape encircling the tube close to the bulb. A convenient pipette rack may be made by cutting holes of the proper size in a small cardboard or wooden box.

17. Keep your staining jars covered when they are not in use. Filter their contents frequently, to remove precipitates and foreign matter. Immediately discard any reagents which become contaminated or otherwise unfit for use.

18. Keep bottles of material well stoppered, and dishes tightly covered.

19. Accurately label all receptacles containing materials being processed. It is best to use labels which are large enough to provide space for brief records. Write the number of the corresponding record card upon each label. For methods of storing material for considerable periods of time, see page 140.

20. Dispose of all solid waste matter in earthenware crocks or other appropriate receptacles, *not* in the sink. A good way to get rid of melted paraffin, or paraffin saturated with toluene, is to pour it into cans or paper cups which can be thrown away as soon as the paraffin solidifies.

21. Never hurry! The old saying that "Haste makes waste" is particularly true where delicate microtechnical procedures are concerned. These are seldom successful unless they are carried out deliberately and with precision.

CHAPTER 5

Compound Microscope and Its Use

Careful study of biological specimens and intelligent evaluation of the quality of preparations being made in the microtechnique laboratory require some knowledge of the optical system of the microscope and methods of illumination. This chapter will be limited to a discussion of the compound microscope as it is used routinely with bright-field illumination. The subjects of dark-field illumination, phase contrast and interference microscopy, polarizing microscopy, and other more specialized methods will have to be omitted. To be useful at all, instructions for the application of any of these methods, and explanations of the optical principles according to which they function, must necessarily be quite detailed, and should be sought in comprehensive works on microscopy or in manuals supplied by manufacturers of the equipment itself.

CARE OF THE MICROSCOPE

When the instrument is not in use, keep it in its case, or at least cover it with a plastic microscope cover, in order to protect it from dust or moisture. Do not allow it to stand in direct sunlight, in a hot place, or near reagents whose fumes are likely to have a corrosive effect or which may react in the atmosphere to deposit a film on optical elements.

Always keep the mirror and lenses of the oculars, objectives, and condenser clean. Particles of dust which may have settled on the optical elements should be removed with a soft camel's hair brush reserved for this purpose. For removing oily deposits, moisture, or other contaminants, use lens paper moistened when necessary by toluene (or xylene) or distilled water. It is best to use a small piece of lens paper only once. The optical elements should never be rubbed with lens paper more often than necessary, and never without first brushing away hard particles which may scratch the soft glass or anti-reflective coatings. In general, if gentle application of a brush cleans the surfaces adequately, avoid the use of lens paper.

When using a microscope to examine wet slides, cover the stage with a glass plate to protect it and the optical parts mounted below the stage. Be careful not to wet any of the objectives with reagents. If, by accident, some part has come in contact with a reagent, sea water, or even distilled water, wipe it dry immediately, using lens paper for the optical elements, and a soft absorbent cloth for mechanical parts. If there is any danger that a corrosive residue will be left, attempt to remove it as soon as possible with lens paper or a cloth moistened with a safe and appropriate solvent (distilled water, toluene, or alcohol), then quickly dry it again.

Do not attempt to examine a thick whole mount or material evidently mounted under a thick coverglass with objectives of high magnification and short working distance. This may not only injure the preparation, but also the objective.

ILLUMINATION

Illumination is a most critical factor in all work with the microscope. To bring out structures which are distinguished by only slight differences in refractive index, color, or density, and to take advantage of a lens system capable of revealing very fine detail, careful attention must be given to the quality of the light (in terms of the wavelengths of which it is composed) as well as its intensity and focus.

In most work with the compound microscope, a strong beam of light is caused to pass through and around the object. This is termed bright-field illumination, because the object is viewed on a light field. If it is to be seen by this method of illumination, a structure must absorb or refract some light, thereby creating a contrast between itself and the surrounding medium.

Sources of Light

Daylight. Direct sunlight is too bright and glaring, and too rich in longer wavelengths of light. Daylight from the northern sky, especially when the sky contains some white clouds, is of favorable quality for microscopy. Unfortunately, the color of daylight changes continually, and its intensity is inadequate much of the time; even at its best, daylight is not strong enough for work with the oil immersion objective.

Electric Microscope Lamps. A good electric illuminator provides light which can be controlled to meet the requirements of the investigator. The more elaborate lamps (Fig. 2) have a system of condensing lenses, an iris diaphragm, and holders for filters. The lens system is necessary if the Köhler method of illumination (p. 35) is to be used, because it permits one to focus the coil or ribbon filament of a tungsten bulb at the level of the iris diaphragm of the substage condenser. The iris diaphragm of the lamp is used to narrow or widen the beam so that light extraneous to the field being examined may be cut off.

For most purposes, a very adequate system of illumination is one in which a ground glass serves as a secondary light source. A Corning Daylight filter ground on one side is particularly good, because it transmits

FIG. 2. A microscope illuminator with holders for filters, iris diaphragm, and condensing lenses. A lamp of this type may be used for illumination by the Köhler system or from a ground glass surface. (Courtesy of American Optical Corp., Buffalo, N.Y.)

light which is well balanced with respect to the various wavelengths. The same effect may be achieved by use of a ground glass coupled with a clear Daylite filter. If a Daylite filter is not available, a blue bulb behind ground glass will provide moderately good illumination.

If a small substage lamp is used, it is centered under the condenser. If the illuminator is of the type placed a few inches in front of the microscope, its tilt is adjusted so that the beam of light coming from it may be reflected by the mirror directly into the condenser.

With any system of illumination employed for critical work with objectives of superior quality, it is an advantage to be able to adjust the intensity of the light by use of neutral density filters, or by a rheostat or transformer. Reducing intensity by closing down the iris diaphragm of the substage condenser impairs the efficiency of an objective, because the full numerical aperture cannot be realized unless the angular cone of light is wide enough to fill the back lens of the objective. Ideally, then, intensity should be controlled at the source, and the iris diaphragm should be closed down only when it is necessary to increase optical contrast or depth of focus.

OPTICAL SYSTEM OF THE MICROSCOPE

In the compound microscope, the lenses of the objective and the lower lens (field lens) of a Huyghenian ocular collectively operate to form an image of the object at the level of the shelf-like diaphragm within the ocular. This image is called the real image, and the orientation of its parts is exactly as in the specimen itself. It is of course small, and furthermore it cannot be seen unless a disk of ground glass is inserted upon the diaphragm of the ocular. The function of the upper lens (eye lens) of the ocular is to act with the lens of the eye to magnify the real image and to form an image on the retina. The retinal image is inverted.

Substage Condenser. A condenser is used to illuminate the object in such a way that a cone of light is brought into focus on the specimen. The cone is then extended so that it just fills the back lens of the objective. The iris diaphragm of the condenser is used to cut off light which is extraneous to that actually being used by the lens system of the objective, and to adjust the angular cone of light. It is a rule of optics that the best resolution with a given objective will be achieved when the iris diaphragm is closed down to the point that the back lens of the objective, as viewed down the body tube after the ocular has been removed, is just filled with light.

Many structures, especially in living cells and organisms, are difficult to see because the optical contrast between them and other structures is weak. In such situations, closing down the iris diaphragm will improve the refractive images of certain elements. This technique does reduce the efficiency of the objective in terms of its capacity to reveal detail, because the full aperture of the objective is not used. However, the compromise may have to be made in order to bring out the refractive image more sharply. The depth of focus is also increased as the aperture is made smaller. Therefore the actual aperture which is ideal for the study of a particular specimen must be determined by observation. When a good refractive image is seen at the full numerical aperture of the objective, and if the depth of field is adequate, reduction of light intensity is better achieved by use of neutral density filters, a rheostat, or a transformer than by closing down the iris diaphragm of the condenser.

Substage condensers are designed to accept parallel light rays. It is therefore best to use the plane surface of the mirror in all operations which involve the condenser. However, if the condenser is removed—this may be necessary if an objective of very low power (as 3.5×) is in optical alignment and there is no auxiliary condenser to illuminate the field evenly—the concave mirror should be employed.

Microscope Objectives and Their Characteristics. Most microscopes are equipped with objectives having an equivalent focus of 16 mm. (10×; "low power") and about 4 mm. (approximately 40×; "high power"). However, for cytology, protozoology, bacteriology, and many aspects of invertebrate and vertebrate histology, an oil immersion objective with an equivalent focus of about 2 mm. (90 to 100×) is essential. A very low power objective, magnifying about 3.5×, will prove useful for scanning larger whole mounts or sections, and for locating objects which are scattered over a considerable area on the slide.

Achromatic objectives are standard equipment on microscopes of moderate price. They are corrected chromatically for two colors (red, blue) of the spectrum, and spherically for one color. The images which they yield are nearly free of undesirable color fringes, and they are adequate for most purposes.

Apochromatic objectives are corrected chromatically for three colors (red, blue, violet), and spherically for two colors, and therefore produce an image with sharper contrast and resolution than achromatic objectives. They are, however, relatively expensive, and to take full advantage of the improved color corrections of apochromatic objectives, it is necessary to use special compensating eyepieces and an achromatic substage condenser.

Resolving Power and Numerical Aperture. The capacity of an objective to reveal fine detail is referred to as its resolving power. It is upon resolving power that effective magnification depends, for if an object is merely made to appear larger, the increase in image size does not guarantee that details of structure will be faithfully revealed.

The resolving power of a lens depends upon its angular aperture, which may be defined as the width of the angular cone of light which it is capable of bringing to a focus. However, the aperture of an objective is usually expressed as the *numerical aperture* (N.A.), which is the sine of one-half the angular aperture multiplied by the refractive index of the medium in which it is designed to operate. Thus, a 43× dry objective having an angular aperture of 82° 30′ has a numerical aperture of .66 ($\sin \frac{82° 30'}{2} =$ 0.65935; 0.65935 × refractive index 1.00 = N.A. .66).

If a drop of oil having the same refractive index as glass is introduced between the front lens of an objective and the coverglass, the light rays issuing at an angle from the coverglass will enter the objective without being deflected or reflected to any appreciable extent. The same thing will happen if the oil is placed directly on the object. Even the more oblique rays proceeding from the object continue their course until they reach the back surface of the front lens of the objective. This system of homogeneous immersion permits more effective use of the angular aperture of the objective. Certain of the light rays which might enter an immersion objective could not enter a dry objective of the same working distance and lens diameter, owing to the fact that the angle at which they are deflected as they pass into air is too wide.

An oil immersion objective with the same angular aperture as the dry objective whose numerical aperture was calculated above would have a N.A. of slightly less than 1.00. And an oil immersion objective having a shorter working distance and an angular aperture of 110° 38′, which is typical of achromatic objectives magnifying about 97×, has a N.A. of 1.25 ($\sin \frac{110° 38'}{2}$ = 0.8223; 0.8223 × refractive index 1.52 = N.A. 1.25).

Most objectives are made to be used with a tube length of 160 mm. and coverglasses having a thickness of 0.18 mm. If the thickness deviates as much as 0.04 mm. from this figure, spherical aberration will be appreciable, and the image will be of lower contrast than might be achieved by use of a coverglass of appropriate thickness. If a good deal of work is to be done with dry objectives of higher power and superior numerical aperture, it will be desirable to make all preparations with coverglasses which are known to be close to the recommended thickness (about 0.16 to 0.20 mm.). The thicker No. 1 coverglasses and thinner No. 2 coverglasses generally fall within this range; in theory, all No. 1½ coverglasses should be suitable, but some are not. Special calipers are manufactured for those who really must have coverglasses that do not deviate appreciably from the recommended thickness.

When a homogeneous immersion system is being used, the thickness of the coverglass is of no consequence, provided the tube length (if adjustable) is set correctly and the coverglass is not so thick that it shortens the working distance severely.

OCULARS. The most widely used oculars are of a type called Huyghenian oculars. These consist of two plano-convex lenses; the convex surfaces of both face downward. The lower lens of a Huyghenian ocular contributes to the formation of a real image at the level of the diaphragm which is within the ocular; the upper lens and lens of the eye work together to magnify the real image and bring an image in focus on the retina.

An ocular magnifying 10× enlarges the diameter of the real image (produced by the objective and lower lens of the ocular) 10 times. In other words, when a 10× ocular and a 40× objective are in alignment, the diameter of the object is enlarged 400 times.

Oculars of 10× (or 8×) are usually standard equipment. Oculars of 5× are helpful in many situations when a lower magnification is desirable. Good 15× oculars are useful for increasing the size of the image if the optical system formed by the objective and substage condenser is a decidedly superior one. It must be remembered that the function of most microscope oculars is to contribute to the formation of a real image and then to magnify this. Obviously, no amount of magnification by an eyepiece can reveal detail not resolved by the objective itself.

USE OF THE VARIOUS OBJECTIVES WITH BRIGHT-FIELD ILLUMINATION

The following instructions apply to a compound microscope—monocular or binocular model—equipped with an adjustable Abbe condenser, Huyghenian oculars, and objectives magnifying 10×, about 40× to 45×, and about 90× to 100×. This combination of optics is the one most likely to be found on microscopes of moderate price used in advanced biology courses, in medical schools, and in clinical laboratories.

Illumination with a Secondary Light Source (Ground Glass) or Daylight

The system of lighting most likely to be employed is one in which a ground glass serves as a secondary light source. As noted previously, a Corning Daylite filter ground on one side, or a clear Daylite filter combined with a ground glass, will provide light which is well balanced with respect to various wavelengths. The directions given for use of each objective are applicable to this system, and also to illumination with daylight, and if carried out properly will fulfill the following conditions:

1. After the objective is focused on the specimen to be studied, the field should be evenly illuminated. It may be necessary to raise or lower the condenser, but in general this should be close to its highest position.

2. The back lens of the objective, when this is viewed with the ocular removed, should be just filled with light. The iris diaphragm of the sub-

stage condenser should therefore be closed or opened until its aperture coincides with the back lens of the objective.

3. If the lamp is equipped with an iris diaphragm, this should be closed or opened so that the beam of light is sufficient to illuminate the field evenly, but it should not be so wide that the image is degraded by glare from excess light.

4. If the intensity of light is too strong, this should be adjusted by using a neutral density filter, a transformer, or a rheostat. The iris diaphragm of the substage condenser should be closed down only if it is necessary to improve optical contrasts or to increase the depth of field.

Ordinarily, it is a good idea first to study a preparation with the 10× objective. After locating the area of interest and centering it in the field of view, swing into alignment whichever of the objectives of higher power seems to be more appropriate. In examining blood for malarial parasites or trypanosomes, fecal material for small amoebae and flagellates, or exudates likely to contain bacteria, it is conventional to examine the preparation directly with the oil immersion objective, or to begin with the 40× dry objective.

USE OF THE 10× OBJECTIVE. Place the slide on the stage of the microscope and center the object above the opening. Swing the low-power objective into alignment, and lower it to within about 5 mm. of the slide. Then, while looking through the ocular, raise the body tube with the coarse adjustment until the specimen is in focus. For superior illumination, raise or lower the condenser until the field is evenly illuminated. Then remove the ocular and adjust the iris diaphragm until the cone of light just fills the back lens of the objective. Replace the ocular and proceed with examination of the material. If the depth of focus is too shallow, or if contrasts between structures (especially in living material) are not strong enough, close down the iris diaphragm. This may also be done to reduce the level of illumination. However, as previously noted, it is desirable, if at all possible, to control intensity by use of neutral density filters, a rheostat, or a transformer.

USE OF THE 40 TO 45× ("HIGH DRY") OBJECTIVE. As the specimen is being examined with the 10× objective, center that portion which is to be examined under higher magnification. If the high-power dry objective is parfocal with the 10× objective, simply swing it into alignment and sharpen the focus with the fine adjustment. If it is not parfocal, raise the body tube, if necessary, to swing the high-power objective into alignment, then lower it until the front lens of the objective is about 1 mm. away from the coverglass. Raise the body tube, while looking through the ocular, until the specimen is in focus.

Adjust the condenser until the illumination of the field is uniform. Remove the ocular and manipulate the iris diaphragm until the margin of its aperture coincides with the margin of the back lens of the objective.

Then replace the ocular and proceed with observation of the specimen. To increase depth of focus or contrasts, reduce the aperture of the iris diaphragm.

USE OF THE OIL IMMERSION (90 TO 100×) OBJECTIVE. Unless the organisms or structures to be examined are more or less evenly distributed, or so small that they are not likely to be recognized except under the oil immersion objective, it will probably be more convenient to locate them under one of the dry objectives first. After carefully centering them in the field, swing the dry objective out of alignment and place a small drop of oil on the area which is to be studied. If the oil immersion objective is parfocal with the other objectives, click it into position and adjust the focus with the fine adjustment. If it is not parfocal with the others, its front lens may be either too close to the object or too far away when it is brought into alignment. If it is necessary, raise the body tube to swing the oil immersion objective into position without striking the coverglass or slide, then lower the body tube with the coarse adjustment until the front lens of the objective appears to be nearly touching the coverglass (or the slide itself, in case the preparation is not covered). Look into the ocular and slowly raise the body tube with the fine adjustment until the specimen comes into sharp focus.

In working with an oil immersion objective, the most favorable resolution will be obtained when the condenser is nearly touching the slide. On many microscopes, the substage assembly is built in such a way that the condenser will be in the correct position when it is raised as far as it will go. Ideally, a drop of immersion oil should be placed between the upper lens of the condenser and the slide. This practice is not often followed except in photomicrography or very critical work with an apochromatic objective and an achromatic condenser. It is the only way to take full advantage of the high numerical aperture of an oil immersion objective.

As with the dry objectives, the iris diaphragm should ideally be opened only enough to fill the back lens of the objective with light, and the intensity of light should be controlled at the source. However, closing down the iris diaphragm may be necessary to increase depth of focus or optical contrast.

If the oil immersion objective is not to be used again for a time, it is best to wipe off the oil with lens paper, and then to apply lens paper moistened with toluene or xylene in order to remove traces of oil which remain.

For many years cedarwood oil was favored for use with oil immersion objectives. Its refractive index of 1.515 makes it ideal for this purpose. However, cedarwood oil left on slides or coverglasses eventually hardens, although it can be removed with toluene or xylene. A number of nondrying synthetic immersion oils with optical properties comparable to those of cedarwood oil have been developed for microscopy and are now in wide use. Nevertheless, it is best to remove all oil from microscope slide preparations as soon as possible. Even if the oil does not harden, it may erode the mounting medium at the edge of the coverglass, and it is also likely to trap dust which may be detrimental to uncovered preparations such as blood films. Slides with oil left on them are a nuisance if books, laboratory notes, or drawings come into contact with them.

To remove immersion oil from an uncovered preparation, do not wipe it off. Add some toluene or xylene to thin it, then draw it off with a piece of lens paper laid flat on the slide. This process should be repeated, each time using a clean piece of lens paper, until it is apparent that little if any oil remains.

USE OF OBJECTIVES OF EXTREMELY LOW POWER. Standard Abbe condensers will not illuminate the entire field when objectives of about 3.5× are used. Auxiliary condensers, attachable to Abbe condensers, are manufactured to overcome this problem. However, for ordinary purposes it is generally acceptable simply to swing the condenser aside (or to remove it from the mount) and to use the concave mirror to reflect light toward the objective.

Illumination by the Köhler Method

This system provides the superior illumination required in exacting research work. The Köhler method requires a lamp which has provisions for control of intensity (by neutral filters and a transformer or rheostat), an iris diaphragm for reducing or enlarging the width of the beam, a system of condensing lenses, and a tungsten lamp with a coil or ribbon filament. Built-in illuminators found on high-grade microscopes fulfill these requirements.

Adjustment of the lamp and optical system of the microscope should be made to meet the following conditions:

1. When the substage condenser is raised until it nearly touches the slide, the image of the lamp filament should be visible on the leaves of the iris diaphragm. In the case of a separate illuminator, this requires adjusting the tilt of the illuminator, using the central portion of the mirror, and focusing the system of condensing lenses in the lamp. The iris diaphragm may be inspected by looking at the mirror, or by holding an auxiliary mirror below the condenser and to one side of the beam of light.

2. When the objective is focused upon the object to be studied, the margins of the aperture of the iris diaphragm of the lamp should be in focus. The condenser may have to be raised or lowered slightly. The aperture should then be closed or opened until it coincides with the field of view of the objective.

3. When the ocular is removed and the back lens of the objective inspected, it should be just filled with light. The aperture of the iris dia-

phragm of the condenser should be increased or decreased as necessary. The ocular is then replaced.

4. The intensity of the illumination should be adjusted by use of neutral density filters, or by a transformer or rheostat. The iris diaphragm of the substage condenser should be closed down only when it is necessary to increase optical contrasts or the depth of focus.

MEASURING WITH AN OCULAR MICROMETER

An ocular micrometer rests upon the shelf which is about half way between the upper and lower lenses of a microscope ocular. Make certain before purchasing a micrometer that it will fit properly in the particular ocular with which it is to be used. To insert the micrometer, the upper element of the ocular is unscrewed and the micrometer is dropped from the opening onto the shelf. The engraved surface must be uppermost. In handling the micrometer, touch it only at its edges. The micrometer may be cleaned with lens paper whenever necessary, but great care should be taken to avoid scratching its surfaces.

The lines on the ocular micrometer, although they are regularly spaced, cannot themselves be used for measurement, because they will be the same distance apart regardless of which objective of the microscope is in optical alignment. Obviously, then, an object having a diameter equal to 10 units when it is viewed with a low-power (10×) objective will have a diameter of about 40 units when it is examined with a high-power (40×) objective. It is necessary, therefore, to establish the value of each unit of the micrometer for each objective with which it is to be used. This is done by partially superimposing the scale of the ocular micrometer on that of a stage micrometer, in which each unit has a definite value. After a given ocular micrometer in a particular ocular is calibrated for use with each of the several objectives on the microscope, the values should be recorded. The stage micrometer may then be dispensed with, unless the body tube is adjusted to a different length for some reason.

A stage micrometer resembles an ordinary microscope slide, and is treated as such. The scale is generally 2 mm. long, and divided into 200 units of 0.01 mm. (10 μ) each. The various objectives, including the oil immersion objective, are focused on this in the usual way. Once a sharp image is obtained, the stage micrometer is manipulated and the ocular is rotated until the first line of the ocular micrometer (marking 0) is set either at the back edge of whichever line of the stage micrometer is to be taken as 0, or directly upon it. It is not necessary to start with the first line (0) of the stage micrometer, although if this is done it may be easier to count the units without making a mistake.

For calibrating the ocular micrometer, it is advisable to use the whole scale (generally 50 or 100 units), or the longest portion of it which can be made to coincide perfectly with any two lines on the stage micrometer. In the example shown in Figure 3, 31 units of a 50-unit ocular micrometer scale equal 18 units (180 μ) of the stage micrometer. The value of each unit of the ocular micrometer is therefore 5.8 μ $(\frac{180\ \mu}{31})$. The entire scale of 50 units appears to occupy the equivalent of 291 μ. Again, the value of each unit of the ocular micrometer is about 5.8 μ $(\frac{291\ \mu}{50})$. In using this method, an estimate must usually be made of the number of microns between the last line of the ocular micrometer and the last line of the stage micrometer just before it. However, if the entire 50-unit (or 100-unit) scale is involved in the calibration, a slight error in the estimate is not likely to affect the value of each unit to any significant extent.

When an ocular micrometer is used in making measurements of organisms or structures, it is superimposed upon the object to be measured, and the number of units occupied by the length or width or other dimension is counted. An estimate will have to be made if a fraction of a unit is involved. The total is then multiplied by the known value of each

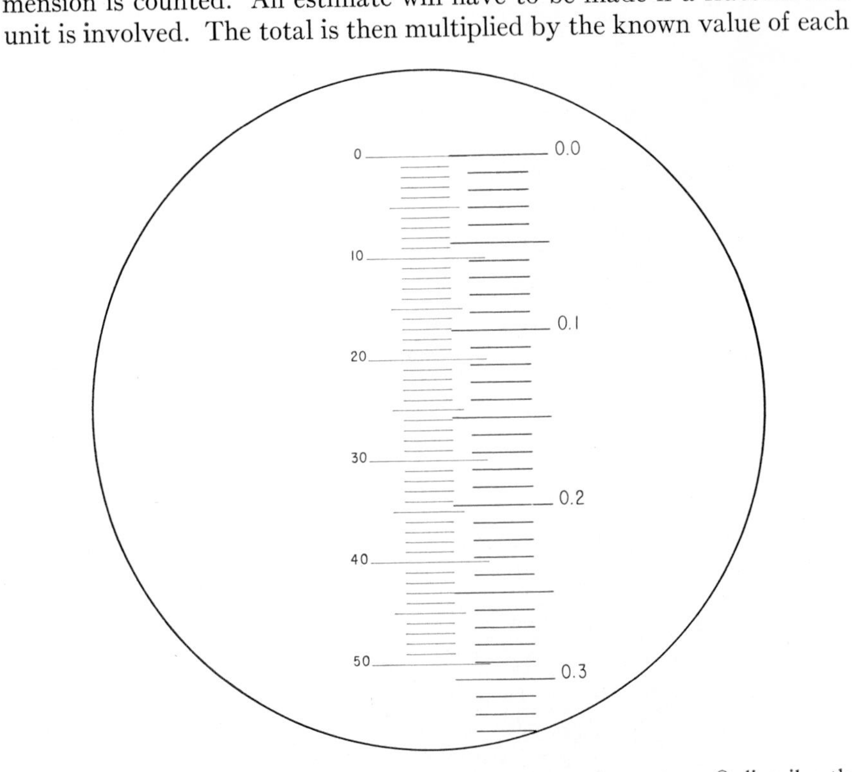

FIG. 3. Ocular micrometer superimposed upon stage micrometer. Ordinarily, the stage micrometer would be more nearly centered than shown here; in order to show all of the lines of both micrometers clearly, the stage micrometer has been displaced slightly to the right. The method for calibrating the ocular micrometer is explained in the text.

unit. Suppose, for instance, that the value of each unit of the ocular micrometer, as used with a certain objective, is 2.1 μ, and the diameter of the nucleus of a cell is equal to 4.5 units. The diameter of the nucleus is therefore 9.45 μ, or close to this. Owing to the difficulty of making an absolutely precise measurement with an ocular micrometer, especially when a fraction of a unit is involved, a number such as 9.45 is obviously too refined. It is perhaps more realistic to say that the diameter of the nucleus, as established at that particular moment, was about 9.5 μ, or between 9 and 10 μ. When the oil immersion objective is used in making measurements, the value of each unit of the ocular micrometer is smaller, so that a more precise measurement of smaller structures may be made with this objective than with those of lower magnification. Conversely, measurements made with a low-power objective are likely to be in error, one way or another, to the extent of at least 1 or 2 microns. In measurement of larger objects, however, an error of no more than a few microns will probably be of no consequence.

MECHANICAL STAGE WITH VERNIERS

A mechanical stage facilitates study of serial sections, and is almost indispensable in any work which regularly involves rapid or systematic scanning of preparations. If the mechanical stage is equipped with verniers, the observer can record the exact position of a small organism or

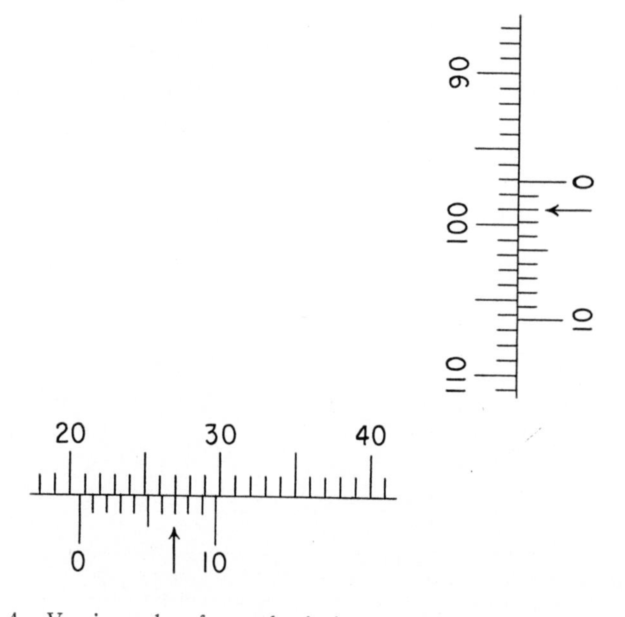

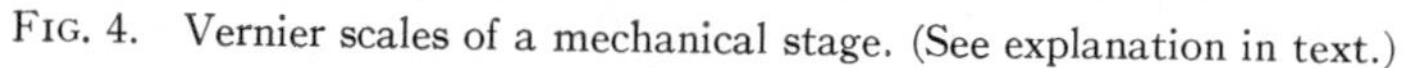
FIG. 4. Vernier scales of a mechanical stage. (See explanation in text.)

a particular structure, so that it can easily be found again if the same microscope is used. Separate verniers are used to establish the position of the object with respect to the tracks of the mechanical stage which move the slide forward or backward and from left to right. The numerical ranges of the verniers do not overlap, so that no confusion can arise over which of the two numbers recorded applies to one track and which to the other.

Each vernier consists of two scales, which are side by side. The more extensive of the two scales is graduated in millimeters. The other scale is 9 mm. long, but is divided into 10 units, so that not more than one of its lines can coincide with any line on the longer scale. When the object whose position is to be recorded is in the exact center of the field of view, look at one of the verniers and find the point on the longer scale where the line marking 0 on the shorter scale falls. This will establish the whole number. The reading is complete if the line marking 0 on the shorter scale coincides exactly with a line on the longer scale. However, if the coincidence is not perfect, the decimal is determined by finding the one line on the shorter scale which most closely coincides with any line on the longer scale. The same procedure is followed to obtain a reading with the other vernier, and the two numbers are recorded with drawings or notes concerning the specimen. In the example shown in Figure 4, the readings should be recorded as 20.7 × 97.2.

In the case of a removable mechanical stage in which neither of the two tracks along which it slides is actually built onto the stage, it is sometimes difficult to relocate a small object whose position was recorded while it was in the field of view of an objective of higher magnification. Once the two readings are reproduced on the verniers, some hunting in the immediate vicinity may be necessary before the object is found.

CHAPTER 6

Methods for the Study of Living and Fresh Material

All branches of the biological sciences are directed toward the understanding of living organisms, and must therefore be based to a large extent upon living specimens. Unfortunately, extended observation of living cells may be difficult for a number of reasons. In the majority of metazoan animals, the cells are rather firmly united in dense masses. In order to examine the cells of such organisms with the microscope, they must first be separated, and it may be nearly impossible to do this without injuring them. We must also be able to keep the isolated cells alive, or at least delay the onset of death and postmortem changes. Even if we could succeed in these efforts, or if we choose to study one-celled or noncellular organisms, eggs, embryos, or other small and transparent specimens, we are confronted by another difficulty. The contents of cells are mostly colorless and transparent, so that they can only be distinguished from one another by slight differences in the extent to which they absorb or refract light.

By various methods, these difficulties can be minimized to permit more intensive study of living or fresh specimens. (The term *fresh* is used here to designate material which is removed from an animal during life or shortly after death, and which is examined at once, without being subjected to any preservative or other strong chemical.) These methods fall into two general categories. First, there are techniques for the preparation and mounting of the specimens so that they will remain for some time in a condition favorable for observation. It is with these techniques that the subject matter of this chapter is primarily concerned. Second, there are various methods for illuminating and examining the preparations. Standard bright-field illumination is discussed at some length in Chapter 5, so that only a few suggestions especially applicable to study of living and fresh material will be given here.

Phase contrast microscopy, which takes advantage of the fact that structural elements of cells and tissues modify the velocity of light waves passing through them, is of great usefulness in study of living and fresh material. Other specialized methods such as dark-field illumination, interference microscopy, fluorescence microscopy, and polarization microscopy may contribute to a still more complete understanding of particular types of cells, tissues, or small organisms. Suitably detailed accounts of these methods of microscopy will be found in modern works on the microscope, as well as in publications prepared by manufacturers for users of the special equipment which is necessary.

Observations of living and fresh material, by one or more methods of microscopy, and of fixed and stained specimens, provide information which is complementary. For this reason, frequent observation of fresh specimens is often advantageous during the course of investigations on permanent preparations.

IMPORTANT GENERAL PRECAUTIONS. Slides and coverglasses must be strictly clean. Methods for cleaning are found on page 19.

Pipettes, glassware, and instruments used for living specimens should never be interchanged with those used for chemicals. Use a separate pipette for removing small organisms from each culture, in order to avoid contaminating the cultures. Likewise, use a separate pipette for each reagent. In work with pure cultures, it is of course necessary to sterilize all glassware and media before using them.

BRIGHT-FIELD ILLUMINATION FOR LIVING AND FRESH MATERIAL. Illumination must be carefully controlled, because many structures of living cells are only distinguished by slight differences in refractive index or color. To improve the refractive images of flagella, cell boundaries, and various discrete bodies within the nucleus or cytoplasm, the aperture of the iris diaphragm of the substage condenser should be adjusted until the best contrasts are obtained. It has been pointed out (p. 29) that when the iris diaphragm is closed down below the point at which the full numerical aperture of the objective is being used, the resolving power of the objective is decreased. However, the compromise may have to be made in order to make the structures clearly visible.

The color images of naturally colored bodies, or structures which have been stained, are often degraded if the iris diaphragm is closed down enough to obtain good refractive images of these same elements. The color image of a particular structure is favored when the intensity of the light is strong, and when light impinges on the structure from many angles. The quality of the color image therefore becomes poorer when the angular cone of the light supplied by the condenser is narrowed, or when the intensity is reduced. To differentiate structures which are colored from those which are not, and to bring out both the color images and refractive images of particular structures, it will probably be necessary to manipulate the iris diaphragm frequently.

MOUNTING METHODS AND MOUNTING MEDIA

Mounting in Water

SMALL AND TRANSPARENT ORGANISMS. The simple and familiar method of mounting living specimens in a drop or two of water is applicable to many forms, including free-living protozoa, small coelenterates, rotifers, ectoprocts, flatworms, nematodes, annelids, and crustaceans, as well as eggs, embryos, and larvae of many invertebrates and lower vertebrates. Obviously, inhabitants of fresh water or soil should be mounted in fresh water, and marine or brackish-water species should be mounted in water of appropriate salinity. In all cases, the water should be free from contamination by metals, chlorine, or other poisons. Water from the culture dish, or pond water, will be satisfactory for fresh-water organisms. Sea water should come in contact only with glass, porcelain, plastics of certain nontoxic types, or well-seasoned wood. Pumping sea water through metal pipes will render it toxic to most organisms.

Supporting the Coverglass. In mounting such delicate organisms as small flatworms, rotifers, or larvae of many invertebrates, it is advisable to support the coverglass at least to some extent. Debris, small algae, and other firm material present in cultures or samples collected in the field will usually serve this purpose. Fine glass rods, which may be drawn out in a flame to any desired diameter, pieces of broken coverglasses, and fine threads or hairs may also be convenient.

Depression slides or small, shallow dishes are more suitable than standard microscope slides for examination of objects which are relatively large and require a considerable amount of liquid.

Flattened organisms having a firm consistency (such as planarians, tapeworm proglottids, and flukes) are less likely to be injured by the weight of the coverglass, which may serve a useful purpose by holding them flat and restraining their movements. In the case of very muscular tapeworms or flukes, it is often advisable to use a thick coverglass and to tie this down so as to compress the specimen somewhat.

Quieting Active Organisms

This may be accomplished either by mechanical restraint or by the use of narcotics. The following suggestions apply to quieting certain active animals for study in a living condition. Methods for narcotizing contractile organisms so that they can be fixed in an expanded state will be considered in Chapter 9.

Ciliates. *Paramecium*, *Euplotes*, *Stylonychia*, and other rapidly swimming ciliates may be slowed by increasing the viscosity of the medium. Methyl cellulose has proved to be very good for this purpose. It seems not to be toxic to any appreciable extent, and does not seriously disturb the osmotic relations of protozoa. A 10% solution of methyl cellulose (viscosity rating 15 centipoises) is a convenient stock mixture. Generally speaking, when a drop of the methyl cellulose solution is stirred into an approximately equal amount of the medium, the viscosity will be raised sufficiently. Some experience will enable one to decide what proportion of the stock solution to use for different types of organisms.

A weak solution (about 0.01%) of nickel sulfate in fresh water or sea water is effective in slowing ciliates, including *Paramecium*. It is convenient to keep a 1% solution on hand, and to dilute this, as needed, with the water in which the specimens are to be observed. Some experimentation may be necessary to determine a concentration that is effective without being a fatal dose.

Another method for slowing ciliates is to mount them in a tangle of fine cotton threads, fibers from lens paper, or filamentous algae.

Sessile or semi-sessile ciliates such as *Vorticella* and *Stentor* can usually be studied without special treatment, but if necessary they can be paralyzed by adding a drop of a freshly made 1% solution of novocaine or cocaine hydrochloride to a few milliliters of water containing the organisms.

Many of the larger ciliates may be studied to advantage by first getting them into a drop of water upon a coverglass, withdrawing water with a pipette or filter paper until the organisms are held against the glass by the surface film, and then inverting the coverglass over the concavity of a depression slide. The thin layer of water holds the specimens in the focal plane, and they are often nearly immobilized.

By adding a suspension of some finely divided, non-toxic substance to the water used for mounting, it is possible to demonstrate the ciliary currents and the discharge of the contractile vacuoles. Powdered carmine or carbon (in the form of India ink) generally serves very well. The colored particles are sometimes ingested in the same way as food, thus making it possible to trace the formation and movement of food vacuoles.

Ciliates in a hanging drop can sometimes be killed and temporarily preserved in a nearly life-like condition by exposing the drop to the fumes of osmium tetroxide for several minutes. *Euplotes* and *Stylonychia* are among the genera which may be treated in this way.

FLAGELLATES. Use the methods recommended for ciliates.

COELENTERATES. Living hydroids (including hydras) and very small anthozoans may be studied without special treatment. To stain the nematocysts in living animals, place them, for 2 hours, in a dish of water (sea water in the case of marine species) which has been tinted light blue by addition of the following solution: methylene blue, 1 gm.; Castile soap, 5 gm.; distilled water, 300 ml. Nematocysts may often be caused to discharge by mechanical stimulation.

Small medusae or ephyrulae may be narcotized by adding a few crystals of ethyl urethane to a small dish of water which contains the material. Another good method for quieting marine medusae is to add, little by little, a solution of magnesium chloride which is approximately isotonic with sea water ($MgCl_2 \cdot 6H_2O$, 75 gm.; distilled water, 1000 ml.).

FLATWORMS. The weight of a coverglass will be sufficient to hold planarians and other small specimens flat and more or less extended. In mounting cestodes, large trematodes, or other muscular forms, it may be necessary to use a thick coverglass and to tie it down with a thread. Tapeworm strobilae begin to relax after they have been in cold water for some time. Some trematodes will discharge eggs from the uterus within an hour or two if they are placed in distilled water. Because the masses of eggs in the uterus often obscure other structures, this method is quite useful for those species which respond to the treatment.

A solution of magnesium chloride, isotonic with sea water (see above), is excellent for relaxing marine acoel, rhabdocoel, and alloeocoel turbellarians.

NEMATODES. Small roundworms are generally immobilized by the weight of the coverglass. If they are not, mount them in a tangle of cotton threads or narcotize them by slow addition of 20% ethyl alcohol.

ROTIFERS. Small rotifers may often be slowed sufficiently by adding some 10% methyl cellulose (p. 42) to the medium. However, it is generally advisable to narcotize rotifers, especially larger types such as *Asplanchna*. Addition of a few drops of a freshly made 1% aqueous solution of cocaine hydrochloride to a small dish of water containing the organisms is a particularly good method, although often impractical because of the problems associated with obtaining and keeping the narcotic. Other anesthetics such as eucaine should be tried. Chlorobutanol (Chloretone) works with some species, and 2% aqueous solutions of butyn, hydroxylamine hydrochloride, and benzamine lactate have also been used with success.

ANNELIDS. To immobilize small fresh water oligochaetes such as *Chaetogaster*, *Nais*, and *Dero*, add 10 to 15 drops of a 1% aqueous solution of chlorobutanol to a Syracuse dish of water containing the specimens. In 5 or 10 minutes, mount the worms on a slide, supporting the coverglass so that they are not crushed. To immobilize marine polychaetes, add, little by little, an isotonic solution of magnesium chloride (p. 43). Addition of fresh water may be helpful in relaxing some species.

DELICATE LARVAE OF INVERTEBRATES. Planulae, trochophores, bipinnaria or pluteus larvae, veligers, and other larvae may sometimes be safely quieted by mounting them in a shallow hanging drop, as recommended for ciliates. Another method of immobilizing larvae is to support the coverglass just enough to avoid heavy pressure upon them. Larval stages of marine gastropods (including nudibranchs), which are enclosed by the gelatinous masses in which the eggs are laid, are quite easily studied. If a small piece of egg mass is mounted, the gelatinous material will usually support the coverglass enough to protect the embryos from being crushed.

SMALL CRUSTACEA AND INSECT LARVAE. These may often be restrained successfully by using supports of such thickness that the coverglass presses on the specimen very slightly. Carbonated water, used for mixing beverages and available at grocery stores, is quite useful for quickly immobilizing many small aquatic arthropods. Mix 1 part of carbonated water with 1 or 2 parts of the water in which the organisms have been living. Once a bottle of carbonated water is opened, its potency declines rapidly unless it is very tightly stoppered and stored in a refrigerator.

Addition of about 4 parts of a 1% aqueous solution of chlorobutanol to 10 parts of water containing the material also works rather well. Another method is to introduce a drop or two of 20% ethyl alcohol every few minutes until the animals are quieted.

Examination of Thin Portions of Larger Organisms

Thin portions of the tail of a small fish (such as a guppy), the web of a frog's hind leg, the tail or external gills of a tadpole, a portion of mesentery, and similar materials may be studied while the animal is alive. The specimen must be kept moist with water or one of the isotonic saline solutions to be described later in this chapter. Active animals must be physically immobilized in some way. In the case of adult amphibians, this can be accomplished by laying the animal upon a thin board, a part of which may be placed over a glass plate on the stage of the microscope, and winding a strip of cloth about both the animal and board. The board must have a hole cut in it beforehand at an appropriate place so that the web of the foot or tissue of the mesentery may be spread out over the hole and examined with transmitted light. Keep the animal as a whole well moistened, especially the portion being studied. Water will be adequate for intact animals, but an isotonic saline solution should be used if internal tissues are exposed.

If mesentery is to be examined, first decerebrate the animal, then wrap it so as to leave part of the abdomen uncovered. Make an incision into the abdomen, draw out a loop of intestine with its mesentery, and fasten it down over the hole, damaging the tissue as little as possible with the needles or pins used to hold it in place.

Amphibian larvae and small fish may be immobilized by placing them in glass cells made especially to fit their bodies, or they may be lightly narcotized. For narcotizing, place them in water which has been shaken with a little ether and then separated by decanting, or add some 1% chlorobutanol to the water.

Dry Mounting

Dry mounting is only useful for study of the external anatomy or activities of organisms which have a thick cuticle or shell, or which otherwise resist rapid desiccation. It is applicable to insects, arachnids, millipedes, centipedes, and terrestrial crustaceans and molluscs. A dry mount is generally observed as an opaque object. Place light-colored specimens over a dark background, and dark specimens on a light background, and illuminate them from above or from the side by means of an illuminator which provides a concentrated beam of moderate to strong intensity. The specimens may be placed on a standard microscope slide, in the concavity of a depression slide, or in a shallow dish. If it is necessary to immobilize the animal without recourse to treatment with ether or fumes of cyanide, supports of suitable thickness may be mounted on a slide so that when a coverglass is applied it will press upon the animal just enough to hold it in position. The supports may be in the form of strips of paper or glass slides, pieces of coverglasses, glass rods, or droplets or rings of mounting media or cements.

Mounting Blood, Lymph, and Coelomic Fluid

Making a Temporary Preparation of Blood. Sterilize the skin of a finger-tip or lobe of an ear by washing it with alcohol, then prick it with a needle that has been sterilized in a flame, or with a disposable lance that has been sterilized before being packaged. Blood from other animals may be obtained in the manner which appears to be most appropriate. In the case of a mouse, for instance, the very tip of the tail may be snipped off and the tail squeezed gently.

As soon as a small drop of blood exudes, touch it with a coverglass. Quickly place the coverglass, blood downward, on a clean slide. Dip a small brush in melted vaspar (a mixture of equal parts of paraffin and

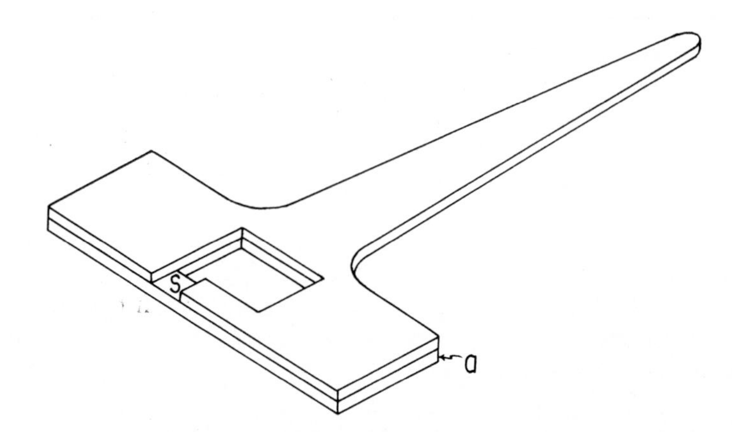

Fig. 5. A simple and inexpensive warming stage. The stage is cut out of sheet copper about 2.5 mm. thick. Extending from the aperture in the center, there is a slot (*s*) into which the bulb of a small thermometer will fit tightly. A sheet of asbestos (*a*) is cemented to the underside of the stage, but does not extend to the tongue. An alcohol lamp or micro gas burner is placed under the end of the tongue, and the flame and distance adjusted to keep the stage at the desired temperature (usually about 37° C.).

petroleum jelly) and draw it around the edge of the coverglass. Mounts sealed in this way are less likely to show annoying drifting currents than those whose edges permit evaporation to take place. If active movements of leukocytes are to be observed, it will be necessary to place the preparation on a warm-stage. An electrically heated stage with an accurate thermostatic control is desirable when work requiring a warm-stage is regularly being done in the laboratory. However, these are expensive and must be handled with great care. A simple and inexpensive warm-stage may be improvised, as shown in Figure 5.

BLOOD FROM POIKILOTHERMOUS ANIMALS. This may be studied at room temperature, although very slight warming is sometimes useful to speed the activity of leukocytes.

LYMPH. This may be withdrawn from the thoracic duct with a hypodermic syringe. It will be easier to secure material from a dog or other mammal of considerable size; the thoracic duct is difficult to find in small mammals.

COELOMIC FLUID. Obtain coelomic fluid with a hypodermic syringe and needle. Some invertebrates such as earthworms provide interesting material.

Mounting in Isotonic Saline Solutions

In order to examine the cells in small pieces of tissue, it is advisable to place them in a fluid whose composition resembles the plasma which normally surrounds them. They may then remain in a life-like condition for a short time. Natural media used for this purpose include blood serum, amniotic fluid, and fluid from the aqueous humor of the eye. These are not always readily available and hence are used only for a few special purposes. The fluids ordinarily employed are dilute solutions of sodium chloride and other salts.

Although ordinarily designated as isotonic, isosmotic, normal, or "indifferent" fluids, they are not entirely without effect upon cells. All of them are toxic to a slight extent, but some are less so than others, as will be explained presently. In addition to being nearly non-poisonous, these solutions have approximately the same salinity as normal blood serum. This means that they will be in osmotic equilibrium with the cells and other components of tissues, and will not cause these structures to shrink or swell. Supravital staining (see p. 52) is usually carried out in isotonic solutions.

Formulas of Isotonic Fluids. "PHYSIOLOGICAL" SALINE SOLUTION. This solution is suitable for washing tissues, rinsing out blood vessels before injection, and for making mounts of a very temporary sort. It is also employed to immerse the eggs of chickens and other birds or reptiles while the embryos are being removed.

Solution for Lower Vertebrates

Sodium chloride	6 to 7 gm.
Distilled water	1,000 ml.

Solution for Birds and Mammals

Sodium chloride	7.5 to 9 gm.
Distilled water	1,000 ml.

RINGER'S SOLUTION. The salt content of this mixture resembles that of blood serum more closely than does the simple saline solution described above. Therefore, it is less likely to produce distortion in cells. It is generally satisfactory for washing and mounting tissues of both invertebrates and vertebrates, as well as for observation of symbiotic protozoa. The composition has been modified by various investigators. The formula given below will prove suitable for most work.

Sodium chloride (for mammals and birds)	8 gm.
(for lower vertebrates)	6.5 gm.
Calcium chloride, anhydrous	0.2 gm.
Potassium chloride	0.2 gm.
Sodium bicarbonate	0.2 gm.
Dextrose (may be omitted)	1 gm.
Distilled water	1000 ml.

Dissolve the calcium chloride last, in order to minimize formation of insoluble calcium carbonate. The solution does not keep long. If sterilization is necessary, it is best done by filtration through a sterile Berkefeld or membrane filter, rather than by autoclaving.

LOCKE'S SOLUTION. The following formula is perhaps better than Ringer's solution for the tissues and embryos of mammals and birds.

Sodium chloride	9 gm.
Potassium chloride	0.42 gm.
Calcium chloride, anhydrous	0.24 gm.
Sodium bicarbonate	0.2 gm.
Distilled water	1000 ml.

Dissolve the calcium chloride last. The solution does not keep well and must not be boiled. If sterilization is necessary, this should be done by filtration.

TYRODE'S SOLUTION. This medium corresponds more nearly to blood serum than do any of the preceding mixtures. It is used for washing cultures of living cells and for other more critical work. However, it requires more care in preparation, and it is not necessary for temporary mounts. The following modification can be sterilized by autoclaving, whereas the original formula requires filtration.

Solution A

Sodium chloride	8 gm.
Potassium chloride	0.2 gm.
Calcium chloride, anhydrous	0.2 gm.
Monosodium phosphate	0.05 gm.
Magnesium chloride	0.1 gm.
Glucose	1 gm.
Distilled water	750 ml.

Solution B

Sodium bicarbonate	0.5 gm.
Distilled water	250 ml.

Prepare and autoclave the solutions separately. Mix 3 parts of solution A with 1 part of solution B, as needed. Some workers may find it convenient to sterilize small quantities of each of the two solutions in small flasks or test tubes, so that mixtures may be made without risk of contaminating larger volumes.

Methods of Preparation and Mounting. BLOOD. Place a drop of Ringer's solution or Locke's solution on a clean slide. Obtain human blood by pricking the skin of a finger-tip or lobe of an ear with a sterile needle or special blood lancet; obtain blood from other animals by a method which appears appropriate. As soon as a drop of blood exudes, touch it with a clean coverglass and quickly place this, blood downward, on the drop of saline solution. If the blood has been taken from a warm-blooded animal, examination on a warm-stage will be necessary to observe favorably the amoeboid movements of the leukocytes. In mounting blood of amphibians having especially large erythrocytes, it may be a good idea to support the coverglass with a fine hair in order to prevent it from crushing them.

EXUDATES AND EXCRETA. Pus, scrapings from the gums, and fecal material to be examined for parasites or their eggs may be mixed with a small amount of an isotonic saline solution before the coverglass is applied. Methods for concentrating parasites and eggs may be found in standard textbooks of parasitology or clinical technique.

THIN MEMBRANES. Peritoneum and subcutaneous tissue should be carefully stripped off with fine forceps. Spread a piece of the tissue upon a clean slide in a drop of Ringer's solution or Locke's solution, and apply a coverglass. It is often advantageous to spread a rather large piece of membrane upon a slide, keeping its center moist by breathing upon it frequently, and allowing its edges to dry so that it will adhere to the slide. Then place a small drop of an isotonic saline solution in the center of it and add a coverglass. The central portion will remain almost perfectly flat. A method for studying the mesentery of a living animal has been described on page 45.

OTHER TISSUES. These may be prepared by either teasing, scraping, or smearing. Each of these methods is adapted to certain kinds of tissue.

Teasing, or shredding apart with needles, is especially useful for the preparation of mounts of nerve, muscle, and other fibrous tissues, or tissues in which cellular portions are interspersed with dense connective tissue. Clip off a very small piece of tissue and place it in a drop of saline solution on a clean slide. Take a clean, sharp needle, mounted in a handle of wood or other suitable material in each hand. Hold them as you would hold pencils. Place the point of one needle firmly upon the tissue, so as to hold it in place. With the other needle, gently pull the fibers apart. Separate the structures carefully; do not crush or break them up. The process of teasing is best carried out under a wide-field binocular microscope. When the structures have been separated sufficiently, add a little more of the saline solution if necessary, and apply a clean coverglass. Examine the preparation at once with the compound microscope, controlling the illumination carefully to obtain suitable optical contrasts.

Scraping is applicable to organs composed of soft cellular matter bound rather firmly together by connective tissue. The liver, spleen, and oral and gastric mucosas are good examples. Scrape a freshly-cut surface very lightly with a sharp scalpel blade, held at a considerable angle to the surface. Place the scrapings in a drop of saline solution on a clean slide, and add a coverglass. The tissue elements may be further separated by drumming lightly upon the coverglass, or by gently raising and lowering one edge of it.

Smearing is a useful method for separating soft tissues, as those of bone marrow, testis, and lymph nodes. Place a small bit of tissue on a slide, in a drop of isotonic saline solution, and apply a coverglass. Slight pressure on the coverglass may assist in spreading the material. This technique may also be applied to tissues from brain, spinal cord, liver, spleen, gastric glands, and other organs, but it will probably be necessary to tease the material with needles before applying the coverglass.

MICRODISSECTION

Precise microdissections, injections, and other operations on cells or embryos have been made possible by invention of mechanical devices which manipulate very fine needles and pipettes under objectives of the compound microscope. The use of these devices, called micromanipulators, will not be discussed here. The investigator whose research requires a micromanipulator should obtain the very detailed information necessary for choosing and operating an instrument from the manufacturers of such equipment.

It is possible, however, to dissect some types of small organisms, including embryos, without a micromanipulator. Work of this type is generally performed under a wide-field binocular microscope, which provides the advantages of low magnification, wide field, and an erect image. Such operations may have as their purpose the removal of some organ (as the salivary glands of *Drosophila* larvae) for detailed study under high magnification, transplantation of tissue from one individual to another, and the separation of blastomeres formed by cleavage of a fertilized egg. The following paragraphs deal briefly with the instruments required, and with some suggestions for handling the biological material.

INSTRUMENTS. These include small knives, needles, forceps, and scissors of high-quality steel, carefully sharpened on fine hones. Fine watchmaker's forceps, with well-ground tips, will serve a variety of purposes. Small sewing needles (sizes 5 to 8), fixed in handles of wood or metal, are useful in their original form and can also be fashioned into a variety of small instruments. A tiny knife can be made by carefully grinding away the opposite sides of the needle close to the tip. Hooks of various shapes can be made from needles, by softening them in a flame, bending them as desired, then re-tempering and sharpening them.

The very fine pins known to entomologists as "Minuten Nadeln" will be useful in some techniques. These pins, now made of stainless steel, are usually about 12 mm. in length and 200 μ in diameter. They can be cemented into a piece of glass tubing with an epoxy cement. It is possible to sharpen them further by honing them, or dipping them for a moment in strong nitric acid. Size 00 insect pins are nearly as slender, and the varnish with which they are regularly coated can be honed off.

Delicate glass needles and hairs are widely used for operating upon very small objects, such as eggs of echinoderms. To make glass needles, take a glass rod about 5 mm. in diameter and 15 to 20 cm. long. Hold it by both ends and heat the middle portion in a gas flame. When the glass becomes soft, remove it from the flame and quickly draw it out as far as you can reach. Break off the thicker ends, for use as handles, and divide the thread into lengths of about 10 cm. To form the needles, draw out pieces of the thread in the flame of a micro burner. By varying the extent to which the thread is heated, and the strength and duration of the pull, one can develop a technique for making both straight and curved needles, with points of any size, taper, or degree of sharpness. When a needle of suitable quality has been obtained, break it off to a length of about 2 or 3 cm. and weld it to one of the handles by heating it in the flame of a micro burner.

To make a hair loop, which is useful for holding eggs and other delicate objects, draw out a piece of glass tubing in the manner just described. Separate the two halves, leaving about 1 or 2 cm. of the capillary portion attached to each of them. Into this insert the ends of a looped human hair, pushing them in far enough to reduce the loop to the desired size. Then seal the hair in place by applying a bit of melted paraffin to the mouth of the capillary portion.

Fine pipettes are also needed for moving small objects from one dish or slide to another. The flow of liquids into and out of such pipettes may be controlled by a rubber bulb, or by means of a small rubber tube held in the operator's mouth.

CONTAINERS FOR SPECIMENS. Shallow embryological dishes are convenient for handling objects which measure a millimeter or more. Some of these dishes are made with a central slot to hold the specimen. Small eggs and similar objects may also be handled on clean slides, which may be scratched or grooved to aid in holding the specimens in place. A layer of wax or agar on the bottom of a dish may be grooved in any way desired and will serve admirably as a working surface.

If the specimens are markedly opaque, it will probably be advantageous to illuminate them from above, using a small spotlight. A piece of black paper or other dark material should be placed beneath the dish if the specimens are light in color; a layer of black wax inside the dish may also prove suitable. If the specimens are dark, a light background is generally preferable.

QUIETING SPECIMENS. By placing the specimens in a very small amount of liquid—just enough to keep them immersed—their movements may often be restrained somewhat. Strips of glass may help to keep them still. However, it is often necessary to employ anesthetics. In such cases, use a substance that is promising in terms of its having been used successfully with organisms of the same general type. See pages 42 to 45 and pages 114 to 118 for suggestions. The choice of an anesthetic for a particular kind of organism, and determination of the amount to use, usually will have to be arrived at through experimentation.

VITAL STAINING

Various structures in living cells can be colored by certain dyes which are relatively non-toxic, and such staining is loosely designated as vital staining. However, there are two very different techniques by which dyes may be applied for vital staining. The term *supravital staining* is used to indicate coloration of isolated cells, small organisms, or pieces of tissue. *Intravital staining* refers to differentiation of cellular constituents by injection of a dye into a living animal.

Supravital Staining

Dyes used for this purpose penetrate the cells in the form of solutions and are taken up as *true stains* by the cellular constituents. In low concentrations they do not kill the cells as a whole. Some of the structures they stain (such as secretion granules) are not actually living entities; in the case of others, it is likely that their life processes cease before or during absorption of the dye. Supravital staining is nevertheless a valuable method for the study of certain structures such as nerve fibers, nerve endings, mitochondria, and various granules in the cytoplasm. It will presently be seen that some dyes (for example, methylene blue) may color a variety of substances, while others (as Janus green B) may affect only one type of cellular constituent.

Methylene Blue for Nerve Tissue. Methylene blue has proved most valuable for the supravital staining of nerve cells, nerve fibers, and nerve endings, although it stains other structures as well. This was one of the early methods of staining introduced by Paul Ehrlich.

QUALITY OF THE DYE. The chemistry of methylene blue is discussed on page 378. For supravital staining, it is necessary to use pure methylene blue chloride, free from zinc.

STAINING ENTIRE SMALL ORGANISMS BY IMMERSION. This method is especially valuable for specimens which are more or less transparent, including hydras and other hydroids, medusae, small flatworms, nematodes, and annelids, as well as young embryos and larvae. The dye solution is made in fresh water or sea water, depending upon the habitat of the species, and a suitable concentration for each type of organism must be determined by experimentation. In general, concentrations ranging from 0.001 to 0.01% are likely to be effective.

When intact organisms are stained by immersion, their various structures take up the dye after different periods of time. Glandular cells are usually stained first; epithelial cells, blood cells, and muscle fibers often become colored in a sequence characteristic for particular organisms. Each structure gradually attains a maximum coloration and then rapidly loses the color. For this reason, it is essential to watch the specimens carefully until the desired structures are sharply stained. If necessary, stained specimens can then be fixed by one of the methods given in a later paragraph. *Nerve cells and fibers generally remain uncolored when intact organisms are stained.*

A slightly more complicated method, which may give good results in staining elements of the nervous system of entire small organisms, is the following: Prepare 100 ml. of a 0.5% solution of methylene blue in distilled water and add 0.05 ml. of concentrated hydrochloric acid. To each 10 ml. of this stock solution, in a small beaker, add 0.2 gm. of powdered sodium formaldehyde sulfoxylate (formaldehyde sodium sulfoxylate), or 2 ml. of a 12% solution of this substance. Heat the preparation, preferably on an electric hot plate or in a water bath, to about 70° or 80° C. (definitely not to boiling), stirring it constantly. The dye soon will become decolorized. Allow the mixture to cool. After a pale yellowish precipitate forms, remove this by filtration, collecting the clear liquid in a small bottle (a dropper bottle will be convenient) that can be tightly stoppered and that is small enough so that the solution will nearly or quite fill it. The decolorized staining solution will deteriorate within a few days, whereas the merely acidified stock solution of methylene blue will keep indefinitely; it is advisable, therefore, to make up only small amounts of the decolorized solution when needed.

For staining, add 1 part of the decolorized methylene blue solution to each 20 to 50 parts of water. In a few minutes, the diluted solution will turn blue. If an objectionable precipitate forms (this regularly happens with sea water), filter the solution before putting in the animal or tissue to be stained. Some nerve elements may take up the stain quickly, but others may require several hours or may remain refractory.

Beginners usually are disappointed with the results of their attempts to bring out elements of the nervous system by any of these methods. Some experimentation and considerable patience almost certainly will be required, as each organism and tissue is apt to have its own peculiarities. Staining with methylene blue has not yet been brought to a point of refinement that will enable one to apply it to a wide variety of subjects with almost certain success.

The report of Spangenberg and Ham (1960) will be helpful to those who have not been successful with the methylene blue techniques described here. It points out factors that influence staining of nerve elements in one particular organism (*Hydra littoralis*), and thus may give the investigator some ideas about how to modify these methods to get better results.

Staining Nerve Tissue after Removal or Exposure. This is a simple and usually effective method for demonstrating ganglion cells nerve fibers, and nerve endings. It is very commonly applied to membranes of the inner ear. The method has yielded exceptionally clear preparations of sensory nerve endings in crustaceans and cephalopods. As soon as possible after killing or decerebrating the animal, remove the desired parts and cut them into small pieces, or expose the nervous strucsures by removing overlying tissues. If necessary, wash them with a suitable saline solution to remove blood or debris, but do not allow any other chemical to come into contact with them. Place the tissue upon a slide or in a shallow dish and bathe it well with dye solution. A 0.1% solution of pure methylene blue in a saline solution is satisfactory for most purposes, though stronger or weaker solutions are preferable in special cases. Do not immerse the tissue completely. Unless they are exposed to the air, the tissue elements may reduce the dye to its leucobase, which is colorless. Observe the preparation at frequent intervals. The time required for staining varies from a quarter of an hour for end-bulbs in the conjunctiva to 3 hours or longer for ganglion cells. Incubation at about 37° C. will hasten the process. Soon after attaining maximum intensity, the color begins to fade. Methods of fixing the stain are described in a later paragraph.

The method described above for using a decolorized stock solution of methylene blue should be tried if this simple technique does not work.

Staining Nerve Tissue by Injection (Perfusion). This procedure insures a thorough penetration of the stain and is, therefore, indicated for the central and sympathetic nervous systems, as well as for nerve endings in the internal organs, muscles, and deeper layers of the skin. In the case of small animals it is best to inject the entire body, as directed in the following paragraph. Organs of larger specimens are removed and injected through their principal arteries.

Immediately after the death of the animal expose the aorta, or one of the carotid or femoral arteries. Into the blood vessel, insert a cannula (see Fig. 9, p. 131) to which a syringe of appropriate size is attached. Also cut the large vein that parallels the artery to be injected and encircle it with a string, proximal to the cut, but do not tie the string. Fill the syringe with saline solution, which for a warm-blooded animal should be heated to 37° C., and inject this in order to wash blood from the vessels. Then fill the syringe with a 0.1% solution of pure methylene blue, warmed if necessary. A somewhat stronger solution may be preferable in certain instances, but 0.5% seems to be the strongest concentration which will give favorable results. Inject slowly, until the colored solution begins to run out of the cut vein. Then ligature the vein tightly and continue to inject slowly until a slight resistance to the plunger indicates that the vessels are well filled. Tighten the string proximal to the cannula and withdraw the latter. A similar procedure is used for injecting individual organs removed from larger animals.

Allow the injected animal or organ to stand for about 30 minutes. Then remove the desired tissues and cut them into small pieces. Place them on slides or in shallow dishes, bathe them well with more of the dye solution, and observe them frequently. Continue as in the preceding method. Preparations made by this method generally show less intense staining, and the finer portions of the nerve elements are likely to be less well-differentiated than if they have been stained in isolated pieces of tissue.

FIXATION OF THE STAIN. Stains obtained by any of the three preceding methods can be made semi-permanent in several ways. However, their usefulness is not likely to last more than a few weeks or months, so that it is desirable to make photographs or drawings while they are still well stained.

Dogiel's method is limited to the preparation of whole mounts, teased material, or sections made by the freezing method. Place the stained tissue in a saturated aqueous solution of neutral ammonium picrate for 2 to 24 hours, and then transfer it to a mixture consisting of equal parts of glycerol and the ammonium picrate solution. Mount small pieces in this mixture. Picro-carmine (p. 347), which provides a good counterstain, may be used in place of ammonium picrate, and the material may be mounted in glycerol.

Bethe's method renders the stain insoluble in alcohol and so makes it possible to prepare sections by the paraffin method, or to mount specimens, with or without sectioning, in a resinous medium. Transfer pieces of tissue 2 to 3 mm. thick from the stain to a saturated aqueous solution of ammonium picrate. After 10 or 15 minutes, transfer them to the following mixture:

Bethe's Fluid

Ammonium molybdate	1 gm.
Chromic acid, 1% aqueous solution	20 ml.
Hydrochloric acid, concentrated	1 drop

Allow the tissue to remain in this fluid for 1 to 2 hours. Wash it in distilled water for an hour and then proceed either with the paraffin method of sectioning (Chapters 14 and 15) or with the usual procedure for mounting entire objects in a resinous medium (Chapter 11).

Janus Green B for Mitochondria. This is a rather specific stain for mitochondria. It was first used by Michaelis about 1900, and subsequently became well known through the work of Bensley and Cowdry in America.

To stain mitochondria in protozoa, small invertebrates, embryos, larvae, and cells in smears, scrapings, and teased material from vertebrates or larger invertebrates, simply mount them in a freshly prepared 0.01% solution of Janus green B. It is convenient to have on hand a 1% solution of the dye in distilled water, and to dilute 1 part of this with 100 parts of the most appropriate medium (water from a culture, pond water, sea water, or an isotonic saline solution).

Another good method of supravital staining with Janus green B is to prepare a dry film of the dye on a clean slide, and then to add the organisms or cellular material in whatever medium matches their osmotic requirements most closely. The procedures for making dry films with neutral red, or a mixture of neutral red and Janus green B, are described on page 58. If Janus green B alone is to be used, the same general directions may be followed. The solution of the dye is most conveniently prepared by diluting 1 drop of a saturated solution in absolute alcohol with 10 ml. of absolute alcohol.

Injection through blood vessels (perfusion) is applicable to the pancreas and other dense organs of vertebrates. Prepare a 0.01% solution of Janus green B in a saline solution, and inject this into the arteries, using gravity as the source of pressure. Allow the dye solution to run slowly through the vessels for 15 minutes. Then ligate the vein draining the area, and continue to inject for a minute before withdrawing the cannula. Allow 10 minutes to pass before removing small pieces of the organ to be studied, and tease them out in the dye solution on a slide.

Neutral Red. This dye, if sufficiently pure, is almost non-toxic in low concentrations. Always take care to obtain neutral red certified for vital staining.

Neutral red colors several types of material in living cells. These include the following: (1) secretion granules in the cytoplasm, including the secretion of the islet cells in the pancreas; (2) specific granules (neutrophil, basophil, eosinophil) of the leukocytes; (3) digestive vacuoles of phagocytic cells in the connective tissue, lymph glands, bone marrow, and various other organs; (4) the system of canaliculi in certain cells such as the parietal cells of the stomach; (5) the reticulum of immature erythrocytes (reticulocytes). Some of these properties of neutral red make it more or less possible to follow the activities of well-defined groups of cells in inflammations and other reactive processes. Hence this dye has been applied extensively to pathological investigations, chiefly concerning the blood and connective tissues.

The living nucleus may show some affinity for neutral red, although this is not always the case. Apparently the coloration is diffuse, resulting from temporary mechanical retention of the dye, and does not involve actual staining of the living chromatin.

Neutral red is a rather sensitive indicator of acidity and alkalinity. Its color changes from red to yellow between pH 6.8 and 8.0. For this reason it is useful in studying changes in acidity and alkalinity, as during the course of digestion in food vacuoles of protozoa.

A 1% solution of neutral red in distilled water keeps well, and it is suggested that this concentration be kept on hand for making dilutions with whatever medium is most appropriate.

Small living organisms may be immersed in a 0.001 to 0.01% solution of the dye in fresh or sea water, depending upon the habitat of the species. They may remain in it for several hours or even days. Certain tissues may assume a dark red color, and within some cells various granules and food vacuoles will be stained bright red.

Tissues of vertebrates are generally stained by injecting, into the blood vessels, a 0.01% concentration of the dye in a saline solution. Subcutaneous injection may be preferable in work on the skin or subcutaneous tissues, and intraperitoneal injection is applicable to mesenteries. When the organs to be studied show a faint rosy tint, remove small portions and tease them out in the dye solution, permitting free access to air. In the pancreas, the islets of Langerhans will appear an intense yellowish red.

Thin membranes, smears of dissociated cells, and protozoa may also be stained to good advantage by the dry film method, described below.

Dry Film Method. This is a convenient technique for supravital staining with neutral red, or with a mixture of this dye and Janus green B. It has been widely used in connection with the study of blood cells, but is also quite suitable for other types of material, including smears of various tissues (such as bone marrow) and protozoa and other small organisms.

If necessary, clean the slides so as to remove every trace of grease. Then place them in 95% alcohol. Wipe each slide with perfectly clean gauze before using it. Flood a cleaned slide with a 0.05% solution of neutral red in *neutral* absolute alcohol. Drain off the excess and touch one edge of the slide to filter paper or a blotter. Then lay the slide horizontally in a dust-free place until the alcohol evaporates completely. The film should be uniform. Films of this type keep quite well and may be stored for use whenever desired. It is a good idea to mark the side of the slide which bears the film.

If it is desired to use neutral red and Janus green B simultaneously, add about 1 or 2 drops of a saturated solution of Janus green B in absolute alcohol to each 10 ml. of the 0.05% solution of neutral red.

To stain protozoa and other small organisms, simply place a drop of the culture medium, intestinal contents diluted with an appropriate saline solution, or other material directly on the slide, and apply a coverglass. Tissue smears should also be prepared in a drop of an isotonic saline solution on the slide.

To make a preparation of fresh blood on a dry film, place a drop of blood on the film and add a coverglass, being careful not to include air bubbles. The drop must be just large enough to fill the space between the coverslip and slide, without spreading beyond. After the blood has spread to the edge of the cover, seal the mount by applying melted vaspar with a small brush. Preparations made only with neutral red will generally keep for about 2 hours at room temperature. The mixture with Janus green B will usually kill the cells within an hour, but in the meantime should differentiate the mitochondria. To observe the motility of the leukocytes, it is advantageous to examine the preparations on a warm-stage at 37° C. Avoid letting the temperature rise any higher, as this will kill the cells within a very short time. The results of staining blood with a mixture of neutral red and Janus green B are as follows: Erythrocytes appear faint yellow, except for reticulocytes in which the reticulum is red. Heterophil (neutrophil) leukocytes show salmon-pink granules. Small mitochondria in these are brought out by Janus green B. Eosinophil leukocytes show the large, refractile granules stained bright yellow. In basophil leukocytes the granules are stained scarlet. Lymphocytes show sparse granules stained light red and numerous rod-like mitochondria stained blue-green by Janus green B. Monocytes contain numerous pink-stained bodies, which may be either vacuoles or granules, often arranged in a rosette. Abundant minute mitochondria lie at the edge of this rosette.

In smears of bone marrow, myelocytes show various types of granules, corresponding in their staining reactions to those of the mature corpuscles. In the connective tissue, the large macrophages (histiocytes) include no specific granules, but often contain phagocytosed material which stains various shades of red or yellow. Fibroblasts have a dense, homogeneous cytoplasm which stains yellowish.

Intravital Staining

The process of coloration to which this term applies is usually not true staining, but depends upon the ingestion of colored particles by phagocytic cells. A colloidal suspension of carbon or of certain dyes is generally employed, although a suspension of fine particles visible with the light microscope may be suitable. When a colloidal suspension is used, the ultramicroscopic particles are aggregated into larger masses within the cells which accumulate them.

Intravital staining is especially useful for identifying phagocytic cells throughout the body. These include: (1) fixed or wandering macrophages of the connective tissue, spleen, lymph nodes, bone marrow, peritoneum, and other tissues; (2) stellate cells of von Kupffer in the endothelial lining of liver sinusoids, and phagocytic cells in the vascular endothelium in other organs; (3) monocytes in the blood or lymph.

Suspensions used for intravital staining are injected into the living animal, either subcutaneously, intraperitoneally, intravenously, or into any other cavity which may be indicated. The route of injection depends, of course, upon what is to be studied. Intravenous injection involves considerable risk to the animal, but may be necessary for work on the macrophages of the liver, spleen, bone marrow, and deeper lymph nodes. For staining serous membranes, intraperitoneal or intrapleural injection is indicated. In studies of phagocytes in areolar connective tissue, subcutaneous injection will be suitable.

The dyes or other substances to be injected should be suspended in sterile distilled water. Do not use a physiological saline solution, such as Ringer's solution. The presence of salts may cause the particles to aggregate in clumps. In vascular injections, clumping of particles is likely to bring about occlusions and consequent death of the animal. Approximate concentrations and dosages will be stated in connection with each of the various substances.

A syringe, preferably of the all-glass type, and hypodermic needles of appropriate sizes, are required for injection. Before use, the syringe, needle, and colored suspension should be sterilized, at least as completely as this can be done, by boiling them. The animal's skin in the area of operation should be shaved and then swabbed with an antiseptic solution such as weak tincture of iodine. The solution to be injected should first be warmed to body temperature.

The rabbit is a favorite subject for intravenous injection, as the veins of its large ears are readily accessible. Intravenous injection of other animals is generally made through a superficial vein of one of the appendages.

Substances Used in Intravital Staining. SUSPENSIONS OF COLORING MATTER. Any non-toxic insoluble pigment, in a finely divided state, may be employed for investigating the phagocytes of connective tissue and von Kupffer's cells of the liver. Carbon (lamp-black) and carmine are commonly employed. To prepare a suspension of either substance, place 1 gm. of it in a porcelain mortar and grind it for several minutes. Then add distilled water from a measuring cylinder, a little at a time, and continue grinding until a smooth paste is formed. Pour in more distilled water, meanwhile continuing to grind and stir the suspension, until a total of 20 ml. has been added. Filter the suspension through coarse filter paper and sterilize the filtrate by boiling. Before using it for injection, bring it to body temperature and shake it thoroughly. For rabbits of medium size, inject 10 ml. intravenously, or 10 to 15 ml. intraperitoneally. In subcutaneous injection, use as much as the connective tissue will take up readily. Generally, the animal is killed in 24 to 48 hours after injection, and pieces of the liver or other tissues are preserved as suggested below.

COLLOIDAL CARBON (INDIA INK). This substance is concentrated by monocytes and endothelial cells of the circulatory system, particularly in areas of inflammatory response to infection or injury. It is taken up also by fixed macrophages and von Kupffer's cells. Higgin's Waterproof Drawing Ink is a readily available form of colloidal carbon. Simply dilute it with an equal amount of sterile distilled water, warm the dilute suspension to body temperature, and proceed to inject it. For rabbits of medium size, inject 5 ml. into a vein of the ear, repeating this daily for 3 days, and thereafter every 3 days if the experiment is to be prolonged.

LITHIUM CARMINE. Carmine in the colloidal state is taken up by all kinds of phagocytic cells. In time it also colors the epithelium of the kidney tubules, making it possible to locate areas of active secretion and absorption. Frequently its color appears in the cells of liver cords. To prepare lithium carmine, make a concentrated suspension of carmine in a saturated aqueous solution of lithium carbonate. The subcutaneous dose is 0.3 ml. for the mouse, 2.5 ml. for the rat, and 5 to 10 ml. for the rabbit. Several injections may be made, at intervals of 3 or 4 days.

TRYPAN BLUE. This is one of several related azo dyes which have been used as intravital stains. Its toxicity is relatively low. Trypan blue is excellent for coloring macrophages and other phagocytes throughout the body, and also the epithelium of kidney tubules. In the case of phagocytes, at least, the dye is taken up in the form of colloidal aggregates.

For injection, use a freshly prepared 0.5% solution of trypan blue in sterile distilled water, warmed to body temperature. The dosage for subcutaneous injection of mice and rats is 1 ml. of this solution per 20 gm. of body weight. For rabbits weighing about 1,000 gm., inject 10 to 15 ml. either subcutaneously or intraperitoneally, but only 3 to 5 ml. intravenously. The day after injection, the skin will begin to appear bluish. If injections are repeated every 4 or 5 days, the skin and mucous membranes will become dark blue and the animal's urine will show the color.

OTHER COLLOIDAL SUBSTANCES. Colloidal silver and gold have been employed, but they offer no particular advantages.

Fixation and Sectioning of Intravitally Stained Tissues. Formalin of 10% strength should be used for fixation of material stained with trypan blue or lithium carmine. Fix small pieces of tissue (no larger than about 5 mm. thick) in this reagent for at least 48 hours; larger specimens should be left in the fixative for a longer time, in proportion to their size. This treatment is also good for other intravitally stained material. However, material colored with carbon or metals may be fixed in any way desired; Bouin's fluid, Zenker's fluid, or Heidenhain's "Susa" fluid will probably prove to be good. If one of these reagents is likely to prepare the tissue better than another for a certain staining procedure, then this should be considered in making a choice.

Fixed material can generally be studied to best advantage if it is sectioned. Sections may be made by the paraffin method, nitrocellulose method, or freezing method (only formalin-fixed material should be used in the case of the latter). The sections are stained with appropriate dyes to bring out the nuclei or other structures in colors which contrast with that of the intravital stain. In this connection, refer to the chapters on sectioning and staining.

REFERENCE

SPANGENBERG, D. B., and R. G. HAM, 1960. The epidermal nerve net of hydra. J. Exp. Zool., *143*, 195–201.

CHAPTER 7

Fixation

Ideally, the study of cells, tissues, and small organisms should include careful observation of living material, but cells or small fragments of tissue, taken from an animal and sandwiched between a slide and coverglass, are not likely to remain normal for more than a few minutes. Such preparations provide a basis for a permanent record only to the extent that they can be accurately photographed or drawn. Moreover, certain nuclear and cytoplasmic structures which may be invisible or poorly defined in living material, even when studied by recently perfected methods of phase contrast microscopy, may be differentiated by appropriate techniques of staining.

Many ingenious methods of sectioning, staining, and mounting tissues have been devised, but for reasons which will presently be explained, these methods can be applied only to specimens which have been *fixed.* Fixation is intended to terminate life processes quickly, preventing autolysis and microbial action, and preserving the organization of cells with as little distortion as possible. If certain chemical constituents of cells and tissues are to be demonstrated, it is important that these be preserved. Usually, fixation is done by immersing the specimens in one of a number of appropriate chemical fluids, called *fixing agents* or *fixatives.* Less frequently, vapors (for example, of osmium tetroxide or formaldehyde) are employed, and in some cases (as with thin films of blood and bone marrow) desiccation is used to accomplish at least certain of the aims of fixation.

THE IMPORTANCE OF FIXATION

Faithful preservation of structures of cells and tissues is obviously the fundamental step in the preparation of material. All subsequent procedures such as sectioning, staining, and mounting serve only to clarify and to differentiate further the products of fixation. Poorly fixed material is worse than useless, since it may not only waste time, but often leads to erroneous conclusions. Intelligence and experience must therefore be applied (1) to the selection of a fixing agent or agents which will preserve the desired structures satisfactorily and will prepare them for appropriate procedures of sectioning and staining, and (2) to the techniques of applying these agents effectively.

The choice of fixing agents for various types of biological material must be based upon a general knowledge of the principles of fixation. This chapter will be devoted to these principles. Chapter 8 will deal with the specific properties and uses of various fixing agents, and Chapter 9 will explain methods of fixing animal materials.

THE EFFECTS OF FIXATION

Fixation accomplishes simultaneously a number of purposes, which are related to the various characteristics of living matter. These will be discussed under several headings below.

TERMINATION OF LIFE PROCESSES AND PREVENTION OF POSTMORTEM CHANGES. If tissues are allowed to die slowly, they will undergo postmortem changes, and structural features they possessed in life will be lost or modified. As we have seen in the preceding chapter, drying and some other distortional changes may be postponed for a time by keeping the tissues in an isotonic salt solution. Even if bacteria of decay are excluded by strict asepsis, destructive changes will take place through the action of enzymes normally present in the cells. This is especially true of the digestive glands and epithelium of the gastrointestinal tract. At death the cell becomes acid and enzymes begin to split up its protein constituents. This disintegrative process is called *autolysis*. Therefore the first functions of a fixing agent are to penetrate and kill the tissues quickly while they are in a nearly normal state, and to prevent postmortem changes due to autolysis and bacterial activity.

HARDENING. In order to embed, section, stain, and mount tissues, it is necessary to treat them with a variety of reagents. Some of these would distort unfixed structures and dissolve certain of their constituents. A fixing agent coagulates, hardens, and renders insoluble many constituents of cells and intercellular matter, thus preparing them to withstand treatment with reagents to which they must subsequently be exposed. This must be accomplished with the least possible alteration in the form of the tissue elements. Coagulation and hardening depend commonly upon precipitation of the proteins of living matter, but other processes are generally involved, and in some cases no precipitation takes place. The nature of the hardening process and its effects upon tissue structures will be discussed on page 68.

PRODUCTION OF OPTICAL DIFFERENCES. A piece of clear glass is more easily seen in air than if it is immersed in an oil. This is because the refractive index[1] of glass (about 1.52, in the case of crown glass) is very different from that of air (1.00) but close to that of oils. Similarly, if a certain constituent of a cell has a distinctly higher refractive index than

[1] Refractive index is a measure of the relative velocity of light passing through a particular medium.

the cytoplasm which surrounds it, it should be distinguishable with the microscope. Fixation raises the relative refractive indices of various structures to a greater or lesser extent, with the result that some of them are made still more readily visible. Even though some structures may later be selectively stained, it should be noted that the color image dependent upon concentration of a dye in a particular structure is not synonymous with the refractive image of this same structure. You can see for yourself how a piece of colored glass submerged in an oil, or in toluene, yields a poor refractive image even if its color image is good, whereas in air the same piece of glass has a good refractive image.

PREPARATION FOR STAINING. It has been known for a long time that tissues which have been fixed will take up stains more readily than fresh tissues. This was first noted by Hartig in 1854. As a matter of fact, most stains will produce a differential coloration of parts *only* in fixed material. Moreover, the nature of the fixing agent has a considerable effect upon the affinity of the structures for various stains. A fixative which favors staining by some dyes may hinder or prevent the action of others. Thus picric acid in mixtures generally gives excellent results with hematoxylins, carmines, and acid aniline dyes, but commonly leads to poor staining with basic aniline dyes.

SUBSTANCES USED AS FIXING AGENTS

The more important ingredients of fixatives are the following: alcohol; acetic acid; formalin; mercuric chloride; chromic acid; potassium dichromate (potassium bichromate); picric acid; nitric acid; and osmium tetroxide ("osmic acid"). A number of other chemicals are used, though much less frequently. These substances all have unlike chemical properties and, as might be expected, act differently upon the diverse constituents of living organisms. A fixing agent which preserves certain structures admirably may injure or destroy others. For example, acetic acid fixes chromatin, but it causes cytoplasm to swell and it does not preserve mitochondria. No single chemical so far discovered will fix tissues in a manner which is altogether satisfactory for a complete study. The best fixing agents are therefore produced by mixing two or more of the above chemicals, each of which possesses certain desirable properties. In appropriate combinations, the ingredients supplement each other, or one of them may offset unfavorable effects of another. As an example, we have already noted that acetic acid swells cytoplasm. For this reason it is commonly mixed with mercuric chloride, chromic acid, or alcohol, in order to counteract the shrinking effect of these powerful fixatives. In such mixtures, acetic acid also improves the fixation of chromatin. However, acetic acid must be omitted or added in very small amounts if mitochondria are to be studied, because it has a tendency to destroy them. Thus it will be apparent that each type of mixture serves certain definite purposes. A complete study of any kind of cell or tissue requires the use of several fixing agents.

GENERAL ANALYSIS OF FIXATION AND CAUSES OF ARTIFACTS

How much and in what ways are the appearances of tissue structures changed by fixation? The value of any study of fixed preparations depends upon the answer to this critical question. However, it is not often that a conclusive answer can be given. There remains a great need for research on fixation, in order that the actions of various fixing agents may be understood and their results interpreted with more certainty. No doubt, such investigations will lead eventually to the development of better and more reliable reagents for the purpose. Having made this observation, let us discuss the problem as it now stands.

Constituents of Living Matter. The extremely complex and ever-changing physicochemical organization of living cells and tissues will continue to provide challenging problems to biologists for a long time. Cells and intercellular matter are made up of a great many substances—proteins, lipids, carbohydrates, salts of metals, and other compounds, which are combined physically and chemically in various ways. Among these components are enzymes and other secretions, as well as diverse granules which may be either living structures or products of metabolism. Obviously the various chemicals employed as fixatives will react in different ways with the constituents of cells. Most attempts to analyze the action of fixatives have dealt chiefly with their effects on proteins.

Early History of Fixation. Turning briefly to the history of microtechnique, we find that methods of preserving tissues were developed very reluctantly to meet the requirements of staining methods which became increasingly more precise. For a number of years after the practice of staining with carmine became widely known through the work of Gerlach (1858), tissues were hardened almost exclusively in alcohol, then called "spirits of wine." However, an improvement was made by Müller in 1859, when he used potassium dichromate to harden the tissues of the eye. Müller's fluid became increasingly popular in the years between 1860 and 1880. By 1875, several good selective stains had come into use and study of cell structure had become a most promising subject. Biologists began to realize that further progress in this line of investigation could be made only by means of better preservatives for delicate cell structures. Revolutionary developments began with the intensive use of mercuric chloride by Lang (1878) and of picric acid by Kleinenberg (1879). Chromic acid had been employed by Hannover as early as 1840. Remak (1854) and others had tried acetic acid, and osmium tetroxide was used by Schultze (1866). However, these last three substances became really well known through the chromic acid-acetic acid-osmium tetroxide mixtures of Flemming (1880). Formic acid was introduced by Rabl (1884), formaldehyde by Blum (1893), and a considerable number of additional substances were tried by various other investigators before the turn of the century.

In the years following 1880, the term *fixation* was gradually adopted for the killing and initial hardening of tissues; use of the term *hardening* was restricted to the subsequent hardening involved in preparing fixed tissues for sectioning or mounting.

Fixed protoplasm was observed to have the form of a spongework, appearing in sections as a network, or a foamy or fibrillar organization. In early days these appearances were thought to represent exactly the living state, and there was much speculation regarding the functions of the solid and liquid constituents.

THE NATURE OF FIXATION. This subject was first investigated in a scientific manner by Fischer, whose findings were described in his book, *Fixierung, Färbung und Bau des Protoplasmas* (1899). Fischer mixed solutions of proteins with various fixatives and observed whether precipitates were formed. When precipitates were noted, he described them and recorded the extent to which they were soluble in water and other solvents. Fischer also compared the precipitating powers of many fixatives and described the characteristic *fixation image* produced by each of them.

Fischer concluded that the coagulation of liquid and semi-liquid constituents of tissues is always a process of precipitation, although he did admit that relatively firm structures such as nucleoli and fibrils may be fixed without the formation of visible precipitates. He was, of course, mistaken in believing that all fixing agents act primarily by inducing precipitation. This is evident from the fact that formalin and osmium tetroxide are energetic fixatives even though they do not coagulate proteins. In any case, Fischer's work made it plain that fixed protoplasm is a product of chemical action and may appear very different from living protoplasm.

Another important work on the subject appeared in the same year as Fischer's classical volume. This was Hardy's paper entitled *On the Structure of Cell Protoplasm.* Hardy cut thin sections of fixed egg albumen, stained them, and found, as did Fischer, that they showed networks like those seen in fixed protoplasm. However, Hardy went much further into the matter, observing that the size of the meshes in both albumen and cell cytoplasm depended upon the fixative used. He also noted the presence of small spherical masses at the junctions of the strands. Other investigators extended observations of this type to the nucleus, and it became clear that certain meshworks and other discrete structures seen in living cells are *artifacts* produced by alteration of the colloidal proteins. (An artifact is a structure not naturally present, but formed by reagents or by changes taking place in the cell as it dies.) Observers became more cautious in drawing conclusions from fixed material.

Unna (1911) placed great emphasis upon the fact that many fixatives are oxidizing agents. Potassium dichromate and osmium tetroxide are among the stronger oxidizing agents, and apparently they act upon lipids largely by oxidation. Oxidative processes may well enter into the fixation of other substances also, but there is no reason to believe that fixation is primarily due to oxidation. Formaldehyde and alcohol, both of which are strong fixatives, are reducing agents. Mercuric chloride and picric acid are only feeble oxidizing agents. Several good fixing mixtures, including the fluids of Helly and Regaud, contain both an oxidizing agent (potassium dichromate) and a reducing agent (formaldehyde).

The action of various fixing agents upon lipids differs greatly, and will be mentioned in the case of each class of fixatives. Carbohydrates generally are simply dissolved out or left in place, depending upon their solubility in the fixing fluid. For this reason, glycogen, which is soluble in water, can be demonstrated in fixed cells only if strong alcohol or certain other reagents are employed. Most metallic salts in tissues are not greatly affected by formalin or alcohol, but they will enter into chemical reactions with other fixing agents.

To summarize this discussion, we may say that what happens during fixation is a complex physicochemical process in which various structures react in different ways to different fixing agents. Extremists claim that the changes it produces are so radical that the study of fixed material possesses little value. However, what is known from the direct study of living material indicates that the structure of cells and tissues may be usefully demonstrated in fixed preparations, provided care and intelligence are exercised in the selection of fixing agents and in carrying out the processes of fixation and subsequent treatment. In critical work, it is always necessary to use different fixing agents or mixtures for the various constituents of the tissue or cell. A composite of the structural appearances observed in these various preparations cannot be taken to represent the organization of living cells, but it will yield much information which is of value.

Penetration of Fixatives. The rate at which a fixing agent penetrates a tissue is an important factor in accomplishing the purpose for which it is used. Various chemicals penetrate at different speeds, so that in the case of mixtures, certain of the components penetrate more rapidly than others.

The rate of penetration of fixing agents has been tested in two principal ways: (1) by measurement of the speed with which they diffuse through protein gels, causing coagulation, and (2) by measurement of the thickness of the zone of opacity noted after a piece of tissue has been left in a fixing agent for a given period of time. The results obtained by various authors using these methods are not in complete agreement; moreover, the speed with which non-coagulant fixatives penetrate and effectively alter constituents of tissues is more difficult to measure than the speed with which coagulant fixatives act. In general, however, it may be said that of the coagulant fixatives in common use, ethyl alcohol penetrates rapidly, mercuric chloride penetrates with moderate speed, and chromic acid and picric acid penetrate rather slowly. Of the non-coagulant fixing agents, acetic acid, formalin, and potassium dichromate penetrate more rapidly than osmium tetroxide.

THE HARDENING ACTION OF FIXING FLUIDS. This is a matter of interest because it determines to a considerable extent the ease with which fixed tissues can be sectioned. Some fixing agents make certain tissues very hard, while others leave them relatively soft. Differences in the extent to which tissues are hardened by fixatives have often been described in rather vague terms. An attempt to find numerical values which would express the hardening effects of different fixing agents was made by Wetzel in 1920. He measured the extent to which pieces of muscle tissue fixed by various agents were bent by a weight. To this measurement he applied a formula involving the dimensions of the piece of tissue, the weight used, and the amount of bending. By calculation he arrived at a coefficient representing the relative elasticity of the tissue. The larger the figure, the harder and less elastic the tissue. Wetzel's data show that absolute ethyl alcohol and acetone hardened muscle tissue to a greater extent than any other substance he tested; the coefficients for both of these substances exceeded 4500. The coefficients (approximate) of other fixing agents were as follows: 10% formalin, 1700; mercuric chloride (saturated aqueous solution), 1100; 0.5% chromic acid, 230; 3% potassium dichromate, 170; 0.5% osmium tetroxide, 170; picric acid (saturated aqueous solution), 70; acetic acid, 9. Wetzel's experiments were limited to one kind of tissue, and although it is not clear whether elasticity is a good index of cutting quality, his work was definitely a step in the right direction.

In most methods of embedding preparatory to cutting sections, the fixed tissue is hardened further in alcohol. Here also, the results of using various agents differ. Wetzel found that the hardness of fixed muscle, as measured in terms of its elasticity, changed after the tissue had been in 80% ethyl alcohol for 4 days. Material fixed in chromic acid, potassium dichromate, picric acid, mercuric chloride, and acetic acid became harder, although never so hard as tissue fixed initially in absolute alcohol. Comparable material fixed in formalin or osmium tetroxide became softer after 4 days in 80% alcohol.

CHAPTER 8

Fixing Agents

We shall now consider individually a number of fixing agents which are in wide use. Since techniques of desiccation and heating are comparatively unimportant, they will be discussed rather briefly. Proceeding to chemical agents, we shall study the properties of most substances which have been employed, with a view to obtaining a conception of the way in which they affect cellular and intercellular constituents. Following a discussion of each chemical substance, instructions will be given for its use, alone or in mixtures. This information will provide a basis for the selection of appropriate fixatives for various organisms or tissues, or for special purposes.

PHYSICAL AGENTS

Desiccation

Drying brings about extensive distortion in most kinds of cells, but the technique is nevertheless routinely used for smear preparations of blood and blood parasites, lymph, pus, spermatozoa, and some other materials. Desiccation not only brings about at least partial denaturation of proteins, but is a simple way to affix cells to a slide. Ordinarily, fixation of dried smears is completed by application of heat or by treatment with alcohol. Several of the well-known blood stains are applied as solutions in strong alcohol, so that they simultaneously fix and stain the dried cells. Among the reagents which operate in this way is the blood-staining mixture devised by Wright.

Heat

The coagulating effect of heat on proteins can be demonstrated by cooking an egg. But heat ordinarily brings about extreme shrinkage and other forms of distortion in nearly all tissues, and therefore is of no use in general histological work. Fixation by heat alone is now largely limited to only a few techniques.

SMEARS OF BACTERIA, SPUTUM, PUS, AND OTHER EXUDATES. In rapid diagnostic work, such smears are dried in the air and then passed several times above a small flame. They are then stained with methylene blue, carbol-fuchsin, or some other aniline dye or dye combination which will differentiate the organisms or cells being sought.

INSECT LARVAE AND ADULTS. These are sometimes killed by immersing them in hot or boiling water for one to several minutes. They are then preserved in 70 or 80% alcohol. This technique may be acceptable for taxonomic or anatomical material, but is unsuited for histological work.

ACCELERATION OF ACTION OF CHEMICAL FIXATIVES BY HEAT. See page 128.

CHEMICAL AGENTS

All of the really important fixatives come under this heading. Chemicals are used almost exclusively in liquid form, in order that they may penetrate the tissues. These include liquids (alcohol, acetic acid, chloroform, nitric acid), solutions of solids (chromic acid and its salts, mercuric chloride, picric acid, osmium tetroxide), and solutions of a gas (formaldehyde). For reasons discussed in the preceding chapter, the better fixing agents for most purposes are mixtures containing two or more of these substances. Vapors of osmium tetroxide or formaldehyde are often used to fix small organisms.

Our plan for discussing chemical fixatives will be as follows: There will be eight headings, corresponding to eight of the more important chemical substances employed in fixation. We will first consider the general properties and uses of each substance, and will then describe mixtures which owe their characteristics largely to its inclusion.

The literature of biological science abounds with hundreds of formulas for fixing mixtures, but a large proportion of these are superfluous. A great number simply offer no special advantages over the well-proved formulas. The objectionable ones are haphazard concoctions for which it would be difficult to imagine any good use. Some writers of general books on technique have added to the confusion by quoting dozens of formulas, good and bad alike, in vague terms and with little or no reference to their purposes. We shall concentrate our attention upon a small number of formulas which, if properly used, will serve most purposes. The discussion of each will include definite and specific statements regarding its composition, the kinds of structures it preserves well, the way in which it should be used, the after-treatment of materials fixed by it, and staining techniques for which it provides particularly good preparation.

Alcohols

Ethyl Alcohol. Of all reagents used in the histological laboratory, alcohol is the most venerable, having been introduced as an anatomical preservative by Robert Boyle, the great physicist, as early as 1663. In fact, it appears to have been the only tissue preservative employed until 1840, when Hannover reported on his use of chromic acid. With the subsequent introduction of other reagents and mixtures offering definite advantages, alcohol has come to be employed less as a fixing agent. However, it continues to have some applications as a simple fixative; in addition, it is an important ingredient of numerous mixtures. Alcohol is also commonly employed to continue the hardening of tissues, after fixation in other reagents.

For fixation, alcohol is generally used in a high concentration (about 80% or stronger). However, weak alcohol (about 30 or 35%) is used when it is desired to induce the cells of a tissue to dissociate, without at the same time allowing them to deteriorate. The applications of strong and weak alcohols are so different that they will be considered separately.

STRONG ALCOHOL. This reagent coagulates many proteins, but does not have a coagulating effect upon nucleoproteins. This means that alcohol is not a dependable fixative for chromatin, or even nuclei. It fixes fibrin, elastin, mucin, collagen, and zymogen granules and is also a good fixative for tigroid granules in nerve cells.

Most lipids are soluble, at least to some extent, in strong alcohol, so it should be omitted from any fixatives which are intended to preserve myelin, Golgi material, and lipids dispersed as granules in the cytoplasm. However, it should be noted that most of these structures become insoluble after treatment with osmium tetroxide, or after prolonged soaking in potassium dichromate, and they may then be safely dehydrated with alcohol.

Glycogen is preserved by strong alcohol and by mixtures which contain a high concentration of alcohol. In fact, these are the only fixatives which preserve this substance for differential staining by Bauer's technique or other methods. Alcoholic fixation is indicated also for the preservation of other water-soluble substances, including many metallic salts. Alcohol causes less chemical alteration than other fixing agents and, therefore, is very useful in preparation for some microchemical tests.

Alcohol penetrates with moderate speed, but not so rapidly as acetic acid or formalin. It hardens tissues far more than any other fixing agent, causes extensive shrinkage and distortion of the cytoplasm, and is a poor fixative for certain constituents of the nucleus. For delicate cells, strong alcohol by itself is probably the worst of all fixing agents. Cells of the testis and other soft organs are often rendered almost unrecognizable by treatment with strong alcohol. Obviously, alcoholic fixation should not be used for delicate tissues except when it is necessary for some special purpose, such as the preservation of glycogen or metallic salts. Absolute alcohol is employed to fix dry smears of blood, lymph, and pus. It seems to cause no further distortion of cells already dried.

For insects, small crustacea, and other invertebrates which are covered by a tough exoskeleton, strong alcohol is often a useful fixative. Not only does it preserve the exoskeleton, but it also fixes the internal organs of small crustacea and insects so that their structure shows very well in whole mounts. Perhaps the slowing of penetration by the exoskeleton and the dilution of the alcohol by water in the tissues result in a mild action, which brings about relatively little distortion.

Directions for Use. Alcohol having a concentration of 80 to 95% is employed for the fixation and preservation of insects and small crustacea, as well as for material which is to be tested for glycogen. Absolute alcohol has been recommended for glycogen, but it possesses no advantages over 95% alcohol, and is more expensive. The volume of alcohol used should be 15 to 20 times that of the material. At least 12 hours should be allowed for fixation of small crustacea (such as *Daphnia*); 24 hours for small insects in which the exoskeleton is not very thick; several days for larger insects; 24 hours for pieces of vertebrate tissue having a size of about 5 × 5 × 10 mm. Material of most types may remain in strong alcohol for several weeks before its staining qualities show marked deterioration. Material which is to be mounted unstained may remain in alcohol indefinitely.

Obviously, material fixed in alcohol requires no special after-treatment for removal of the alcohol. Specimens which are to be mounted entire, without being stained or bleached, are passed into absolute alcohol and then into terpineol-toluene or some other clearing agent. Similarly, one may transfer material to absolute alcohol and then proceed directly with methods of embedding in paraffin or nitrocellulose. If material is to be stained, it is first passed through the alcohol series to the same strength of alcohol as that in which the dye is dissolved, or, in the case of an aqueous stain, to water.

WEAK ALCOHOL (30 TO 35%). This is a useful fixing agent for tissues whose elements are to be separated by dissociation. It fixes the cell contents with little shrinkage, or even with a slight swelling, and softens the intercellular cementing substances to such an extent that the cells can be shaken or teased apart. Ranvier's one-third alcohol (1 part of 95% alcohol to 2 parts of distilled water) is one of the safer and more convenient reagents for isolation of cells and fibers. Its use is explained in Chapter 12.

Methyl Alcohol (Methanol). Pure methyl alcohol kills, fixes, and hardens tissues in the same manner as ethyl alcohol and may be used in its place, either alone or in mixtures. Methyl alcohol is tax-free and very inexpensive. *However, it is highly poisonous* if taken internally, and the inhalation of its fumes should be avoided as far as possible.

Alcoholic Mixtures. Combinations of alcohol with acetic acid, formalin, and chloroform will be described under this heading. Although it is generally conceded that they are less favorable than alcoholic solutions of mercuric chloride or of picric acid for preservation of many delicate structures, they usually penetrate rapidly. Moreover, for certain purposes, fixatives containing mercuric chloride or picric acid are objectionable.

Alcohol is a reducing agent and, as a general rule, should not be mixed with oxidizing agents such as chromic acid, potassium dichromate, or osmium tetroxide. One such mixture (Perenyi's fluid, p. 85) seems to be acceptable, for it serves to fix the eye without excessively hardening the lens, and also fixes the eggs of molluscs or fish without causing their abundant yolk to become brittle.

Carnoy's Acetic Acid-Alcohol

Absolute alcohol	3 parts
Glacial acetic acid	1 part

(*Mix immediately before use.*)

This mixture penetrates and fixes rapidly, causing somewhat less shrinkage than alcohol alone. The acetic acid partially counteracts the tendency of alcohol to shrink and to harden tissues. Acetic alcohol fixes nuclei and cytoplasm well enough for rough histological work, but is by no means suited for general use in histology. It is strange that some authors suggest that it be used to fix root-tips and similar delicate objects, because it causes great distortion. It should also be noted that this mixture destroys mitochondria, Golgi material, and calcareous structures.

In zoological work, acetic acid-alcohol is a good fixative for the entire bodies or chitinous parts of insects, crustacea, and other arthropods, as well as for arthropod eggs and larvae. Allow 1 to 3 hours for fixation of eggs, small crustacea, or similar objects. The time necessary for fixing increases with the size and hardness of the material, 24 hours being required for parts of large and thick-skinned insects. Transfer fixed material to 70 or 80% alcohol. Change this 2 or 3 times at intervals of several hours, in order to remove the acid. Material may then be embedded, stained, or stored in alcohol for some time.

Carnoy designed this fluid in order to fix the eggs of nematodes, which are enclosed in resistant membranes. For this purpose, however, the mixtures of Carnoy and Lebrun (p. 96) or Duboscq and Brasil (p. 103) are much better.

In histology, acetic acid-alcohol is used principally for preservation of glycogen in tissues. For this purpose, it may be preferable to alcohol alone, due to the fact that it causes less shrinkage. Fix 5 × 5 × 10 mm. pieces of tissue for 3 to 6 hours. Next, place them in absolute alcohol for 24 hours, changing the alcohol once or twice. Then proceed as directed on page 446.

Carnoy's Acetic Acid-Alcohol, with Chloroform ("Carnoy II")

Absolute alcohol	6 parts
Glacial acetic acid	1 part
Chloroform	3 parts

(*Mix immediately before use.*)

This combination penetrates and acts even more rapidly than Carnoy's original mixture. However, it is likely to produce still greater distortion and should be tried only in cases where penetration is especially difficult. It is said to be a good fixative for nematodes. Fix small specimens for 15 to 30 minutes, larger ones for 1 to 2 hours. Transfer them to 95% alcohol; change this two or three times, and proceed with staining, or dehydration and embedding. This fluid, like the preceding, preserves glycogen, but destroys Golgi material and calcareous skeletons. Some writers recommend it for general work on glandular and lymphatic tissues, for which it is quite unsuitable. These tissues do not require a fixative of unusual penetrating power and their delicate cells are badly distorted by this reagent. For many impermeable objects, including acanthocephalan worms and eggs of ascarid nematodes, the addition of mercuric chloride to this mixture is desirable (refer to Carnoy and Lebrun's fluid, p. 96).

Formalin-Alcohol

Alcohol, 85%	90 ml.
Formalin, commercial	10 ml.

This is a convenient and rapid fixative for rough work in zoology, histology, or pathology. Owing to the absence of acetic acid, it brings about more shrinkage than the fixatives to be considered next, but for the same reason does not attack calcareous structures. Formalin-alcohol is quite commonly used to fix bones and teeth which are subsequently to be decalcified. The advantages of fixing such tissues with a reagent which does not decalcify, and later treating them with an acid to dissolve the calcium deposits, are mentioned on pages 145 and 148.

Formalin-Acetic Acid-Alcohol (FAA) Mixtures

These are useful fixatives for rapid histological and pathological work, and can be used with tissues which are to be sectioned and mounted preceding impregnation of nerve elements according to Bodian's Protargol method. The following are representative formulas:

	Author		
	Lavdowsky	*Tellyesniczky*	*Galigher*
Formalin	10 ml.	5 ml.	10 ml.
Alcohol (approximately 95%)	30 ml.	75 ml.	70 ml.
Distilled water	60 ml.	25 ml.	15 ml.
Acetic acid, glacial	2 ml.	5 ml.	5 ml.

(*Add the acetic acid immediately before use. The other ingredients may be mixed in advance, if desired.*)

Lavdowsky's mixture contains less alcohol than the other two mixtures. It penetrates and fixes less rapidly, but at the same time it brings about less shrinkage. It is an adequate all-purpose fixative for invertebrates which do not possess a hard exoskeleton and in which there are no calcareous structures requiring preservation. It is convenient for field work. This fluid is also quite useful for preserving large vertebrate embryos, especially those of mammals.

The other two formulas yield mixtures which penetrate and fix more rapidly. In this respect, the mixture suggested by Galigher is better than that proposed by Tellyesniczky. Both fixatives are suitable for arthropods and arthropod eggs. For such materials they are preferable to Carnoy's fluids, except in cases requiring the extreme penetrating power of the latter.

In any of the above mixtures, a minimum of 3 to 6 hours should be allowed for fixation of *Daphnia*, *Cyclops*, and similiar organisms; 12 to 24 hours for larger crustaceans, small insects, insect larvae and insect eggs, arthropods with thick exoskeletons, and blocks of vertebrate tissue; large vertebrate embryos should be left in the fixative for 2 or 3 days. Material may remain in any of these fluids for at least a week without danger of injury. Transfer specimens from Lavdowsky's mixture to 50% alcohol, and then to 70% alcohol. From either of the other mixtures, transfer material to 70% alcohol.

Nitric Acid-Alcohol

Alcohol, absolute	90 ml.
Nitric acid, 25% aqueous solution	10 ml.

This mixture is useful for fixation of cardiac muscle, because it preserves intercalated disks better than most histological fixatives in routine use. However, it is not a good general fixative, and its applicability is therefore restricted. Leave the tissue in the fixative for about 24 hours, then wash out the nitric acid in at least three changes of 95% alcohol before completing dehydration and embedding.

Acetic Acid

This organic acid is familiar because of its presence in vinegar (to the extent of about 4 to 10%), to which it gives the characteristic sharp taste and smell. Pure anhydrous acetic acid is usually termed *glacial acetic acid*, because it freezes at 16.7° C. (about 62° F.), forming a crystalline, ice-like mass. After cold nights it may be necessary to liquefy this acid by placing the bottle in warm water. Acetic acid mixes with water and alcohol in any proportions, and the mixtures do not freeze so readily.

Acetic acid, in the form of vinegar, has been used since ancient times for the "pickling" of food matter. The earliest record of its application to histology is found in a paper by Remak (1854), who indicated that its utility for demonstrating cell nuclei was already known at that time. In 1874, Auerbach, a Dutch anatomist, described its action on the nuclei in the liver of the carp. A few years later, Flemming included acetic acid in a fixing mixture which rapidly found favor. As Flemming's fluid grew famous, the desirable properties of acetic acid became widely known, and it was incorporated into mixtures with various other substances. Today acetic acid is used frequently in combination with many other fixatives. It is employed as the sole fixing agent in certain techniques of simultaneous fixation and staining, as in the preparation of chromosomes by the aceto-carmine method.

Acetic acid does not cause coagulation or gelation of most proteins, and on this account is not by itself suitable for fixation. Nevertheless, it does have a most important and useful effect upon cells by causing them to swell. As this fact suggests, the inclusion of acetic acid in mixtures will counterbalance to some extent the shrinking effect of other fixing agents, such as alcohol. The swelling brought about by this and many other acids is probably due at least in part to formation of non-diffusible salts with proteins. Because of the osmotic pressure exerted by these salts, water enters the cell and causes it to become swollen.

Acetic acid penetrates rapidly. When it is used in mixtures, it probably penetrates more quickly than the other ingredients. Any swelling which it may cause at first is usually more than offset by the shrinking effect of the other ingredients, so that the cells become smaller than they were originally. Mucin and collagen are also caused to swell by acetic acid, but these substances tend to remain swollen.

Acetic acid precipitates nucleoproteins in the interphase nucleus, and fixes chromosomes with less alteration than most other agents. Therefore, it is included in nearly all mixtures used for fixation of stages in mitosis and meiosis.

In concentrations in which it is normally used in preparation of fixatives, this reagent has no effect upon fats. In general, it is destructive to mitochondria, although the extent to which it dissolves or alters these bodies differs for various species and even for different cells of the same animal. This matter is discussed further in connection with Flemming's fluid. The effects of acetic acid upon Golgi material are not well known, but it is almost never included in fixatives used for cells in which Golgi material is to be demonstrated.

As might be expected from our knowledge that acetic acid does not fix cytoplasm, it hardens less than any other fixative. Not only does it fail to harden tissues, but it lessens the extent to which they are hardened by later treatment with alcohol.

Acetic acid may be included in mixtures with all other fixing agents.

In making acetic acid-alcohol mixtures (including fixatives containing mercuric chloride), add the acid just before use. On standing, it will react with alcohol to form an ester. It is a wise precaution, in all cases, to add the acetic acid at the time of use, as it may react with other ingredients.

No special washing-out process is required for acetic acid itself, as it diffuses out quickly into water or alcohol. Other ingredients of the fixing solution will determine what special procedures for washing must be followed.

Formalin (Formol)

It is important to keep in mind exactly what is meant by *formalin* and *formaldehyde*. Formaldehyde is a gas produced by the oxidation of methyl alcohol. Formalin (also called formol, or formolose) is a saturated solution of this gas in water, and contains approximately 37% of formaldehyde, by weight. A 10% solution of formalin is therefore the equivalent of a 3.7% solution of formaldehyde. Commercial formalin can be conveniently measured volumetrically, and *throughout this book, all figures for quantities or percentages of this substance refer to commercial formalin.*

The introduction of formalin as a fixing agent was made by Blum, in 1893. He had previously discovered its antiseptic properties. Blum first noticed the hardening effect of this substance upon his own fingers. This led him to try it as a preservative, and finally to section and stain several tissues after hardening them in formalin. He found them to be quite well fixed.

For fixation, formalin is very commonly employed in aqueous solutions of about 10%, although weaker or stronger solutions are used for some techniques. It is a convenient and economical fixing agent for certain purposes which will presently be discussed. One special advantage of formalin is that it leaves tissues in favorable condition for many stains and microchemical tests. Formalin also occupies an important place in many of the more widely used fixing mixtures, as will be apparent throughout this chapter.

In the course of a few months, a white solid may slowly accumulate in bottles of formalin. This is paraformaldehyde, a product of polymerization. In order to minimize this chemical change, it is well to keep the stock of formalin in brown bottles or opaque containers, and to store these in a cool place.

Formalin is a strong fixing agent. It renders tissues tough and elastic, but not brittle. It does not coagulate nucleoproteins or other proteins. Acting as an additive fixative, formalin hardens protein gels, and does this without separating the water from the protein of which they are composed. In cells fixed in formalin, the general form is usually well preserved, and the cytoplasm is finely granular, not spongy as after fixation in coagulant fixatives. The nucleus often appears very much as it did in the living cell.

In cells fixed in formalin, most lipids are preserved and are in favorable condition for staining with dyes of the Sudan group or for impregnation with osmium tetroxide, although they are still subject to dissolution by their usual solvents. Certain phospholipids (as lecithin) become less soluble after they have been treated with formalin.

Formalin is employed, along with compounds of chromium, to precede the impregnation of nerve cells and fibers with silver, gold, or other metals. For the fixation of mitochondria, it is employed in mixtures with potassium dichromate.

The rate of penetration of tissue by formalin is rapid. However, fixation in simple formalin solutions should be continued for at least 1 or 2 days, as this substance acts slowly upon proteins. Tissues may safely be left in formalin solutions much longer than in alcohol. For ordinary purposes, their staining properties show little or no deterioration after weeks, or even months, and overhardening does not occur.

Formalin itself produces little shrinkage, but material fixed in formalin shrinks considerably when it is subsequently treated with alcohol. This is not excessive if sections are made by the freezing or nitrocellulose method, or in the case of most objects which are dehydrated and mounted whole. Entire objects fixed in formalin become more transparent after they are cleared than they do when other fixatives have been used. However, fixation in formalin alone is generally not suitable for material which is to be embedded in paraffin. Formalin prevents at least some of the proteins of cells and intercellular substances from being coagulated and hardened by alcohol. Thus when the tissues are exposed to a clearing agent and then to hot paraffin, great shrinkage and distortion may take place.

Valuable fixing mixtures are obtained by combining formalin with alcohol, acetic acid, picric acid, or mercuric chloride. Since formalin is an active reducing agent, it would seem irrational to mix it with strong oxidizers such as chromic acid or potassium dichromate. In practice, however, some very fine fixing fluids are thus obtained; among them are the fixatives of Regaud, Orth, Bensley, Navashin, and various modifications of Zenker's fluid which contain formalin. Obviously, all such mixtures are unstable and should be prepared a short time before use.

Formalin, by itself, requires no special techniques of washing, and storage of tissue in alcohol or passage of tissue through an alcohol series will remove it.

If material preserved in aqueous formalin is to be stained in an aqueous dye solution, first wash it with several changes of water, over a period of one to several hours, depending upon the size of the objects. The washing of material in preparation for metallic impregnation is described in connection with specific methods of impregnation.

Many stains give good results with material fixed in formalin alone. These include hematoxylins and numerous aniline dyes. Among basic aniline dyes which should be mentioned are methylene blue, basic fuchsin, and crystal violet. Acid aniline dyes such as eosin, orange G, acid fuchsin, and azocarmine G also work well. Carmines yield stains of fine brilliance and selectivity after formalin fixation, notwithstanding an old idea to the contrary.

NEUTRALIZING FORMALIN. Formalin solutions generally have a slight acid reaction, owing to the fact that formaldehyde oxidizes slowly to formic acid. In ordinary histological work, this is no particular disadvantage. However, there are certain special cases in which acidity is harmful, and for these it is necessary to neutralize the formalin, or to use only refined grades which have been shown on analysis to be free of acid. Acidity may cause the formation of artificial vacuoles in cytoplasm, which would interfere with the study of vacuoles normally present. Some cytoplasmic granules, such as those of the islet cells of the pancreas, cannot be preserved well unless the formalin is neutral. If formalin is used for the preservation of organisms having delicate calcareous skeletons (for example, calcareous sponges and pluteus larvae), it must be nearly neutral, for even low concentrations of acid may dissolve such skeletal structures either partially or wholly. Acidity is also considered unfavorable for impregnation of tissues with silver according to the methods of Ramón y Cajal, Bielschowsky, and Del Río Hortega.

Neutralization of commercial formalin may be accomplished by placing enough calcium carbonate in the bottle to make a layer 1 to 5 cm. deep (depending upon the size of the bottle). The mixture should be shaken every day or two, for 10 days, after which the carbonate may be allowed to settle and the clear liquid decanted off as needed. Some workers prefer to neutralize the formalin after diluting it with water, by adding a strong aqueous solution of sodium borate or sodium carbonate, a little at a time, until a sample withdrawn for testing shows a red color with phenolphthalein. The effect of several alkaline substances upon the acidity of solutions of formaldehyde is mentioned by Baker (1958, p. 112.)

When formalin is used as the sole fixing agent, it generally is applied as a 10% solution. Such a ten-fold dilution, especially if it is made with neutral formalin, or if it is kept neutral after dilution, will be perfectly suitable for simply preserving organisms. It may also serve acceptably as a fixative for small organisms such as hydroids, ectoprocts, rotifers, trematodes, cestodes, ascidians, and many kinds of larvae which are to be prepared as whole mounts. It has been used for histological material, but for this purpose it is much inferior to Bouin's fluid, Zenker's fluid, and some other standard fixatives, unless certain constituents of the tissue are destroyed by these fixatives. Extensive use of formalin solutions in pathological work is justified by the speed with which the tissue can be washed and made ready for freezing and sectioning. To hasten fixation, the formalin solution may be warmed to about 35° or 40° C.

For contractile invertebrates (which usually are narcotized first), it is a common practice simply to add to the water containing the specimens enough commercial formalin to make about one-tenth of the total volume. Or, one may cautiously withdraw most of the water from the specimens and quickly pour in a quantity of 10% formalin, heated to about 60° C. In this way, fresh-water hydras and some ectoprocts (such as *Bugula*) may be fixed in an expanded state, without prior narcotization.

BUFFERED FORMALIN. Because dilute formalin is apt to become oxidized (and therefore acid) more quickly than strong solutions, it is good to buffer solutions prepared for use as fixatives. The following formula works well, and even specimens with delicate calcareous skeletons may be left in it for some time without risking dissolution of the skeletons.

Buffered Formalin Solution

Formalin	100 ml.
Distilled water	900 ml.
Sodium phosphate, monobasic ($NaH_2PO_4 \cdot H_2O$)	4 gm.
Sodium phosphate, dibasic, anhydrous (Na_2HPO_4)	6.5 gm.

FORMALIN IN PHYSIOLOGICAL SALINE SOLUTIONS AND IN SEA WATER. A number of investigators have noted that certain types of cells or tissues of terrestrial and fresh-water animals are preserved more faithfully by formalin if this is diluted by an isotonic saline solution than by distilled water. The fixation of some marine invertebrates, or their eggs or tissues, is superior if the formalin is mixed with sea water. The idea that the fixation is better because the isotonic solution minimizes distortion is evidently not correct, however.

The osmotic pressure of a 10% solution of formalin in water is already considerably higher than that of vertebrate bloods. The use of a saline solution simply raises this osmotic pressure further. Moreover, what is known about the effects of fixatives contradicts the idea that those fixatives having a high osmotic pressure always cause shrinkage and that those having a low osmotic pressure always induce swelling. When a solution of indifferent salts has advantages, it is likely that these are not so closely related to the osmotic pressure as to the reactions which take place between the fixative and the proteins of the cells.

The possibility that formalin diluted by a saline solution will give better results should definitely be explored if this substance is to be used for fixation of a particular type of material, and in fixation of marine organisms it is certainly safe if not desirable to use sea water more or less routinely.

USE OF WEAK FORMALIN FOR DISSOCIATION. To dissociate epithelia, nerve cells, and some other tissue elements, use a very weak solution, made by adding 0.5 ml. of formalin to 250 ml. of water or physiological saline solution. Detailed directions are given on page 180.

FORMALIN VAPOR. Vapor fixation is used occasionally for isolated cells, thin membranes, smears, and very small bits of solid tissue. Vapor of osmium tetroxide is used more frequently, but it is unfavorable to some stains, including Sudan III and IV (commonly used for lipids), and to methods of impregnation with silver or gold. When these stains or methods of impregnation are to be used for objects such as those mentioned, formalin vapor may be employed. Lay or smear the material upon a slide and invert this over the mouth of a bottle or dish which contains commercial formalin. Place the container in an incubator, at a temperature of about 37° C. Isolated cells or thin smears will be fixed within 15 to 30 minutes, while several hours may be required for small bits of solid tissue. After fixation, proceed with the desired method of staining or impregnation.

FORMALIN-GLYCEROL. When material must be stored in formalin for many weeks or months, it is well to add about 10 ml. of glycerol to each 100 ml. of 5 or 10% formalin. This will keep material soft and pliable.

OTHER FORMALIN MIXTURES. Combinations of formalin with alcohol and acetic acid have been considered under Alcohols. Numerous mixtures of formalin with picric acid, mercuric chloride, and various other chemicals are discussed in connection with these compounds.

RE-FIXATION. This procedure is often advantageous when it becomes necessary to stain formalin-preserved material by a method which gives superior results after a different fixative. For example, re-fixation with Zenker's fluid often leads to better results with Mallory's triple stain.

For Blocks of Tissue

1. Remove formalin by placing the tissue in dilute ammonia water (30 drops of ammonium hydroxide to 100 ml. of water) for 2 days, in an incubator at about 37° C.
2. Wash for 24 hours in running water.
3. Re-fix in an appropriate mixture such as Zenker's fluid, Helly's fluid, or Bouin's fluid.
4. Wash in water or alcohol, depending upon the fixative used; dehydrate, embed in paraffin or nitrocellulose, section, and stain.

For Sections

1. Place sections for 30 minutes in ammonia water or ammonia alcohol (30 drops of ammonium hydroxide to 100 ml. of distilled water or 70% alcohol).
2. Wash for 1 hour in running water.
3. Re-fix for 1 hour in the desired fluid.
4. Wash in water or alcohol, according to fixative used. Iodize sections after Zenker's fluid or other fixatives containing mercuric chloride.

5. Stain, dehydrate, and mount sections.

REMOVAL OF PRECIPITATES CAUSED BY FORMALIN. A fine crystalline precipitate, brown or black in color, is sometimes produced in tissues by the action of formalin upon laked hemoglobin. To remove this, treat sections or blocks of tissue with dilute ammonia and then wash them thoroughly with water, as described in steps 1 and 2 of the preceding methods. After that, prepare and stain them in the usual way.

Chromic Acid

A solution of chromic acid is prepared by dissolving its anhydride, chromium trioxide, in water. The reaction is $CrO_3 + H_2O \rightarrow H_2CrO_4$. The anhydride absorbs moisture rapidly from the air, and for this reason it should be kept in a tightly stoppered container. It is a good idea to keep in the laboratory a 1% aqueous solution of chromic acid ready for use in the preparation of fixing mixtures.

Chromic acid was probably the first metallic compound to be employed for histological fixation. Its use for this purpose was described by Hannover, in 1840. Hannover began fixing various tissues in chromic acid after he had seen it used successfully for the preservation of an eye in the laboratory of Jacobson, at Copenhagen. Chromic acid came to be widely known because of its inclusion in the mixtures introduced by Flemming, and it is still used chiefly in mixtures of the same general type.

Chromic acid is a strong fixing agent and acts by coagulating proteins, with which it apparently forms chemical compounds. The process must involve some oxidation, since the fixed tissues commonly develop a greenish color which indicates the presence of chromium oxide. When used alone, chromic acid coagulates the cytoplasm as a network. It also induces the formation of a coarse network in the interphase nucleus, but fixes chromosomes rather well.

This substance is generally combined with acetic acid and osmium tetroxide, which counteract some of its undesirable effects on cytoplasm and interphase nuclei. Such mixtures include the well-known fluids of Flemming. Chromic acid is an active oxidizer and, on this account, some workers believe that it should not be mixed with reducers such as formalin or alcohol. Nevertheless, some such mixtures (Karpechenko's fluid and Perenyi's fluid) are excellent fixatives for certain purposes. All chromic acid mixtures should be made from stock solutions immediately before use.

Little is known about the effects of chromic acid on lipids, but if the fat of adipose tissue is exposed to prolonged treatment, it will become insoluble in the usual solvents of lipids. Alone, or in combination with acetic acid, it destroys mitochondria. However, chromic acid-osmium tetroxide mixtures containing little or no acetic acid are among the better fixatives for mitochondria.

Chromic acid penetrates slowly, and mixtures which contain it sometimes fix unevenly at different depths. On this account it is customary to use them only for very small pieces of tissue. It is often stated that material fixed in chromic acid is brittle and difficult to section, but perhaps it is more accurate to say that its hardening effect is moderate. The shrinkage which it causes is also moderate. Standard mixtures in which chromic acid is only one of the ingredients generally bring about less shrinkage.

In using chromic acid mixtures, the volume of fixing fluid should be large—25 to 50 times that of the tissue. The time of exposure is also important and differs for each mixture and type of material. It is a wise precaution to keep material in darkness during fixation, since some elements of cells fixed in chromic acid are known to deteriorate under the influence of light.

As soon as fixation is complete (see paragraphs on various mixtures), all traces of chromic acid should be removed by washing the material for at least 12 hours in running water. Otherwise, a precipitate of insoluble chromium suboxide may appear when the tissue is brought into alcohol, especially in the light. Methods of washing are discussed further in Chapter 10, page 133.

Material fixed in chromic acid is often stained very successfully with iron hematoxylin. Alum hematoxylin yields only fair results. Basic aniline dyes (safranin, crystal violet, basic fuchsin) sometimes give unusually brilliant differentiation following fixation in mixtures containing chromic acid. Carmines, however, do not stain well after fixatives of this type.

Chromic Acid Mixtures. FLEMMING'S FLUIDS. In his *Zellsubstanz, Kern und Zelltheilung* (1882) Flemming stated that he based his fixing fluid (the first formula, often referred to as Flemming's "weak" fixative) upon von Flesch's chromic acid-osmium tetroxide mixture. He found that the addition of acetic acid improved fixation of the nucleus and its staining properties.

Flemming's "Weak" Solution

Chromic acid, 1% aqueous solution	25 ml.
Osmium tetroxide, 2% aqueous solution	5 ml.
Acetic acid, 1% aqueous solution	10 ml.
Distilled water	60 ml.

Flemming's "Strong" Solution

Chromic acid, 1% aqueous solution	75 ml.
Osmium tetroxide, 2% aqueous solution	20 ml.
Acetic acid, glacial	5 ml. (or less)

(In both of the above formulas, add the acetic acid immediately before use. The osmium tetroxide-chromic acid combination may be kept ready-mixed in the correct proportions.)

Flemming's mixtures, containing the proportions of acetic acid indicated above, are excellent for the fixation of nuclei and mitotic figures in *very small pieces of tissue*. The osmium tetroxide gives homogeneous fixation of cytoplasm, preserves mitotic spindles, asters, and fine fibrils, and prevents coarse coagulation of proteins; acetic acid and chromic acid fix the chromosomes; and chromic acid hardens the tissue and prepares it for staining with basic dyes. However, these fluids are absolutely unsuitable for general histological work, as they penetrate poorly, harden excessively, blacken the material, and interfere with the action of many dyes. In objects of considerable size, the outer layers are over-fixed by osmium tetroxide and appear homogeneous, while the center portion shows a coarser coagulation characteristic of the influence of chromic acid. Between the two, there is often a layer of beautifully fixed cells.

The "weak" solution is suitable for very small and delicate organisms, such as the larvae of coelenterates, echinoderms, annelids, and molluscs. It may also be employed for tiny fragments of tissue from larger animals. Use a relatively large volume of the fluid (at least 50 times that of the tissue) and allow it to act for 24 to 48 hours.

The "strong" solution may be employed for all but the smallest and most permeable objects. Fix small pieces (2-mm. thick) of testis or other organs for 24 to 48 hours. The volume of liquid need be only 4 or 5 times that of the tissue.

After fixation in either mixture, wash the material in running water, or in many changes of water, for 12 to 24 hours, then pass it up through the alcohols and proceed with the paraffin method. Iron hematoxylin follows these fixatives very well.

FLEMMING'S FLUID WITH LITTLE ACETIC ACID. An excellent fixative for mitochondria is obtained by omitting all or most of the acetic acid from Flemming's "strong" solution. Fixation of chromosomes may still be good enough to permit determination of stages in maturation. This is not the case with Altmann's fluid and some other fixatives for mitochondria. The harmful effects of acetic acid upon mitochondria differ for various kinds of cells even in the same animal. It is desirable to include some acetic acid, if this can be done, in order to improve the fixation of nuclei and chromosomes. In working with a type of cell which has not been investigated, acetic acid may be omitted in first trials. After that, the effects of adding 0.5 to 5% of the acid may be tested.

Fix small pieces of material for 24 to 48 hours, wash them in running water for 12 to 24 hours, and proceed with the paraffin method. For techniques of demonstrating mitochondria, see Chapter 26.

KARPECHENKO'S FLUID. This fluid was designed for work in plant cytology, particularly for the fixation of nuclei and chromosomes. It is an excellent fixative for this purpose. It penetrates rapidly and produces a most delicate and beautiful fixation of both interphase and dividing nuclei

Karpechenko's Fluid

Chromic acid	0.5 gm.
Distilled water	54.5 ml.
Acetic acid, glacial	5 ml.
Formalin	40 ml.

(The chromic acid may be dissolved in the distilled water and kept as a stock solution to which the other ingredients are added immediately before use; or use 50 ml. of a 1% stock solution of chromic acid plus 4.5 ml. of water.)

in root tips, stamens, and pistils. At the same time, it gives a good fixation of cytoplasm. Karpechenko's fixative has been used very little for animal material. It has, however, been found to be quite good for stages in the transformation of spermatids, and deserves more extensive trial.

Fix small objects or thin slices of tissue for 6 to 8 hours, or overnight, keeping the container in darkness or subdued light. Then wash them in running water for about 6 hours, pass them through the alcohols, and proceed to embed according to the paraffin method. Iron hematoxylin and basic aniline dyes (for example, safranin) give fine stains following this fixative.

Perenyi's Fluid

Chromic acid, 1% aqueous solution	15 ml.
Nitric acid, 10% aqueous solution	40 ml.
Alcohol, 95%	30 ml.
Distilled water	15 ml.

(Mix immediately before use.)

This is a special reagent for some objects which become extremely hard and difficult to section after ordinary histological fixatives. Among these is the entire vertebrate eye. Perenyi's fluid fixes the retina with very little separation of its layers, and it does not harden the lens excessively. Fix the entire eye for 12 to 24 hours, depending on its size, keeping the container in darkness or subdued light. Next place it in 70% alcohol for 48 hours or longer, changing the alcohol once or twice. After that, transfer it successively to 80%, 95%, and absolute alcohols, allowing it to remain in each for 2 days or longer. Then proceed to open the eye, embed it in nitrocellulose, and section it, as described on page 497.

This mixture is used also for the yolky eggs of arthropods, molluscs, fishes, and some other animals, in which it leaves the yolk soft enough to be sectioned. Fix such objects for 10 to 12 hours, transfer them to 70% alcohol, and embed by the paraffin method.

Following this fixative, material stains rather poorly. Use alum hematoxylin and eosin for general structure, iron hematoxylin for details.

Potassium Dichromate (Potassium Bichromate)

Potassium dichromate was introduced as a fixing agent by Müller, in 1859. "Müller's fluid" consists of a 2.5% aqueous solution of potassium dichromate, to which is added 1% of sodium sulfate. (It is possible, however, that the sulfate serves no useful purpose.) Müller employed this solution to fix smooth muscle and nerve plexuses in the choroid coat of the eye. Soon it came to be widely used for fixing tissues of all kinds. In later years, mixtures of potassium dichromate with other substances proved superior. Today, simple solutions of potassium dichromate are employed only for hardening tissues (chiefly those of the nervous system) which are to be impregnated with silver or mercury, and in preparation for staining the myelin sheaths by Weigert's method. But even for these methods, it is more common to use mixtures.

Potassium dichromate does not coagulate nucleoproteins or other proteins. Unless it is acidified, it does not fix chromatin, and on this account it is decidedly unsuitable for work with nuclei and chromosomes. However, mixtures of potassium dichromate with mercuric chloride, formalin, or other substances which do fix nuclei include a number of important histological and cytological reagents. A potassium dichromate solution which has been acidified (as with acetic acid) behaves much like chromic acid, coagulating chromatin and cytoplasmic proteins as networks, but fixing chromosomes rather well.

The effects of potassium dichromate upon fats and other lipids are related to its strong oxidizing properties. In a short period of fixation (up to 12 hours), most of these substances are little affected, and are subsequently dissolved by alcohol or other solvents. But if the treatment is prolonged, the lipids are oxidized and resist solution. This reaction with lipids probably is responsible to some extent for the effectiveness of potassium dichromate in fixing cells in which mitochondria are to be demonstrated. Potassium dichromate is also used in connection with certain techniques for impregnating nerve cells and nerve fibers with silver and mercury.

Potassium dichromate diffuses rather rapidly into tissue. However, because it does not cause coagulation of proteins, or even gelation of most of them in the relatively short time it is likely to be applied to tissue, the speed with which it moves through tissue cannot be compared with the speed with which other fixing agents effectively produce changes in the proteins. Potassium dichromate hardens tissues slowly and confers upon them a moderately good cutting consistency. Little shrinkage occurs during fixation itself, but the total shrinkage in the case of delicate cells may be more than 50% by the time the tissue is finally embedded in paraffin.

Since potassium dichromate is a strong oxidizing agent, some writers argue that it should never be mixed with formalin or other reducers. In practice, however, several potassium dichromate-formalin mixtures (the fixatives of Regaud and Helly, for instance) have proved their worth. The formalin should always be added immediately before use.

With potassium dichromate and mixtures that contain this substance, it is advisable to use a large volume of fixing fluid (25 to 50 times the volume of the material). The fixed tissue should be washed in running water for 12 to 24 hours. If fixation is prolonged, the fixing solution should be replaced with a fresh mixture in the event that precipitates appear, or at least once a day in the case of potassium dichromate-formalin mixtures.

In tissues fixed with the following mixtures of formalin or acetic acid with potassium dichromate, nuclei are stained slowly but well by alum hematoxylin or iron hematoxylin. Eosin and other acid dyes for cytoplasm also act rather slowly but satisfactorily. Potassium dichromate mixtures containing little or no acetic acid cause mitochondria to stain well with acid fuchsin and with hematoxylin. Mixtures containing mercuric chloride will be treated under a discussion of that substance.

Orth's Fluid

Potassium dichromate	2.5 gm.
Sodium sulfate (may be omitted)	1.0 gm.
Distilled water.	100 ml.
Formalin, *to be added immediately before use*	10 ml.

This mixture, which is Müller's fluid to which formalin has been added, has a limited use in vertebrate histology. It can be employed with success for rather large pieces of tissue. It fixes the majority of tissues fairly well, preserving cytoplasmic granules and producing, in most cases, very little alteration in the shape of cells. Orth's fluid is quite useful for bulky pieces from the central nervous system. Excellent preservation of human and other mammalian kidneys has been achieved with it, but results with testis and lung tissue have been poor. A serious disadvantage of this mixture is that it often causes nuclei to assume an unnatural, hollow appearance. The fixation of nuclei can be improved by adding 5 ml. of glacial acetic acid to the above formula. In cases where it is practical to use smaller pieces of tissue, a fixative containing mercuric chloride (such as Zenker's fluid) or picric acid (such as Bouin's fluid) will show histological details more clearly.

Slices of tissue may be as large as several centimeters in length and width, but should not be more than 7 or 8 mm. in thickness. Small pieces of tissue are hardened sufficiently in 2 days, but larger ones require 4 or 5 days. During fixation, it is best to keep the containers in darkness or subdued light. Wash fixed material in running water for 24 hours. Then pass it up through the alcohols and embed it in paraffin or nitrocellulose. If sections show a brownish color, bleach them with potassium permanganate and oxalic acid (p. 144).

Material fixed in Orth's fluid stains well, though quite slowly, with alum hematoxylin or iron hematoxylin, and takes acid counterstains such as eosin acceptably.

Smith's Fluid

Potassium dichromate	5.0 gm.
Distilled water	87.5 ml.
Formalin, commercial	10 ml.
Acetic acid, glacial	2.5 ml.

(The formalin and acetic acid should be added immediately before use.)

This mixture has been found to be quite good for fixation of amphibian eggs and embryos, as well as other yolky materials. Remove as much as possible of the jelly surrounding amphibian eggs and place them in a large quantity of the fluid, for 24 hours. Rinse them with several changes of water and preserve them in 3 to 5% aqueous formalin, until needed. If the formalin becomes decidedly yellow in color, change it.

Zenker's Fluid

Stock Solution:	
Potassium dichromate	2.5 gm.
Mercuric chloride	5.0 gm.
Sodium sulfate (may be omitted)	1.0 gm.
Distilled water	100 ml.
Add immediately before use:	
Acetic acid, glacial	5.0 ml.

Zenker's fluid is one of the best and most widely used fixing reagents for histological and embryological materials. It preserves nuclei, cytoplasm, and connective tissue elements quite well, but, owing to the presence of acetic acid, it destroys mucin, mitochondria, zymogen granules, and some other elements of the cytoplasm. It is satisfactory for most routine work in histology and pathology. In embryology it serves very well for the preservation of mammalian embryos. However, this fluid does not fix either nuclei or cytoplasm suitably for cytological purposes. For critical studies on the histology of delicate tissues (such as testis) it is not as good as Bouin's fluid.

Zenker's fluid is suitable for materials which are to be sectioned. Tissues fixed in this mixture have a tendency to become rather hard in paraffin, but when embedded in nitrocellulose they have a better cutting consistency than those fixed in Bouin's fluid. This reagent is definitely unfavorable for objects which are to be mounted entire. Specimens fixed in it are not so transparent after clearing as those fixed in alcohol, formalin, mercuric chloride, acetic acid, or picric acid mixtures. In addition, the carmine stains generally used for entire objects do not give best results following fixation in this mixture.

The time required for fixation ranges from about 3 hours for small and permeable specimens, such as young chick embryos, to about 24 hours for objects as dense as a slice of kidney 5 to 7 mm. in thickness. Do not leave material in this fluid much longer than is necessary, in order to minimize the formation of precipitates of mercury within the tissues. After fixation, wash material in running water for 12 to 24 hours. Then pass it up through the alcohol series to 70 or 80% alcohol and treat it with iodine in order to remove mercurial precipitates (see pp. 93–94). Continue with either the paraffin or nitrocellulose method of embedding and sectioning.

Many stains give excellent results with material fixed in Zenker's fluid or one of its modifications. Alum hematoxylin and iron hematoxylin stain slowly but very well. Eosin, generally used as a counterstain with alum hematoxylin, gives unusually striking and brilliant differentiation after this fixing agent. The eosin-methylene blue combination is also excellent for general histology. In order to obtain a brilliant differentiation of collagenous connective tissue, use Mallory's triple stain, Heidenhain's "Azan" method, or Lillie's modification of Masson's "trichrome" stain. Verhoeff's elastic tissue stain also works well after this fixative. It has previously been pointed out that carmines do not give their most brilliant or selective stains after Zenker's fluid or its modifications.

Helly's Fluid ("Zenker-Formol")

Zenker's fixative (stock solution)	100 ml.
Formalin, *to be added immediately before use*	5 ml.

Maximov's Fluid

Zenker's fixative (stock solution)	100 ml.
Formalin, *to be added immediately before use*	10 ml.

(In order to preserve lipids, 10 ml. of a 2% osmium tetroxide solution may be added to the above.)

In these mixtures, formalin is substituted for acetic acid. Their effects are similar to those produced by Zenker's fixative, but they preserve mucin, zymogen granules, and other cytoplasmic granules. For study of cytoplasmic granules, as those in the islet cells of the pancreas, use neutral formalin in preparing the solutions.

There is very little difference between the two formulas above, though the second may act somewhat more rapidly. With it, good fixations of nearly every organ have been obtained. Even entire eyes of monkeys, fixed without being punctured, show well the structural details of the retina. It seems, however, to be a poor fixative for testis.

The time required to fix specimens in these modifications will vary as stated for Zenker's fluid. During fixation, it is best to keep the solution in darkness or subdued light. Wash fixed material well in running water, bring it gradually to 70 or 80% alcohol, and iodize it according to the procedure generally followed after fixation in fluids containing mercuric chloride.

Iron hematoxylin differentiates zymogen granules well after fixation in these fluids. Various types of cytoplasmic granules, as those in the pituitary gland and islets of the pancreas, will be brought out clearly by staining with alum hematoxylin, Giemsa's stain, and Heidenhain's "Azan" method.

Champy's Fluid

Potassium dichromate, 3% aqueous solution	7 parts
Chromic acid, 1% aqueous solution	7 parts
Osmium tetroxide, 2% aqueous solution	4 parts

(*Mix the three solutions together immediately before use.*)

This mixture has been widely employed for demonstrating mitochondria, but it often prepares cells as a whole quite adequately for staining with iron hematoxylin. Fix very small pieces of tissue for several hours, then wash them thoroughly in running water for several hours or overnight.

Champy's fluid is also a good initial fixative for ciliated protozoa to be impregnated with silver according to the method of Chatton and Lwoff (p. 406). The ciliates are fixed (either in smears on coverglasses or by discharging them into the fixative with a pipette) for a few minutes, then transferred directly to Da Fano's fixative, in which they should remain for at least 2 or 3 hours. A longer period does no harm. It is a good idea to change the Da Fano's fixative at least once to prevent carrying over any of the ingredients of Champy's fluid into subsequent steps.

Bensley's Fluid; Regaud's Fluid. These are mixtures of potassium dichromate and other substances and are designed especially for fixation of mitochondria. Their formulas and applications will be found in Chapter 26.

Mercuric Chloride (Bichloride of Mercury; Corrosive Sublimate)

As early as 1854, Remak employed a dilute solution of mercuric chloride to preserve liver cells of a rabbit embryo. The use of stronger solutions seems to have been introduced by Lane in 1878.

Mercuric chloride is soluble in cold water to the extent of a little over 5%, but in boiling water or 95% alcohol, its solubility is raised to about 30%.

Physiological saline solutions increase its solubility to about 10%, and in sea water a concentration of about 15% can be obtained. However, the belief that such solutions will be isotonic is unfounded, for even if no other salts or formalin are added, the concentration of mercuric chloride itself will raise the osmotic pressure to a level higher than that of tissue fluids. It is convenient to keep on hand a saturated solution of mercuric chloride in distilled water (and also in physiological saline solution, if desired).

A simple solution of mercuric chloride produces a rapid and usually rather fine coagulation of most proteins of the cytoplasm and nucleus, although nucleoproteins are coagulated less effectively by it. In solutions which are more acid than the isoelectric point of the proteins—and this is the case with fixatives containing acetic acid—the mercury is added to the proteins as an ion. This reactivity is facilitated by the presence of sodium chloride. However, when the solution of mercuric chloride is close to the isoelectric point of the proteins, sodium chloride interferes with the reactivity of mercuric chloride with the proteins and may even encourage the dissolution of the coagulum. On this account, mixtures of mercuric chloride and acetic acid are generally preferable to mercuric chloride by itself, except when it is desired to study some substance, such as mucin, which would be affected adversely by the acid. Mixtures of mercuric chloride with other fixing agents will be discussed subsequently.

Mercuric chloride has little effect on triglyceride lipids, although it is said that in cells treated with mercuric chloride these become less stainable with dyes of the Sudan group than is the case with comparable cells fixed in formalin. Phospholipids form compounds with mercuric chloride, but the subsequent fate of these in solvents of lipids has not been studied.

Mercuric chloride penetrates with moderate rapidity, and its hardening effect is also moderate. By itself, it induces little shrinkage, but tissue fixed in it will generally shrink considerably after passage through various solvents and melted paraffin. However, material fixed in mercuric chloride has a fairly good cutting consistency. Brittleness has been said to result from fixation in mercuric chloride, but in general this seems not to be a serious problem.

Mercuric chloride is used in mixtures with many substances: potassium dichromate, osmium tetroxide, chloroform, alcohol, trichloroacetic acid, and acetic acid. Such mixtures are valuable for various types of work in general zoology and histology. Their use in cytology is more limited. Mann's mercuric chloride-osmium tetroxide mixture is employed for demonstrating Golgi material, and the powerful mixture of Carnoy and Lebrun has been widely used for the eggs of nematodes and other objects which are not readily penetrated. A saturated aqueous solution of mercuric chloride, with 5% acetic acid, is commonly used preceding Feulgen's stain for chromatin. As a rule, however, picric acid and chromic acid mixtures are preferable to mercurial reagents for preserving delicate structures such as mitotic figures and flagella.

Mercuric chloride solutions should be allowed to act only until fixation is assuredly complete. Prolonged exposure may result in the formation of excessive mercury precipitates which are difficult to eliminate. All traces of mercury must be removed from the material by subsequent washing and treatment with iodine. This matter is considered fully under Special Precautions (below).

Tissues which have been fixed in mercuric chloride or other mercurial mixtures are stained quickly and with good differentiation by a great many dyes. Alum hematoxylin and iron hematoxylin give very bold images of nuclei. Acid aniline dyes produce a brilliant, precise differentiation of cytoplasmic and intercellular structures. Carmines stain superbly after fixation in mercuric chloride and certain mixtures which contain it, although the results are no better than those sometimes obtained with formalin-fixed material. However, carmines do not give the best stains after fixation in mixtures of mercuric chloride with potassium dichromate (Zenker's fluid and its modifications) or with osmium tetroxide (Mann's fluid). Basic aniline dyes used for staining nuclei generally give good results after mercuric chloride.

Special Precautions to Observe in the Use of Mercuric Chloride. POISONING. Mercuric chloride is highly toxic. Take care that it does not get into the mouth or eyes. It will make dark spots upon the skin.

PREPARING SOLUTIONS. Use distilled water, never tap or spring water, for making solutions. This applies also to the preparation of physiological saline solutions to be used as solvents for mercuric chloride.

INSTRUMENTS. Never bring metallic instruments into contact with mercuric chloride or solutions containing it. If you do, you will soon find out why it is called *corrosive* sublimate. The instrument will be damaged and the solution contaminated. Handle the dry salt with a spatula of porcelain, glass, paper, wood, plastic, or horn. To manipulate material in mercuric chloride solutions, use glass pipettes, rods of wood, plastic, or glass, or forceps tipped with ivory or bone. If tissues must be pinned out before fixation, do this with hardwood splinters, sharp toothpicks, or glass needles.

PERIOD OF FIXATION. In most cases, it is important to leave material in mercurial solutions only until fixation is completed. Prolonged immersion may cause excessive mercury precipitates to form in the tissues. The time required for fixation varies according to the strength of the solution, the temperature, and the size and permeability of the objects. Some suggestions as to the time necessary for fixing objects of various sorts will be given in connection with each formula described. In case of doubt, remove a specimen from the fixing solution, slice it in two, and observe whether or not it has become white and opaque to the center. After some experience of this sort, one will know the approximate rate at which a certain mixture penetrates different types of material.

Washing and Iodizing Material. First it is necessary to wash out free mercuric chloride or other mercury compounds which remain dissolved in the tissues. The manner of washing differs in various cases, as follows:

1. From aqueous mercuric chloride, mercuric chloride-acetic acid, Heidenhain's "Susa" fixative, Gilson's fluid, Petrunkewitsch's fluid, Schaudinn's fluid, or Worcester's fluid, transfer material to 50% alcohol. After a few minutes or hours, depending upon the size of the object, put it into 70% alcohol. If there is a large quantity of material, it is well to discard and replace the 50 and 70% alcohols once or twice during the treatment.

2. From the fixative of Carnoy and Lebrun, transfer material to 95% alcohol. Change this alcohol 3 or 4 times, at intervals of one to several hours.

3. From Mann's fluid, rinse material in distilled water for 30 minutes; if Golgi material is to be demonstrated, proceed with the Mann-Kopsch method (p. 439).

4. From Zenker's fluid and formalin-containing modifications (Helly's fluid, Maximov's fluid), wash material in running water for 12 to 24 hours. Then place it successively in 30, 50, and 70% alcohols, leaving it in each for a minimum period of 1 to 6 hours, according to the size and permeability of the object (see p. 139).

Next it is important to remove mercurial deposits, because the presence of these will spoil otherwise good preparations. The deposits consist of minute blackish particles and some larger needle-like crystals with a blackish mass at either end. The composition of these deposits is still uncertain. Their removal can be accomplished by means of an iodine solution, or a mixture of iodine and potassium iodide. Evidently the mercury deposits are converted to mercuric iodide, which is soluble in alcohol.

Specimens are generally iodized before they are sectioned. In most cases, iodine is introduced when the tissue is in 70 or 80% alcohol, but in the case of material fixed in Carnoy and Lebrun's fluid it is convenient to iodize it while it is in 95% alcohol. Iodizing in water or alcohol weaker than 70% is not advised, because dissociation of the tissue may take place.

It is convenient to keep on hand a saturated solution of iodine in 95% alcohol, or the following mixture: iodine, 2 gm.; potassium iodide, 3 gm.; 95% alcohol, 100 ml. Add one of these, drop by drop, until the alcohol containing the specimen is a deep amber color. As the iodine performs its task of removing the precipitates, the alcohol will be decolorized. It may be a matter of minutes or several hours before this happens, but if decolorization is complete, add more iodine-alcohol. If the process must be repeated several times, it is a good idea to discard the original alcohol to which iodine has been added and to start with a fresh solution. This is especially recommended when the iodizing solution does not contain potassium iodide; otherwise, a precipitate of red mercuric iodide may accumulate on the surface of the specimen. When the rate of decolorization begins to slow down, add iodine in smaller quantities, and discontinue addition of iodine once it is apparent that no more decolorization is likely to take place.

In the case of material which is to be mounted entire, every trace of mercurial precipitate must be removed before subsequent steps are taken. If the material is to be sectioned, however, the removal may be effected—in whole or in part—at a later time, by treating deparaffined sections with iodine-alcohol according to the procedure described above for entire specimens or blocks of tissue. This method may save time, because the iodine solution will act more rapidly on thin sections than on blocks of tissue.

If the iodine stains the tissue, the higher alcohols into which tissue is passed prior to embedding in paraffin or nitrocellulose will usually remove it. If sections or specimens which are to be mounted entire are not readily washed free of iodine by the time one wishes to pass them down the alcohol series prior to staining, leaving them in 95% alcohol for a time may accomplish complete removal. It is also possible to use the following solution:

Sodium Thiosulfate Solution for Removal of Iodine

Sodium thiosulfate	0.75 gm.
Distilled water	90 ml.
Alcohol, 95%	10 ml.

Transfer sections or other objects to this solution from lower-grade alcohols (30 or 35%). After decolorization is complete, wash material in water. If material is to be stained in an aqueous solution, transfer it directly to this. If the staining solution is made up in alcohol, pass material through the alcohol series to the appropriate concentration before staining.

Mercuric Chloride without Other Fixing Agents. A saturated solution of mercuric chloride in distilled water, physiological saline solution, or sea water may be used to fix small invertebrates in which it is desired to preserve calcareous structures. These organisms include calcareous sponges, corals, pluteus larvae, and stages in the metamorphosis of echinoderms. A saturated aqueous solution of mercuric chloride in physiological saline solution is an excellent general fixative for planarians. This solution may be used for salivary glands and other mucous glands, in order to avoid the swelling of mucin which is caused by acetic acid. However, it shrinks the gland cells considerably.

Mixtures of Mercuric Chloride With Other Fixing Agents

Mercuric Chloride-Acetic Acid ("Sublimate-Acetic")

Mercuric chloride, saturated solution in distilled water or physiological saline solution	100 ml.
Acetic acid, glacial (to be added shortly before use)	5 ml.

Owing to the presence of acetic acid, this solution shrinks tissues less than a solution of mercuric chloride by itself. However, it cannot be used if calcareous spicules, scales, or plates must be preserved. It might be characterized as a second-rate general fixative for small, soft-bodied invertebrates and vertebrate tissues. It is, however, rather reliable for fixation of material to be stained by the Feulgen method. Leave specimens in the fixing solution until they are opaque and white to the center. Transfer them to 50% alcohol, and proceed as directed for mercuric chloride-fixed material in general.

Schaudinn's Fluid (with Acetic Acid)

Mercuric chloride, saturated aqueous solution	66 ml.
Alcohol, 95%	33 ml.
Acetic acid, glacial (*to be added immediately before use*)	5 ml.

Schaudinn's fluid has been widely used as a fixative for protozoa, especially intestinal organisms which can be fixed in smears of fecal material on coverglasses. The smears are dropped face down on the fixative for 15 to 20 minutes. The fixative may be warmed to 60 or 70° C., and the time of fixation reduced to 5 or 10 minutes. Intestinal protozoa may also be fixed in bulk, by diluting the fecal material with physiological saline solution, straining it to remove large objects, and adding several times its volume of hot Schaudinn's fluid.

After fixation, transfer the material to 50% alcohol, then to 70% alcohol, and iodize it in the usual way. Iron hematoxylin is generally used for staining material fixed in Schaudinn's fluid.

Worcester's Fluid (Modified)

Mercuric chloride, saturated aqueous solution	100 ml.
Formalin	5 ml.
Acetic acid, glacial	5 ml.

(*Mix immediately before use.*)

This is a good general fixative for many free-living protozoa. It gives excellent results with amoebae, including the large *Amoeba proteus*, as well as with *Euglena* and various other flagellates. It is satisfactory for many ciliates, especially those which have a rather stiff pellicle, such as *Stylonychia* and *Euplotes*. Place amoebae in a shallow dish, with very little water. When pseudopodia have been formed, pour in the fixing fluid as rapidly as possible. Flagellates and ciliates may be concentrated by centrifugation, if necessary, and then discharged into the fixative by means of a pipette. Allow material to remain in the fixing solution for at least an hour—several hours' exposure will do no harm. Then transfer it to 50% alcohol and pass it after a time to 70% alcohol. Iodize it, and stain it according to a suitable method. Iron hematoxylin may be recommended as a good stain to bring out the general morphology.

Carnoy and Lebrun's Fluid

Absolute alcohol	1 part
Chloroform	1 part
Acetic acid, glacial	1 part
Mercuric chloride, sufficient to saturate the above mixture	

(*Prepare immediately before use.*)

This reagent will penetrate and kill even the most refractory objects within a few minutes. Considering the composition of the mixture, it causes surprisingly little shrinkage. However, its use should be restricted to specimens which are difficult to penetrate. Material which is at least moderately permeable can be fixed more successfully with other agents.

Carnoy and Lebrun's fluid is an excellent fixing agent for eggs of ascarid worms and other nematodes after they have formed heavy membranes. Uteri of ascarid worms should be fixed for 30 minutes. Carnoy and Lebrun's fluid is also preferable to other fixatives for many species of adult nematodes, acanthocephalans, and other organisms which are difficult to penetrate. Fix such organisms for 1 or 2 hours. After fixation, transfer specimens to 95% alcohol. In all cases, change the 95% alcohol several times and then add iodine in the usual way. Embed uteri of ascarid worms in paraffin and stain the sections with iron hematoxylin. Nematodes, acanthocephalans, and other organisms which are to be mounted entire may be stained with alum cochineal, borax carmine, or some other stain of this general group.

Gilson's Fluid

Distilled water	220 ml.
Alcohol, 70%	25 ml.
Nitric acid (approximately 80%)	4 ml.
Acetic acid, glacial	1 ml.
Mercuric chloride	5 gm.

Gilson's fluid is one of the better fixatives for entire preparations of several types of invertebrates. It penetrates well, acts rapidly, and does not cause the muscles of some worms to contract so strongly as do other fixing solutions. In addition, it leaves fewer precipitates in the tissues than other mercurial solutions.

This reagent has its greatest use in the fixation of both free-living and parasitic flatworms. It gives good results on cestodes and trematodes, and is also suitable for many turbellarians. These animals may have to be flattened by appropriate methods (Chapter 9, p. 112). Fix small species (planarians, *Dipylidium*) for 3 hours or longer. Large worms (*Fasciola*, *Taenia solium*, large polyclads) should be left in the fixative for 12 to 24 hours. Material may remain in this fluid for several days without suffering any harm, though prolonged exposure is neither necessary nor desirable. After fixation, wash specimens in several changes of 50% alcohol. Pass them into 70% alcohol and iodize them. Any of the carmine stains will give brilliant results. Alum hematoxylin is also a good stain for tissues fixed in this fluid.

Gilson's fluid is not recommended for routine use with materials which are to be sectioned. It coagulates the cytoplasm in a coarse reticulum and induces the formation of unnatural vacuoles. On the other hand, it confers an unusually good cutting consistency upon some refractory objects, including pieces of large nematodes, and preserves them well enough for ordinary work in morphology or histology.

Petrunkewitsch's Fluid

Distilled water	150 ml.
Alcohol, absolute (or 95%)	100 ml.
Nitric acid, 80%	5 ml.
Acetic acid, glacial	45 ml.
Mercuric chloride, sufficient to saturate the above mixture	

This fixative differs from Gilson's fluid in that it contains larger proportions of alcohol and acetic acid, hence it is better for organisms or tissues which are difficult to penetrate. These include nematodes and crustaceans, as well as tissues and eggs of insects. Material fixed in Petrunkewitsch's fluid should be handled after fixation in the same way as objects fixed in Gilson's fluid.

Heidenhain's "Susa" Fluid

Mercuric chloride, saturated solution in physiological saline solution	50 ml.
Trichloroacetic acid	2 gm.
Formalin	20 ml.
Acetic acid, glacial	4 ml.
Distilled water	30 ml.

This is a good general fixative for the tissues of vertebrates and higher invertebrates. It penetrates rapidly and fixes nuclei, cytoplasm, and connective tissue elements well enough for critical histological work. It preserves delicate epithelia with a minimum of distortion, and for vertebrate kidneys it is perhaps superior to any other fixative. Owing to the acids present, "Susa" causes mucin to swell and does not preserve zymogen granules.

Tissues fixed in "Susa" cut somewhat more easily than those fixed in formalin, Zenker's fluid, or Bouin's fluid. For this reason it is especially good for the ovaries of fishes, amphibia, and birds, which contain yolky eggs, as well as the uterus and other organs which contain large masses of connective tissue and muscle fibers.

Fix specimens in the usual manner, by immersing thin slices of tissue (3 to 5 mm. thick) or entire small organisms in a relatively large volume of fixative. In many cases it is advisable first to inject the fluid into the blood vessels or other cavities. The time required for fixation ranges from 3 to 4 hours for small embryos or thin slices of permeable tissue to 24 hours for large embryos or slices of tissue more than 5 mm. thick. Transfer the material from the fixing fluid to 50% alcohol, and change this several times. Place it in 70% alcohol and treat it with iodine in the usual way.

The staining properties of material fixed in this mixture are exceptionally good. Heidenhain's "Azan" stain and Mallory's triple stain yield preparations in which the structures are differentiated clearly and beautifully. Alum hematoxylin and eosin give excellent results, as does iron hematoxylin.

Mann's Fluid

Mercuric chloride, saturated solution in physiological saline solution . .	10 ml.
Osmium tetroxide, 1% aqueous solution	10 ml.
(*Mix shortly before use.*)	

This fixative is used for material in which Golgi material is to be demonstrated by the so-called Mann-Kopsch method. It was not originally devised for this purpose, however. See page 439 for instructions concerning the application of Mann's fluid in this technique.

Picric Acid (Trinitrophenol)

The yellow crystals of picric acid are sold in a slightly moist condition, in order to assure safety in handling. The dry substance burns quietly when unconfined, but explodes when detonated. Picric acid is soluble in water, alcohol, and a number of other solvents. Its solubility in water at warmer room temperatures approaches 1.5%. For preparation of fixatives, it is convenient to keep in the laboratory a saturated aqueous solution with an excess of picric acid in the bottom of the bottle.

Picric acid was evidently first used, in combination with dilute sulfuric acid, by Kleinenberg in 1879. Since then, it has been utilized in many combinations, a number of which are very popular. Picric acid is seldom, if ever, used alone as a fixing agent.

Picric acid acts by coagulating proteins, and probably becomes combined with them. It generally causes formation of a network in both nucleus and cytoplasm. Addition of acetic acid greatly improves the fixation of nuclei. By itself, picric acid induces only moderate shrinkage, but tissue fixed in it becomes badly shrunken after passage through organic solvents and melted paraffin. It causes only very slight hardening, and this is one of its assets. It penetrates rather slowly.

Picric acid seems to have little or no effect upon lipids. It may be that the aggregation of lipid droplets noted after fixation is brought about indirectly, due to distortions in the cytoplasm. Mitochondria are preserved by picric acid, but the best-known picric acid mixtures (such as Bouin's fluid) contain acetic acid or other acids which tend to destroy mitochondria. Picric acid is also combined with various other substances. It is a weak oxidizing agent, so that there is no question as to its compatibility with reducing agents such as alcohol and formalin.

A decided advantage of picric acid mixtures is that tissues may remain in them for many hours or days without becoming over-hardened or spoiled by formation of precipitates.

Hematoxylin stains, carmine stains, and acid aniline dyes give excellent results with material fixed in picric acid mixtures. However, basic aniline dyes often give inferior results.

WASHING AND HARDENING MATERIAL. There is a long-standing controversy as to the correct manner of washing excess picric acid from fixed tissues. According to some authorities, picric acid should always be washed out with alcohol not weaker than 50%, and preferably in 70% alcohol. This procedure is based on the idea that compounds formed by picric acid and proteins are soluble in water. However, contradictory evidence has been obtained to show that proteins coagulated by picric acid are largely insoluble in water.

In spite of the fact that the coagulates seem not to be dissolved by water, washing in alcohol is recommended. It is certainly a safe procedure, and there is some evidence that, in the case of cells from testis, ovary, bone marrow, and certain other delicate tissues, material washed in water yields inferior preparations. It may be that prolonged washing in water brings about undesirable changes of some sort.

Ordinary histological specimens may be transferred from aqueous picric acid mixtures (Bouin's fluid, Kleinenberg's fluid) directly to 50% alcohol. This should be changed once or twice, at intervals of one to several hours, depending upon the size of the objects. The material should then be transferred to 70% alcohol, in which the washing may be continued for as long as necessary. After this, a suitable method of staining and mounting, or of embedding and sectioning, may be carried out.

In the case of cytological material, delicate small organisms, or gonadal tissue, the transfer from the fixing fluid to alcohol should be gradual. A simple way of accomplishing this is to pour off the excess fixing fluid, leaving just enough to cover the specimens, and then add small amounts of 70% alcohol at intervals until the concentration has been raised nearly to 70%. Following this, the specimens may be transferred to clean 70% alcohol.

To remove picric acid from the material, change the 70% alcohol whenever it has become deeply colored, which may be once or several times a day. If the material is to be embedded in paraffin, without first being stained, it is not necessary to remove all picric acid; two or three changes of 70% alcohol will be sufficient. Any remaining picric acid will be dissolved out of the sections when they are brought into alcohol. Neither is it necessary to remove all color from specimens which are to be stained in borax carmine, because borax will complete its removal. However, it is best to remove all picric acid from objects which are to be stained entire with hematoxylin either before being embedded or before being prepared as whole mounts. There is an old idea that material fixed in picric acid is unsuitable for embedding in nitrocellulose. This is not true, although it is a good idea to wash out all picric acid from the material before bringing it into nitrocellulose. Regardless of how material is to be prepared, do not allow it to lie for weeks or months in alcohol which is deeply colored with picric acid, as it may then be extremely difficult to remove the yellow color.

Extraction of picric acid may be hastened by warming the alcohol to 35 or 40° C. The most convenient way of doing this is to place the container of material in an incubator. Extraction is about twice as rapid at 40° C. as at the usual room temperatures. Many technicians hasten the process by adding a few drops of a saturated solution of lithium carbonate or sodium carbonate to the alcohol. This is very effective, and may be necessary in obstinate cases. Such treatment may, however, injure the staining quality of tissues, and therefore is not to be recommended as a routine practice.

Kleinenberg's Picro-Sulfuric Acid

Distilled water	100 ml.
Sulfuric acid (concentrated)	2 ml.
Picric acid, sufficient to saturate the above mixture	

This mixture causes little shrinkage in delicate tissues, and is good for arthropod materials because it penetrates the exoskeleton rather quickly. Another important property of this mixture is that it hardens yolk less than do many other fixatives. On this account it is useful for yolky eggs (as those of molluscs and teleost fishes) and especially for those which are covered by impermeable membranes (as those of crustaceans and other arthropods). The membranes of arthropod eggs should be pricked with a sharp needle soon after they are placed in the fixative, and the same treatment is necessary for teleost eggs which have a thick chorion (for example, those of trout). Mollusc eggs should be freed as far as possible from the surrounding jelly. Allow 24 to 30 hours for fixation of such objects.

Transfer objects from picro-sulfuric acid to 50% alcohol. After a few minutes or longer, depending on their size and permeability, place them in 70% alcohol. Proceed with appropriate methods for preparing whole mounts or for embedding in paraffin. Carmine and hematoxylin stains are often quite effective after this fixative.

Power's Fluid

Kleinenberg's picro-sulfuric acid fixative	1 part
Saturated aqueous solution of mercuric chloride with 5% acetic acid	1 part

This mixture of two well-known fixatives is excellent for many ciliates (as *Paramecium* and *Euplotes*) and larger free-living amoebae which are to be prepared as whole mounts. Staining by the dilute acidulated borax carmine method gives good results after Power's fixative. Perhaps this fixative will also prove useful for various small invertebrates.

Fix protozoa for several hours or overnight, then transfer them gradually to 70% alcohol for removal of picric acid and mercuric chloride. If there is any danger that mercurial precipitates will remain in the organisms, iodize the alcohol in the usual way.

Bouin's Fluid

Picric acid, saturated aqueous solution	75 ml.
Formalin, concentrated	25 ml.
Acetic acid, glacial	5 ml.

The mixture keeps well. Nevertheless, some workers persist in following Bouin's recommendation that it be made up just before use.

It is probable that Bouin's fluid is used more widely than any other fixing reagent. It is a safe, dependable, and convenient fixative for most routine work in zoology, histology, embryology, and parasitology. Bouin's fluid penetrates rapidly, fixes evenly, and preserves many cell structures with a minimum of alteration. It serves as well for specimens which are to be mounted entire as for those which are to be embedded and sectioned. Tissues fixed in this mixture usually have a favorable consistency for sectioning.

Bouin's fluid is a first-rate fixative for nuclei and chromosomes. It is widely used in cytological studies on mitosis, oogenesis, spermatogenesis, and fertilization. In fact, Bouin designed this fluid for his study of cytological changes brought about in the mammalian testis by cutting the vas deferens, which led him to conclude that the interstitial cells secrete the male hormone. Bouin's fluid also fixes rather well some delicate cytoplasmic structures such as cilia, mitotic spindles, and astral radiations.

This reagent should not be used in the study of certain structures or substances. It fails to fix lipids, leaving them soluble in alcohol as well as in toluene, xylene, and related solvents. It does not preserve mitochondria, Golgi material, neurofibrils, zymogen granules, and certain other cell granules. In addition, it causes mucin to swell, and it does not fix glycogen.

Bouin's fluid is an excellent general fixative for many kinds of biological material. Invertebrates for which it is suitable include the following: free-living amoebae; *Ephelota* and some other suctorians; *Hydra* and all kinds of hydroids; small medusae, such as those of *Obelia*; sea anemones and other non-calcareous anthozoans; larvae of trematodes and cestodes; annelids; soft parts of all kinds of molluscs; tissues of crustacea, insects, and other arthropods. It is fairly satisfactory for flatworms in general, although Gilson's fluid or mercuric chloride-acetic acid is preferred for at least some turbellarians and trematodes, and mercuric chloride mixtures or formalin may be better for many cestodes. Bouin's fluid is unsuitable for the following: sponges; large medusae; nematodes; adult echinoderms; crustacea and insects which are to be mounted entire. Among lower chordates, Bouin's fluid is recommended for small tunicates (such as *Perophora*), tunicate larvae, *Branchiostoma* ("amphioxus"), and the ammocoetes larvae of lampreys. In embryology, it serves well for eggs, embryos, and larvae of coelenterates, flatworms, annelids, echinoderms, tunicates, hemichordates, and *Branchiostoma*. It is also excellent for vertebrate embryos. In fixing chick embryos and early stages in mammalian embryology, it is best to dilute Bouin's fluid with an equal volume of water. Tissues of adult vertebrates, including human tissues, are well-fixed in most instances. However, Bouin's fluid should not be used for salivary glands or other organs in which it is desired to preserve mucin. It is unfavorable to the staining of striations in skeletal or cardiac muscle and gives poor results with mammalian kidneys. It should not be employed when it is necessary to preserve calcium deposits; it is, in fact, a useful decalcifying agent.

Bouin's fluid is a rapid fixer. Young embryos and small larvae are fixed within a few minutes, but it is advisable to leave them in the solution for several hours, in order to insure thorough hardening. In fixing delicate invertebrate larvae, particularly contractile organisms (such as advanced larvae of annelids), add to the water containing the organisms about one-third its volume of Bouin's fluid. After the larvae have settled to the bottom, draw off the liquid and replace it with full strength Bouin's fluid. Embryos of considerable size (72-hour chick embryos, 5 to 8 mm. mammalian embryos) or thin slices of vertebrate tissue should remain in the fluid for at least 3 to 6 hours. Fix larger embryos or pieces of tissue over 5 mm. thick for 24 hours or more. Materials may safely remain in Bouin's fluid for a considerable length of time, which makes this a convenient reagent to take into the field. However, it is best not to leave material in the fixing solution for more than 4 or 5 days. Longer exposure may affect the staining qualities adversely.

Transfer specimens from Bouin's fluid to 50% alcohol (use intermediate mixtures or add alcohol gradually if delicate structures are involved), then to 70% alcohol. Wash with repeated changes of 70% alcohol, not water. This matter has been considered fully in the general discussion of picric acid.

For staining objects which are to be mounted entire, borax carmine (with or without counterstaining) generally gives excellent results after fixation in Bouin's fluid, although other carmine stains also work well. In routine histological and pathological work, stain sections with alum hematoxylin and eosin. For strong differentiation of collagen use Lillie's modification of Masson's "trichrome" stain. Mallory's triple stain and Heidenhain's "Azan" stain give fair results after fixation in Bouin's fluid; better results can usually be achieved with these two methods if the material has been fixed in Zenker's fluid. Iron hematoxylin stains nuclei, chromosomes, spindles, and asters very well after Bouin's fluid.

Duboscq and Brasil's Fluid

Alcohol, 80%	150 ml.
Picric acid	1 gm.
Formalin	60 ml.
Acetic acid, glacial	15 ml.

This modification of Bouin's fluid was designed to penetrate refractory objects such as small crustacea, insects, and nematodes. It will penetrate the cuticle and preserve the internal organs well enough for detailed study. This mixture is especially valuable for fixation of eggs of *Parascaris* from the time of fertilization to the end of the second meiotic division. As soon as the fertilization membrane has been raised, aqueous solutions fail to penetrate rapidly enough. During meiosis, the cytoplasm of the egg is so delicate that it would be badly shrunken by Carnoy and Lebrun's fluid, which is ordinarily used for later stages. However, Duboscq and Brasil's fluid will penetrate the thin shell which is present during the maturation stages, and will fix the egg with minimum distortion. This reagent will not satisfactorily penetrate the thicker shells of eggs in pronuclear and later stages.

Fix *Parascaris* uteri for 6 hours or longer, and crustacea and insects for 12 to 24 hours, according to size. Specimens may safely remain in the fluid for 2 or 3 days. Transfer material to 70% alcohol, and change this several times to remove the picric acid.

ALLEN'S MODIFICATIONS OF BOUIN'S FLUID. In an attempt to prevent clumping of the metaphase chromosomes and shrinkage of the cytoplasm, Allen experimented with several modifications of Bouin's fluid. The modifications involved changing the proportions of the three original ingredients and adding other substances. The following two of Allen's formulas have proved very useful.

Allen's PFA 3 Fluid

Picric acid, saturated aqueous solution	75 ml.
Formalin	15 ml.
Acetic acid, glacial	10 ml.
Urea (*to be added immediately before use*)	1 gm.

Carothers used this modification (but with only 0.5 gm. of urea) in her striking cytological study of the segregation of chromosomes in Orthoptera. Fix testes of grasshoppers, squash bugs, or other insects for 24 hours, and pass them by way of 50% alcohol to 70% alcohol. After washing has been continued long enough, proceed with the paraffin method. Stain sections with iron hematoxylin.

Allen's B 15 Fluid

Bouin's fluid	105 ml.
Heat to 37° C. and add	
Chromic acid	1.5 gm.
Stir until dissolved, then add	
Urea	2 gm.
Continue to stir until this is dissolved. Use at once.	

The foregoing mixture is used for critical work on mammalian chromosomes, especially in the cells of the testis. Fix small pieces of tissue immediately after the death of the animal. Still better, inject the warm fixing fluid into the genital arteries or aorta and, after a few minutes, remove the testes, cut them up with a razor blade, and place them in an adequate volume of the fixative. Keep the liquid warmed to 37° C. in an incubator during fixation. Approximately 2 hours are required to fix pieces of testis 4 mm. or less in thickness. Larger or denser objects require 3 to 4 hours, or slightly longer. Replace the fixative by adding 70% alcohol, a little at a time. Change the alcohol several times and proceed by the paraffin method. For cytological purposes, stain sections with iron hematoxylin. Staining is slower than after ordinary Bouin's fluid. Flemming's triple stain may also be used. Alum hematoxylin and eosin will stain fairly well, but with less striking color differentiation than is obtained after Bouin's fluid or Zenker's fluid.

Hollande's Fluid

Picric acid	4 gm.
Cupric acetate	2.5 gm.
Distilled water	100 ml.
Formalin	10 ml.
Acetic acid, glacial	1.5 to 5 ml.

Dissolve the cupric acetate, then the picric acid. Add the formalin and acetic acid later.

This mixture has proved to be one of the better fixatives for certain kinds of protozoa, but in recent years it has been used very little for other types of material. Its suitability as a general histological and cytological fixative should be explored.

Hollande's fluid is one of the better fixatives for flagellates and ciliates which are to be impregnated by activated silver albumose (Protargol) according to the method of Bodian (p. 419). However, iron hematoxylin also gives good results after Hollande's fluid. The Feulgen nucleal reaction may sometimes be used with at least moderate success after this mixture, although a saturated aqueous solution of mercuric chloride (with 5% acetic acid) or Schaudinn's fluid will probably give better fixation for this technique.

The amount of acetic acid specified in the original formula is 1.5 ml. to each 110 ml. of the other liquid ingredients. However, increasing the amount of acetic acid to 5 ml. seems not to affect the utility of Hollande's fluid for silver impregnation or staining, and may lessen the extent to which the organisms shrink during fixation.

After fixation, transfer the material gradually to 70% alcohol and leave it in this until the picric acid and cupric acetate have been washed out rather thoroughly.

Osmium Tetroxide (Osmic Acid)

Osmium tetroxide is commonly called "osmic acid," but this is quite incorrect. It is not an acid, as it is a non-electrolyte, forms no salts, and is neutral to indicators. Unfortunately, this name will be difficult to abolish altogether, because it has been widely used in biological publications for a century.

Osmium tetroxide is sold as pale yellow crystals. It dissolves rather slowly in water and is highly volatile. For this reason, it is generally kept in sealed glass ampules containing 0.5 or 1 gm. *Do not inhale the fumes of osmium tetroxide or expose the eyes to them, since they are extremely irritating and destructive.* In aqueous solution this oxide becomes reduced if any organic matter is present, so that special precautions are necessary in preparing and storing solutions. This matter will be discussed under a separate heading.

Osmium is a heavy metal related to platinum. It occurs, along with other rare metals, in platinum ores from various countries. At extremely high temperatures, osmium forms the tetroxide. This is one of the more expensive substances used in microtechnique, and currently sells for about $20.00 per gram.

Schultze introduced osmium tetroxide as a fixative in 1866, employing it in his studies on the luminescent dinoflagellate, *Noctiluca*. Later, von Flesch tried it in a mixture with chromic acid, and in 1882 Flemming improved the combination by adding acetic acid. In 1890, Altmann observed mitochondria in cells fixed in a mixture of potassium dichromate and osmium tetroxide, and subsequently combinations of this type came to be widely used in routine techniques for demonstrating mitochondria.

Osmium tetroxide does not coagulate proteins, but it causes them slowly to form homogeneous gels. In addition, this substance affects proteins in such a manner that they are not coagulated by subsequent treatment with alcohol. Cells and organelles are preserved very well by it. These are among the reasons for its importance as a fixative for electron microscopy. The nature of its action is not definitely known, but it operates as an additive fixative. Tissues fixed in it become brown or black and lose their affinity for dyes, especially acid dyes. Thorough washing helps to make the tissues more stainable.

Osmium tetroxide is soluble in lipids, and blackens unsaturated lipids; it therefore also blackens Golgi material and mixtures of unsaturated and saturated lipids. Because of its solubility in lipids, subsequent reduction of it (as by alcohol) may result in blackening, and this must not be construed as a test for presence of only unsaturated lipids.

Osmium tetroxide hardens tissues only slightly, and it prevents them from being hardened much by subsequent treatment with alcohol. The presence of chromic acid or potassium dichromate in mixtures overcomes the lack of hardening. The shrinking effect of osmium tetroxide is negligible. However, the fixed tissues often shrink, crack, or crumble during the process of paraffin embedding.

Osmium tetroxide is a strong oxidizing agent, and on this account should not be mixed with reducing agents such as alcohol and formalin. In cytological work it is commonly combined with chromic acid and acetic acid (as in Flemming's fluids) in order to obtain a fairly homogeneous fixation of cytoplasm instead of the coarse coagulation caused by chromic acid alone. Osmium tetroxide-potassium dichromate mixtures (such as Bensley's fixative) are also widely used for mitochondria. For this purpose the acetic acid is reduced (generally to less than 1%) or eliminated entirely. Solutions of osmium tetroxide alone are now employed chiefly in special methods which utilize its property of blackening lipids and Golgi material, and rendering them less soluble in alcohol, oils, and resins. When used for such purposes, osmium tetroxide may serve as both fixative and stain. However, better results are obtained by fixing material first in formalin or in a special fixative. Some methods for demonstrating Golgi substance and lipids by the use of osmium tetroxide will be found in Chapter 26.

Osmium tetroxide penetrates rather slowly, and also acts slowly on proteins. When used as the only fixative, and when the time of exposure must be short, as in preparation of material for embedding in Epon for either electron microscopy or light microscopy, the pieces of tissue must be very small. Even when it is used in combination with other fixing reagents, the pieces of tissue should be at least moderately small, and adequate time should be allowed for the osmium tetroxide to penetrate. Blackening of the tissue is not likely to be of consequence in material fixed for embedding in Epon, because the sections will be so thin; however, the blackening may be offensive in the case of thicker sections or whole mounts of small organisms, and it should be minimized by keeping the container in the dark during the period of fixation.

Sections of material fixed in osmium tetroxide, or in mixtures of this substance with chromic acid and potassium dichromate, stain well with aniline dyes (safranin, crystal violet, orange G, acid fuchsin, and a number of others) provided they have been properly washed and bleached. Iron hematoxylin gives good results, but alum hematoxylin usually yields poor preparations. As a general rule, it is well to avoid osmium tetroxide in fixing material which is to be mounted entire, except in the case of protozoa and other very small organisms, or when it may serve a useful purpose by blackening certain structures.

PREPARATION OF OSMIUM TETROXIDE SOLUTIONS. Osmium tetroxide can be kept conveniently as a 2% or 4% aqueous solution. Most older formulas specify a 2% solution, but fixations for electron microscopy and for related techniques of light microscopy (as the Epon method) generally call for a 4% solution. Thus, if embedding in Epon is to be done, it may be well to keep a 4% aqueous solution on hand, and to dilute small quantities of this, as the need arises, to make a 2% solution. Solutions of osmium tetroxide must be prepared, stored, and used with special care, because the tetroxide of osmium is highly volatile and also because the presence of any organic matter will reduce it to an inert, dark-colored hydrated dioxide ($OsO_2 \cdot 2H_2O$). It is advisable to make a rather small quantity of solution at a time, for osmium tetroxide is very expensive and its solutions are subject to spoilage. Because osmium tetroxide is ordinarily kept on hand as a 2% or 4% stock solution, a 25- or 30-ml. bottle is the most appropriate size for preparing a 2% solution from 0.5 gm., or a 4% solution from 1 gm. A smaller bottle can be used for preparing a

4% solution from 0.5 gm. Select an absolutely clean bottle (treatment with sulfuric acid-potassium dichromate cleaning solution, followed by repeated washing with distilled water, may be advisable) which has a tight-fitting glass stopper (never use a cork!). Make sure that the neck is wide enough to admit the glass ampule in which the osmium tetroxide has been sealed. Fill the bottle with the required amount of distilled water.

Soak the ampule in hot water to remove the label and other foreign substances, and to melt the osmium tetroxide. Let the liquid run to one end, then let the ampule cool until the osmium tetroxide solidifies in one mass. Wash the ampule thoroughly in distilled water, then dry it with a clean cloth or laboratory tissue, taking care not to contaminate the outside with fingerprints or any other organic matter. After this point, it is a good idea to wear clean rubber gloves or disposable plastic gloves. Holding one end of the ampule with a laboratory tissue, score it for about half its circumference, near the middle, with a triangular file, as you would a piece of glass tubing. Wipe away the filings, wrap the ampule in a sheet of clean paper, and break it. Drop the end with the solidified osmium tetroxide into the bottle of distilled water. (The other half may be discarded after it has been soaked a while in hot water in the sink, to prevent any residue of osmium tetroxide from being freed in the laboratory or in a waste receptacle.) Stand the bottle in a bath of water warmed to about 60° C. The osmium tetroxide will melt again and start to dissolve, and if the bottle is kept warm and shaken periodically, solution will soon be complete. (If, when the bottle is cooled, fine particles become evident, resume the warming and shaking.)

Another good way to liberate the osmium tetroxide from the ampule, after this has been thoroughly washed, is to drop the ampule into a clean bottle of distilled water, then break it with a clean glass rod. The bottle may be warmed, as directed above, to hasten solution of the osmium tetroxide. This method is a little more messy because of the broken glass at the bottom of the bottle.

Always keep the bottle stoppered. Because the bottle will probably be a rather small one, and therefore one which can easily be knocked over, it is a good idea to store it within a wide-mouthed plastic jar, or within a glass jar lined with paper toweling to minimize the chance that it will be broken. The solution should be kept out of the light, and preferably in a refrigerator, to slow down reduction of the osmium tetroxide catalyzed by the presence of organic matter. Darkening of the liquid indicates that reduction has been taking place.

Fixation by Osmium Tetroxide Vapor. This is a useful method for isolated cells, thin membranes (such as the retina), and smears of various kinds. The material to be fixed is placed on a slide or coverglass, and this is inverted over the mouth of a bottle containing a 2% solution of osmium tetroxide and left there until it begins to turn brown. Isolated cells or thin smears are fixed within 1 to 15 minutes, while several hours are required for fixation of the retina.

Small bits of solid tissue may be suspended in a bottle of osmium tetroxide by means of a thread, or in a little bag of cotton gauze held in place by the pressure of the stopper. The bottle, with suspended tissue, should be kept at about 40° C. for 1½ hours. Upon removal, the tissue may be placed in 50% alcohol and subsequently processed for embedding in paraffin.

It has sometimes been stated that osmium tetroxide vapor penetrates more rapidly than solutions. This seems very unlikely, because the vapor can only penetrate beyond the surface of the tissues by dissolving in their fluids, and thus becoming a solution. An apparently valid claim for the vapor method is that nothing can be dissolved out of the cells during fixation.

Washing and Bleaching Material. Materials fixed in osmium tetroxide or in mixtures containing it should be washed thoroughly in water, preferably running water. Wash for at least 2 hours in the case of thin smears or isolated cells, and as long as 24 hours for objects as thick as the retina. The material may then be passed up the alcohol series.

Materials are darkened as the reduction of osmium tetroxide takes place. The extent of blackening depends upon the strength of the solution, the length of time the tissues remain in it, and the nature of the tissues themselves. Blackening may be minimized by keeping the tissue in darkness during fixation, and washing it very thoroughly in running water after fixation. However, unless it is desirable to retain an osmium impregnation for demonstration of lipids, the material should be bleached.

Bleaching is generally carried out upon thin paraffin sections which have been affixed to slides (Chapter 15, p. 241). After removing paraffin from the sections, pass them down the alcohol series to 50% alcohol and transfer them to the bleaching agent. Nitrocellulose sections may be bleached as loose sections, or after they have been affixed to slides. Bleaching may be accomplished with a 5% solution of hydrogen peroxide in 50% alcohol. This is a very safe reagent, but it requires many hours to bleach material which has been blackened to a moderate extent, and it may fail altogether in obstinate cases. An equally safe and much more effective bleaching solution can be made by adding a commercial hypochlorite solution, such as "Clorox," to alcohol. One drop of "Clorox" for each 10 ml. of 50% alcohol is a suitable proportion. This solution will usually bleach sections in from 10 to 45 minutes. The potassium chlorate-hydrochloric acid bleaching method (p. 144) may be used in difficult cases, though it is more likely to injure delicate sections.

Objects which are to be mounted entire may be bleached in the same way as sections.

After sections or other materials have been bleached, wash them in running water for one hour or longer, in order to remove every trace of the bleaching agent.

Osmium Tetroxide Mixtures. See Flemming's fluids (p. 83), Champy's fluid (p. 90), Bensley's fluid (p. 438), and Mann's fluid (p. 98).

REFERENCE

BAKER, J. R., 1958. *Principles of Biological Microtechnique.* New York: Barnes and Noble, Inc.

CHAPTER 9

Technique of Fixation

In order to fix and preserve animal structures without altering their appearance any more than necessary, it is important not only to use a suitable fixing agent, but also to apply this agent in an effective manner. This means that the fixative must (1) penetrate and act upon the tissues rapidly, since destructive postmortem changes begin immediately after the cessation of life, and (2) it must accomplish this without causing extensive distortion. *Failure to consider the importance of these matters is responsible for many disappointments in preparing microscopical material.*

The methods of fixation applicable to a particular kind of organism or tissue depend upon the physical character of the structures to be preserved, especially their bulk, permeability, and reactivity. All fixatives are strong chemicals and will cause many organisms to contract or otherwise become distorted, unless this is prevented by appropriate means. The penetrating power of the fixative is also very important. Fixatives which penetrate rapidly (such as Bouin's fluid, Carnoy's fluid, and alcohol), can be used for large or relatively impermeable objects, but only small pieces of tissue should be fixed in mixtures containing osmium tetroxide and chromic acid. The procedures employed for various kinds of animals are necessarily great in number and variable in detail. Nevertheless, they are all adaptations of a few basic methods. These methods will be outlined in a general way in this chapter, and reference will frequently be made to specific directions given in other parts of the book.

PREPARATION OF SPECIMENS

Be careful to use healthy, normal organisms, except when the object is to study some pathological condition. Receptacles, pipettes, and other instruments used for handling living material should be labeled conspicuously and kept separate from those used for chemicals. Otherwise, unintentional destruction or degeneration of living material is likely to occur.

Although invertebrates of many kinds may be cultivated by simple methods, various other delicate organisms are difficult to keep alive in the laboratory, requiring special provisions for proper water, food, aeration, and control of temperature. This is certainly true of numerous marine animals, including sponges, hydroids, certain medusae, ctenophores, siphonophores, ectoprocts, and many others. In the absence of special facilities, such organisms must be fixed as soon as possible after they are taken from their natural habitat. When carrying specimens back to the laboratory in large jars, plastic pails, or plastic bags, keep them cool. Material which has been crowded into a small volume of water will seldom, if ever, be of any use. In many cases it is safer to transport marine animals with wet seaweed rather than in water, but when the specimens are immersed in water there should be plenty of it and it should be cold and well aerated. Delicate marine organisms will usually not live in aquaria if the sea water has been in contact with metal.

Fixation Without Previous Preparation

Some small organisms may be killed and fixed in a satisfactory manner by simply immersing them in a fixing solution. This is successful only with species which are small enough to be quickly penetrated and which are not so sensitive as to contract strongly or to suffer pain when brought into contact with strong chemicals. These include protozoans, sponges, and some coelenterates, helminths, echinoderms, and arthropods, as well as the eggs, embryos, or larvae of most invertebrates and vertebrates. In case there is foreign matter adhering to the specimens, this should be washed or brushed off before fixation.

Exposure to the fixative must be brought about as suddenly as possible, especially in the case of organisms which are likely to contract. Terrestrial animals and embryos and larvae of many kinds may simply be dropped or squirted from a pipette into the fixative. Organisms which are usually in contact with a solid substrate (fresh water hydras, for instance) should be placed in shallow dishes containing very little water. When they are extended, the fixing fluid is quickly poured over them.

Hot fixing fluids penetrate and act very rapidly. Bouin's fluid or any other strong fixative, heated to about 60° C., will kill amoebae, hydras, and some other animals so quickly that they do not have time to contract. Hot fluids are also used for fixing entire nematodes, insects, and other invertebrates possessing a resistant cuticle. A very penetrating solution, such as Carnoy and Lebrun's fixative, need not be heated. A discussion of the use of heat in connection with the application of fixatives will be found on page 128.

Preparation of Contractile Animals (Invertebrates and Lower Chordates)

Physical Restraint. The application of pressure during fixation will serve to prevent many invertebrates from shrinking, curling, twisting, or retracting their appendages. This method is employed chiefly for flattened organisms, as illustrated by the following three paragraphs.

STRONG PRESSURE. This is necessary to hold large and tough trematodes (such as *Fasciola*), leeches, centipedes, ticks, and flattened insects outspread. A convenient method is illustrated in Figure 6. Place the specimen between two slides or other small pieces of glass and, with a string, tie these together tightly enough to hold it properly extended. Drop the whole preparation into the fixing fluid. After an hour or so, cut the string to admit the fixing fluid more rapidly, being careful not to separate the pieces of glass or to disturb the specimen until it shows signs of being well-hardened. In many cases, a rubber band may be used for the same purpose, provided it is not stretched too tightly.

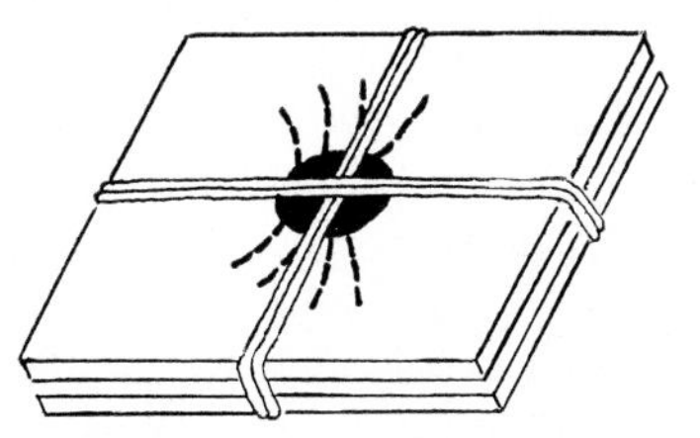

FIG. 6. An arthropod immobilized between pieces of glass in such a way that the appendages will remain spread out when the animal is fixed. The same general method is useful for keeping leeches, flukes, and some other animals slightly flattened during fixation.

SLIGHT PRESSURE. A light weight can safely be applied to delicate animals such as turbellarians, small trematodes (such as *Clonorchis* and *Pneumoneces*), cestodes, and small arthropods. To fix turbellarians (rhabdocoels, polyclads, and most triclads) or small trematodes in a flattened condition, proceed as follows: Lay the specimens on a slide or plate of glass and cover each of them with a small piece from a broken slide, or in the case of very delicate forms, with a coverglass. With a pipette, place a small quantity of fixing fluid at the edge of the coverglass, and allow it to flow under. After 20 to 30 minutes remove the glass and carefully transfer the specimen to a dish of fixing fluid.

THE WEIGHT OF THE ATMOSPHERE. This alone is sufficient to prevent delicate embryos and larvae from curling or twisting. The specimen is placed on a moist glass plate or on the bottom of a shallow dish, and is gently straightened out. Then sufficient fixing fluid is added to wet it thoroughly, but not enough to float it off the glass. A few more drops of fixing fluid are added after a minute or two. Within a few minutes the specimen will be hardened enough so that it can safely be immersed in the fixing agent. A very important application of this method is the fixation of chick embryos (see p. 508). Among other objects for which it is used are tapeworm scoleces, small specimens of *Branchiostoma* ("amphioxus"), and the ammocoetes larvae of lampreys. The last two should be anesthetized with magnesium sulfate or some other suitable agent before they are laid on the glass.

TENSION. Tension, instead of pressure, is sometimes employed to hold tapeworms in a normal state of extension (see p. 481).

Anesthetics for Aquatic Invertebrates and Lower Chordates. A great many organisms can be fixed in an expanded state only after they have been rendered insensitive or paralyzed by an anesthetic. Unfortunately, few systematic studies of the suitability of various anesthetic substances in common use have been attempted. Methods which work well with certain organisms may be inferior for anesthetization of others.

The following brief selection of anesthetics, and of methods for their use, is based upon practical experience. Nearly all of them were suggested by published accounts, but a considerable amount of work was required to segregate a few really useful methods from a large number of less reliable or useless ones. Likewise, many details concerning these methods were discovered by experiment. The list is obviously incomplete, but it does provide a few reliable methods which may be used with a good many types of invertebrates.

Remember that it is always essential to fix the organisms as soon as they are sufficiently paralyzed, since death and postmortem changes almost immediately follow complete anesthesia. This is particularly true in the case of protozoans, hydroids, small medusae, and delicate larval forms.

MAGNESIUM SULFATE (EPSOM SALTS) AND MAGNESIUM CHLORIDE. These are useful anesthetics for many aquatic organisms, including the following: hydroids; medusae (except very small ones, such as those of *Obelia*); sea anemones, corals, and other anthozoans; nemerteans; small fresh-water annelids and tubicolous marine annelids (terebellids, sabellids, serpulids); leeches (which should subsequently be flattened between slides); holothurians and ophiuroids, echiuroids and sipunculids; certain phoronids; nudibranchs, pteropods, and heteropods; ascidians; hemichordates; "amphioxus"; and ammocoetes larvae of lampreys.

Place fresh-water organisms in clean water and allow them to stand until fully expanded. Then introduce the salt in such a manner that it diffuses slowly and evenly through the water, without disturbing the animals. This may be done by suspending one or more little cloth bags, filled with crystals of magnesium sulfate or magnesium chloride, so that their tips just touch the water containing the specimens. Another method of introducing the salt is to add a few drops of a solution at intervals. Whatever method is used, allow the salt to become gradually more concentrated, until the organisms no longer contract when touched.

Probably the best way to anesthetize marine organisms with either magnesium chloride or magnesium sulfate is to prepare a solution which is isotonic with sea water, and to add this gradually until the organisms no longer respond to stimulation. In the case of magnesium chloride, a

7.5% solution of the crystalline salt ($MgCl_2 \cdot 6H_2O$) is approximately isotonic with sea water; in the case of magnesium sulfate ($MgSO_4 \cdot 7H_2O$), a 20% solution is approximately isotonic.

The concentration of magnesium chloride or magnesium sulfate required to bring about paralysis differs according to the size and nature of the organisms. Hydroids and small medusae are generally paralyzed within 30 to 40 minutes after the addition of salts is begun. The process must be continued at the same rate for 1 to 3 hours or somewhat longer in the case of scyphomedusae, small anemones, annelids, and phoronids. Holothurians and large anemones may require as long as 10 or 12 hours. As soon as the organisms are paralyzed, rapidly siphon off most of the water. Then introduce the fixing solution, pouring it gently down the side of the container in order to disturb the specimens as little as possible. When the material is fairly well hardened, transfer it to fresh fixing fluid.

A less refined method may be used for some organisms, such as leeches, "amphioxus," and ammocoetes larvae of lampreys. Simply add the salt with a spoon or spatula and accelerate diffusion by gentle stirring or agitation. When the animals lie motionless and do not react to prodding, they may be fixed. "Amphioxus" and ammocoetes larvae are simply laid on a glass plate or on the bottom of a shallow dish and fixed as described on page 113. Leeches should be placed between two slides, which are then tied together and immersed in a fixative.

MENTHOL-CHLORAL HYDRATE. Even though menthol is very slightly soluble in water, it acts rapidly as an anesthetic for small and permeable organisms such as hydroids and some ectoprocts. This substance does not cause even the most sensitive organisms to contract. Some hydroids (as *Obelia* and *Plumularia*) react to it by expanding to the very limit. Quite often, however, anesthesia is not deep and the fixing agent irritates the organisms so much that they contract to some extent before death. On this account it is advantageous to mix a small quantity of chloral hydrate with the menthol, which will paralyze the contractile structures very effectively. Chloral hydrate by itself is not satisfactory, since it causes many organisms to contract.

The menthol-chloral hydrate mixture is an excellent anesthetic for many of the lower invertebrates. For hydroids it is incomparable, enabling one to fix practically all species of certain groups (Campanulariidae, Sertulariidae, and Plumulariidae) with the hydranths completely expanded. It gives good results with fresh-water ectoprocts such as *Plumatella* or *Cristatella* and with the marine *Membranipora*, but it cannot be recommended for *Bugula*. It is not as good as magnesium sulfate or magnesium chloride for the larger gymnoblastic hydroids such as *Tubularia* and *Syncoryne*. Menthol-chloral hydrate is very satisfactory for narcotizing the cestodarian *Gyrocotyle* and some monogenetic trematodes.

To prepare the mixture, place approximately equal amounts of menthol and chloral hydrate in a mortar or porcelain evaporating dish. Add enough water (fresh or sea water, according to the habitat of the animals) to moisten the crystals. Grind the mixture, meanwhile adding a little water, until it forms a thin paste.

Place the organisms in a culture dish or some other fairly shallow vessel, with sufficient water to cover them to the extent of about 1 cm. Many hydroids, *Obelia* among them, will expand better if the dish is placed in bright sunlight. Now float upon the water enough of the menthol-chloral hydrate mixture to form a very thin layer over the surface. At intervals of 2 or 3 minutes observe the material through a low-power microscope or hand lens, and when the organisms no longer show expanding or contracting movements, gently prod some of them with a needle. As soon as they fail to contract when so stimulated, fix the material immediately. Colonies of *Obelia*, *Campanularia*, and many other hydroids may be picked up by the base with forceps and dropped into Bouin's fluid or other fixative. Ectoprocts and very delicate hydroids such as *Hydractinia* should be killed by pouring the fixative into the dish in which anesthetization was carried out. Introduce a quantity of Bouin's fluid, or some fixative containing mercuric chloride, to equal approximately the volume of water in the dish. In about 5 minutes, the material may be transferred to full-strength fixative. To fix with formalin, simply introduce enough strong formalin to make a 5 or 10% solution.

COCAINE. This alkaloid is expensive and difficult to obtain because of the legal restrictions that pertain to narcotics; hence it is not often used except when other anesthetics fail. It will paralyze rotifers and some ectoprocts (as *Bugula*) in an expanded condition. Place the organisms in a small dish of clean water, under a dissecting microscope. When they are expanded, introduce a little of a *freshly prepared* 1% aqueous solution of cocaine hydrochloride with the least possible disturbance. Use about 1 ml. of the cocaine solution for each 100 ml. of water (or 1 drop to about 5 ml.). If the organisms continue to show movements after 10 minutes, add a little more cocaine, repeating this if necessary. As soon as movement appears to have ceased, prod a few individuals to see if they react. When the organisms appear to be completely paralyzed, withdraw as much liquid as possible and pour in the fixative; or, if the organisms can be picked up with a small pipette without carrying over too much fluid, squirt them into a dish or vial of fixative. With mercuric chloride and some other fixatives, cocaine forms a white precipitate; however, this is dissolved by alcohol.

EUCAINE. This synthetic drug is more readily available than cocaine, because it is not, in the legal sense, a narcotic. It works well on leeches, which may not respond favorably to some of the other commonly used relaxing agents, and it may prove to be as good as cocaine for many aquatic organisms. To the water containing the animals, add as much of a fresh 1% aqueous solution of eucaine hydrochloride as seems to be necessary. For large leeches, a strong mixture (about 1 part of the solution to 5 parts of water) may be required, but for smaller leeches and most other organisms, a much weaker mixture should suffice.

NOVOCAINE. This is another synthetic preparation that can be obtained without much difficulty. Novocaine hydrochloride may be used as a 1% solution in the same way as cocaine and eucaine. It gives good results in narcotizing tapeworm scoleces. In *Taenia pisiformis* and *Dipylidium caninum*, for instance, the rostellum generally remains completely extended.

PROPYLENE PHENOXETAL. This viscous fluid has bactericidal properties, and has been used for preserving vaccines and cosmetics. It is a valuable narcotizing agent for many kinds of organisms to be prepared for dissection, as well as for histological work. It works well with such diverse groups as fishes, molluscs, annelids, and nemerteans. A 1% solution (by volume) is made up in warm water, and this is diluted as needed with from 10 to 100 parts of fresh water or sea water. Appropriate concentrations for particular kinds of organisms will have to be worked out by trial.

Propylene phenoxetal should be handled with care. It is toxic, and may also cause temporary skin ulcerations.

CHLOROBUTANOL. This substance (1,1,1-trichloro-2-methyl-2-propanol; "Chloretone") has been highly recommended for a variety of invertebrates. It is not, however, very useful for narcotizing hydroids and other coelenterates, rotifers, or ectoprocts. When completely immobilized by the effects of chlorobutanol, such organisms are contracted and often somewhat disintegrated. Similar results are obtained with *Stentor*, *Vorticella*, and other ciliates. Chlorobutanol is fairly satisfactory for immobilizing small oligochaetes, such as *Chaetogaster*, *Limnodrilus*, and *Dero*. Place the worms in a small quantity of water and add a 1% aqueous solution of chlorobutanol, a few drops at a time, until they no longer react to stimulation. Chlorobutanol has also been used successfully for immobilizing some of the more active and contractile tapeworms. This reagent may be recommended for some organisms which produce a great deal of slime, because it seems to inhibit the secretion of slime. For slugs, a saturated solution of chlorobutanol in boiled (air-free) water has been found to be useful.

PENTOBARBITAL SODIUM. This drug is widely prescribed as a sedative by physicians. It has been applied with much success to relaxing prospective museum specimens of gastropod molluscs, so that they do not contract and withdraw into their shells when dropped into a preservative. For this purpose, a 10% aqueous solution is used. Pentobarbital sodium should be tried more extensively for narcotizing small organisms to be fixed for histological or cytological studies, but at considerably lower concentrations than those used for organisms which are simply to be preserved in an uncontracted state.

ETHYL ALCOHOL. Gradual addition of ethyl alcohol, to the extent of about 10%, is an excellent method for immobilizing earthworms (p. 485). Some acanthocephalans may be fixed in a relaxed state, with the proboscis well extruded, if this technique is used.

ETHYL URETHANE. This is a good anesthetic for small medusae, such as those of *Obelia* and *Sarsia*, and for the ephyra stage of scyphozoan medusae. Simply sprinkle a few crystals upon the surface of the water. When the medusae have settled to the bottom and no longer react to stimulation, either pipette them into a fixing agent or withdraw most of the water and replace it with a fixing fluid.

CHLOROFORM-WATER. Chloroform has been used with success for killing medusae and echinoderms with their appendages extended. Shake up some water with a small quantity of chloroform and then separate the water by decantation. It is also possible to pour a very small amount of chloroform on the surface of the water containing the animals.

Other Methods for Aquatic Organisms. FRESH WATER FOR KILLING MARINE ORGANISMS. Ophiuroids (brittle stars) and active marine annelids (such as *Nereis* and *Eunice*) become greatly contorted and frequently rupture or break into pieces when placed in a fixing fluid. If these organisms are placed in fresh water, they soon become limp, and may then be fixed in good condition. However, anesthetization in an isotonic solution of magnesium chloride will probably be preferable, as it will cause less histological alteration than an abrupt transfer to fresh water.

NITRIC ACID METHOD FOR PLANARIANS. See page 479.

ASPHYXIATION. Gastropods, both aquatic and terrestrial, die in an extended state when asphyxiated. The method is described below, in connection with terrestrial organisms.

Methods for Killing Terrestrial Invertebrates. ETHYL ALCOHOL FOR EARTHWORMS. This method has been mentioned in the preceding section, since it is also applicable to aquatic animals. For detailed directions, see page 485.

ASPHYXIATION. This is the best way to kill snails and slugs in an extended state. Excellent results are obtained with *Helix*, *Ariolimax*, and similar gastropods. Fill a jar completely with water which has been boiled (to free it from dissolved oxygen) and cooled. Put the animals into the jar and close it air-tight with a lid or plate of glass. Some species die within 6 hours, others require 12 to 24 hours. The effect may be hastened by adding a pinch of tobacco to the water, or the water may be saturated with chlorobutanol.

FUMES OF ETHER, CHLOROFORM, OR CYANIDE. Insects, arachnids, and myriapods are commonly killed by this means. Select a wide-mouthed bottle with a screw cap or tight cork closure. Cut a piece of screen wire to form a tight-fitting false bottom. If ether or chloroform is to be used, place a wad of cotton, upon which the anesthetic may be poured, under the screen.

If potassium cyanide is used, a few lumps of this substance may be placed under the screen. Keep the jar tightly stoppered except when introducing specimens or taking them out. Remove specimens as soon as they are dead.

If specimens are to be mounted whole, it is generally desirable to spread their appendages. After removing the animal from the jar, lay it on a glass plate and arrange the appendages. Then place a small piece of glass over the specimen and run alcohol or some other fixing fluid between the two pieces of glass. As soon as the animal's joints have stiffened, remove it to a dish or bottle of the fluid. Small insects and other small arthropods may simply be placed between two microscope slides, to keep their appendages outspread, and killed in alcohol or Carnoy's fixative, without first anesthetizing or poisoning them.

In case any parts of the animal are to be sectioned, remove and fix them at once. With small organisms, it may be better simply to open the body and place the entire organism into the fixative. For cytological work, kill animals by decapitation, not by anesthesia or poisoning. The reason for this will be discussed presently.

Methods for Killing Vertebrates and Large Invertebrates

The tissues are least affected if the animal is killed by decapitation, bleeding, or a blow upon the head. Anesthetics should be used only to the extent required by humane considerations. Lethal quantities of ether and chloroform, especially the latter, are likely to injure the cells. These substances should be employed only for anesthesia, until the animal can be killed by bleeding. Otherwise, the tissues may be spoiled for critical work.

Methods for Various Animals. FISHES. These are commonly killed by a blow upon the head, or by making an incision into the heart.

FROGS, SALAMANDERS, REPTILES. To kill small species, cut off the head quickly with stout scissors or cartilage shears. Kill large animals by a blow upon the head, or anesthetize and bleed them.

BIRDS. Traditionally the neck is wrung or the head is cut off. If this seems inhumane, treat them in same way as mammals.

CATS, DOGS, RABBITS, RATS AND OTHER SMALL MAMMALS. Put the animal into a gas-tight box, jar, or can, and introduce a cloth or sponge well saturated with ether. As soon as the animal becomes unconscious, remove it and cut its throat or open some large blood vessel.

LARGE MAMMALS. Tissues from freshly-killed sheep, cattle, and pigs may be obtained at almost any meat-packing plant. Human tissues can be obtained in strictly fresh condition only from surgical operations, and are likely to be pathological. The value of autopsy material depends upon how long after death the autopsy is performed and the temperature at which the body has meanwhile been kept. The physical condition of the subject at the time of death is another important factor. Very soon after death the digestive glands show signs of autolysis and the epithelial lining of the digestive tract begins to slough off. At ordinary room temperatures the brain, spinal cord, and most other organs undergo decided changes within 6 hours. However, under the same conditions, skin, muscles, cartilage, and bone may remain in rather good condition for 24 hours or even longer.

Dissection and Subdivision

Dissection is the careful separation of anatomical parts, generally with reference to natural boundaries. It is not indiscriminate cutting, hacking, or mangling. In microtechnique, dissection is ordinarily employed for the removal of organs or parts of organs from the body. *Subdivision*—as the term is used here—designates the cutting of an organ, region, or entire animal body into segments or slices, in which the various structures retain their natural relations. The purpose of both processes is to facilitate the penetration of fixing solutions and other reagents, essential for the preparation of thin sections.

Dissection. GENERAL INSTRUCTIONS. Plan operations in advance. Have instruments, other necessary equipment, and fixing agents ready. Begin to dissect as soon as possible after the death of the animal. Work carefully but rapidly, so that there will not be time for postmortem changes to occur. Use sharp, clean instruments, but do not employ fine instruments for heavy work which might dull or bend their points. Avoid jabbing or pinching the tissues and cut only when necessary. Never put metal instruments into solutions containing mercuric chloride or other corrosive substances. Handle material in such liquids with wooden, plastic, or glass instruments which can be fashioned to suit various purposes. Before placing tissues into a fixative, it is generally advisable to wash them with physiological saline solution, in order to remove adhering blood or foreign matter.

DISSECTION APPLIED TO VARIOUS ANIMAL TYPES. *Small Invertebrates and Embryos.* Permeable organisms as large as 10 mm. (or even 15 mm., if they are slender) can best be fixed and stained or sectioned without subdivision. However, dissection is frequently useful in the study of small invertebrates, especially arthropods. In preparing large embryos for sectioning, cut them into several pieces after they have been hardened for a short time. When embryos are to be studied by means of dissection, they should first be thoroughly hardened.

Larger Invertebrates. For critical work in histology and cytology, remove the organs from perfectly fresh specimens, subdivide them if their size makes this necessary, and place them in an appropriate fixing solution.

Animals with Hard Shells or Chitinous Cuticles. Arthropods, which are to be dissected and mounted, or which are to be sectioned for the study of general histology, can be fixed in a satisfactory manner if several small apertures are made in the body wall. Another method is to inject the fixing fluid. Large nematodes may be treated similarly. Gastropod molluscs should be killed by asphyxiation, pulled out of their shells, their bodies dissected or injected with a fixative, and then placed in the fixing fluid. Pelecypods and brachiopods can be fixed well enough for general study by propping apart the valves of their shells and placing them in a fixative.

Fishes, Amphibia, Reptiles, Birds. Methods of killing these animals have been discussed in a previous paragraph. The procedure for dissection is much the same as that which will be described for mammals.

Plan for Dissecting a Mammal. If a variety of tissues is to be obtained from one animal, begin with the organs in which postmortem changes occur most rapidly. Remove, subdivide if necessary, and fix the various organs in the order they are subject to postmortem changes. The following outline is designed for small mammals, such as rats and guinea pigs. It can readily be modified to fit other vertebrates or to meet the requirements of special work.

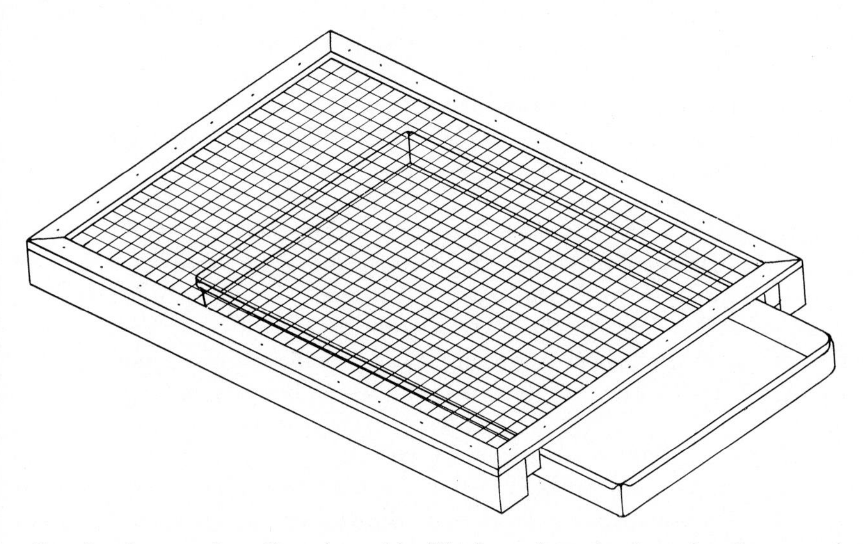

FIG. 7. A convenient dissecting table. The frame is made of wood, and coarse wire netting ("hardware cloth") is stretched over it. An enameled pan under the netting catches drippings. The animal's appendages are simply tied to the netting, and flaps of skin or muscle may quickly be fastened to the wire by clips or hooks. A dissecting table about 18 by 24 inches will serve for mammals the size of a rabbit or smaller.

Killing. Anesthetize the animal with ether. Remove the skin from one side of the neck, locate the jugular vein, and cut it. This method of killing leaves only a little blood in the vessels of the various organs. Ordinarily this is not a disadvantage. Should it be important for the vessels to remain filled with blood, in order that they may be seen more readily, kill the animal by cutting the spinal cord close to the base of the skull.

Lift the animal by its hind legs and hold it over a sink or other receptacle until bleeding nearly ceases. Then lay it, ventral side up, on a dissecting table or pan, spread its appendages, and fasten them in place. An inexpensive and easily made dissecting table is shown in Figure 7; this type will serve well for animals which are too large or too muscular to pin out in a dissecting pan of the conventional type. Make a mid-ventral incision through the skin from pubes to sternum. Reflect the skin and make a similar incision through the muscular wall of the abdomen. Spread or fasten down the halves of the abdominal wall. Remove the great omentum, in order to expose the viscera.

1. *Digestive System.* First remove the pancreas, and cut from it several slices not more than about 4 or 5 mm. thick. For general histological work, Helly's fluid will serve as a good fixative. Remove the stomach, by cutting across the esophagus some distance anterior to its junction with the stomach, and across the duodenum posterior to the pyloric sphincter. Slit open the organ and rinse it with warm physiological saline solution. Cut strips about 1 by 2 cm. from the fundic and pyloric regions, pin them to wood or cork (Fig. 8*A*), and place them in the fixative. Either Bouin's fixative or Zenker's fixative may be recommended. The latter is favored when material is to be stained by Mallory's triple stain. The transitional region where the esophagus and stomach are joined, and the region of the pyloric sphincter, make interesting and instructive preparations, and should be fixed separately. Cut a 3 to 5 cm. length from each major region of the intestine (duodenum; ileum; colon) and any special regions (such as the ileocecal valve) which one may wish to study. Unless the tissue is pinned out, as directed above for stomach, a pipette should be used to force the fixative through the pieces of intestine. Bouin's fluid or Zenker's fluid will be suitable for most purposes. From the liver cut several slices 4 or 5 mm. thick. Heidenhain's "Susa" fixative is good for liver tissue, but Zenker's, Helly's or Bouin's fluid may also be used.

2. *Gonads.* Remove testes or ovaries and place them in a fixative such as Bouin's fluid or Allen's B 15 modification of it. If the organs are of considerable size (as large, for example, as the testis of a rat) cut them into thin slices after they have been in the fixative for about half an hour. Use a sharp razor blade for this purpose.

3. *Kidneys, Adrenal Glands, Spleen.* Remove the adrenal glands and fix them in Zenker's fluid or Helly's fluid. Cut the kidneys lengthwise

or transversely, into at least 3 parts. Heidenhain's "Susa" fixative is excellent for kidney tissue. Cut several transverse slices from the spleen; for general histological work, fix them in Bouin's fluid or Zenker's fluid.

4. *Central Nervous System; Eyes; Ears.* Special directions for these structures will be given in subsequent paragraphs.

5. *Salivary Glands.* Dissect out the parotid, submaxillary, and sublingual glands. Cut each into several slices and fix these in Helly's fluid or a saturated aqueous solution of mercuric chloride. Avoid the use of fixatives containing an acid.

6. *Lung.* Cut through the ribs and muscles on either side of the sternum, and remove the ventral wall of the thorax. Then cut through one of the two primary branches of the trachea, just back of the bifurcation of the latter, and carefully remove the corresponding lung. Attach a large rubber bulb to a piece of glass tubing and fill the apparatus with a fixing solution, preferably Zenker's fluid or Bouin's fluid. Insert the tube into the primary bronchus entering the lung, and tie it in place with a string. Then distend the lung by exerting gentle pressure on the bulb. When the lung is moderately distended, slip the tube from the bronchus, at the same time pulling the string tight so as to prevent the escape of liquid, and place the lung in the fixative.

7. *Cardiac Muscle.* Cut slices from walls of the ventricles and auricles and fix these in Zimmermann's nitric acid-alcohol mixture.

8. *Artery, Vein, and Nerve.* Remove the skin from one hind leg. On the medial surface of the thigh, locate the femoral artery, vein, and nerve, which lie parallel and close together in a depression called the iliopectineal fossa. With fine forceps work two strings around all three structures, at points separated by about 1 or 2 cm. Tie knots in the strings, cut through the blood vessels and nerve outside of the segment thus tied up, and then carefully separate the segment from the surrounding tissues. Lay it on a narrow piece of smooth wood, stretch it to normal length, pin down its ends, and place the preparation in Bouin's fluid or Zenker's fluid.

9. *Sciatic Nerve.* Bisect the biceps femoris and other superficial muscles on the lateral aspect of the thigh, so as to expose the sciatic nerve. Cut through the nerve close to the hip joint, dissect it down to the knee, and cut it off at this point. Lay the nerve on a narrow strip of wood and pin it down at either end, being careful not to stretch it. Immerse it in 10% formalin or some other suitable fixative. After a few hours, remove the pins and cut the nerve into lengths of about 1 cm.

10. *Other Tissues.* Fix small pieces of esophagus in the way suggested for intestine. The tongue should be cut in slices not over about 4 or 5 mm. thick, and in such a way that the surface layers will eventually be sectioned transversely. Bouin's or Zenker's fluid will be good for fixation of tongue; the latter is recommended if staining by Mallory's triple stain is contemplated.

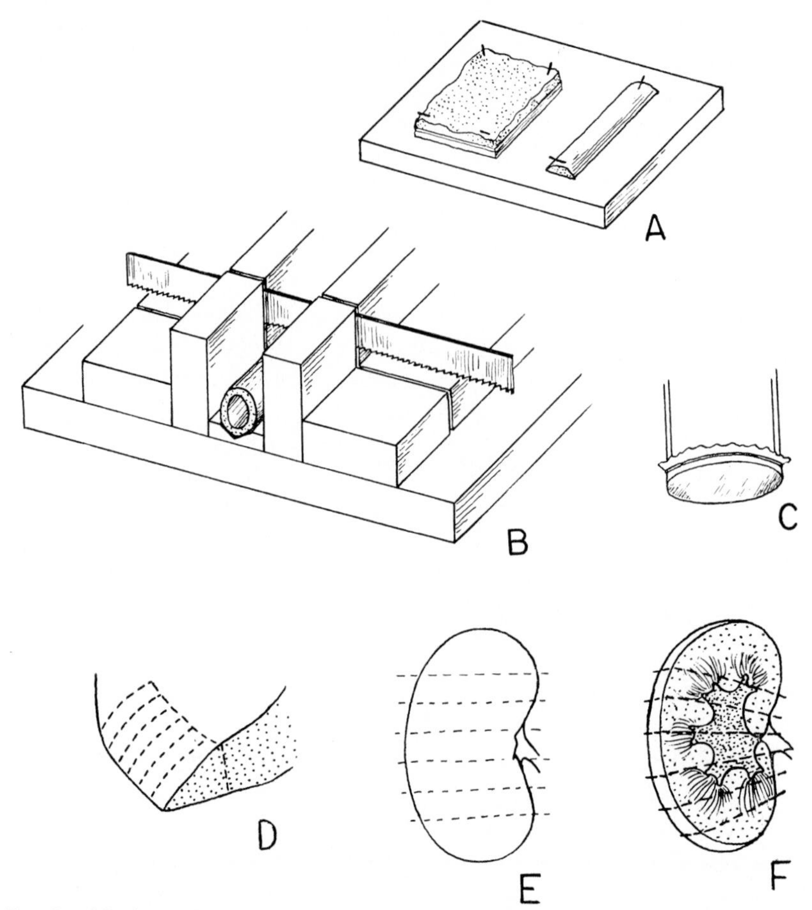

FIG. 8. Methods of preparing various types of material for fixation. *A*, Portions of the digestive tract and muscle pinned down to prevent them from becoming distorted by contraction. The same technique is applicable to pieces of nerve, skin, and blood vessels. *B*, A simple miter box used for sawing bones into short lengths or thin sections. *C*, Stretching a thin membrane over a vial or glass tube to prevent it from wrinkling. *D*, Cutting thin slices from the liver, spleen, and other dense organs. *E*, Subdivision of a kidney in preparation for making transverse sections. In the case of a large kidney, it will be advisable to divide these slices into sectors to facilitate penetration by the fixative. *F*, Subdivision of a kidney in preparation for making partial sagittal sections.

If areolar tissue, pieces of mesentery, and similar materials are to be prepared, see suggestions on page 168 for spreading these before fixation.

11. *Tendon.* Dissect out one or two of the flexor tendons from the ventral surface of the tarsus and metatarsus. Tendons are usually quite well fixed by 10% formalin.

12. *Skin.* Cut off several of the pads of thickened skin from the soles of the feet. Fix these in Bouin's fluid or some other suitable reagent.

13. *Skeletal Muscle.* From one of the leg muscles cut a thin slice

about 2 cm. long, 5 mm. wide, and 3 mm. thick. Choose a portion in which the muscle fibers run nearly parallel. Pin the slice to a piece of wood (Fig. 8*A*), but do not stretch it to more than its normal length, and immerse it in Helly's fluid or Bouin's fluid.

14. *Bone.* Disarticulate the hip-joint and remove the leg from the body. Cut away all of the tendons and muscles from the femur and tibia. Using a miter box and a very fine saw blade (Fig. 8*B*), cut the tibia into several pieces 4 or 5 mm. long. Fix them in 10% formalin for several days.

Dissection of the Central Nervous System, Eyes, and Ears. As a rule, it is best to use a separate animal for these organs, because they are likely to undergo postmortem changes rather rapidly. Normally a considerable amount of time is required to remove them. Strip off the skin from the top and sides of the head. With bone cutters, carefully cut and break away the roof and sides of the skull.

1. *Eyes.* Cut the extrinsic muscles of the eye and the optic nerves. Remove and fix the eyes in Helly's fluid or Perenyi's fluid.

2. *Removal of the Brain.* When the brain has been exposed, cut through the tentorium cerebelli, which is the bony partition between cerebrum and cerebellum, and carefully remove this. Cut through the dorsal arch of the atlas, and then transect the spinal cord just posterior to the foramen magnum. Tilt the head forward, and lift the posterior end of the brain, cutting the cranial nerves as they are encountered. In this way, carefully remove the brain. If the brain is to be fixed entire, place it upon a layer of cotton, in a large dish or bottle of fixative. Ten percent formalin is generally used for fixation of the entire brain. If parts of the brain are to be processed individually, proceed as follows:

3. *Cerebrum.* Cut across the brain stem between the cerebrum and cerebellum. Carefully remove the dura mater and place the cerebrum in a large volume of 10% formalin for a few hours. Then, with a very sharp blade, cut out several slices, vertical to the surface of the organ and across its long axis. Make the slices about 5 mm. thick, 6 or 7 mm. wide, and 1 cm. deep. Return them to the fixative.

4. *Cerebellum.* Remove the dura mater and place the cerebellum in a fixative which is suitable for the method of staining or impregnation contemplated. (See especially Ramón y Cajal's method of impregnation, p. 411). After the cerebellum has been in the fixative for about an hour, slice off its lateral lobes and the lateral portions of its median vermis. The slab which remains, and which includes the median parts of the cerebellum and medulla oblongata, should be about 6 or 8 mm. thick.

5. *Ears.* Cut out the temporal bones. Then cautiously slice away the surface of each, in order to make a small opening into the internal ear. Place them in a suitable fixative. Bouin's fluid gives good fixation and at the same time begins the process of decalcification.

6. *Spinal Cord and Ganglia.* Make a median longitudinal incision

through the skin at the back of the neck, to the level of the first thoracic vertebra. Reflect the skin, and cut away the muscles so as to expose the neural arches of the vertebrae. Cut through the pedicles of the neural arches, beginning with the axis, and working back to the first thoracic vertebra. Gently lift the end of the spinal cord, free it from its bed, and locate the first pair of spinal ganglia and nerves. Dissect out the ganglia, which lie in the intervertebral foramina, and cut the nerves a short distance lateral to them. Work backwards in the same way, gently freeing the cord and ganglia and cutting the nerves. Remove the cord and lay it on a long, narrow strip of wood. Spread the spinal nerves at right angles to the cord and pin them down. Place the spinal cord in 10% formalin for several hours or days. Then cut from it a number of segments 6 to 8 mm. long, each including a pair of spinal ganglia and nerve roots. If only the ganglia are wanted, do not take time to remove the cord. Simply dissect out the ganglia and fix them by the desired method. For demonstrating Golgi material in cells of the spinal ganglia, refer to the method of Da Fano and the so-called Mann-Kopsch technique (Chapter 26).

Subdivision. After dissecting out the organs of vertebrates and larger invertebrates, it is necessary in most cases to subdivide them, because the entire organs may be too large and dense to be penetrated rapidly enough by fixatives and other reagents. Subdivision is also necessary in preparing invertebrates or embryos of considerable size.

Subdivision is carried out by cutting the specimen into segments or slices, in which the component parts retain their natural relationships. Cutting must be done with reference to some important plane or other landmark, so that the locations of structures may be determined correctly from the sections (see Fig. 8*D*, *E*, *F*). It is necessary to use a razor blade or very sharp knife, in order to avoid crushing the tissues. The size of pieces to be cut depends largely upon the penetrating power of the fixative. Slices of tissue should not be much more than 5 mm. in thickness for fixation in Bouin's fluid, Zenker's fluid, 10% formalin, and comparable fixatives used for general histological purposes; they should be no thicker than 2 or 3 mm. if cytological fixatives containing chromic acid and osmium tetroxide are to be used. For general histology, the pieces may be moderately large (up to 1 by 1.5 cm.) in the intended plane of sectioning.

Before placing pieces of tissue in a fixative, rinse them briefly with a physiological saline solution, in order to remove adhering blood or foreign matter. Very soft organs such as the testis, brain, and spinal cord should be hardened to some extent before they are subdivided. A testis as small as that of a mouse may be cut into slices after it has remained in a fixative such as Bouin's fluid or Allen's B 15 modification for about 30 minutes, but a large brain being fixed in 10% formalin should be left in this for several hours or overnight before it is cut into smaller pieces.

Preparation of Contractile Tissues

PINNING DOWN MUSCULAR ORGANS AND NERVES. If portions of the digestive tract or urinary bladder are simply placed in a fixing fluid, contraction of the muscle layers will cause the specimens to twist and curl. This will make it impossible to prepare sections in which the relationships of the various tissue layers to one another are clear. Such distortion may be prevented by pinning the pieces of tissue to a smooth strip of soft wood or cork before placing them in the fixing agent (see Fig. 8*A*). The mucosa of the organ should be turned away from the wood or cork. This method is applicable to other contractile tissues, including skeletal muscles and blood vessels. It is also useful for some non-contractile specimens, such as nerves and tendons, which are likely to become bent or twisted. The tissue should be fastened down in a *normal* state of extension, but *not stretched.* In case the fixing fluid contains mercuric chloride or is strongly acid, use splinters of hardwood or glass needles instead of metal pins. Sharp toothpicks can be stuck into sheet cork, and sometimes serve very well.

SPREADING THIN MEMBRANES. To prevent mesenteries, subcutaneous areolar tissue, and other membranes containing elastic fibers from contracting into a useless mess, they should be spread on a glass or plastic ring or over the end of a glass or plastic tube. Place the ring or end of the tube against the membrane *in situ.* Bring the tissue over the edge, tie it in place with a string, and cut around it with sharp scissors. This method is illustrated at *C* in Figure 8, and described more fully on page 168.

METHODS OF APPLYING FIXATIVES

The method to be used depends upon the size and density of the objects and upon the penetrating power of the fixative.

Fixing by Immersion

This is the usual method of fixation and suffices for small invertebrates, embryos, and small pieces of tissue. It is often necessary to inject the fixative into large or dense specimens, but even then the process is completed by immersion.

BRINGING SPECIMENS INTO THE FIXATIVE. Tissues of vertebrates and larger invertebrates are removed by dissection, subdivided, and fastened down if this is necessary. They are rinsed with physiological saline solution, to free them from adhering blood or foreign matter, and then dropped into the fixative. Other objects which can be handled in air, without being injured, are treated in the same way. Very small invertebrates, embryos, and larvae may be taken up, with a little water, into a pipette and expelled gently into the fixative. In order to fix delicate contractile invertebrates (such as hydroids and sea anemones), it is best to narcotize them, remove most of the water from the organisms, and then pour the fixing fluid on them. The use of hot fixing fluids is often advantageous. When removing the water, take care to disturb the organisms as little as possible.

Volume of the Fixing Fluid. This must always be many times the volume of the material. The proportion differs for various fixatives, but in general the volume of fixing fluid should be 25 to 50 times that of the objects. In case the fixing solution is considerably diluted by water brought in with the material, allow the specimens to settle and harden for a few minutes, then draw off most of the liquid and replace it with a new fixing solution.

Position of Objects in Fluid. In order to obtain uniform penetration, bulky objects should be suspended in the fixative. If this is impractical (for example, in the case of entire brains), a quantity of cotton may be placed on the bottom of the container to support the specimen. If tissues have been pinned to a strip of wood, cork, or cardboard, float the strip on the fixative, with the tissue downwards.

Duration of Treatment. This depends upon the size and permeability of the objects, as well as upon the properties of the fixing solution. The approximate time required to fix objects of different types and sizes has been stated for various fixing solutions in Chapter 8.

Washing and Hardening. The way in which these steps must be carried out depends upon the fixative used. A general discussion of this subject will be found in the next chapter, and suitable methods to follow the use of various standard fixatives are described in Chapter 8.

Effects of Temperature. Warm or hot fixing solutions penetrate more rapidly, reducing the possibility that unfavorable postmortem changes may occur. There may be other advantages to their use also. For example, some contractile organisms (as hydras, various flatworms, nemerteans, and annelids) are killed so quickly by hot fixatives that contraction is minimized.

In the case of those substances, including mercuric chloride, which coagulate proteins, it is likely that their effectiveness as coagulants is increased as the temperature is raised. If this is so, the dilution of a fixative as it penetrates the deeper layers of a piece of tissue may be offset to some extent by an increase in its coagulative capacity when heated.

What has just been said should not be taken to mean that fixatives should routinely be heated. Some fixing mixtures definitely give better results, at least with certain types of material and for certain purposes, when applied at lower temperatures. Flemming's fluid, for instance, seems to bring about superior fixation of nuclei in certain tissues at temperatures around 0° C. The preservation of specific substances, such as glycogen, is often carried out at lower temperatures in order to reduce the solubility of these substances in alcohol or water.

It is quite likely that even moderate heating may be unfavorable for fixation of many histological and cytological elements. Good judgment, based on experience, will have to be used in deciding whether heat should be applied, and to what temperature the fixative may advantageously be raised.

If heat is to be used, the fixative should be brought to the desired temperature before the specimens are dropped into it. Should it appear desirable to maintain the higher temperature after the specimens have been immersed, a water bath or incubator is to be preferred for keeping it warm. It is not a good idea to allow specimens to rest on the bottom of a container which is being heated directly over a hot plate.

Fixing by Injection (Perfusion)

A fixative may be caused to penetrate dense or bulky tissues with maximum speed and thoroughness by injecting it into the blood vessels or other natural cavities. Tissues destined to be used in cytological work are frequently fixed in this manner, which is especially valuable for the gonads and central nervous system.

Injecting Apparatus. For very small animals such as mice and frogs, it is most convenient to use a glass or plastic syringe of 10 to 25 ml. capacity. For larger animals, including cats, rabbits, and rats, a syringe holding at least 100 ml. is necessary, and it may be advisable to use some form of constant low-pressure injecting apparatus. Gravity will supply the necessary pressure if the container of fluid is raised to a height of 2 or 3 feet. One may use a funnel, supported by a ring-stand, and connected to the injecting cannula by a rubber tube. A pressure equal to 10 or 12 mm. Hg is enough for small animals, and 15 to 20 mm. Hg is sufficient for cats and animals of similar size. High pressure must not be used, because it will lead to rupture of smaller vessels. Avoid metal syringes and cannulas, because most fixing solutions will corrode them.

Injection through Blood Vessels. If tissues are to be taken from various parts of the body, it is best to insert a cannula through the left ventricle, into the aorta, and to inject the entire body. Otherwise, injection may be made by way of the thoracic aorta, directing the cannula backward in order to fix the testes, kidneys, and other organs located in the same general area, or forward to fix structures of the head and neck region. Even in an animal the size of a cat or rabbit, the arteries of individual organs are rather small to be easily handled, and it is advisable to inject through a larger artery such as the aorta, one of the carotid arteries, the coeliac artery, anterior mesenteric artery, posterior mesenteric artery, or one of the femoral arteries, depending upon the location of the organs one desires to fix. The following directions may be applied to injection either of the entire body or of any region of a small animal.

Preparation of Equipment. Assemble and test the injecting apparatus. Have the dissecting instruments conveniently at hand. These should include sharp scissors, scalpel, and coarse and fine forceps. Cartilage shears are also convenient for use on large animals. Some absorbent cotton or gauze will be needed. Have ready several hundred milliliters of an isotonic saline solution (such as Ringer's solution), and a like quantity of fixing fluid. When injecting mammals or birds, keep these fluids in a water bath at 40° C. until they are needed, and also have available several cloths which can be wrung out in warm water and used to keep the animal at body temperature.

Preparation of the Animal. Anesthetize the animal with ether. As soon as it becomes unconscious (the heart still beating), lay the animal upon its back, on an operating table or dissecting pan, and tie down its legs. Make a longitudinal incision through the skin, in the mid-ventral line, from the level of the first rib to the genitalia. Separate the skin and reflect at right and left. Beginning at the diaphragm, cut forward through the muscles and ribs at a short distance from each side of the sternum. Carry the cuts up to the first rib, but not through it. Wind a stout string around the sternum, just back of the first rib. Tie it tightly and then cut off the sternum behind this point. Cut off the tip of the heart. Turn the animal over and hold it above a sink or pan until bleeding ceases. Then wash out the thorax with warm water. Open the abdomen by making a median longitudinal incision through the muscles and underlying omentum.

Inserting the Cannula. Locate the artery which is to be injected. If the entire body is to be injected, pass two strings around the aorta, close to the heart. If only a part of the animal is to be injected through a portion of the aorta or some other peripheral artery, pass three strings around the vessel as shown in Figure 9. Unless a hypodermic needle is used for injection, it will probably be necessary to make a small incision (*i*) in the vessel, at a point proximal to the center string (*b*). Fill the injecting apparatus with an isotonic saline solution which, for mammals or birds, should have been warmed to about 40°C. Addition of 0.1% novocaine to the saline solution will facilitate injection by causing the arterial walls to relax. Discharge a small amount of the saline solution through the cannula into the artery, so that it points away from the heart. Tighten the knot in the string nearest the heart (*c*) and, if three strings are used, also tie the center string tightly. In any case, the string toward which the cannula points (*a*), should loosely encircle the artery, *beyond the end of the cannula.*

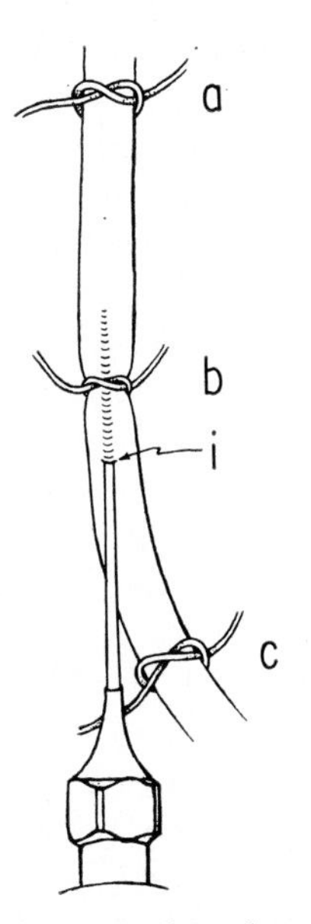

FIG. 9. Method of inserting a cannula or hypodermic needle into a blood vessel. (See text for explanation.)

Washing Out the Vessels. Start the flow of saline solution with a *very gentle but steady pressure.* Use just enough pressure to distend the artery. Too much pressure will rupture the vessels. Continue the injection until no blood can be seen in the desired part and the vein coming from it is filled with clear liquid. If the testes are of special interest, slit open the scrotum and examine them. In case the entire body is being injected, the liver and intestines should be pale and the liquid running from the right ventricle should be practically free of blood.

Injecting the Fixative. Detach the connecting tube from the syringe or receptacle and hold or fasten it up in order to prevent the escape of liquid and entry of air. Fill the syringe with fixing fluid. For mammals and birds, this should be warmed to about 40°C. Attach the connecting tube again, being careful to exclude air. Now inject the fixing fluid with gentle pressure, until it has filled the desired organs and runs from the heart. Toward the end of the process, the pressure may safely be increased to some extent. If the fixative is colored (as Bouin's fluid, for instance), its progress can easily be observed. Fluids containing mercuric chloride cause the tissue to become white and opaque. A small amount of some dye may be added to colorless fixatives, in order to show their presence. When the injection is complete, ligature the base of the heart or, if only one area has been injected, ligature the vein draining it. Then tighten the string distal to the cannula (*a*), loosen the one which holds it in place (*b*), and withdraw the cannula.

Removing and Immersing Tissues. Allow the animal to remain in position for 10 to 15 minutes. Then remove the desired parts, and place them in a quantity of the fixing solution. Use old instruments for this purpose, as most fixatives corrode metal. After another half hour, cut the tissues into thin slices with a razor blade. Return them to the solution until fixation is completed. The time required depends upon the tissues and upon the fixative used. Then wash the tissues in the manner which is appropriate, and proceed to dehydrate them preparatory to embedding in paraffin or nitrocellulose.

DISTENTION OF HOLLOW ORGANS. By filling saccular organs with a fixing solution, it is possible to fix them in a nearly normal state of distention and at the same time to bring the fixing fluid quickly into contact with the inner layer of tissue. This applies both to organs with muscular walls, such as the urinary bladder, and to delicate collapsible structures such as the lungs. Insert a cannula of appropriate size into the orifice of the organ and tie a string tightly around the tissue encircling the cannula. Attach the cannula to a rubber bulb or syringe filled with the fixing fluid. Inject a sufficient amount of fluid to distend the organ to its natural size, *but no more.* Withdraw the cannula, at the same time tightening the string to prevent escape of liquid, and immerse the organ in a quantity of fixing fluid. After the tissue has become hardened, pieces of the desired size may be cut from it with a sharp, thin blade.

Fixation by Vapors

This method is limited to isolated cells, thin membranes (such as the retina), smears, and very small bits of solid tissue. The advantage claimed for vapor fixation is that it can dissolve nothing from the cells. Only two substances are regularly used in this manner.

OSMIUM TETROXIDE VAPOR. This substance is the one most frequently employed. Directions for its use will be found on page 108.

FORMALDEHYDE VAPOR. This is used in a few cases when osmium tetroxide would interfere with the stain (as Sudan IV, for lipids) or impregnation with silver or gold. The procedure used for fixation in vapor of formaldehyde is described on page 81.

CHAPTER 10

Washing, Hardening, Preservation, Bleaching, Decalcification, and Desilicification

WASHING

As a rule, it is necessary to wash the fixing agent out of the tissues before proceeding with dehydration, infiltration preparatory to embedding, or staining. Some fixatives are conventionally washed out with water, others with alcohol. The fluid which should be used for washing and the approximate time required for the process have been given for each fixing solution described in Chapter 8. For convenience, this information will be briefly summarized, and some procedures used for washing will be described.

Certain ingredients of fixatives should be washed out as soon as fixation is complete. These include chromic acid (especially when it is used in combination with a reducing agent), some other acids which may bring about destructive changes within a matter of hours, and osmium tetroxide, which will blacken the tissue. Chromic acid also tends to discolor tissues and to form precipitates. Mercury compounds may become injurious within hours or days, depending upon other substances present, and sometimes continue to form precipitates which are difficult to remove. Picric acid and potassium dichromate do not cause destructive changes for many days, but should not be allowed to act much longer than is necessary, in order to limit the discoloration of tissues. A fact of paramount importance is that any of the foregoing reagents will interfere with the action of stains, if they are present in the tissues or if they have been allowed to act for too long a time. Formalin and alcohol are relatively harmless, and both are easily washed out when this is necessary. Although they are among the better preservatives, these substances bring about, in the course of time, a slow deterioration in staining properties.

In certain special methods of metallic impregnation, and particularly in connection with demonstration of structures in the nervous system, fixing agents such as potassium dichromate, mercuric chloride, or cobalt nitrate are allowed to remain in the tissues, so that they may enter into chemical combination with other substances to be introduced later.

Washing is therefore omitted or reduced to mere rinsing. Washing is shortened in some methods for mitochondria and other structures if the fixative acts also as a mordant. However, these are exceptions.

General Rules for Washing

Aqueous Fixatives. 1. Aqueous solutions containing any of the following: chromic acid; potassium dichromate; osmium tetroxide; formalin—wash in water. This group includes the fixatives of Flemming, Orth, Regaud, and Zenker, and many other mixtures.

2. Aqueous solutions containing picric acid or mercuric chloride (potassium dichromate-mercuric chloride combinations, such as Zenker's fluid, belong under 1)—wash in 70% alcohol. This group includes Bouin's fluid and many of the mercuric chloride mixtures in general use. There is much difference of opinion regarding the advisability of washing out picric acid with alcohol, as recommended here. This matter has been discussed at some length on page 99.

Alcoholic Fixatives. These are always to be washed out with alcohol. Proceed as described below under Washing in Alcohol.

Containers for Material. Protozoa, very small invertebrates, embryos, and larvae are best handled in vials. Allow the objects to settle before making each change of liquid. Cautiously withdraw the liquid with a pipette and pour the new fluid down the side of the vial. The settling of material may safely be hastened by centrifuging at low speeds.

Larger specimens may be placed in wide-mouthed bottles, stender dishes, or screw-capped jars. The latter are especially convenient.

Washing in Water

Running Water. This is the most efficient and least troublesome method for objects which are of sufficient size and firmness to withstand such treatment.

Various kinds of apparatus are employed for this purpose. If a piece of cotton netting or plastic screen is tied over the mouth of a bottle, and a small stream of water is focused on it, diffusion will be accelerated to some extent. It is not a good idea to increase the flow of water to the point that the tissue is thrown about.

There are better ways to accomplish a thorough but gentle washing. One simple method is to place the specimens in a glass or plastic tube of appropriate length and diameter, tie cotton netting or plastic screen over both ends of the tube, and lay it in a pan through which water is running. Large or firm specimens may be tied in bags of cotton netting and suspended in a pan of running water. Some workers make or purchase little tissue baskets of wire or plastic screen.

Another efficient washing apparatus is shown in Figure 10. The unique features of this device are that *all* of the water passes over the tissue and that the flow is *downward*, so that gravity aids rather than hinders the removal of heavy salts. To make this apparatus, take a 10 cm. length of glass tubing having a diameter of about 1 cm., or make such a tube by removing the bottom of a shell vial. Fit a cork to a wide-mouth bottle or jar of appropriate size. Make two holes in the cork: one of sufficient size to hold the large tube, and a smaller one for the U-shaped outlet tube. Assemble these parts as shown in the illustration and tie a piece of cotton netting over the bottom of the large tube. Place the tissue in the tube and allow a small stream of water to flow gently into it. In some laboratories, a long pipe, with numerous small outlet tubes, is installed over the draining board of the sink. This makes possible the use of several washing units at the same time.

Repeated Changes of Water. Protozoa, small and delicate invertebrates, embryos, and larvae must be washed in this way. Place material in tall vials, test tubes, or centrifuge tubes. The volume of water must be many times that of the objects. Every few minutes, or as often as the objects settle to the bottom, carefully remove as much water as possible with a pipette, and refill the containers. The settling of small objects may safely be accelerated by slow centrifugation.

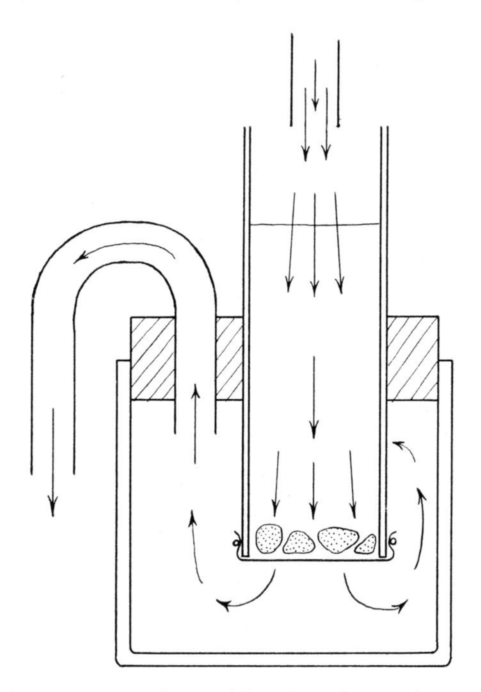

Fig. 10. An efficient apparatus for washing tissue in running water. (See text for explanation.)

In some localities the tap water contains a good deal of air and may form bubbles which lift small specimens to the surface. If this is the case, the water should be boiled and then cooled, or at least allowed to stand for a while before it is used.

Duration of Washing. Washing is generally continued until all or nearly all of the uncombined fixative has been removed.[1] However, tissues should not remain in water longer than is necessary, since they may become softened or even partially disintegrated by prolonged soaking. If the fixative is deeply colored, as after fixation in compounds of chromium, the progress of washing can be estimated from the color of the tissue and wash water. The approximate time required for a thorough washing in running water is about 3 to 6 hours for small and permeable objects (such as chick embryos), 6 to 12 hours for slices of vertebrate tissues up to 5 mm. thick or for embryos up to 8 mm., and 18 to 24 hours for larger or less permeable objects. These estimates refer to the removal of chromic acid, potassium dichromate, or osmium tetroxide. After fixation in formalin, a few minutes in water will suffice for most purposes, or the alcohols used for dehydration can be counted on to remove the fixative. In washing chromium compounds from small and delicate objects, make repeated changes of water over a period of 6 to 24 hours, depending upon how soon the water no longer shows discoloration. *As a general rule, do not wash objects in water for more than 24 hours.*

After they have been washed in water, tissues are transferred gradually to 70% alcohol, in which they may be stored unless they are to be stained or dehydrated and embedded as soon as possible. A series of alcohols consisting of 15, 30, and 50% alcohol will be sufficiently gradual for all but very delicate specimens.

Washing in Alcohol

Following Alcoholic Fixing Solutions. The general rule is to transfer material to alcohol of about the same percentage as that contained in the fixing fluid, or to the next lower or next higher step in the conventional alcohol series. Then change this several times before replacing it with stronger or weaker alcohol.

1. From Carnoy's acetic acid-alcohol, or the modification containing chloroform ("Carnoy II"), transfer material to 95% alcohol. Change this two or three times. From Carnoy and Lebrun's fixative, which contains mercuric chloride, transfer material to 95% alcohol, change this several times, and proceed to add iodine as directed on page 93. After iodizing the material, change the alcohol once or twice in order to remove the excess iodine.

[1] It has already been pointed out that washing is omitted or shortened in some special methods for mitochondria, Golgi material, and nerve tissue, in order that the fixative may act as a mordant.

2. From most formalin-acetic acid-alcohol mixtures, or the fixatives of Duboscq and Brasil or Perenyi, transfer material to 70% alcohol. Following Duboscq and Brasil's fluid or Perenyi's fluid, change the alcohol several times, until color no longer comes freely from the tissue. Formalin and acetic acid will come out in subsequent treatment by almost any method.

3. From Gilson's fluid or Petrunkewitsch's fluid, transfer material to 50% alcohol. After some minutes or hours, according to the size of the objects, place material in 70% alcohol. Then add iodine to remove precipitates of mercury, as described on page 93. After iodizing material, change the alcohol at least once or twice in order to wash out excess iodine.

FOLLOWING AQUEOUS PICRIC ACID OR MERCURIC CHLORIDE SOLUTIONS. Material fixed in these reagents is washed with alcohol in order to avoid the dissociating effect of water. There has been a great deal of discussion about this procedure, especially in regard to picric acid. The matter is reviewed on page 99.

4. From aqueous mercuric chloride, mercuric chloride-acetic acid, Heidenhain's "Susa" fixative, and Worcester's fluids, transfer material to 50% alcohol. With very delicate objects, the change should be made gradually. Treat material further as directed under 3 (above).

5. From Bouin's fluid, modifications (aqueous) of Bouin's fluid, or picro-sulfuric acid, transfer histological material to 50% alcohol. Change this once or twice, depending upon the size and quantity of the objects, and then replace it with 70% alcohol. In the case of cytological and embryological material, or delicate invertebrates, the transfer from the fixing agent to alcohol must be gradual. A method for gradual introduction of alcohol is described on page 100.

To remove picric acid from the tissues, change the 70% alcohol whenever it has become deeply colored, which may be once or several times daily. If the material is to be embedded in paraffin, before being stained, it is not necessary to remove all picric acid, and 2 or 3 changes of 70% alcohol will be enough. Neither is it necessary to remove every trace of picric acid from specimens which are to be stained with borax carmine or picro-carmine. On the other hand, it is best to remove all yellow color from objects which are to be stained with hematoxylin, and then mounted entire or eventually embedded and sectioned. Complete removal of picric acid is recommended also in the case of material which is to be embedded in nitrocellulose.

The extraction of picric acid may be accelerated by warming the alcohol in an incubator to 35 or 40° C. Some workers add a few drops of a saturated solution of lithium or sodium carbonate to the alcohol. This practice is not recommended, because it may impair the staining qualities of the tissue.

HARDENING

It has been pointed out in Chapter 8 that fixing agents differ a great deal with respect to their hardening action. Ethyl alcohol hardens tissues more than any other fixative in common use. This is one of several reasons why it is rarely used for fixation unless it is diluted with water or mixed with other fixing agents.

According to a study published by Wetzel in 1920 (see Baker, 1958), muscle tissue stored for 4 days in 80% alcohol after initial fixation in 10% acetic acid, 0.5% chromic acid, 3% potassium dichromate, or a saturated solution of picric acid or mercuric chloride was hardened further, though not excessively. Muscle fixed initially in formalin or 0.5% osmium tetroxide was apparently softened at least slightly by the 80% alcohol.

At the same time that it hardens tissues, alcohol has other important effects upon them. The first of these is *dehydration*, or the replacement of water by alcohol. The extent of dehydration depends upon the strength of alcohol employed. Alcohol also removes lipids, unless the solubility of these has been decreased by a fixative such as potassium dichromate or osmium tetroxide. The solvent action of alcohol is utilized to wash out certain fixatives (for example, picric acid) as described in the preceding section. Hardening, dehydration, and washing are considered as separate processes, but the first two of these, and often the third, are accomplished simultaneously by treating material with alcohol.

Shrinkage is the most undesirable effect of ethyl and methyl alcohols. The amount of shrinkage varies according to the fixative used. This is very clearly illustrated by some experiments in which the volumes of whole livers were measured before and after fixation in several reagents, and again after dehydration with alcohol and infiltration with paraffin. For instance, the volume of a liver fixed in 3% potassium dichromate was about the same as it was before fixation, but by the time it had been completely dehydrated its volume was reduced by 36%; by the time it had been infiltrated with melted paraffin, its volume was reduced by 51%. On the other hand, a liver fixed in a saturated solution of mercuric chloride lost 9% of its original volume, but dehydration caused an additional decrease of only 11%; after infiltration, the total decrease in volume was about 30%.

Alcohol has been blamed for the excessive hardness sometimes exhibited by material embedded in paraffin, but it is likely that the use of xylene for clearing or overheating during infiltration is to a large extent responsible for this difficulty.

Hardening is omitted in the case of material which is fixed in a weak fixing agent with the idea of softening the tissue so that it will dissociate. It is unnecessary, though not injurious, to use alcohol in the case of objects which are to be mounted in glycerol or other aqueous media, or those which are to be sectioned by the freezing method. Alcohol, if present, should be washed out before either of these steps is taken.

Ethyl, Methyl, and Isopropyl Alcohols. Ethyl alcohol is ordinarily used for hardening and may be obtained at a reasonable price by educational institutions and hospitals. Pure methyl alcohol (synthetic methanol) and isopropyl serve equally well and are within the reach of laboratories which cannot obtain ethyl alcohol without paying a high tax.

Method of Hardening with Alcohol. Alcohol should be allowed to diffuse into the tissues very gradually. This is necessary in order to avoid violent diffusion currents which may shrink and distort delicate structures if they are suddenly transferred from an aqueous medium to strong alcohol. The usual procedure is to place the fixed tissues successively in alcohols of increased strength. In most histological and zoological work, the series may consist of 15, 30, 50, 70, 80, 95, and 100% alcohols. For extremely delicate structures, the series should begin with 10% alcohol and be increased by steps of 10%. Some workers recommend intervals of 5%, but when such extreme caution is necessary, it is safer and more convenient to introduce alcohol by means of a dropping or diffusion apparatus. Hardening progresses throughout the series of alcohols, but the tissues will certainly be rather well hardened by 70% alcohol, and the principal function of the stronger alcohols is that of *dehydration*. Therefore we will now consider only the steps up to 70% alcohol. Complete dehydration necessary for embedding or mounting specimens will be dealt with in later chapters.

Tissues Which Have Been Washed in Water. After washing specimens thoroughly in water, place them in the lowest grade of the series, which is ordinarily 15% alcohol (or 10% alcohol, in longer series for more delicate material). At proper intervals, transfer them to the second, third, and subsequent members of the series. The length of time the objects must remain in each strength of alcohol depends upon their size and permeability. A small, permeable object (such as a young chick embryo of not over 48 hours' incubation) need be left in each alcohol for only about 30 to 45 minutes. A 3 or 4 mm. slice of testis or spinal cord may require about 2 hours. A relatively large and dense object, such as a piece of mammalian skin, liver, or kidney about 5 or 6 mm. thick, should probably remain in each of the lower grades for 2 to 3 hours, and in 70 and 80% alcohols for at least 3 hours. It is inadvisable to leave material more than about 6 hours in alcohol under 50%, because weak alcohol is likely to dissociate tissues upon longer exposure. Specimens may remain for several days in any of the stronger alcohols. Concerning the storage of material in alcohol, refer to Preservation (p. 140).

Tissues Which Have Been Washed in Alcohol. When alcohol is used to wash the tissues, it hardens and dehydrates them at the same time. Specimens are treated in this way after fixation in alcoholic reagents, or in aqueous solutions of picric acid or mercuric chloride. The procedure has been described on page 136.

PRESERVATION

It is often necessary to preserve specimens for a long period of time. Such is the case when materials have been collected in the field, or when time cannot be given to complete the preparations at once. In clinical work, specimens are often saved to supplement records. Stocks of fixed material are commonly kept on hand for research and teaching purposes.

Preservatives

A good preservative will prevent disintegration and will not dissolve or otherwise alter important constituents of the fixed tissue. It must neither shrink nor swell the tissue elements excessively. It should not soften the tissues, but should give them a firm consistency, without causing them to become very hard or brittle. In addition, the tissues must not become discolored and their staining properties must not be seriously impaired. No ideal preservative is known. Each of the better methods of preservation has advantages for certain kinds of material or for use under certain conditions. The following is a good general rule: *If specimens are to be sectioned, embed them in paraffin or nitrocellulose as soon after fixation as circumstances permit. If they are to be stained as whole specimens, do this at the first opportunity; then embed them for sectioning or store them in a resinous medium for mounting entire, as the case may be.*

PARAFFIN. Material for sectioning may be kept for many years after it is embedded in paraffin. No deterioration occurs. Paraffin sections, affixed to slides, also keep well. Store the specimens in a cool place and protect them from dust.

NITROCELLULOSE. When embedded specimens are to be preserved for future sectioning, embed them on fiber blocks. Some woods give off pigments which discolor the tissues. Embedded material may be preserved in glycerol-alcohol or soaked in pure glycerol and kept in a covered container, without liquid. In either case, use bottles having caps of glass or plastic. Avoid corks and metal caps.

ALCOHOL. A solution of 70% ethyl, methyl, or isopropyl alcohol is a convenient preservative and much used. However, after some time it may unfavorably affect the staining properties of material. Specimens which are not to be stained may generally be kept in alcohol for months or years, but the alcohol should be replaced whenever it becomes much discolored. Specimens tend to become brittle and crumbly after a time. The addition of 1 part glycerol to 9 parts of 70% alcohol will lessen this effect.

The staining qualities of material begin to deteriorate after it has remained in alcohol for a few weeks. The effect varies, but most stains give poor results with material which has remained in alcohol for several months. If circumstances make it impossible to go on with the preparation of material for many weeks or months, consider the following suggestions. Specimens which are eventually to be stained and mounted entire may advantageously be transferred to glycerol (below) for storage. Formalin or formalin-glycerol (below) affect the staining properties of some types of material less rapidly than does alcohol, and may also be recommended. Formalin should not be used when water or a reducing agent would be harmful, as in the case of tissues fixed in picric acid or chromic acid, or in the case of tissue in which glycogen is to be demonstrated. Absolute or 95% alcohol is recommended for preserving glycogen, but this may cause the tissue to become quite hard.

Methods of washing out various fixatives and transferring tissues to alcohol have already been described in this chapter. Thorough removal of chromic and mercurial fixatives is necessary to avoid discoloration and formation of precipitates, especially when material is kept in alcohol for some time. In the case of material fixed in picric acid, change the alcohol at suitable intervals, depending upon the size and quantity of the objects, until it no longer becomes deeply colored. If material is stored in alcohol discolored by picric acid, it is undesirable to use cork stoppers for vials or bottles containing the specimens (see Containers, p. 142).

FORMALIN. A 5 to 10% aqueous solution of formalin is generally used. This preservative affects the staining properties of tissues less rapidly than alcohol. Stains of at least fair quality may generally be obtained with carmines, hematoxylins, eosin, and a number of other dyes when material has remained in formalin for as long as 6 months. After that, results are uncertain. Specimens do not become as brittle in formalin as in alcohol. If desired, the softness of the tissues may be increased by adding 10 ml. of glycerol to each 90 ml. of 5 or 10% formalin. If yolky eggs are placed in formalin after fixation in Smith's fluid (which contains potassium dichromate), they do not become as crumbly as they might otherwise. Materials fixed in agents other than formalin should be thoroughly washed before they are placed in formalin. If formalin is used to preserve organisms having delicate calcareous skeletons (as spicules of calcareous sponges or of pluteus larvae), it must be buffered (p. 80) or neutralized (p. 79) and *kept* neutral by placing a small amount of calcium carbonate in the container with the specimens. Concerning the removal of precipitates which sometimes occur in formalin-preserved tissues, see page 82.

GLYCEROL. This is an excellent preservative for specimens which are to be stained and mounted without sectioning. It keeps the material soft and pliable. Good stains have been obtained with carmines, hematoxylins, and aniline dyes after material has been stored in glycerol for 10 years. Glycerol is also a fine preservative for specimens which have been stained. In fact, it is the only medium in which methylene blue and some other delicate stains keep at all well. For this purpose, glycerol must be *neutral*. Transfer material from 70% alcohol to a relatively large volume of 10% glycerol in 70% alcohol, in a loosely covered dish. Allow this to concentrate by evaporation until it reaches a syrupy consistency, or simply make gradual additions of glycerol to replace the alcohol. Specimens may be transferred from water or aqueous formalin in a similar manner, by using 10% glycerol in water. Add a small crystal of thymol to discourage molds. Evaporation may be hastened by putting the dish on top of a paraffin oven, or in an incubator. In order to avoid discoloration, do not store glycerol-preserved material in cork-stoppered bottles or metal-capped jars. Use bottles with covers of glass or plastic.

CANADA BALSAM AND OTHER RESINS. It is often very convenient to keep stained or unstained material in bottles of balsam, damar, thick cedar oil, or synthetic resins until it is needed for preparation of whole mounts. Unfortunately, specimens grow more and more brittle in these media. It may become very difficult to mount delicate invertebrates, such as hydroids and small worms, without breaking them.

Screw-capped vials or jars are the most suitable containers. To protect stained specimens from the action of light, use bottles of dark glass or keep them in a very subdued light. Carmines and some aniline stains may keep perfectly for years. Hematoxylins are more likely to fade after a time, probably due to formation of acid in the resin. Sensitive stains such as methylene blue and methyl green fade quickly in resins, but may be preserved for some time in glycerol.

Containers

Vials, bottles, and jars with screw caps of molded plastic are very satisfactory for general use. Metal caps (cardboard-lined) are not objectionable when material is stored in alcohol or resins, but are subject to rust if used with formalin, and to corrosion by glycerol and some other substances. Glass-stoppered bottles may be used, but they are relatively expensive and the stoppers may become "frozen" in place. For large quantities of material, glass-capped fruit jars are excellent and inexpensive containers. *Avoid the use of cork stoppers.* This is especially important when material is preserved in alcohol, glycerol, or any mixture containing either of these. Specimens will become badly discolored by the pigments and tannin which these liquids extract from cork.

A convenient method for preserving numerous specimens in small lots is the following: Put them into vials of appropriate size, filling each vial with the preservative and enclosing a pencil-written label. Stopper the vials with cotton or cloth plugs. Pack them into a large, wide-mouthed bottle or jar provided with a glass or plastic lid. Fill the jar with the same preservative and cover it tightly. A similar method for larger pieces of tissue is to tie them in individual cheesecloth bags, attaching to each of these a tag bearing appropriate data. A number of bags may be stored in one large container of preservative, but they should not be packed in tightly.

Labels

Label each container fully, showing the kind of material it contains; record the number, preservative used, the date, and other pertinent information. Labels to be pasted upon bottles should be written with waterproof ink, pencil, or typewriter. After the label has dried in place, it is well to give it a coat of shellac or clear lacquer; otherwise, the label may drop off at some future time. Another precaution worth taking is to place in the container a second, pencil-written label, bearing essential data. This cannot get lost.

BLEACHING

Bleaching is necessary if the specimens contain much natural pigment, or if they have been darkened by osmium tetroxide or some other fixative. Upon occasion, it may be desirable to bleach tissues which have been stained unsuccessfully, or in which a stain has faded. Restaining will be possible in a good many cases.

Methods for Sections

When sections are to be made, it is best to section the material first and then to bleach the sections. Fix and wash specimens in the usual way. Make sections by the paraffin, nitrocellulose, or freezing method.

Removal of Osmium Stains or Natural Pigments. HYDROGEN PEROXIDE METHOD. A 5% solution of hydrogen peroxide in water or 50% alcohol is a very safe reagent, but it requires many hours to bleach moderately colored material, and it may fail altogether in obstinate cases.

SODIUM HYPOCHLORITE METHOD. This is a safe and rapid technique. Transfer sections from 50 or 70% alcohol to the following mixture:

Sodium Hypochlorite Solution for Sections

50% alcohol	10 ml.
Commercial sodium hypochlorite solution ("Clorox" or other brand)	1 drop

Bleaching will usually be accomplished within 10 to 45 minutes. Then wash the sections in running water for an hour, or in repeated changes of 50% alcohol. In very difficult cases it may be necessary to use a stronger hypochlorite solution or to use Mayer's potassium chlorate-hydrochloric acid method (p. 144), although either of these may injure delicate sections.

Removal of Stains Caused by Potassium Dichromate, Chromic Acid, or Dyes. Potassium Permanganate-Oxalic Acid Method. Bring sections by way of the alcohol series to water. Place them in a 0.25% aqueous solution of potassium permanganate for 5 to 10 minutes. Rinse them thoroughly in distilled water and place them in a 5% aqueous solution of oxalic acid for 5 to 10 minutes. If decolorization is not complete, rinse the sections in water and repeat the process. After bleaching, wash sections in running water for 5 to 10 minutes and then proceed to stain them.

Methods for Objects to be Mounted Entire

Sodium Hypochlorite Method. The following is a good all-purpose bleach for entire objects.

Sodium Hypochlorite Solution for Entire Objects

Alcohol, 70% or 80%	10 ml.
Commercial sodium hypochlorite solution ("Clorox" or other brand)	3 or 4 drops

Transfer specimens from 70 or 80% alcohol to the bleaching solution and leave them until they are decolorized. If this is not accomplished within 2 days, renew the solution. Deeply pigmented specimens may require several changes, at intervals of 2 or 3 days. When bleaching has progressed to the desired extent, wash material in 3 or 4 changes of 70% alcohol, in order to remove all traces of chlorine.

Mayer's Potassium Chlorate-Hydrochloric Acid Method. This is rapid and thorough, but it may prove injurious to delicate invertebrates.

Put a few crystals of potassium chlorate into a bottle or test tube and add 2 or 3 drops of concentrated hydrochloric acid. When greenish chlorine gas evolves, add 5 to 10 ml. of 70% alcohol.[1] Transfer material from alcohol of similar strength to the chlorine solution. The time required for bleaching varies from a few minutes to a day or longer. Wash the bleached specimens in several changes of 70% alcohol and proceed to stain them, or to dehydrate and mount them unstained.

Sulfurous Acid. This is very good for removing discoloration caused by potassium dichromate. To prepare a solution, add 2 to 4 drops of concentrated hydrochloric acid to 10 ml. of a 2% aqueous solution of sodium bisulfite. Sulfurous acid is formed and goes into solution. Transfer objects from water to a freshly prepared mixture. Bleaching requires 6 to 12 hours. After treatment, the material should be thoroughly washed in water.

[1] Water should be used in place of alcohol for bleaching the exoskeleton of arthropods.

DECALCIFICATION

Some animal tissues contain deposits of calcium salts which are too hard to be cut with a microtome knife. It is necessary to dissolve these calcium compounds with an acid before the usual methods can be employed to make sections for study of the soft parts. Vertebrate tissues which require decalcification are bones, teeth, calcified cartilage, dermal scales, and tissues in certain pathological states such as arteriosclerosis, tuberculosis, and several kinds of tumors. Among invertebrates, decalcification is necessary preparatory to sectioning entire molluscs, some crustaceans, echinoderms, brachiopods, calcareous coelenterates, calcareous sponges, and foraminiferans.

Decalcification may be accomplished at any one of three points in the process of preparation, and various acids may be employed. The decalcifying agent to use, and the time to apply it, will depend upon the size and nature of the specimens.

Simultaneous Fixation and Decalcification of Small Objects

Many fixing solutions contain sufficient acid to decalcify small objects during the course of fixation. These include the fixatives of Bouin, Zenker, Carnoy, Gilson, Petrunkewitsch, and Heidenhain ("Susa"), as well as mercuric chloride-acetic acid. Any of these will dissolve the spicules and small calcareous plates of invertebrates, or the deposits of calcium in cartilage and the developing bones of embryos or larvae up to several centimeters in length. Specimens may then be embedded and sectioned according to the paraffin method or nitrocellulose method. Bouin's fluid may be allowed to act for many days and thus will serve to decalcify large embryos and similar objects. A serious disadvantage of simultaneous fixation and decalcification is that the accumulation of carbon dioxide gas may inflate and distort the tissues before they become hardened.

Decalcification of Fixed and Hardened Tissues

Teeth and bones of adult animals, or other large and heavily calcified objects, should be fixed and hardened before they are decalcified. After being hardened in alcohol, the specimens are usually treated with an acid, then dehydrated, embedded in nitrocellulose, and sectioned. However, the tissue relations in pieces of bone are best preserved by embedding specimens in nitrocellulose and then decalcifying them (see p. 148).

Fixation

Trim away muscles from bones before fixing them, but be careful not to injure the periosteum. In general, bones and teeth should be sawed into pieces 2 to 5 mm. thick. However, the thickness of the slices must be varied according to the structures included. In order to study the inner ear, for example, a much thicker mass of bone must be prepared. For sawing, use a fine-toothed blade and a miter-box (see p. 124). The tissues may be fixed in various reagents. Ten percent formalin and formalin-alcohol are convenient general fixatives. Material should remain in either of these for 2 to 4 days, but may be left in aqueous formalin for many weeks. Helly's or Maximov's modifications of Zenker's fixative will give the tissues excellent general staining properties; Zenker's fluid with acetic acid is most favorable if Mallory's triple stain, especially suitable for developing bone, is to be used. When using any of these formulas, fix tissues for 24 hours and wash them in running water for 12 to 24 hours. In all cases, bring specimens to 70% alcohol before putting them into one of the following liquids.

Decalcifying Fluids

Hydrochloric acid, formic acid, and nitric acid have proved more satisfactory than others. These acids should be greatly diluted so that they do not swell or otherwise injure the soft parts. This also makes their action slower, thus avoiding rapid production of CO_2 which might accumulate in the tissues and distort them. Alcohol is generally agreed to be the best medium for dilution, because it protects the soft tissue elements from the acid.

Hydrochloric Acid. A 3% solution of hydrochloric acid in 70% alcohol is a very effective decalcifying agent, and causes no serious damage to tissues. After fixing and hardening bones or teeth, transfer them from 70% alcohol to a relatively large volume of the acid solution, in a loosely covered container. The time required for decalcification varies from 1 to 15 days or longer, according to the amount of calcium present. Every day add about 20 drops of concentrated hydrochloric acid for each 100 ml. of liquid; but whenever the liquid has become markedly cloudy, replace it with a completely fresh mixture instead of adding concentrated acid. Warmth and occasional stirring of the fluid will hasten decalcification. When bubbles fail to appear upon the surface of the material, shortly after addition of acid, make a test by sticking a fine needle into some unimportant part of the specimen. If decalcification is complete, a needle will pass easily into the tissue. An x-ray examination provides a more accurate test, but the equipment may not be available.

Following decalcification, place the tissue in neutral 70% alcohol. Change this once or twice daily, until it remains neutral (a test with blue litmus paper is accurate enough). Neutralization may be accelerated by adding a few drops of a saturated aqueous solution of lithium carbonate to the alcohol. After neutralizing, proceed to embed and section the tissues. For adult bones and teeth, or large invertebrates, the nitrocellulose method (Chapter 16) is used. Small invertebrates, fetuses, or young animals may be embedded and sectioned in paraffin after decalcification.

Nitric Acid. Half a century ago, an aqueous solution of nitric acid ranging from 1 to 10% was the favorite decalcifying agent. A 5% aqueous solution of nitric acid decalcifies rapidly, but injures the staining properties of material, commonly rendering nuclei incapable of taking up hematoxylins and some other stains. Nitric acid of 1 to 3% strength, in 70% alcohol, is much less harmful in this respect. If decalcification requires several days, 5% formalin may be added to aid in protecting the tissues. On the whole, better results are ordinarily obtained with hydrochloric or formic acid.

Formic Acid. The following mixture is a very good decalcifying reagent for bones and teeth. It is recommended especially for large specimens.

Formic acid, 85%	50 ml.
Alcohol, 95%	50 ml.
Sodium citrate, 20% solution in 1% aqueous solution of trichloroacetic acid	50 ml.

Fix the tissues in 10% aqueous formalin or formalin-alcohol. After fixation in aqueous formalin, specimens may simply be rinsed in water, but if formalin-alcohol was used they should be brought gradually to 30% alcohol. Place the tissues in a relatively large volume of the decalcifying solution and cover the jar loosely. Action is more rapid if the liquid is kept at 40 to 50° C. Change the liquid after a day or two; do this again several times in the case of large objects. Complete decalcification may be ascertained by testing the object with a sharp needle, or by x-ray examination. One advantage of this solution is that tissues show no damage and exhibit good staining properties after remaining in it as long as several months. After decalcification, wash specimens in running water for 24 hours or longer, according to their size. Thorough washing is necessary to avoid formation of precipitates. Then proceed to embed and section the tissues in nitrocellulose (Chapter 16).

Chelating Agents. Decalcification with the aid of strong acids may spoil tissues if certain methods of staining or metallic impregnation are to be used. A technique that takes advantage of a chelating agent to bind the available calcium may be preferable, even if it works slowly. EDTA, the disodium salt of (ethylenedinitrilo)-tetraacetic acid, serves this purpose well. Any preparation of high purity (99%) may be used. Make up a 5% solution in 10% formalin or buffered formalin (p. 80). The formalin will operate as a fixing agent if the tissue is fresh, or for keeping the tissue from deteriorating if it has already been fixed. The volume of fluid, in proportion to the tissue to be decalcified, should be substantial—perhaps 100 ml. for a few pieces 5 mm. in diameter, decidedly more than this for an intact animal 1 or 2 cm. long—and should be renewed each week until decalcification is believed to be complete (three weeks will probably suffice). The tissue is then washed in water, dehydrated and embedded, or stored in 70% alcohol for embedding at a later time.

Decalcification After Embedding in Nitrocellulose

This procedure is designed to preserve the natural relationships of delicate tissues supported by bone, calcareous plates, or spicules. It is very useful for cancellous bones, such as the temporal bone, and for demonstrating red marrow within the bones. The method has been used successfully for calcareous sponges, corals, and echinoderms.

Fix, wash, dehydrate, and embed specimens in nitrocellulose (Chapter 16). Put the blocks of embedded tissue into a relatively large quantity of 3% hydrochloric acid or nitric acid in 70% alcohol, and stopper the container tightly. Keep the jar or bottle in a warm place, shake it occasionally, and replace the liquid after a day or two. When decalcification is complete, as evidenced by trial cutting or x-ray examination, transfer the blocks to 70% alcohol. Change the alcohol once or twice daily, until it no longer causes blue litmus paper to change to red. Neutralization may be accelerated by adding a few drops of a saturated solution of lithium carbonate to the alcohol. Gas bubbles may have formed in the tissue during decalcification. If so, remove them by subjecting the block to negative pressure. This is easily produced by means of a water-faucet aspirator. Preserve the block in glycerol-alcohol (p. 261) or in 70% alcohol until it is to be sectioned.

DESILICIFICATION

This step is necessary in preparing to make sections of sponges and radiolarians which possess siliceous spicules or skeletons.

Fix, wash, and bring material to 80% alcohol. Then place it in a 5% solution of hydrofluoric acid in 80% alcohol. This acid dissolves glass, so a coating of paraffin should be applied to the inner surface of the dish or bottle used for treating the specimens. (For the same reason, it is unwise to leave containers of hydrofluoric acid in the vicinity of optical equipment.)

Desilicification usually requires only a few hours, or a day or two at the most. When microscopic examination proves the absence of siliceous bodies, place the material in 80% alcohol. Change this two or three times, at intervals depending upon the size and quantity of the objects. Then proceed to embed according to the paraffin or nitrocellulose method.

CHAPTER 11

Preparation and Mounting of Entire Objects

Small organisms, or portions of somewhat larger ones, can often be studied to advantage after they have been mounted entire in a suitable medium. Such preparations, ordinarily called "whole mounts," are valuable for study of the general anatomy of organisms which cannot be dissected. It is sometimes advisable to work out the anatomy of specimens in whole mounts before attempting to interpret sections.

Entire objects are generally mounted in one of the synthetic resins or balsam. Glycerol, glycerol jelly, and other non-resinous media are used for mounting a few types of organisms which resist infiltration by resins or to which resins are injurious. Hard structures are frequently mounted dry.

General Directions for Handling Material. In treating specimens with alcohol or other reagents, they should be covered with sufficient fluid to equal several times their bulk. It is generally not advisable to use higher concentrations of alcohol (especially absolute alcohol) or clearing agents for more than one lot of material, as these reagents lose strength by evaporation, and alcohols may absorb moisture from the air. The reagents also become slightly diluted by fluids diffusing out of the material itself. Some economy may be achieved by diluting higher concentrations of alcohol for re-use in lower concentrations, where slight deviations in percentage are of little consequence.

As a rule, specimens should be transferred from one liquid to another by pouring off the first liquid, and then quickly covering them with the second. If the specimens are very small or delicate, carefully withdraw the liquid from the dish with a pipette, and gently pour the succeeding liquid down the side of the container. Although it is possible to lift robust specimens from one container to another with a paper spatula, this technique is not recommended for any materials which are likely to be damaged by repeated manipulation or by even short periods of exposure to air.

Stender dishes, with accurately fitted covers, are the best containers in which to handle specimens of moderate size. Extremely small objects, such as protozoa and microscopic larvae, can best be handled in shell vials or short test tubes. Such material must be allowed to settle before each change of liquid, and care must be taken not to stir it up and accidentally draw it off with the pipette used to remove the liquid.

Pieces of glass tubing, inserted into rubber bulbs of various sizes, are useful for quickly drawing off amounts of liquids larger than those which can be pipetted off with ordinary medicine droppers.

Always keep the containers covered or stoppered, except when changing fluids or examining material.

PREPARATION OF UNSTAINED SPECIMENS TO BE MOUNTED IN RESINOUS MEDIA

Certain organisms, especially arthropods, which possess considerable natural color, may often be studied to advantage without any staining. Calcareous or siliceous structures, as the skeletons or tests of some protozoa, are routinely brought into the mounting medium as directly as possible.

1. FIXATION. Insects and other organisms, which are to be mounted unstained and used chiefly for the study of hard parts, are commonly fixed and hardened in 95% alcohol, although concentrations as low as 80% may serve as well. Aqueous fixing solutions are usually unsuitable for such material because of their poor penetrating power. It is not necessary to use the more refined fixing reagents for material of this sort. In order to keep the appendages spread apart, put a small piece of glass on each specimen. If the weight of this is not sufficient, tie the specimen between two pieces of glass. Small flies, mosquitoes, and other insects which usually fold their wings above their bodies when they are killed may most conveniently be mounted so that they are viewed from one side. Small and relatively permeable specimens such as aphids and small flies should be left in 95% alcohol for at least 8 or 10 hours; larger types, or those with heavy exoskeletons, should be fixed for at least a day or two. Material of this type can be stored indefinitely in tightly stoppered vials of alcohol. In case only appendages or other portions of the specimen are to be mounted, carefully dissect off the desired parts after fixation.

2. HARDENING. Specimens fixed in 95% alcohol are usually well hardened, and most of the water in them is replaced by alcohol. It is now necessary to complete the process of dehydration. Unless the specimens are deeply pigmented and require bleaching as directed in step 3, go directly to step 4.

3. BLEACHING. Chlorine, derived from sodium hypochlorite, is suitable for bleaching many kinds of insects, arachnids, and other arthropods which are to be studied with transmitted light. The commercial product known as "Clorox" is convenient and inexpensive. A safe and efficient bleaching fluid can be made by adding 3 or 4 drops of Clorox to 10 ml. of 70% alcohol. The time required for bleaching varies from a few hours to

several weeks, depending upon the size of the organisms and the extent to which they are pigmented.

Transfer the material to 70% alcohol for a time, and then to the bleaching solution; leave it until it is sufficiently bleached. If it is not adequately bleached in two days, renew the liquid. In the case of specimens which are very deeply pigmented, this may have to be done several times, at intervals of 2 or 3 days. Then wash the material with several changes of 70% alcohol to remove all traces of chlorine. After passing the specimens through 95% alcohol for as long as appears advisable considering their size, proceed to step 4.

4. DEHYDRATION. Complete the dehydration of the specimens by placing them in absolute alcohol. It is essential that the material remain in absolute alcohol until all traces of water have been removed from it, otherwise it may not be possible to clear it properly.

Change the absolute alcohol once or twice, depending upon the amount and nature of the material. The length of time required for complete dehydration depends upon the size of the specimens and their permeability. Small and permeable insects (aphids, for example) should remain in absolute alcohol for at least 3 or 4 hours; larger forms should be left in it for a longer time, in proportion to their size. Material of this type is not likely to be harmed by prolonged exposure to absolute alcohol, so if in doubt about how long to leave it, allow a little more time than might be necessary.

5. CLEARING. Now that all traces of water have been removed from the specimens, they are in condition to be permeated by some substance

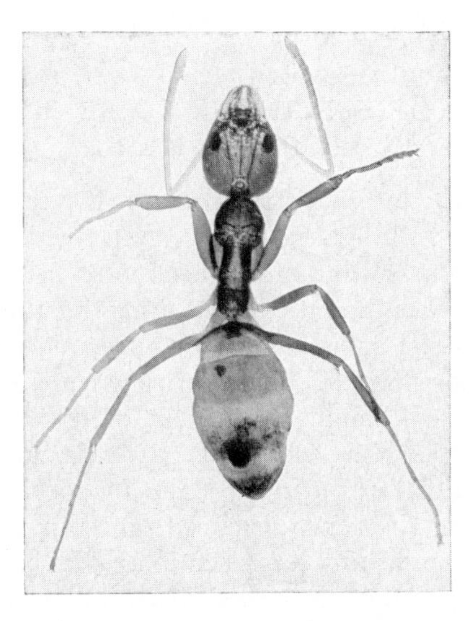

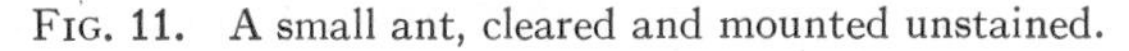

FIG. 11. A small ant, cleared and mounted unstained.

which will render them transparent and make it possible to mount them in a resinous medium. Fluids to be used for clearing must therefore be miscible with alcohol and with the resinous medium. Toluene and xylene fulfill these requirements, but will not clear material which contains traces of water. Either of them—especially xylene—is likely to make specimens brittle. Terpineol-toluene (a mixture of 1 part of terpineol with 3 parts of toluene) will clear very completely, absorbing traces of water if any should be present, and will cause less hardening. A mixture of terpineol with xylene may also be used, as well as creosote-toluene, creosote-xylene, carbol-toluene (phenol-toluene), or carbol-xylene (see p. 16).

Pour off the absolute alcohol and quickly add a mixture of equal parts of absolute alcohol and the clearing agent. Leave the specimens in this for an hour or two after they sink to the bottom, or overnight, keeping the container tightly covered at all times.

Replace the mixture with the clearing agent. After several hours, or when the specimens appear to be thoroughly cleared, change the clearing agent. There should be no opaque spots in the material after it has remained for several hours in this liquid. If such spots are present, it is probable that dehydration was not complete enough, and the material should be returned to absolute alcohol.

6. Infiltration With a Resinous Medium; Mounting; Completing Preparations. The method to be used in transferring the specimens to balsam or some other resinous medium and mounting them depends upon their size and nature. The procedures and media suited to various types of material, either stained or unstained, will be explained under step 9 of the next section. When the mounts have been prepared, consult Chapter 28 for suggestions for drying and labeling them.

STAINING OF SPECIMENS TO BE MOUNTED IN RESINOUS MEDIA

In order to study the internal structure of many types of organisms to be mounted entire, it will be necessary to stain them with suitable dyes. Certain staining mixtures containing carmine or hematoxylin are most commonly used for whole mounts. They bring out the nuclei of cells, so that organs which consist of tissues in which the nuclei are close together are often clearly differentiated. This is why gonads, ducts of the reproductive system, and epithelium of the gut of flatworms, for instance, show up quite well among the cells of the parenchyma. Other dyes may be used as counterstains, or as primary stains for special purposes.

Briefly summarizing the general plan of procedure, the specimens are fixed, washed, and then usually hardened in alcohol before being stained. After staining, they are dehydrated, cleared, infiltrated with a resinous medium, and mounted. The directions for carrying out these steps in the procedure will be made sufficiently general to apply to various types of material. In all steps where the technique for different types of material varies considerably, several alternatives will be given and the particular uses of each of them will be explained.

1. FIXATION. Care must be exercised to select a fixing fluid which will preserve the specimens in nearly natural form, and which will also fix the tissues in such a way that the various structures will be well differentiated after staining. Directions for preparing and using fixatives are given in Chapters 8 and 9. The following suggestions may prove helpful in choosing a fixative.

a. Use alcoholic fixatives when great penetrating power is required.

b. Avoid acid fluids when calcareous structures are to be preserved.

c. Fix soft, watery, or vesicular organisms (such as small jellyfishes, *Volvox*, and thin-walled rotifers) in aqueous fixatives. Subsequent dehydration of such material must be carried out very gradually.

d. Fix specimens which are active and evidently normal. In the case of aquatic organisms, avoid specimens covered with foreign matter.

e. Fix contractile organisms in an expanded condition, by using anesthetizing agents or mechanical restraint.

2. WASHING. When the specimens have been thoroughly fixed (see remarks given in connection with various fixing reagents), wash them by the method which is most appropriate. This and subsequent steps can be carried out most conveniently with the material in a stender dish with an accurately fitted cover. Except for destaining, however, the material can be handled nearly as well in small wide-mouth bottles or vials. Wash out aqueous fixatives, except picric acid, by placing the material in running water for several hours. Methods of washing material in running water are explained in Chapter 10. If the material is very delicate, simply use many changes of water. Wash out alcoholic fixatives with several changes of alcohol; begin with a concentration of alcohol which is not too different from that in the fixing fluid, and then pass the material gradually to 70% alcohol. Wash out picric acid (as in the case of Bouin's fluid) with 70% alcohol until practically all of the yellow color has left the material. The process may be hastened by warming the alcohol, but addition of an alkali is not recommended. If strong alcohol alone has been used for fixation, washing and hardening are omitted; proceed directly from step 1 to step 5, unless the material needs to be bleached.

3. HARDENING. If the material was washed in running water, transfer it successively through 15, 30, and 50% alcohols to 70% alcohol. Leave small and permeable objects such as protozoans, hydras, small larvae, and young embryos in each strength of alcohol for at least an hour; specimens of the size of a 72-hour chick embryo should be left in each change for at least 2 hours; leeches or large flukes should be left in each

change for at least 3 hours. Delicate organisms (such as rotifers or *Volvox*) which have been fixed in formalin may simply be rinsed in water and transferred to an aqueous staining mixture without being hardened in alcohol.

Specimens which were washed (or washed and iodized) in 70% (or 80%) alcohol will already have been hardened sufficiently.

Material may be stored in 70% alcohol until needed. However, the staining properties may begin to deteriorate after a few months, so that it is best to stain and mount the specimens as soon as possible.

4. SPECIAL TREATMENT REQUIRED BY CERTAIN TYPES OF MATERIAL. If the material was fixed in a fluid containing mercuric chloride, it must be treated as follows: Transfer the specimens (by way of a series of alcohols, if necessary) to 70 or 80% alcohol. Add to this enough of an alcoholic iodine or iodine-potassium iodide solution to give it a light straw color. Add more of the iodine solution if it becomes decolorized. When decolorization no longer takes place, transfer the specimens to fresh alcohol of the same strength, so that they will give up any iodine they have absorbed. (Read the remarks concerning mercuric chloride on p. 93.) If the specimens are deeply pigmented, such as some planarians, they should be bleached as suggested on page 144.

5. STAINING. Having been fixed, washed, and hardened, the material is now in condition to be stained. It is passed up or down the graded series of alcohols to about the same percentage as that in which the stain is dissolved; if the staining solution is aqueous, the material is transferred gradually to water.

Unless the object of staining is simply to make superficial structures more sharply defined by tinting them with a rather non-specific stain, one of the following methods is likely to be a good choice:

Borax carmine (p. 335). Commonly used for many types of material, including flukes, tapeworm scoleces and proglottids, hydroids, ectoprocts, and chick embryos.

Lynch's precipitated borax carmine (p. 336). Excellent for smaller flukes, hydroids, ectoprocts, and many other specimens of small or moderate size; not suitable for forms with a relatively impermeable cuticle or those with large cavities (as echinoderm larvae).

Dilute acidulated borax carmine (p. 340). A useful method for small organisms, such as larvae of invertebrates; not suited to larger organisms or those having a relatively impermeable cuticle.

Picro-carmine (p. 347). Good for tapeworm proglottids and larger flukes, as well as for other organisms in which the parenchyma is rather dense; also well suited to some delicate specimens, such as rotifers, which may not take up dilute acidulated borax carmine, or which cannot safely be treated by Lynch's method.

Alum cochineal (p. 344). Penetrates well, and therefore is good for nematodes, small tunicates, larger turbellarians, and some flukes; may be applied also to chick embryos and similar materials, but other carmine methods which are more selective will probably be favored.

Dilute alum hematoxylin (p. 353). Suitable for small and transparent organisms in general; echinoderm larvae containing calcareous structures may be stained progressively without danger of dissolving these.

More detailed information concerning the types of material for which each of these methods, as well as other techniques, is suited or not suited will be found in Chapters 21 and 22. If directions for staining, destaining, and neutralizing are followed, the material will then be ready for counterstaining and final stages of dehydration (step 6).

6. Dehydration and Counterstaining. The stained material is now in 80% alcohol. The next step is to gradually remove all remaining water from it and—if desirable—to apply a counterstain which will give a contrasting color to certain structures not brought out by the nuclear stain.

Material stained with dilute alum hematoxylin is not often counterstained. However, it is conventional to counterstain certain types of specimens stained by a carmine method, especially if they have tentacles, nematocysts, a ciliated body covering, or exoskeletal bristles. A suitable counterstain will also bring out the perisarc of hydroids and the zooecia of transparent ectoprocts.

Among the more effective counterstains for carmine are the following: fast green (p. 390); light green (p. 390); indigo-carmine (p. 390). Any of these may be applied as a dilute solution in alcohol. In the case of fast green, light green, and indigo-carmine, a 0.1% solution in 80% alcohol will give good results. Some experience with the material at hand may lead one to use a weaker or slightly stronger solution. All three of these dyes tend to act rather rapidly and may overstain the material, spoiling the effect of the primary carmine stain. When in doubt, it is better to start with a solution no stronger than 0.1%. If the stain comes out too readily when the material is transferred to alcohol, it is likely that the staining solution is not acid enough. In this case, add a drop of very dilute acetic acid (a 0.5 or 1% solution will probably be strong enough), to the staining dish, and repeat the process of counterstaining. If too much acid is added, the counterstain may be very dark and removable only by the addition of a little alkali (as sodium bicarbonate) to the alcohol in which the specimens are washed. In general, a more uniform and attractive counterstain is likely to be obtained by staining progressively, rather than by first overstaining and then removing some of the dye. For this reason, use of a dilute, very weakly acidified solution over a longer period of time is to be preferred over use of a more concentrated or more strongly acidified solution for a shorter time.

After counterstaining has progressed to a point which appears to be optimum, replace the dye solution with clean 80% alcohol. If the stain is too light, put the specimens back in the staining solution. If a higher concentration of dye or acidification seems to be required, follow suggestions given above.

Replace the 80% alcohol with 95% alcohol, and leave the specimens in this for an hour or longer, depending upon the size and nature of the material.

Any of the counterstains discussed above may also be applied as solutions in 70, 90, or 95% alcohol. Staining in 90 or 95% alcohol may be favored if the dye tends to wash out too quickly in spite of slight acidification; material may be transferred, after a brief rinse in clean 95% alcohol, to 100% alcohol. This method shortens the time required for dehydration after staining.

Complete the dehydration of the material by placing it in absolute alcohol. It must remain in this until every trace of water has been removed from it, otherwise it may be impossible to clear and mount the specimens properly.

Change the absolute alcohol once or twice, depending upon the amount and nature of the material. The length of time required for complete dehydration must be determined by the size and nature of the specimens. Very small and permeable forms should remain in absolute alcohol for at least 30 minutes. Specimens the size and consistency of a 72-hour chick embryo or a fluke about 1 cm. long should be left for at least an hour or two; larger specimens should remain in absolute alcohol for a proportionately longer time. Too much time in alcohol does no harm, but too short a time may lead to trouble.

7. Clearing. Once all traces of water have been removed from the specimens, they are next cleared by transferring them to a suitable clearing agent such as terpineol-toluene or toluene. (See p. 16 for advice on choice of a clearing agent.) The transfer from absolute alcohol to the clearing agent must be made gradually, in order to avoid the formation of violent diffusion currents which may distort delicate structures. Unless the specimens are unusually delicate, pour off the absolute alcohol and quickly add a mixture of equal parts of absolute alcohol and the clearing agent. Leave the specimens in this for an hour or more after they sink to the bottom, or overnight. Keep the container tightly covered. Then pour off the mixture and replace it with the clearing agent. After several hours, or when the specimens appear to be thoroughly cleared, change the clearing agent again, to insure that no traces of alcohol remain in the specimens.

For extremely delicate organisms, such as *Volvox* and rotifers, begin with a mixture containing 10% of the clearing agent in absolute alcohol, then follow this with mixtures containing gradually increasing amounts (20%, 30%, and so on) of the clearing agent.

8. INFILTRATION WITH RESINOUS MEDIA. After the specimens have been saturated with a clearing agent in which the resinous medium of choice is soluble, the specimens are ready to be permeated by the resin in which they are to be mounted. If they are at all delicate, the introduction of the resin must be brought about very gradually. The difference in density between clearing agents, such as terpineol-toluene or toluene and a synthetic resin or balsam, is so great that an abrupt transfer to the resin is likely to shrink and distort delicate tissues, and to collapse vesicular structures. This point has often been completely overlooked in works on zoological technique. A great deal of otherwise well-prepared material has been needlessly shrunken and ruined by being transferred directly from a clearing agent to a thick resin.

The rapidity with which the transfer to a resin can safely be made depends upon the delicacy and permeability of the specimens. The appendage of an arthropod, such as the leg of an insect, is rather firm and can safely be taken from the clearing fluid and mounted in thick resin. However, in order to mount safely a hydroid such as *Obelia* without collapsing the perisarc, the resin should be slowly introduced over a period of several days. Tapeworms, flukes, and small annelids should be similarly treated. Delicate larvae, such as those of echinoderms, must be infiltrated still more slowly. Extremely delicate organisms and relatively impermeable specimens, such as hookworms, must be infiltrated with the greatest care by adding resin over a period of 1 to 2 weeks.

The gradual introduction of the mounting medium can best be carried out as follows: Place the material in a stender dish of suitable size, together with a sufficient amount of the clearing agent to cover it to a depth of about 5 mm. Fold a piece of coarse filter paper into the shape of a cone, and adjust the last fold so that the cone will rest on the edge of the dish and its tip will dip down into the clearing agent but not touch the material.

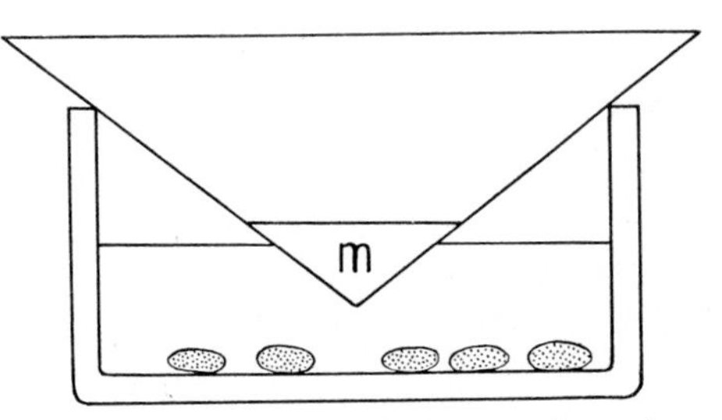

FIG. 12. Method of gradually infiltrating material with a resinous mounting medium. The thick medium (*m*) is placed in a cone of filter paper which dips into the clearing agent containing the specimens.

This arrangement is illustrated by Figure 12. Wet the filter paper with the clearing agent before placing it into the dish, otherwise the paper may soak up the fluid so rapidly that the specimens will be drawn into tight contact with the paper. The amount of balsam or other resin to be put in at one time depends upon the amount of the clearing agent in the dish and the character of the material. Unless the specimens are unusually delicate, it is safe to start with about 5 large drops of thick resin for approximately 10 ml. of the clearing agent, and to add the same amount twice a day until the contents of the dish reach the consistency of a medium syrup.

Occasionally examine the material. In case the specimens have shrunk or have partially collapsed, stir in some more of the clearing agent until they resume their normal form, and then add resin more slowly.

Remove the filter paper and leave the cover ajar to allow the mixture to concentrate by evaporation until it becomes a thick syrup. The material may then be mounted, or it may be stored for a while, provided the container is covered rather tightly. Specimens may be stored in well-stoppered bottles of most mounting media, including balsam and the synthetic resins, for a long time. It is advisable, however, to mount material as soon as possible, as it may become more brittle (eventually so much so that it cannot be handled safely), and the stain may fade, especially if a rather large surface of the medium is exposed to air.

If the specimens are hard and tough, as is the case with appendages of insects, they may be mounted without preliminary infiltration. Go from step 7 to step 9.

9. Mounting in Resinous Media. For most types of stained and unstained specimens which are to be mounted entire, either a synthetic resin or neutral Canada balsam may be recommended. However, balsam is more likely to become yellow with age, and it also hardens much more slowly than synthetic resins dissolved in volatile hydrocarbons such as toluene. For some unstained specimens which have setae, fine bristles, or sculpturings that are important but difficult to resolve in these media because of similarities in the refractive indices, the possibility of using Hyrax or some other medium of high refractive index should be considered. A discussion of the advantages and disadvantages of various synthetic and natural resins will be found in Chapter 27.

In mounting, care must be exercised to (1) properly orient the specimens on the slides, (2) use an appropriate amount of the mounting medium, (3) use a coverglass of suitable size, (4) support the coverglass if the specimens are delicate, and (5) prevent the inclusion of air bubbles in the mounts.

Orienting the Specimens. If the specimens are not large, it is best to place them in the center of the slide. A marker may be made for this purpose by placing a slide on a piece of white paper, marking around it with a pencil, and then ruling diagonal lines across the outline. In mounting, lay the slide on this marker and place the specimens over the intersection of the diagonals.

If a large number of small specimens, such as eggs or larvae of echinoderms, are to be mounted on a slide, place a drop of resin containing the material in the center of the slide. When the coverglass is lowered, the resin will flow out and evenly distribute the material. If only a few small specimens are to be mounted on a slide, it is desirable to concentrate them in a group under the center of the coverglass. To do this, place a very small drop of resin in the center of the slide, pick out the specimens with a fine pipette or needle, and put them into this, making sure that they are completely covered by the medium. Then let the drop of medium harden for a time in a dust-free place. More resin, some supports, and the coverglass may then be added without disturbing the specimens to any great extent.

A single large specimen, such as the fluke *Fasciola*, should be laid lengthwise near the right-hand end of the slide, thereby leaving room for a label at the left end. If several specimens representing a succession of stages (young, mature, and gravid tapeworm proglottids, for instance) are to be mounted on one slide, lay them in order.

When the specimens are large enough to be handled individually, lay them in positions which will show their structures to the best advantage. It is often desirable to mount two or more specimens on one slide, placing some of them with the dorsal side up, others with the ventral side up. In mounting small specimens, put as many on a slide as the supply of material will permit, but be careful that they are not clumped together and that none of them run out from under the coverglass.

Choice of a Coverglass. Circular coverglasses are best for mounting specimens of small or moderate size. The resin will flow evenly to the edges of circular coverglasses, but it is sometimes difficult to make it fill in under the corners of square ones. Rectangular coverglasses must be used for very large specimens. Select a coverglass which will overlap the specimens at least 2 or 3 mm. on all sides. If the specimens come any closer to the edge of the coverglass, the stains may soon fade from them. Very small or thin objects which are to be examined with lenses of higher power should be mounted under coverglasses of No. 1 thickness. Use No. 2 coverglasses for thick specimens, as they are much less likely to be broken when the finished preparations are handled.

Supporting the Coverglass and Mounting in Cells. In mounting organisms which are delicate, the coverglass should be supported by some object of suitable thickness. Otherwise, the shrinking of the resin as it dries may draw the coverglass down upon the specimens and crush them. Fine glass rods (Fig. 13*A*) are useful for this purpose, and may be drawn out to any desired diameter. Simply heat the middle of a short length of glass rod about 3 or 4 mm. thick over a Bunsen burner until it is soft, and then quickly pull apart the ends. The fine rods thus prepared may be broken into convenient lengths and sorted according to diameter. Pieces of broken coverglasses may also be used for this purpose, although these yield less attractive mounts. When the material has been placed on the slide and properly oriented, select 2 or 3 pieces of glass rod which are just large enough to hold the coverglass clear of the specimens. Place them on the slide close to the material, and then add more resin and a coverglass (see Fig. 14). Even when the specimens are not sufficiently fragile to require protection, it is often advisable to use supports to hold the coverglass level.

Thick specimens, particularly if they are delicate (such as chick embryos and small tunicate larvae), should be mounted in cells of proper depth. Cells of glass, metal, and cellulose acetate may be secured from dealers; they are made in various sizes and thicknesses. Very satisfactory and economical cells can be stamped out of sheet aluminum by a machinist. The type of cell illustrated in Figure 13*B* is made by first stamping out a square of the desired size, and then punching a large circular hole in it. For small and thin objects use 22-gauge aluminum and make the cells 18 mm. square. For somewhat larger and thicker objects, use 18- or 20-gauge aluminum and make the cells 22 mm. square. For still larger and thicker ones, such as 72-hour chick embryos, use 14-gauge aluminum and make the cells 25 mm. square. Each cell should be carefully smoothed on a grindstone and wiped perfectly clean before it is used. Regardless of their composition, the cells should be cemented to slides with a small amount of the mounting medium and allowed to stand until this hardens before the specimens are placed in them.

Another good way of mounting thicker specimens, especially those

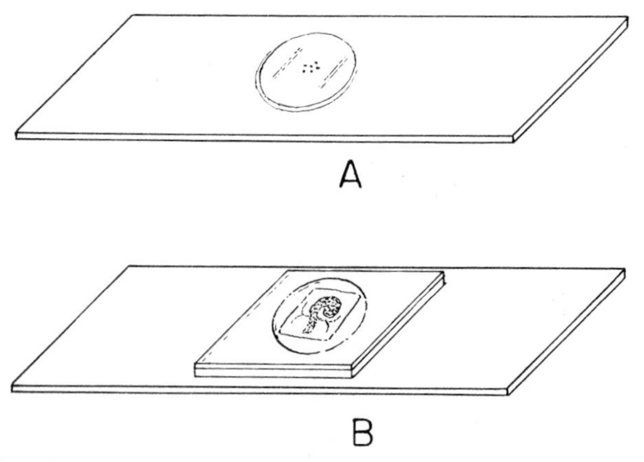

FIG. 13. Use of fine glass rods (*A*) and prefabricated cell (*B*) for supporting the coverglass in the preparation of whole mounts.

which are elongated (such as chaetognaths, many worms, ammocoetes larvae, and "amphioxus") employs strips of glass about 2 or 3 mm. wide, and of appropriate length, cut from microscope slides or slightly thicker stock. Two of these strips, cemented down on a slide, will support a square or rectangular coverglass after the medium fills in the trough between them. This method has at least one advantage over mounting in cells: air bubbles, if trapped in the medium, will eventually emerge from one side or the other.

Amount and Consistency of the Mounting Medium. Entire objects should be mounted in the thickest medium in which they can conveniently be handled. Good judgment in this respect can only be acquired through experience. If a very thick resin is to be used, it will probably be necessary to keep it warm so that it can be applied more easily.

Use enough mounting medium to spread somewhat beyond the edge of the coverglass. The medium ordinarily shrinks as the solvents in it evaporate, and if one uses barely enough to run to the edge of the coverglass it will eventually recede and leave the edge of the coverglass unsupported, or will withdraw unevenly and suck in bays and bubbles of air. In making relatively thin mounts, use enough mounting medium to form a very narrow rim (0.5 mm.) around the coverglass. In thicker mounts the rim of medium should be somewhat wider, perhaps 1 mm. or a trifle more than this. Here again, experience is the best teacher.

Placing the Coverglass. All that now remains to be done is to lower the coverglass into position. Put the slide bearing the material and the mounting medium in a convenient place on the table in front of you, or hold it between the thumb and forefinger of the left hand. Pick up the coverglass by one edge with a pair of fine-pointed forceps, warm it over a small flame, and quickly move it to a position immediately over the area of the slide upon which it is to be placed (Fig. 14). Lower the edge of the coverglass which is opposite to that held by the forceps until it rests on, or is very close to, the slide or supports. As the coverglass comes

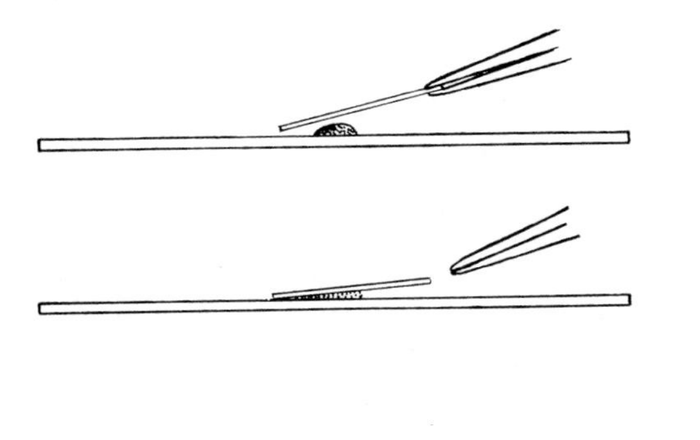

FIG. 14. Correct technique for application of coverglass. (Explanation in text.)

in contact with the mounting medium, slowly but steadily let down the edge held by the forceps so that the medium will spread evenly and without disturbing the material. When the tips of the forceps almost touch the slide, slowly withdraw them. If the medium is very thick, gently warm the slide high above a very small flame until it spreads very slightly beyond the edge of the coverglass. If there is not enough medium to form a slight rim, add a little with a needle or small glass rod. In case a considerable excess of medium has unintentionally been used, wipe off as much as you can without moving the coverglass.

Only through experience can the technician acquire the skill and judgment which will enable him always to make neat mounts. Profit by your mistakes, and you will soon become proficient.

Alternative Methods for Delicate or Collapsible Specimens

Many organisms which are vesicular (as *Volvox*) or which possess a cuticle which is relatively impermeable (as acanthocephalans) are likely to collapse and to remain that way if they are dehydrated in the usual series of alcohols. It may be possible, however, to prepare good mounts by dehydrating the specimens very gradually in glycerol and then replacing this with strong alcohol.

PROCEDURE. Transfer fixed, washed, and stained organisms from water to a dish of 10% glycerol. Add a small crystal of thymol to discourage the growth of molds. Cover the dish with lens paper and leave it in a warm, dust-free place until nearly all of the water has evaporated, and the consistency of the fluid approximates that of pure glycerol. (Glycerol, like alcohol, absorbs water from the air, so it will not become completely anhydrous.)

Transfer the specimens to pure glycerol and leave them in this for a few hours. Replace the glycerol with 95% alcohol. Organisms which may have collapsed will usually expand to their original shape. Complete the process of dehydration in 100% alcohol (two changes).

The specimens may now be cleared and mounted in one of two ways:

a. Clear them very gradually in terpineol-toluene, toluene, or some other standard clearing agent by transferring them through successively more concentrated mixtures of the clearing agent with absolute alcohol. After the specimens have remained for a sufficient time in the clearing agent without any alcohol, transfer them to a fresh change. Then begin to infiltrate them very slowly with the mounting medium, either by placing them in a dilute solution of the mounting medium and letting this thicken by evaporation, or by adding the medium gradually to the clearing agent by the method described on page 157. They may then be mounted as directed in step 9, page 158.

b. Transfer them to a mixture of 10% Venetian turpentine in absolute alcohol. Cover the dish with lens paper and place it immediately in a desiccator containing an ample amount of soda lime, because both the alcohol and Venetian turpentine are hygroscopic. In a few days or weeks, after all of the alcohol appears to have evaporated, the specimens are mounted in the Venetian turpentine with which they have been infiltrated.

Although the Venetian turpentine method has some serious disadvantages (see p. 456), it may enable one to produce acceptable mounts of specimens which are likely to collapse or become otherwise distorted during infiltration with any of the resinous media in routine use.

Dehydration by the usual series of alcohols may of course precede infiltration and mounting in Venetian turpentine. However, the advantages of dehydrating certain types of material first in glycerol and then in alcohol have already been pointed out.

MOUNTING IN NON-RESINOUS MEDIA

Before Canada balsam was introduced for making permanent preparations, microscopists mounted objects in various kinds of more liquid media such as formalin, syrup, chloral hydrate, and glycerol. The mounts were generally made somewhat more permanent by application of several coats of cement to the edge of the coverglass and the adjoining parts of the slide.

As methods were developed for making mounts of nearly all types of specimens in resinous media, the utility of more fluid media diminished. However, a number of aqueous mounting media made firm by inclusion of gelatin or gum arabic in the formula are still favored for certain types of material, such as nematodes, mites, tardigrades, small insects, and tissues stained to bring out lipid inclusions. Glycerol by itself is commonly employed for making temporary preparations, especially of unstained specimens, and is an excellent medium in which to tease apart or dissect materials, since it renders them very soft and pliable.

Instructions will be given here for using glycerol (glycerin, which is a cheaper commercial preparation of glycerol, not quite pure, is perfectly suitable) in making temporary or semi-permanent preparations, and also for mounting specimens in glycerol jelly. The latter is one of the more important aqueous media, and will serve to make rather durable mounts if the edges are sealed with a suitable cement and care is exercised in handling them. After some experience in working with glycerol and glycerol jelly, one should be able to modify the general procedure in whatever way is necessary to prepare mounts in any of the other aqueous media described in Chapter 28.

MOUNTING IN GLYCEROL OR GLYCEROL JELLY. 1. Fix and wash specimens by a suitable method. Nematodes, which are among the more common forms to be mounted in these media, are generally fixed in hot (steaming) 70 or 80% alcohol, and then stored in alcohol of the same strength. If an aqueous solution of formalin has been used, material may simply be left in the fixative, if this is not objectionable, or it may be washed in water and gradually brought to 70% alcohol. In the case of material fixed in other aqueous or alcoholic media, it is best to store it in 70% alcohol after appropriate washing.

2. If a stain is to be used, apply it in the conventional manner. Alum cochineal is a rather good stain for nematodes and some other invertebrates of the type likely to require mounting in aqueous media because of a cuticle which is relatively impermeable to resins. The refractive index of glycerol is low, and as a rule specimens mounted in glycerol or glycerol jelly are not stained.

3. Gradually replace the alcohol (or water, in the case of material which has not been stored in alcohol, or which has just been stained in an aqueous mixture) with glycerol. This may be accomplished by either of two procedures:

a. Make a 10% solution of glycerol (glycerin) in 70% alcohol or water. If water is used for dilution, add a small crystal of thymol to discourage the growth of molds. Transfer the material to a moderately large volume of the mixture in a shallow dish. Cover the dish with a sheet of lens paper, taping the edges of the paper tightly to the dish, or simply leave the lid ajar; place the dish in a dust-free chamber to allow the alcohol or water to evaporate. If 70% alcohol is used in making the mixture, the liquid will reach the consistency of nearly pure glycerol after several days; a longer time will be required if the mixture is prepared with water.

b. Make a solution of 10% glycerol in 70% alcohol or water. Place the specimens in just enough of this mixture to cover them. Over a period of about 24 hours, add more glycerol until the mixture consists almost entirely of glycerol. Then transfer it to pure glycerol. This procedure of adding glycerol at intervals may of course be coupled with the technique of allowing the alcohol or water to evaporate.

4. After the specimens have been in pure glycerol for a time, mount them in glycerol or glycerol jelly, following the suggestions below.

If only temporary mounts are to be prepared in glycerol, simply transfer material to a drop of glycerol on a slide and apply a coverglass. To prepare a somewhat more permanent mount, place a clean slide on a turntable (Fig. 15) and spin a ring of asphalt cement, about 3 mm. wide and of appropriate diameter, upon it. The diameter of the ring should be such that when the coverglass of choice is applied, it will rest upon the ring close to its inner margin. Allow the ring to dry, and if necessary apply additional coats of the cement until a cell of proper depth for the specimen has been built up. Such cells may be prepared in quantities and kept on hand in a dust-free slide box. Place the specimen in a drop of glycerol in the cell and orient it as desired. Add more glycerol if necessary; the total

FIG. 15. A turntable used for making shallow cells on slides and for ringing mounts with sealing cements. (Courtesy of Williams, Brown, and Earle, Philadelphia.)

amount should be barely enough to fill the cell after the coverglass is applied and its edges pressed down against the cement. Lower the coverglass slowly, taking care not to include bubbles of air. If a trifle of glycerol exudes, wipe it off with a cloth moistened with strong alcohol. If much exudes, it will probably be best to transfer the specimen to another cell and try again. After the coverglass is seated against the cement, proceed with step 5.

If glycerol jelly is used for mounting, immerse the bottle in warm water (about 50° C.) until the medium is fluid. Transfer a drop of it to the slide, and place the specimen which has been cleared in glycerol into it. It will be best to work with the slide on a warming table maintained at a temperature of about 40° C. Apply the coverglass, wipe away any excess medium, and allow the preparation to cool. The mount may be sealed, if desired, as directed in step 5. However, it is best to let the exposed edges of the medium dry for at least a day before applying cement.

It may be desirable to build up a cell of asphalt cement by ringing the slide on a turntable, as directed above for making thicker mounts in glycerol. If this is done, the precautions about using an appropriate amount of medium and taking care to exclude air bubbles must be observed. The slide should be kept warm while the coverglass is being lowered, so that the medium flows easily. With the tips of forceps or some other suitable instrument, gently press the edges of the coverglass down onto the cement, which will be sticky if the slide is warm enough. Wipe away any small excess of medium which may have exuded, and then apply a seal of cement, as directed below.

5. Place the slide on the turntable, and with a small camel's hair brush paint a ring of asphalt cement about 2 mm. wide over the edge of the coverglass. Begin the application outside the margins of the coverglass, on the ring of cement if a cell was built up before mounting. Gradually move the brush up over the edge of the coverglass. Spin the turntable at moderate speed, and apply the tip of the brush very lightly to the preparation. Let the first coat dry, then add at least two more coats.

Additional details concerning the suitability of various other cements for sealing mounts in aqueous media, and a description of the technique of double mounting in both an aqueous and a resinous medium, are given in Chapter 27.

DRY MOUNTING

This method of mounting is used principally for hard structures of two general types:

1. Shells or other objects too opaque to be cleared and studied by transmitted light.

2. Small scales, hairs, or spicules that would become almost invisible if mounted in a resinous medium or one containing glycerol. Some such structures will take up stains, and therefore may also be mounted in resinous media; others cannot be stained, and must be studied dry or in water. In either case, however, dry mounts usually show the structures more satisfactorily.

Opaque Mounts

The calcareous tests of some larger foraminiferans, and other completely opaque objects, can most advantageously be studied by reflected light. The antennae and other appendages of insects can be treated in the same way, but it is usually best to bleach them and mount them in a resinous medium. If the objects are white or nearly so, as tests of foraminiferans, they should be mounted against a black background. If the objects are dark, the background should be light. Hard rubber cells with bottoms have been manufactured for this purpose, but are not generally available. Nearly the same effect can be obtained by mounting the specimens in a glass, plastic, or metal cell and applying a dull black or white paint to the back of the slide.

If the preparations are to be handled extensively, it is best to put a very thin coat of a resinous mounting medium or glue on the area of the slide within the cell. This will hold the specimens in place and prevent them from being broken if the slide is handled roughly by accident. Apply the adhesive by means of a camel's hair brush, being careful not to put on so much that it soaks into the material. Then orient the specimens as desired. If they are small and abundant, simply strew them over the bottom of the cell. The specimens must be thoroughly dry, otherwise moisture evaporating from them will cloud the coverglass. After the adhesive has dried, place a tiny drop of a resinous mounting medium on each corner of the cell and gently press a clean coverglass into place. Finally, apply black paint (or white paint if the specimens are dark) to the back of the slide.

Transparent and Semi-Transparent Objects

The membranous wings of many insects, scales scraped from the wings of moths and butterflies, and delicate calcareous or siliceous spicules of sponges are sufficiently transparent so that they may be mounted dry and studied by transmitted light. The structure of such objects can often be observed much better in dry specimens than in those embedded in a mounting medium, as many of their details become invisible in resins or in media containing glycerol.

The process of mounting such specimens is very simple. Place a clean slide on a turntable and spin a ring of a resinous medium upon it, or a cement such as gold-size, about as thick as the specimens which are to be mounted. The inside diameter of the ring should be slightly less than the diameter of the coverglass which is to be applied. Allow the ring to dry. Such cells may be prepared in quantities and kept on hand for use as needed. Thoroughly dry the specimens in the air, or in an incubator if this is necessary, and then place them in the shallow cell. Gently warm the slide until the ring becomes sticky. Then quickly warm a coverglass of suitable size, and gently press it down upon the cell until it adheres at all points. When the preparation has cooled, seal it with one or two coats of resin or cement.

CHAPTER 12

Spreading, Smearing, and Dissociation

SPREADING METHOD

Thin membranes such as the mesenteries, subcutaneous areolar tissue, or the amnion can be mounted and studied to best advantage by this method. The preparations show all of the structures in almost natural relations, and are thin enough to be examined with transmitted light.

1. SPREADING THE TISSUES. In order to prevent the membranous tissue from shrinking and wrinkling, spread it upon a frame or a flat surface. A convenient device for this purpose consists of two stiff rubber or polyethylene rings, one of which fits rather snugly inside of the other. Place the smaller ring under the desired area of mesentery, amnion, or subcutaneous tissue,[1] preferably without detaching this from the surrounding parts. Press the larger ring over it. Then cut the membrane around the rings. A single glass or plastic ring may also be used, if a thread is tied around the tissue where it overlaps the ring (see Fig. 8*C*), p. 124). Small bits of subcutaneous tissue are sometimes stretched upon a coverglass and allowed to dry just enough to adhere to the glass, and then fixed. This method is much less satisfactory than the preceding.

2. FIXATION AND STAINING. Fix the stretched tissue by immersing it (or floating the coverglass to which it is attached) in a suitable fixing fluid. For general study, including the differentiation of elastic and collagenous fibers, fix the material in formalin, stain it by Verhoeff's method for elastic tissue, and counterstain it with eosin. To outline squamous epithelial cells or lymph plexuses, use the silver nitrate method (p. 403). For a critical study of connective tissue cells, including the differentiation of intracellular granules, fix and stain the tissue by Giemsa's method (p. 396). To preserve fats, fix material in 10% formalin and stain it with Sudan IV (p. 443). Specific directions for staining are given under each method.

3. MOUNTING. After all stains (except stains for lipids), the tissues are dehydrated, cleared in toluene, xylene, or some other clearing agent, removed from the rings or other supports, cut into pieces, and mounted in balsam or synthetic resin. After staining lipids, mount the material in glycerol jelly.

[1] Areolar tissue can be taken most easily from the axillary or inguinal region.

PREPARATION OF SMEARS

The cells of blood, lymph, and spermatic fluid, as well as various normal and pathological exudates, can be studied in a satisfactory manner in thin smears. This method is valuable also for spreading apart the cells of soft organs such as testes, lymph glands, spleen, or bone marrow, and for microorganisms suspended in intestinal contents or in media such as water, blood, cerebrospinal fluid, and coelomic fluid. Some kinds of material are simply spread out by pressure, rather than being actually smeared.

Smears or films are classified as *wet* or *dry*, according to whether they are immersed in a fixing reagent while they are still moist, or allowed to dry and then fixed more completely with heat or alcohol.

In making smears, the material should be spread evenly and as thinly as is required by the size and number of structures present. When the cells are small and very numerous, as in blood, the films must be thin in order to show only one layer of structures. When they are larger or less numerous, the smears should be correspondingly thicker. The process of spreading must not injure or distort any of the structures which are to be studied. The exertion of excessive pressure in making smears is especially to be avoided. No pressure whatever is necessary in smearing blood and other very fluid materials.

Slides or coverglasses upon which smears are to be made must be completely free from dirt or grease. It is nearly impossible to spread material evenly upon glass which bears any trace of grease.

Preparation of Smears on Coverglasses. A small quantity of material is placed upon a coverglass, and spread out into a thin film by means of a platinum loop, tips of fine forceps, or a toothpick or glass rod. The film may then be dried, if this method of fixation is desirable. Otherwise it should be dropped face down on the surface of a fixing reagent in a shallow container (such as a Petri dish). Some experience will help the laboratory worker to decide how viscous the smear must be in order for it to adhere properly.

With certain types of material, the adhesion of wet-fixed smears to coverglasses may have to be facilitated by addition of some compatible coagulable material, such as fresh horse serum, fresh egg albumen (use this sparingly!), or tissue which can be smeared out in order to supply a matrix of protein. In working with symbiotic protozoa and other small organisms, if the immediate habitat (for example, the lumen of the gut) is not of a consistency which will adhere well when it is fixed, the judicious addition of some tissue from another organ of the same host may solve the problem without introducing any objectionable contaminants.

Another method of preparing smears on coverglasses is to place the material on one coverglass and then apply another on top of it. If the material is moderately solid, the coverglasses may be pressed together just enough to disperse it more uniformly. The coverglasses are then quickly slid apart and fixed by drying or by dropping them on the surface of a fixing reagent. This technique for preparing films is less desirable for most purposes than the one described above, as it may cause larger cells or organisms to be crushed.

A disadvantage of making smears on coverglasses is the difficulty of handling them without some breakage or loss. However, by using Columbia dishes (which are miniature Coplin jars made to hold coverglasses having a diameter of 22 mm.), or staining racks designed for coverglasses, and taking care not to twist or bend the coverglasses while placing them into or removing them from the grooves, one can soon learn to keep damage to a minimum. Much smaller quantities of reagents are required for processing smears on coverglasses than on slides. In transferring a coverglass from one container or rack to another, hold it with forceps, close to the edge, in such a way that the smear itself is not damaged.

PREPARATION OF SMEARS ON SLIDES. In many cases it is better to spread the material upon a slide, rather than upon a coverglass. With blood and some other very fluid materials, this is best done by using one slide to draw out the contents of a drop in a very thin film on another slide. This method is described in detail below. Otherwise, the techniques for preparing smears on coverglasses may be applied also to making smears on slides, except that those to be wet-fixed are plunged with a steady motion into the fixing reagent instead of being dropped on the surface.

Specific Methods for Various Materials

Dry-Fixed Smears of Blood. 1. PREPARATION OF FILMS. Lay several absolutely clean slides in a row, upon a piece of white paper. At hand have some bits of sterilized absorbent cotton or gauze. Sterilize the skin of a fingertip or lobe of an ear by washing it with alcohol, then prick it with a sharp needle which has been sterilized over a flame, or by means of a sterile lancet manufactured for this purpose. If the blood flows rapidly at first, wipe it away with cotton or gauze.

Take a clean slide—preferably a new one—in the right hand, holding it by its edges near one end, between the thumb and first finger. As soon as a small drop of blood exudes from the puncture, touch the free end of the slide to it in such a way that it is collected on the under surface, close to the edge (Fig. 16*A*). Instantly lower the slide onto a second slide, near its right end. The slide bearing the blood should rest lightly and evenly on the slide upon which the smear is to be made. As soon as the blood makes contact with the latter, push the first slide steadily and with

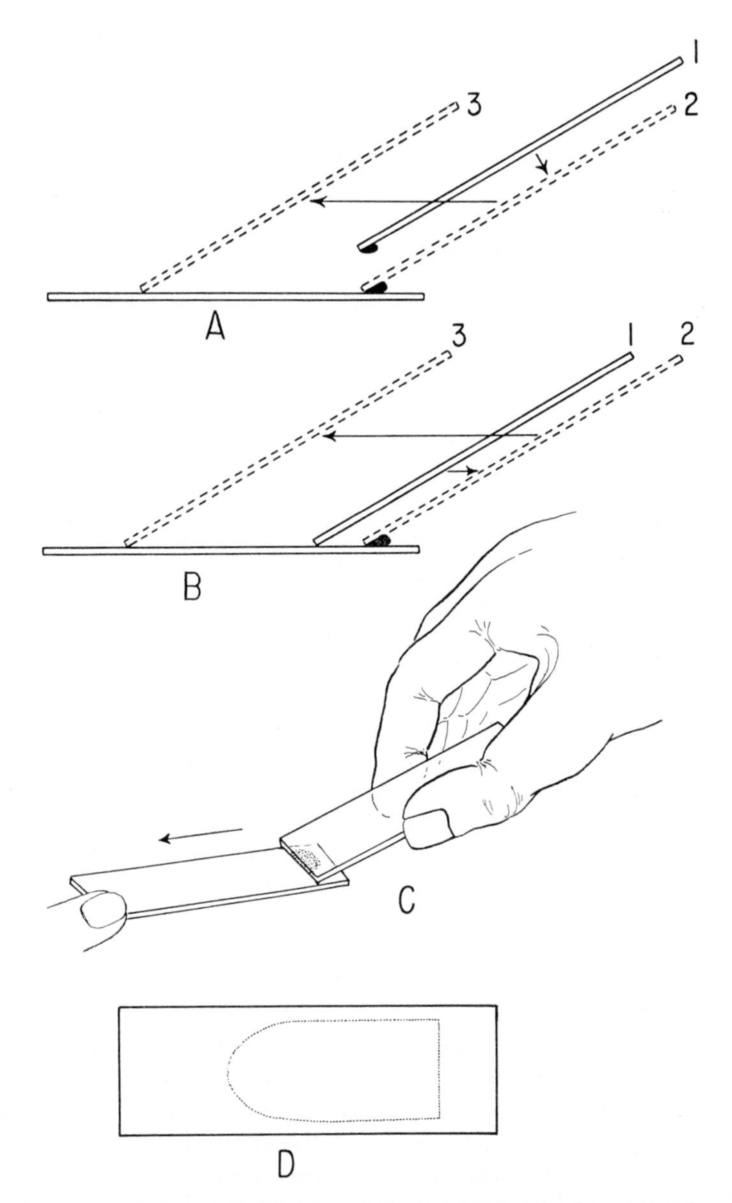

FIG. 16. Methods of making thin films of blood. *A*, The blood is collected near the edge of the slide used for spreading the blood (*1*). This is lowered into contact with the horizontal slide (*2*), and then moved across the slide with a steady motion (*3*). *B*, The blood is applied to a slide near one end. The slide used for spreading the blood is brought down just to the left of it (*1*), then moved to the right until it touches the drop of blood (*2*). It is then moved across the slide (*3*). *C*, The proper way to hold the slide used for spreading the blood. *D*, Outline of a correctly prepared film.

moderate speed toward the left end of the second. The horizontal slide will have to be immobilized by a finger of the left hand (Fig. 16*C*); some workers use both thumb and forefinger. If this procedure is executed properly, the blood will be drawn out into a thin film, without crushing the corpuscles. It is essential that the blood be collected upon the slide as soon as it has exuded from the puncture, and that the whole process be carried out very rapidly, so that the blood has no opportunity to coagulate. The drop of blood should not be so large that the motion of spreading must be continued as far as the left edge of the slide where it is being held. Ideally, the last bit of blood is spread out before the moving slide reaches the point where the horizontal slide is being held, so that there is no slowing down or uncertainty of movement as the slide reaches the end of its course and is finally lifted off. The width of the film normally narrows as the sweep of one slide across the other is nearing a close, so that the outline of the film as a whole is somewhat tongue-shaped (Fig. 16*D*).

Some workers prefer to collect the drop of blood near one end of the slide on which the film is to be made. Then the slide used to draw out the film is lowered and moved backward until the blood spreads out along the vertex of the angle formed by the two slides (Fig. 16*B*). The action of moving one slide across the other is carried out as directed above.

Note that in either of these methods the slide which spreads the blood is pushed, so that the blood follows it. The slide must not be dragged over the blood, and it should rest as lightly as possible on the horizontal slide.

The thickness of blood smears may be regulated by the angle and speed at which the slide used for spreading the blood is drawn across the horizontal slide. An angle of about 30° generally permits the preparation of ideal films in which most of the corpuscles are not quite touching one another. A smaller angle or more rapid motion will produce a thinner film, whereas a greater angle or slower motion will produce a thicker film.

Do not use the end of a slide for making more than one smear before washing off the dried blood. If care is taken in handling the slides, each end may be used once for making a smear. When several smears are to be made in close succession, a convenient system is to lay down each slide as soon as it has been used for drawing out a film, and then preparing a film on it. If this procedure is followed, caution must be taken to avoid getting any dust or oil from the fingers on the slides as they are held.

Irregularity in a blood smear indicates that the slide on which the film was made was not clean, that the end of the slide used for drawing out the film was dirty or nicked, that the drop of blood may have begun to coagulate, or that smearing was not carried out with a steady motion.

After making each film, wipe the blood from the wound and wait until another small drop exudes. If blood coagulates about the opening, wash it off with a piece of cotton or gauze moistened with water. Then dry the wound from which blood may again exude. The flow of blood can be hastened by pinching or rubbing the skin near the puncture.

2. Fixation and Staining. In most histological and clinical work, the films are allowed to dry in the air. They are then fixed with absolute methyl alcohol (acetone-free) and stained by Giemsa's method (p. 395), or simultaneously fixed and stained by Wright's method (p. 393).

Wet-Fixed Smears of Blood. In order to avoid the distortion of corpuscles which occurs in drying, the films may be immersed in a fixative while they are still moist. A saturated aqueous solution of mercuric chloride or Schaudinn's fluid will serve well for such preparations. After 10 to 30 minutes in the fixative, the preparations are washed in several changes of 70% alcohol and are then treated with iodine-alcohol. They may be stained with alum hematoxylin and eosin, iron hematoxylin, Giemsa's stain, or by some other method chosen to bring out particular structures. After they are stained, the preparations are dehydrated by passing them through the alcohols (or acetone, in the case of Giemsa's stain) and cleared in toluene or xylene. Finally, they are mounted in thin balsam or synthetic resin under No. 1 coverglasses.

Lymph. 1. Obtain material from the thoracic duct of a dog or some other animal of considerable size, since it is very difficult to find the duct in small species. Kill the animal and skin the left side of its thorax. Remove the left thoracic wall, cutting through the ribs to the left of the sternum and about an inch from the vertebral column. Use bone cutters or cartilage shears for this purpose. Find the thoracic duct, which lies at the left side of the aorta. Into it insert a needle which is connected to a small hypodermic syringe, and draw out some lymph.

2. Place a drop of lymph upon a clean slide and spread it by 2 or 3 sweeps with a piece of platinum wire.

3. Allow it to dry, and fix and stain it as directed for dry blood films.

Spermatozoa. 1. Obtain fluid containing sperm from the testis, epididymis, or vas deferens of the animal.

2. If the fluid is extremely dense with spermatozoa, dilute it with a physiological saline solution. Put a drop of the fluid on a clean slide and draw it out into a thin film in the same manner as described for making a blood smear, or disperse a little of the fluid on a coverglass with a platinum loop until it forms an even film.

3. If the film is on a slide, fix it while it is still moist by immersing the slide in mercuric chloride-acetic acid, Bouin's fluid, Flemming's weak solution (p. 83), or in some other suitable fixative. If the preparation is on a coverglass, this should be dropped face down on the fixative. (The coverglass will float if the fixative is an aqueous one.) Leave the smear in the fixative for 10 to 30 minutes. Then wash it by whatever method is most appropriate. Iron hematoxylin and alum hematoxylin and eosin are good stains for spermatozoa; impregnation with Protargol,

following the method recommended for protozoa (p. 419), often yields interesting preparations.

4. Dehydrate, clear, and mount the film in balsam or a synthetic resin.

Smears of this type may also be allowed to dry, fixed in absolute alcohol, stained by Giemsa's method, then dried again and mounted.

Pus. Make smears in the manner directed for blood, or spread the pus upon a slide by means of a loop of platinum wire. Allow the smears to dry, then either pass them several times above a flame or place them in 95% alcohol for 2 or 3 minutes. Stain them by whichever of the following methods will best show the structures in which you are interested.

STAINING WITH LOEFFLER'S METHYLENE BLUE. This stain is good for bacteria in general, and for nuclei.

1. Cover the smear with several drops of Loeffler's methylene blue (p. 379) and leave it for 15 seconds to 1 minute.

2. Rinse it several times with water. Dry it in the air, waving it high over a flame if rapid drying is important.

3. Add a drop of thin balsam or synthetic resin and mount under a coverglass, or simply place a drop of immersion oil on a thin part of the smear and examine it with an oil-immersion objective.

STAINING BY GRAM'S METHOD. Some bacteria (as streptococci, staphylococci) stain by this method, and accordingly are designated "Gram-positive." Those which do not retain the primary dye are called "Gram-negative." The method is much used, together with other diagnostic procedures, in determinative bacteriology.

1. Make smears, allow them to dry, and fix them with heat.

2. Cover them with a solution of crystal violet (p. 380) for 1 or 2 minutes.

3. Rinse with water and cover for 30 seconds to 1 minute with the following solution:

Gram's Iodine Solution

Iodine	1 gm.
Potassium iodide	2 gm.
Distilled water	300 ml.

4. Place the slide in 95% alcohol until the color ceases to come out of the smears.

5. Counterstain them for 30 seconds with a solution of safranin prepared as follows:

Safranin 0, 2.5% in 95% alcohol	10 ml.
Distilled water	100 ml.

Safranin will bring out Gram-negative species which were decolorized by treatment with alcohol.

6. Rinse off the safranin in water. Dry and examine the smears with an oil-immersion objective, or pass them through absolute alcohol and toluene, and mount in resin under a coverglass.

STAINING WITH GIEMSA'S SOLUTION. Fix and stain the dried smears as directed for smears of blood. This stain is excellent for demonstrating the structure of leukocytes, and the methylene blue in Giemsa's solution will stain any bacteria present.

Sputum. Sputum is examined most commonly for *Mycobacterium tuberculosis*. The method of preparing it for this purpose depends upon the "acid-fast" characteristic of this organism.

ZIEHL-NEELSEN'S METHOD FOR TUBERCLE BACILLI. 1. Pick out small, yellowish-white cheesy masses from the sputum of a tubercular patient. Place one or two of them on a slide, break them up, and spread them into a smear with a platinum loop.

2. Dry the smear and fix it by passing it quickly above a flame several times.

3. Cover it with Ziehl-Neelsen's carbol-fuchsin stain (p. 378) and heat the slide until the liquid steams for 1 or 2 minutes.

4. Rinse the slide in water and place it in 80% alcohol containing 2% hydrochloric acid. Leave it in this until the film has lost nearly all of its color.

5. Wash in water for a minute or two.

6. Counterstain it for 10 seconds with Loeffler's methylene blue solution (p. 379).

7. Wash the preparation again. Then dry it and mount it in balsam or synthetic resin, or examine the dry film with an oil-immersion objective. *M. tuberculosis* will appear red on a blue background. Most other bacteria present will be stained blue.

Fecal Material and Intestinal Contents. The following method is designed especially for the identification and study of symbiotic protozoa.

1. *a.* To prepare dense fecal material, as that of mammals, first mix the material thoroughly with a small stick or glass rod. If it is thick and hard, dilute it with a physiological saline solution. Then transfer a small amount of material to a slide or coverglass and smear it out with a platinum loop or tips of clean forceps, or with a small stick or glass rod.

b. To prepare films of the soft intestinal contents of invertebrates such as termites, annelids, and molluscs, transfer the material quickly, and by the method which appears to be most appropriate, to a clean slide or coverglass. Spread it gently with a loop or tips of clean forceps.

2. If the smears are too fluid to adhere when placed in the fixative, allow them to stand for a minute or two, until they are only slightly moist. As a rule, however, one should make an effort to prepare smears which are sticky enough to fix immediately. Smears on slides are lowered gently

into the fixative; coverglasses are dropped face down on the surface of the fixative. For intestinal amoebae and flagellates to be stained with iron hematoxylin, Schaudinn's fluid has long been a standard fixative. Bouin's fluid is also quite good. Both Bouin's and Hollande's fluids are superior for ciliates, flagellates, and gregarines to be impregnated by Bodian's Protargol method (p. 419).

3. After appropriate post-fixation treatment, stain or impregnate the smears by the chosen method. If iron hematoxylin is used, it is best to destain the preparations one at a time on the stage of the microscope, and to watch them carefully until the majority of the organisms appear to be differentiated as desired. Keep the slide or coverglass well flooded with the destaining agent. It may either be dipped occasionally into the destaining solution, or the solution may be placed upon it with a pipette.

4. After completing the procedure of staining or impregnation, and subsequent washing, dehydrate and clear the preparations. Mount them in thin balsam or synthetic resin.

Bone Marrow, Lymph Nodes, Spleen. 1. Split open a piece of femur, rib, or other bone containing red marrow, from a freshly killed animal. Place a bit of red marrow upon a clean slide and lay a second slide upon it. If the marrow does not spread between the two, press gently to bring about this result. Then slip the slides apart with a steady and rather rapid motion. In the same way make smears of small pieces of pulp from the spleen or a lymph node. Another convenient method for preparing smears is to touch the freshly cut surface of an organ to a clean slide. A considerable number of cells will adhere to the slide and may be treated as a smear.

2. *a.* The smears may be allowed to dry, then fixed and stained by Giemsa's method. Dry and mount them in balsam or synthetic resin. Or,

b. The smears may be fixed on slides or coverglasses while still moist, by immersing them for 10 to 30 minutes in a saturated aqueous solution of mercuric chloride or some other suitable fixative. After appropriate post-fixation treatment, stain the smears with alum hematoxylin and eosin; then dehydrate, clear, and mount them.

Testis. Smear preparations of the testis are often valuable in cytological work carried out with a view to establishing chromosome numbers or the condition of the testis with respect to stages in spermatogenesis.

TEMPORARY PREPARATIONS. These are valuable for rapid identification of the stages of spermatogenesis in the gonad.

1. Place a bit of testis tissue (or an entire testis or testicular follicle of an insect or other small animal) upon a clean slide. Cover it with a drop of aceto-carmine or iron aceto-carmine (p. 341), or aceto-orcein (p. 342).

2. Place a coverglass upon it. Press upon the coverglass until the cells are spread apart. Pry up the cover a little, to bathe the tissue with stain, and gently press it down again. Staining takes place within a few minutes.

3. If it is desirable to preserve the preparation for several days, seal it with petroleum jelly, melted paraffin, or warmed "vaspar" (a mixture of equal parts of petroleum jelly and melted paraffin).

PERMANENT PREPARATIONS. 1. Hold a small bit of fresh testis tissue between the tips of a pair of forceps, and smear it back and forth over the surface of a slide or coverglass. Follow a zig-zag path, so the same area is not covered twice. Smears of testis may also be made by the methods described for bone marrow.

2. Fix the smears in a mixture which is particularly suitable for the staining method to be used. Bouin's fluid and Flemming's weaker solution (p. 83) are generally good before iron hematoxylin, and a saturated aqueous solution of mercuric chloride with 5% acetic acid is regularly used before the Feulgen nucleal reaction. Smears fixed in Flemming's mixture should be washed in running water for 30 minutes, then brought gradually to 70% alcohol; an hour in this will harden them somewhat.

3. After staining smears by the chosen method, dehydrate, clear, and mount them in balsam or synthetic resin.

Other Materials. SALIVARY GLANDS OF DROSOPHILA. See page 500.

DISSOCIATION METHODS

By isolating the cells and intercellular structures of a tissue, it becomes possible to examine these elements from every angle. In the case of liquid or soft tissues, separation of cells is easily accomplished by the smear method. The procedures required to separate the cells, fibers, and other structures of firmer tissues depend upon the nature of these structures and the means by which they are naturally joined together. There are four general methods of dissociation, as follows:

Teasing

This term is applied to the process of separating structures by the skillful use of sharp needles. Fibrous tissues such as nerves and tendons, which are not extremely fragile and are not held firmly together by a cementing substance, can be teased apart without special preparation. The majority of other tissues must be dissociated by chemical means before they can be teased apart successfully.

Preparation of Material to Be Teased without Previous Dissociation. FRESH MATERIAL. This may be teased apart in physiological saline solution and studied without being stained, or a stain may be drawn under the coverglass with the aid of a piece of bibulous paper or filter paper. The preparation may be preserved for a time by placing a small drop of glycerol in contact with one edge of the coverglass and allowing it to run under the coverglass. Such preparations, like all liquid mounts, are difficult to preserve and to handle, even if they are sealed in some way.

When permanent preparations are desired, it is much better to mount the tissue in balsam or synthetic resin, by the method to be described in a later paragraph.

GLYCEROL METHOD FOR FIXED MATERIAL. Either unstained or stained material may be put into a dish of 10% glycerol in water or 70% alcohol. If water is used for dilution, add a small crystal of thymol to discourage growth of molds. Cover the dish with lens paper, or leave its lid slightly ajar, and keep it in a dust-free place. When the glycerol has thickened by evaporation of the water or alcohol, tease apart the structures on a slide and apply a coverglass. Formalin (10%) is a suitable fixing agent for material which is to be prepared in this manner. To preserve the mounts, seal them with asphaltum cement. Glycerol mounts are, in any case, essentially temporary, and should be avoided if possible.

MATERIAL TO BE MOUNTED IN A RESINOUS MEDIUM. For permanent preparations, this is the best method. The following outline is based upon the preparation of nerves and tendons, but anyone possessing a little ingenuity can adapt it to other materials or other methods of staining.

1. Pin several 1 or 2 cm. lengths of nerve and tendon to a small piece of wood in order to keep them straight, but do not stretch them. Fix them in 10% formalin for at least 12 hours. The material may remain in formalin for weeks or months.

2. *a. Osmium Impregnation for Nerves.* Rinse, with several changes of distilled water, one or more small pieces of nerve which have been fixed in formalin. Place the material in a small glass-stoppered bottle containing sufficient 2% osmium tetroxide solution to cover it to a depth of about 4 mm. (See p. 107 for the method of preparing this solution.) Put the bottle away in a dark place for a week or 10 days. Then pour off the solution, rinse the tissue with several changes of distilled water, and wash it in running water for 2 hours. Dehydrate it with the usual graded series of alcohols, leaving it in each grade for 1 hour or longer. When the material reaches absolute alcohol, proceed to step 3.

b. Alum Hematoxylin and Eosin Stain for Tendons and Other Tissues. Wash the material fixed in formalin in running water for 2 or 3 hours. Then tear it or cut it into coarse shreds and place it in 4 or 5 times its volume of dilute alum hematoxylin (p. 353). After 30 minutes, tease apart a shred of tissue and examine it under the microscope. If the nuclei are deeply stained, pour off the staining solution. Otherwise, allow the staining solution to act for a longer period of time. After the stain has been poured off, rinse the material with several changes of distilled water. Then wash it in running water for 2 hours.

If the fibers become heavily stained, which is unlikely, place the material in a half-saturated solution of aluminum ammonium sulfate until inspection shows that they have given up the stain. This may require several hours. Then wash the material in running water.

Transfer the pieces of tissue up the alcohol series to 80% alcohol, and complete dehydration with 95% and absolute alcohols, adding enough eosin to each of these to give the solution a deep pink color. Leave the material in the absolute alcohol for an hour or two. Then transfer it to absolute alcohol without eosin, and leave it for 20 to 30 minutes.

3. Transfer the material from absolute alcohol to a mixture of equal parts of absolute alcohol and terpineol-toluene. Leave it in this for 15 to 30 minutes.

4. After an hour, pour off the mixture and add terpineol-toluene. Renew this after another hour or more. *Material which is to be teased must not be cleared in toluene alone, as this substance toughens connective tissue fibers and renders nerve fibers very brittle.*

5. Add Canada balsam or synthetic resin to the terpineol-toluene containing the material, putting in a little more every few hours until the liquid reaches the consistency of a medium syrup.

6. Tease apart the material as instructed in the following paragraphs. Then place a drop of moderately thick mounting medium and a circular coverglass upon it. Be careful to use the proper amount of resin and to put the coverglass down gently, so that the material does not mat together or run to the edge.

METHOD OF TEASING. Use clean, sharp needles, mounted in handles of wood, bone, or other suitable material. There should be two pairs of needles: a pair with rather stout points which can be used to pull apart masses of tissue; and a pair ground to fine, delicate points suitable for the final separation of minute structures. Always inspect the needles critically before beginning work and, if necessary, sharpen them on a carborundum hone. Wipe off the needles after you have finished each preparation, or whenever they have become the least bit gummy with resin.

Place a small bit of tissue, with a drop of mounting medium from which it has been taken, on the center of a clean slide. Take one of the coarse needles in the left hand and put its point firmly down upon the tissue, far enough from the edge so that you get a good hold upon it. Take the other coarse needle in the right hand and gently but firmly shred apart the tissue. Begin at the edge opposite to that held by the other needle and work gradually into the tissue. *Separate the structures very carefully, but do not crush or break them. Avoid squeezing the tissue between the needles and the slide.* This procedure can be carried out to best advantage under a binocular dissecting microscope. When the tissue has been separated into small shreds, take the fine-pointed needles and continue the process of teasing until the individual fibers or other structures have been isolated. When you think this has been accomplished, examine the material under a compound microscope. If necessary, continue to work carefully with the sharp needles. Finally, remove any thick pieces of tissue from the slide, add a drop of mounting medium and put on a circular coverglass. To apply the coverglass, hold it over the material, at an angle of about 30° to the surface of the slide. Lower it until one edge touches the slide, then slowly let down the opposite edge, so that the mounting medium spreads evenly and no bubbles of air are trapped in the medium.

Dissociation in Special Fluids

The cells of most tissues are held firmly together by intercellular cementing substances, and these are generally much harder and tougher than the cells themselves. This makes it impossible to separate the cells and other structures by the simple process of teasing. Any mechanical force sufficient to pull them apart would crush them or tear them to pieces. In order to overcome this difficulty, tissues are dissociated in a liquid which softens or dissolves the intercellular cements and, at the same time, fixes the cells and fibers. After this treatment, the structures can easily be teased or shaken apart.

Dissociating Fluids and Their Use. The reagents which have been recommended for dissociation include nitric acid, caustic soda, osmium tetroxide, picric acid, sodium chloride, chromic acid, salts of chromium, potassium permanganate, chloral hydrate, formaldehyde, and alcohol. Most of these agents do not preserve cell structures as well as the better fixatives in common use, and they do not give them the best staining properties. These disadvantages are greatest in the case of strong acids, alkalis, and oxidizing agents. Of the dissociating agents mentioned, weak alcohol and formalin are by far the least harmful. They serve most purposes quite well and are the only ones whose use will be discussed in detail.

Ranvier's One-Third Alcohol

Alcohol, 95%	1 part
Water	2 parts

This is a generally useful dissociating fluid and is particularly recommended for epithelia, glands, and smooth muscle fibers. It acts rather slowly and preserves the structures with a minimum of alteration. Simple columnar epithelia may easily be shaken apart after being treated with it for 36 hours; cells of stratified epithelia which are not keratinized are usually thoroughly dissociated after 48 hours. Gastric glands require about 3 days, and smooth muscle fibers about 4 days. It seems not to be favorable for nerve cells.

Gage's Formaldehyde Solution

Formalin	2 ml.
Physiological saline solution	1,000 ml.

This mixture is excellent for dissociating simple epithelia and nerve cells. Epithelia are dissociated thoroughly in 24 hours, and nerve cells of the spinal cord in 40 to 60 hours.

General Directions for Dissociation. Regardless of which dissociating agent is used, employ *small* pieces of tissue. Cut slices 3 or 4 mm. thick, or tear off shreds of fibrous tissue, and place them in stoppered vials about half full of the dissociating agent. Shake the vials at least once a day, and note the amount of tissue which breaks up. When the extent of dissociation appears to be considerable, examine a drop of the preparation in order to ascertain if the desired elements have been freed.

Suggestions Concerning Material. To demonstrate smooth muscle and simple columnar epithelium, slit open a piece of small intestine. Wash the mucosa, then scrape it lightly with a scalpel and dissociate the scrapings in Gage's formaldehyde solution for 24 hours. Remove the mucosa, cut the muscular layers into thin strips, and dissociate the tissue in one-third alcohol for 4 days. To demonstrate ciliated epithelium, strip off the lining from a piece of the trachea, and dissociate it in Gage's formaldehyde for 24 hours. For nerve cells, the spinal cord of an ox is very favorable and easy to obtain. Dissociate it in Gage's formaldehyde solution for 3 to 5 days, and make mounts of the gray matter as directed under Permanent Preparations of Nerve Cells (p. 182).

Methods of Mounting Dissociated Material. Epithelia, glands, and smooth muscle fibers, when sufficiently dissociated, can be shaken apart. Skeletal and cardiac muscle fibers must be teased apart on slides, and nerve cells are shown best in smears of dissociated gray matter.

Temporary Mounts. The simplest method of examining dissociated material is to shake or tease it apart, mount a very small quantity of it on a slide, and apply a coverglass. The cells may be dissociated further by drumming lightly upon the cover with a pencil or other small instrument. Material which has been dissociated in either of the fluids mentioned above may be mounted temporarily in the same fluid. If an acid, alkali, or colored salt has been used, the material should be washed and mounted in water. Such preparations can be stained by placing a drop of acidulated methyl green, aceto-carmine, dilute alum hematoxylin, borax carmine, or aqueous eosin at one edge of the coverglass and drawing it into the preparation by means of a piece of filter paper held in contact with the liquid at the opposite edge of the cover. It is hardly worthwhile to mount such preparations in glycerol.

Permanent Preparations of Dissociated Epithelial Cells or Smooth Muscle Cells. 1. Apply a thin film of Mayer's albumen affixative (p. 241) on a clean slide.

2. Place upon this a drop of the sediment containing cells which have been dissociated by shaking. Tilt the slide back and forth in order to spread the material somewhat.

3. When the material is almost dry, place the slide in absolute alcohol for a few minutes. This will coagulate the albumen, and cause it to hold the material on the slide.

4. Place the slide for a minute or longer in a 1% solution of Parlodion. Drain it and expose it to the air until the coating sets to form a film. Then place it into 70% alcohol.

5. Pass the slide down through the graded series of alcohols to water. Place it in dilute alum hematoxylin (p. 353) until the nuclei of the cells are deeply stained, which will occur ordinarily within 30 to 45 minutes. Other stains can be used, if desired.

6. Wash it in running water for 10 or 15 minutes.

7. Pass it up through the alcohols and counterstain it for 2 minutes in 0.5% eosin in 90% alcohol.

8. Rinse the preparation in 95% alcohol, and place it in absolute alcohol until the counterstain is differentiated.

9. Place it in terpineol-toluene for 1 or 2 minutes, then transfer it to toluene and leave it for a short time. Mount it in balsam or synthetic resin.

Permanent Preparations of Skeletal or Cardiac Muscle. 1. Dissociate bits of muscle in a 20% aqueous solution of nitric acid for 24 hours. Replace the nitric acid with water and shake the material vigorously in a vial or test tube. Place a small amount of the dissociated tissue on a slide and tease apart the individual fibers.

2. Drain the water from the slide and cover the teased material with several drops of absolute alcohol. After a minute, renew the alcohol.

3. At the end of another minute, pour off the alcohol and replace it with a drop of a 1% solution of Parlodion. Tilt the slide back and forth so as to spread the Parlodion. When this substance has congealed to form a film, place the slide in 70% alcohol.

4. Stain the fibers with alum hematoxylin and counterstain them with eosin, as in the preceding method. Alum cochineal is also a fairly good stain for such material.

5. Place the preparation in 95% (*not* absolute) alcohol until the stain is properly differentiated.

6. Clear and mount it as in the preceding method.

Permanent Preparations of Nerve Cells. These mounts show the nerve cells and portions of their processes, free from other structures.

1. Cut slices 4 to 5 mm. thick from the brain or spinal cord. The cervical region of the spinal cord of cattle provides excellent material for demonstration. Dissociate the tissue in Gage's formaldehyde solution for 40 to 60 hours.

2. Pick out bits of gray matter and place them upon a slide. Lay a second slide upon it and slip the two apart, thereby spreading the tissue into a smear.

3. Place the smears, while they are barely moist, in 70% alcohol. Leave them for 10 minutes or longer.

4. Stain the smears with alum hematoxylin, counterstain with eosin, dehydrate, clear, and mount. These steps are described under the method for epithelial cells and smooth muscle cells (p. 181).

PRESERVATION OF UNMOUNTED MATERIAL. Dissociated material may be preserved indefinitely in glycerol. Place it in a relatively large volume of 10% glycerol in water or 70% alcohol in a shallow dish covered with a sheet of lens paper, and allow it to stand in a warm, dust-free place (as in a box on top of a paraffin oven). If water is used to dilute the glycerol, it is a good idea to add a small crystal of thymol to the mixture in order to prevent the growth of molds. When the mixture has thickened, by evaporation of the alcohol or water, to the consistency of a light syrup, store the material in a bottle with a glass stopper or plastic lid. Do not use a cork or rubber stopper, because glycerol erodes rubber and extracts enough pigment from cork to stain the preserved material.

Dissociation by Digestive Enzymes

With the use of certain enzymes, it is possible to remove certain structures of tissues or of cells while leaving others unaltered and more easily studied. This method is applicable to fresh tissues or to those which have been lightly fixed, as in the lemon juice (or formic acid) and gold chloride technique (p. 425). Material fixed in alcohol or formalin may be used if it is first washed in running water for 24 to 48 hours. The enzyme preparations usually employed, and the nature of their action, are described below.

PEPSIN. This digestive enzyme of the stomach dissolves albumins, collagen, mucin, and elastin. Elastin is dissolved very slowly, however. Pepsin has no effect upon fats, carbohydrates, chitin, keratin, and neurokeratin. Its capacity to digest collagen makes pepsin useful for the isolation of elastic fibers (such as those of the nuchal ligament) and for softening the connective tissue of muscles when making teased or pressed preparations to show nerve endings.

1. Make an acidulated solution of pepsin, as follows: powdered pepsin, 1 gm.; distilled water, 1,000 ml.; hydrochloric acid, 3 ml.

2. Incubate small pieces of tissue or frozen sections of unfixed material in this liquid, at 37° C., for 30 minutes to 2 hours, depending upon the effect desired.

3. Wash the material with several changes of distilled water.

4. *a.* Make temporary mounts in water, or

b. Stain the material by a method appropriate to its nature. Then dehydrate, clear, and mount it in resin. However, because alcohol and clearing agents toughen or shrink muscle and elastic tissue, it is best to transfer these tissues to glycerol, tease or press them apart, and mount them in glycerol jelly.

PANCREATIN. The inexpensive dried preparations of pancreatic enzymes found in commerce contain trypsin (a proteinase), amylopsin (an amylase), and steapsin (a lipase). Such preparations effectively digest albumins, mucin, elastin, starches, and fats, but do not attack collagen and chitin. The following procedure is a useful application of the proteolytic capacity of pancreatic extracts, which results in the removal of cells from among connective tissue fibers:

1. Prepare an alkaline solution of pancreatin, as follows: pancreatin, 5 gm.; distilled water, 100 ml.; sodium bicarbonate, 1 gm.

2. Prepare frozen sections, 50 to 80 μ in thickness, of fresh tissue. (Spleen, lymph node, and liver are recommended for practice.)

3. Incubate the sections in the above solution, at 37° C., for 24 hours. This will destroy the cells, but leave the reticular fibers of connective tissue intact.

4. Wash the sections in water, shaking the dish gently in order to dislodge cellular debris.

5. Pass one end of a clean slide into the water and under a good section. Lift the section from the water, and affix it to the slide by Wright's nitrocellulose method (p. 277).

6. Stain it for a few minutes in a 0.5% solution of eosin in 90% alcohol.

7. Rinse it with water and place it in 95% alcohol for 2 or 3 minutes. Clear it in terpineol-toluene, followed by toluene, and mount it in balsam or synthetic resin.

Corrosion

Corrosion is the process of separating and cleaning hard structures by destroying, with strong chemicals, the soft parts that naturally surround them. This method is important in the study of shells, spicules, ossicles, bristles, scales, and chitinous skeletons. Another noteworthy application of corrosion is that of injecting blood vessels, air passages, or other internal cavities with wax, latex, or some other resistant substance, and then corroding away the soft parts by immersion in a strong acid or alkali. Beautiful casts of the cavities are thus made.

CALCAREOUS AND SILICEOUS STRUCTURES. The shells of foraminiferans, spicules of sponges, skeletal plates or pedicellariae of echinoderms, and other calcareous or siliceous structures may be cleaned by means of a strong alkali. Place the material in an adequate volume of a 10% or 20% aqueous solution of sodium or potassium hydroxide, or in a commercial preparation of sodium hypochlorite (such as Clorox), until the organic material has been removed. (This may take only a few minutes, or it may take several days, and it may be necessary to replenish the alkali solution once or twice.) The sediment can then be washed in several

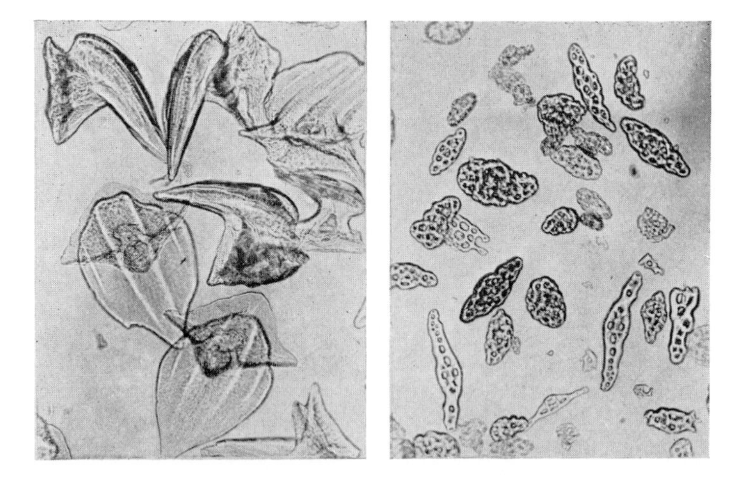

FIG. 17. Scales of a dogfish shark, *Squalus suckleyi* (left), and calcareous plates from the skin of a sea cucumber, *Eupentacta quinquesemita* (right), isolated by corrosion of soft parts in sodium hydroxide.

changes of water (a conical centrifuge tube and a pipette may be helpful in this operation) and mounted under a coverglass for observation. If permanent preparations are desired, the calcareous or siliceous structures may be dried and mounted (p. 166), or they may be dehydrated in several changes of absolute alcohol, passed through at least two or three changes of toluene, and mounted in a resinous medium.

INSECTS AND OTHER CHITINOUS STRUCTURES. The bodies of insects which are too large or opaque to be mounted successfully by the method given on page 150 may be treated as follows: Kill the insect with cyanide fumes and soak it in a 10% aqueous solution of sodium hydroxide until the internal parts have disintegrated and the exoskeleton has become soft and transparent. Make a small slit in the exoskeleton and remove remnants of tissue with fine forceps. Place the specimen again in the alkali for a day or two. Then wash it thoroughly in water, and flatten it between two slides which are tied together. Place the slides in absolute alcohol for one to several days, then in oil of wintergreen until the animal is cleared. Separate the slides and mount the specimen in balsam or synthetic resin. Should the exoskeleton remain dark after the soft parts have been removed, bleach it with the aid of any commercial preparation of sodium hypochlorite, such as Clorox (2 or 3 drops to 10 ml. of 50% alcohol), then flatten it between slides and proceed as directed above.

The radula of a snail may be prepared by soaking the buccal mass in a 10% aqueous solution of sodium hydroxide until all of the soft parts are destroyed. Then wash the radula in water, flatten it between two slides, dehydrate, clear, and mount it.

CHAPTER 13

Methods of Sectioning, Microtomes, and Microtome Knives

Cutting sections is the most important way by which organisms or their tissues are separated into thin, translucent portions for microscopic study. One great advantage of sections is that they show the microscopic structures in nearly natural relations. Furthermore, standard methods of sectioning permit the use of strong and effective fixing agents. Dissociation methods offer neither of these advantages, and their use is accordingly limited. A serious disadvantage of sectioning is that individual cells, fibers, and other structures are cut through in one plane, regardless of their natural boundaries, with the result that only fragments of many structures are present in any given section. The proportion of complete structures to fragments in a particular section depends upon the size of the structures and the thickness of the section.

Sections must be cut thin enough so that light can pass through them readily. Generally, sections to be used for histological or cytological studies should be no thicker than one or two layers of cells. Very thick sections may be useful for following the course of blood vessels through the tissues, or for demonstrating the course of nerves and their processes, but they are not apt to be suitable for study of the cells or tissues themselves because of their relative opacity and the fact that out-of-focus images of overlying and underlying layers make it difficult to obtain clear views of any particular structures. Extremely thin sections may not provide sufficient perspective for histological studies, although they may be indispensable for understanding certain details.

The thickness of sections is measured in microns, 1 micron (abbreviated μ) being one-thousandth of a millimeter (0.001 mm.). For ordinary histological purposes, sections 8 to 12 μ thick are most useful. For cytological work, sections about 8 μ or slightly thinner will probably be suitable, but sections as thin as 1 μ can be used to advantage in understanding some aspects of the organization of cells. Electron microscopy dictates a rather different attitude toward the analysis of histological and cytological details because it requires sections that are only a fraction of a micron thick.

A high degree of precision in section cutting is accomplished through the use of several special instruments and methods. It is known that botanists commonly sectioned plant tissues as early as 1770. It is difficult to ascertain when sections were first employed in animal histology, but this certainly took place at a much later date. Zoologists in the days of Schleiden and Schwann (around 1840) made sections of tissue by holding the specimen in one hand and cutting it with a sharp razor. Sections obtained in this manner, even by the most skillful operators, were always quite thick and much distorted. Better sections could not very well be obtained because of the softness of most animal tissues, which caused them to be crushed and torn by the knife, and because of the irregular spacing of sections and the wavering stroke of the unguided knife. For these reasons, improvements in methods of sectioning were pursued along two lines: (1) the development of techniques for hardening and supporting the tissues, now called *embedding*, and (2) the invention of machines, called microtomes, for cutting thin and uniform sections. Advances along these lines took place simultaneously and were dependent upon one another. However, to simplify the discussion, each will be considered separately.

EMBEDDING METHODS

PRIMITIVE METHODS. The earliest aid to hand sectioning consisted of using pieces of pith or cork in order to support the tissue. Eventually, tissue to be sectioned was commonly hardened in alcohol. Softer tissues were sometimes supported by surrounding them with alcohol-hardened liver. Methods as crude as these are little used today, but alcohol continues to play an important part in hardening and embedding procedures.

THE PARAFFIN METHOD. In 1869, Klebs, an assistant to the celebrated pathologist Virchow, began to coat his specimens with paraffin or wax, which he found supported them better than pith or cork. A few years afterward, other workers developed a method of *infiltrating* or saturating tissues with paraffin. Tissues so treated are supported at every point, both externally and internally, and can be cut into extremely thin sections without damaging or disarranging any of their parts. This and other processes in which the tissue is saturated as well as surrounded by an embedding medium (*interstitial embedding*) have survived as the methods in common use today.

Use of the paraffin method increased greatly before the turn of the century, following the invention of a rotary microtome which would cut paraffin sections in such a manner that each of them adhered to the preceding section. The *ribbons* or *series* thus formed contain the sections in natural order and make it possible to *reconstruct* any part by following it through successive sections. Today the paraffin method, the details of which are given in Chapters 14 and 15, is used more than any other method of sectioning. It is unsatisfactory, however, for either hard or watery tissues, for large objects, and for the cutting of very thick sections.

THE NITROCELLULOSE ("CELLOIDIN") METHOD. Ten years after Klebs began to support specimens with a coating of paraffin, Duval and Latteaux accomplished the same purpose by dipping objects repeatedly in a solution of nitrocellulose. This method also was soon developed into a process of infiltration and interstitial embedding. Subsequently, a special form of nitrocellulose for this process was marketed as "Celloidin." Through repeated use of this trade name, the nitrocellulose method came to be called the "Celloidin method." This method is more successful than the paraffin method for embedding hard tissues, watery tissues, large objects, or any that are to be cut into very thick sections (over 20 or 30 μ). It is not well adapted to the preparation of serial sections. Techniques for embedding material in nitrocellulose are explained in Chapter 16.

SYNTHETIC POLYMERS. Electron microscopy takes advantage of a number of synthetic media which, although they are of relatively low molecular weight and penetrate tissues readily, become very hard when polymerized. Their hardness and other properties make them suitable for embedding material that must be sectioned at about one-tenth of a micron. These media can also effectively be used in preparation of sections about 1 μ thick, for study with the light microscope. Chapter 18 gives detailed instructions for embedding in Epon, which is perhaps the most important of the synthetic polymers, and for cutting sections with the ultramicrotome.

OTHER EMBEDDING METHODS. Gelatin is sometimes used for embedding, in a single block, numerous small objects which are to be sectioned by the freezing method. In earlier times, soap masses were extensively employed, and soap-wax mixtures have also been devised. Certain water-soluble waxes (carbowaxes, polyethylene glycol waxes) have recently found favor for special purposes. Those who contemplate using waxes of this type will find helpful suggestions in the contributions of Giovacchini (1958), Jones, Thomas, and O'Neal (1959), Riopel and Spurr (1962), and Wade (1952).

THE FREEZING METHOD. Freezing is the simplest and most rapid process by which tissues can be hardened for sectioning. The process does not require treatment of the tissue with solvents which remove fats, and it can be carried out with unfixed material. This method is described in Chapter 17. It is used primarily in clinical and microchemical work.

MICROTOMES

Microtomes, or machines for cutting thin sections, were used by English microscopists about 200 years ago. It appears that the first device of this sort was invented by Cummings, in 1770. Improved models were developed soon afterward. These machines were employed chiefly by amateurs who used them to cut sections of plant tissue. They were not adopted by scientists in Germany or in other European countries, and cannot be regarded as the direct ancestors of modern microtomes. In fact, the existence of these early machines was not generally known until rather recently, so for a long time the microtome was believed to be of relatively modern origin.

The first microtome invented and used in Germany was that of Oschatz (1843). However, some German scientists persisted in cutting sections by the freehand method, maintaining that microtomes were needed only by clumsy persons. Between 1865 and 1868, the development of better staining methods emphasized the importance of thin, perfect sections. Improved microtomes began to appear, first in Germany and later elsewhere. Beginning with the primitive instrument of Oschatz, several distinct patterns, each with numerous variations, evolved. The most important of these are described below.

Simple Microtomes

HAND MICROTOMES. Oschatz' microtome consisted of a metal tube, into one end of which a screw was threaded while the other end of the tube was fitted into a hole in the center of a circular metal plate. The specimen was inserted into the open end of the tube, and could be forced out short distances by turning the screw. After each movement of the screw, a section was cut by drawing a razor across the circular plate. This simple apparatus involved the essential mechanisms needed for accurate section cutting: a feeding device to advance the specimen, and a guide for the knife. The term *simple microtome* is applied to instruments in which the knife, though guided, is held in the hand.

Instruments of the type just described were used by Ranvier (Professor in the Collége de France), and because of his influence, they grew in favor. A modernized form of Oschatz' or Ranvier's microtome is the hand microtome (or well microtome) which botanists sometimes still use for sectioning stems.

TABLE MICROTOMES. These instruments have a cast-iron frame provided with clamps for attachment to a tabletop. Modern table microtomes (Fig. 18) generally have two glass tracks, over which a straight-edge razor or standard microtome knife may be drawn, and projections of the casting which will guide a broad, chisel-like knife. The calibrated feed screw passes through the bottom of the casting. Above this, it is threaded into a block which bears an object clamp. The lower end of the screw bears a ratchet wheel, and when this is turned the block is pushed upward a known distance for each tooth of the ratchet. Thirty or 40 years ago, these instruments were used extensively for making frozen sections, and occasionally for some other purposes. They have been almost completely superseded by other types.

FIG. 18. Table microtome. (Courtesy of American Optical Corp., Buffalo, N.Y.)

Sliding Microtomes

In 1868, Rivet invented a microtome in which the knife was firmly clamped to a heavy block. This block rested upon a long, horizontal track, and sections were cut by sliding it back and forth. Before each cutting stroke, the object carrier was raised by pushing it up an inclined plane. The use of accurate mechanical devices for holding and manipulating both the specimen and knife made it possible to cut sections with a degree of precision previously unattainable. Rivet's instrument, originally made of wood, was the progenitor of all *mechanical* or *double-acting* microtomes.

TYPICAL SLIDING MICROTOMES. The instruments found in nearly every laboratory follow Rivet's pattern in a general way. Of course, the entire mechanism has been brought to a high degree of precision. These instruments have, with few exceptions, automatic feeding devices to advance the object a measured distance before each cutting stroke of the knife. The feeding mechanisms are of two distinct types, and the instruments are classified on the basis of these differences. The *inclined plane*

FIG. 19. Sliding microtome. (Courtesy of American Optical Corp., Buffalo, N.Y.)

models retain this feature of Rivet's microtome and employ an accurately calibrated feed screw to push the object carrier up the incline. The *vertical feed* type, illustrated by Figure 19, has the feed screw and track for the object carrier placed at right angles to the horizontal knife track. Good instruments of this class are offered by several makers. They have an advantage over inclined plane models, because the operation of the vertical feeding mechanism does not change the position of the object with reference to the length of the knife track, so that the same long sweep of the blade can be used throughout the cutting of a large specimen. In microtomes of both classes, the knife is ordinarily clamped at one end only, and if the blade is thin or the object hard, the knife is likely to spring, therefore cutting sections of unequal thickness. In some larger vertical feed models this fault is eliminated by means of a heavy clamp which holds the knife at both ends.

The oblique position of the knife, and its long sweeping stroke, are favorable to the cutting of a more or less flexible embedding matrix such as nitrocellulose. Hence the sliding microtome is regularly used in the nitrocellulose method. A sliding microtome equipped with a freezing device (p. 271) is excellent for cutting frozen sections. The ordinary sliding microtome with a movable knife, set at an oblique angle to the object, is *not* suited to the production of ribbons or series of sections in paraffin.[1] However, the oblique orientation of the knife makes this type of microtome useful for cutting paraffin sections of dense connective or muscular tissues.

SPECIAL SLIDING MICROTOMES. Several types have been developed particularly for the sectioning of large or hard objects. In some kinds, the knife is clamped at both ends and remains stationary, while the heavy object carrier slides upon a broad track. In others, the knife is clamped at both ends to a rectangular frame which slides upon tracks located at opposite sides of the microtome. Some of these large instruments are operated by the turning of a wheel, but nevertheless are not rotary microtomes in the sense that this term is usually employed. All of them are too large and cumbersome to be recommended except for the special purposes named.

Rotary Microtomes

In 1883, Pfeifer (a mechanician at Johns Hopkins University) invented a microtome in which the turning of a wheel caused the object carrier to move up and down a vertical track. The knife was clamped in place, with its edge upwards and its long axis at right angles to the face of the object clamp. An automatic device turned the feed screw a measured distance before each downward stroke of the object carrier. Three years later, another American, Charles Minot of Harvard University, independently invented a similar but better machine. A modern version of Minot's "rotary microtome" is illustrated by Figure 20. The feeding mechanism of this instrument consists of a ratchet wheel mounted upon one end of the feed screw, which is turned by a pawl during the upward motion of the object carrier. By varying the position of a cam, it is possible to regulate the point where the pawl engages the ratchet wheel. This determines the extent to which the ratchet wheel is turned and, consequently, the distance the object is advanced for each section. Several features of this microtome make it particularly suitable for sectioning tissues embedded in paraffin. The stationary knife holder prevents the blade from springing under the impact of the hard paraffin block containing the specimen, and so insures sections of uniform thickness. The knife is placed transverse to the path of the paraffin block, which is trimmed to a rectangle and placed with its lower edge parallel to the knife's edge. Each time the block strikes the knife it makes contact with the edge of the preceding section. This pressure, and a slight amount of heat created by impact of the block against the knife, weld the sections together, so that successive sections form a ribbon. Since the knife is nearly vertical, gravity causes the ribbon to leave its edge

[1] Models in which the knife remains stationary can be used for this purpose, but they are not as satisfactory as regular rotary microtomes.

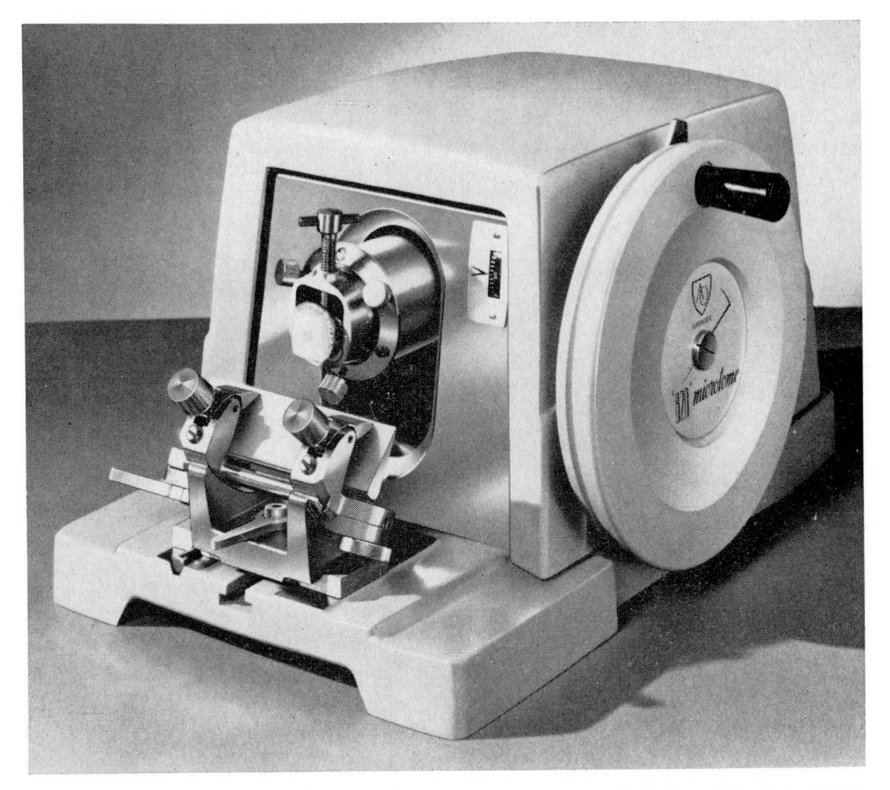

FIG. 20. A rotary microtome. (Courtesy of American Optical Corp., Buffalo, N.Y.

quickly; this facilitates handling the sections. For these reasons, the rotary microtome and the paraffin method grew rapidly in favor and are now used more extensively than any other techniques for sectioning.

Minot's rotary microtome has been greatly improved and modified by various makers. Modern rotary microtomes have holders which allow the knife to be shifted so that any part of its blade may be brought into use, and which permit adjustment of the tilt of the knife. Most models have a convenient ball-and-socket or ball-and-flange mounting for the object holder, which makes it possible to lock the specimen into any desired position by tightening one or more screws. Some of the more elaborate instruments are driven by electric motors, but there seems to be nothing better than hand operation to provide the much-needed control of the cutting stroke.

Some manufacturers have offered special attachments for holding the knife obliquely, for cutting sections in nitrocellulose. However, the vertical position of the blade and object make a rotary microtome always unsatisfactory in the conventional nitrocellulose method.

Swinging ("Clinical") Microtomes

In these, the knife holder is pivoted at one end and swings in an arc, or is provided with a double swinging device making the blade move in a flattened curve. Such microtomes (Fig. 21) are much used for cutting frozen sections.

FIG. 21. Clinical microtome equipped with freezing attachment designed to use liquid carbon dioxide. (The same type of freezing attachment may be used in a sliding microtome.) (Courtesy of American Optical Corp., Buffalo, N.Y.)

Ultramicrotomes

For cutting the extremely thin sections required for electron microscopy, special microtomes, called ultramicrotomes, have been devised. In some of these, the system advancing the specimen is a mechanical one; in others, it operates on the principle of thermal expansion. The tissue is embedded in a hard synthetic polymer, and cut on a very sharp knife made of broken glass or a polished diamond. The usefulness of ultramicrotomes extends to light microscopy, because the combination of certain methods of fixation with embedding in plastic polymers of low molecular weight permits one to make preparations that show some cytological details beautifully. Chapter 18 is devoted to embedding in one of these polymers (Epon), and to sectioning with the ultramicrotome.

Selection of Microtomes

According to the facts already brought out, a vertical rotary microtome is needed for almost all paraffin sectioning. A well-made sliding microtome is required for sectioning material embedded in nitrocellulose. The latter, with a good freezing attachment, will serve well for cutting frozen sections.

Features to be considered in choosing a microtome are: (1) General strength and rigidity of construction. (2) Precision and wearing qualities of the feeding mechanism. A good test is to operate the microtome at high speed and observe whether sections of uniform thickness are obtained. A device which lifts the pawl from the ratchet wheel, upon its return stroke, is found in certain microtomes. This prevents wear upon the teeth, which will eventually result in uneven cutting. (3) Convenience and rigidity of the mechanisms for holding and orienting the specimen. The ball-and-socket or ball-and-flange types are excellent. (4) Rigidity and easy manipulation of the knife holder. There should be convenient arrangements for lateral movement of the blade and adjustment of its tilt. In sliding microtomes, the knife block should be very heavy, and in all types it should rest upon broad, accurately ground tracks. A clamp which holds the knife at both ends will eliminate uneven cutting caused by springing of the blade.

Use and Care of the Microtome

Detailed directions for operating the rotary microtome are given in Chapter 15. The sliding microtome is discussed in Chapter 16. The cutting of frozen sections is explained in Chapter 17. The following general instructions, however, are important for all methods of sectioning and all types of microtomes.

1. A sharp knife is the first requisite of good section cutting, and there

is no substitute for it. The selection and care of microtome knives occupies the next part of this chapter.

2. The knife must be tilted to the correct angle, which differs for paraffin and nitrocellulose. (See pp. 233 and 263.)

3. The object must be properly oriented with respect to the knife edge, so that it may be cut with reference to axes or planes of symmetry. A discussion of this important subject will be found on page 207.

4. *When working about a microtome, be extremely careful not to cut yourself on the sharp knife, or to injure the cutting edge by striking it with hard objects.*

5. Before beginning to cut sections, make certain that every adjustment screw in the knife holder and object holder is tightened. Looseness may result in injury to the knife, to the operator, or to the embedded material.

6. Cut sections with a steady, even stroke, making this fast or slow according to the type of material. Serial sections are generally cut with a rapid stroke. Yolky, crumbly, or brittle structures may require a very slow motion.

7. At intervals, or whenever scratches appear upon the sections, lightly wipe the cutting bevel of the knife with a camel's hair brush or a bit of cotton soaked in toluene or xylene, but avoid touching the actual cutting edge. Also wipe the upper edge of the block with a brush. If scratches persist, the knife is probably nicked; use a different part of the blade. Other sources of trouble and methods for their correction are discussed on page 237.

8. Keep the microtome clean. Cover it when it is not in use.

9. Lubricate the instrument frequently, following the manufacturer's directions. In general, use petroleum jelly upon sliding surfaces and typewriter oil upon other bearings.

MICROTOME KNIVES: THEIR SHARPENING AND CARE

The microtome knife is the *sine qua non* of section cutting. Good embedding procedures and accurate microtomes are essential to the precision of modern methods of making sections, but they cannot serve well unless used in conjunction with a knife of the proper steel and design, sharpened to a straight, smooth cutting edge. These matters are best discussed under several headings.

Types of Microtome Knives and Razor Blade Holders

Design of the Knife. A microtome knife should be sufficiently rigid that it will not bend under the impact of the hardened and embedded specimen. Bending will result in sections of uneven thickness, or in ridges being formed upon the sections. On this account, microtome knives are wedge-shaped in cross section, as shown in Figure 22. The stoutness of the cutting edge is further increased by grinding the apex of the wedge at an angle (*f-e-f′*) less acute than the general angle of the blade (*b-a-b′*). This produces two narrow, beveled surfaces, *the cutting facets* (*or bevels*), which meet at an angle of about 30° to form the cutting edge (*e*). The secondary wedge (*f-e-f′*) formed by these bevels is much stouter and less likely to bend than would be the case if the sides of the blade were continued straight to *a*. In sharpening a knife, the proper angle is maintained by means of a tubular steel honing *back* (*hb*) obtained from the manufacturer. This back should either fit the blade tightly enough so that it does not slip, or it should be held in place by an inner spring. Setscrews installed in some honing backs are a nuisance, since they prevent the blade from being rotated easily during the process of honing.

The stoutest and most serviceable knives are those in which both surfaces are flat, as shown at *A*. Blades which have been slightly hollow-ground on one side, as at *B*, satisfy most purposes, though they are perhaps a little weaker on account of the decreased width at all points from *f* to *b*. A few makers offer blades which are hollow-ground on both sides, as shown at *C*; these are satisfactory only for cutting small and soft objects. Note the narrowness of such a blade, particularly at *f-f′*. The only excuse for hollow grinding is the claim that it eliminates the need for a honing back. Even this appears to be erroneous, because the

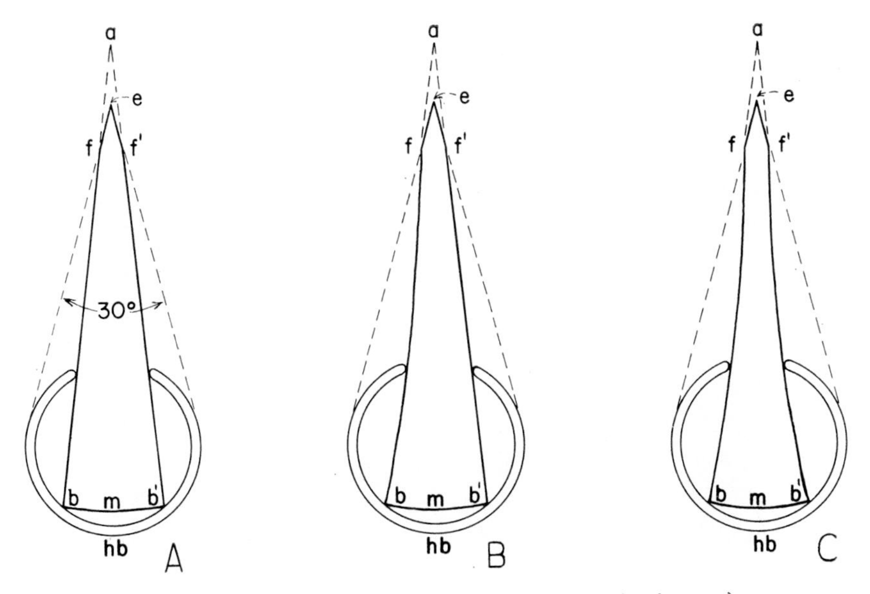

FIG. 22. Types of microtome knives. (Explanation in text.)

thickness of the base of the blade shown at *C* would have to be greatly increased in order to provide the necessary angle for grinding the cutting edge. So wide and cumbersome a blade probably does not exist. A straight-sided blade, fitted with a correct honing back, is the choice of most workers.

A "plain" microtome knife is a rectangular blade not provided with a shank for attachment to the microtome. Such knives are the easiest to obtain and to sharpen. They are used with all rotary microtomes and most other types. Some sliding microtomes are made to hold knives with curved shanks, but adapters make it possible to use plain knives with them. Certain other very large microtomes require specially designed knives.

QUALITY OF STEEL. A microtome knife should be made of hard and slightly resilient steel. If the metal is too soft, the edge will soon be dulled by use. If it is extremely hard or brittle, the edge will become chipped and broken in cutting hard objects. Knives of different makers, or even individual knives produced by the same manufacturer, vary in hardness. Actual trial is the only basis upon which a good knife can be selected with certainty.

SAFETY RAZOR BLADE HOLDERS. Safety razor blades have been used for many years as substitutes for regular microtome knives. Several firms supply a type of holder (Fig. 23) into which razor blades may be clamped in order to give them the needed rigidity. Only 1 or 2 mm. of the edge projects. The attractive feature of such a contrivance is that it eliminates the "time-consuming and tiresome labor of sharpening microtome knives." This, like most Utopian schemes, unfortunately meets with practical obstacles, which will be briefly commented upon. (1) Safety razor blades are too short to be employed successfully in the sliding microtome, the chief advantage of which lies in its long, sweeping stroke. (2) The cutting

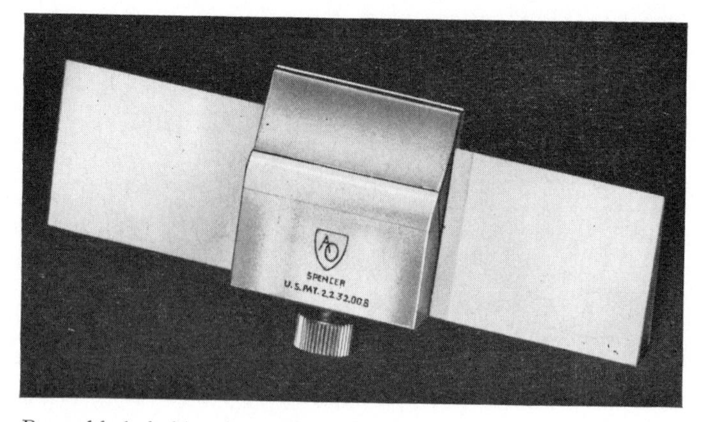

FIG. 23. Razor blade holder, for use with rotary microtomes. (Courtesy of American Optical Corp., Buffalo, N.Y.)

edges of most safety razor blades show, under the microscope, a multitude of small nicks, and the cutting facets are marked by corresponding deep scratches running at right angles to the edge. The pattern of this saw-toothed edge is plainly discernible upon sections cut at 5 μ or less. A well-sharpened microtome knife has, on the contrary, a relatively smooth, clean-cutting edge. (3) The thin, unsupported edge of a razor blade will spring under the impact of the block, if the latter is large or hard, or in any case if sections are being cut at thicknesses greater than 15 μ.

For these reasons, the usefulness of razor blades is limited. In the case of paraffin blocks which are small and in which the embedded tissue is not very hard, razor blades may cut acceptable sections within the range of about 7μ to about 15 μ. However, even with such ideal material, a sharp microtome knife gives smoother, more uniform, and less compressed sections. Once the operator has learned to sharpen and care for a knife properly, he will find that this involves very little actual drudgery. He will be amply repaid for the time he takes to keep his knives in good condition.

Although safety razor blades and holders cannot replace microtome knives if high-quality work is to be done regularly, there are times when their use may be recommended. In cutting tissues which are thought to contain grit or hard calcareous or siliceous material, substitution of a razor blade may prevent damage to the cutting edge of a microtome knife. Razor blade holders are also used a great deal in laboratories in which students practice cutting sections before they have learned how to sharpen a microtome knife.

DISPOSABLE MICROTOME KNIVES. Far better than razor blades are high quality disposable blades made expressly for microtomy. DISPO blades (sold by Scientific Products, a division of American Hospital Supply Corporation, Evanston, Illinois) can be recommended highly. They are 12 cm. long, and therefore have a cutting edge as extensive as that of a small microtome knife. The metal stock from which they are manufactured is about 1 mm. thick, so they have little tendency to vibrate. These disposable blades cost about one dollar each, and are used in connection with a holder that adapts them to a microtome and provides additional support. They are perhaps the answer to the problem encountered in a teaching laboratory, where one might wish for something better than a razor blade but less expensive in cost as well as maintenance than a good microtome knife. DISPO blades are cheap enough that one may be assigned to each student. If properly cared for, they should serve for several weeks.

Sharpening Microtome Knives

Each worker should use, sharpen, and care for his own microtome knives. The old adage "everybody's business is nobody's business" is applicable to the microtechnique laboratory. A knife used by several persons is almost certain to get into bad condition. It is quite as important for a technician to sharpen his knife skillfully as it is for the player of a stringed instrument to tune it correctly.

Appearance of a Good Cutting Edge. In order to cut thin sections in a thoroughly satisfactory manner, the edge of the blade must be straight and sufficiently smooth to show no considerable nicks or irregularities under a magnification of 100 to 200 diameters. The cutting facets should be polished until they show none but very shallow scratches.

The Hone and Strop Method of Sharpening. This consists of grinding the knife lightly upon one or more hones, to remove nicks from its edge, and then polishing the edge upon a leather strop. This method is fairly rapid and, when carried out properly, is adequate for all ordinary purposes. It requires a minimum of equipment, but this must be selected with care. Another technique of honing, the abrasive-plate glass method, is preferred by some workers, and will be discussed subsequently.

Hones. A fine-grained yellow Belgian hone is needed for the first grinding. In the hands of a skilled worker it may be made to serve for the entire process of honing. It is advisable to have also a very fine, smooth blue-green or Arkansas stone upon which to complete the grinding. In selecting these hones, examine carefully the grade and uniformity of their texture. A hone must be large enough so that the entire length of the blade may pass over it when the knife is moved in long, sweeping, diagonal strokes. For knives up to 185 mm. in length, hones $10 \times 2\frac{1}{2}$ inches are satisfactory. Avoid small hones, very thin ones, or any which have been spliced together.

In use, the center of a hone becomes worn down more rapidly than the ends. Eventually the surface of the hone becomes so concave that it will no longer grind a straight bevel upon the knife. Whenever the surface appears to be markedly concave, rub it upon a flat piece of glass (preferably plate glass) which is flooded with water and sprinkled plentifully with fine emery powder, until it is again flat. After this treatment, wash every trace of emery powder from the hone.

Strops. It is advisable to have two strops, both of which should be of fine calfskin, mounted *smoothly* upon wood. They may be mounted upon separate strips of wood, or upon the opposite sides of the same strip. Strops mounted upon sponge rubber cannot be recommended. The softness of this substrate allows the strop to bend under the blade and then to spring back across the cutting edge. This removes more metal from the edge than from the center of the cutting facet, and consequently produces a blunt edge. The strops may advantageously be 18 to 20 inches long. The surface of one strop should be impregnated with fine jeweler's "rouge" (ferric oxide). A rouged strop may be purchased, or it may be prepared by applying a thin film of light mineral oil to any strop and then rubbing in a light dust of fine rouge. The second strop requires no special treatment. Both should be kept *scrupulously clean.*

METHOD OF HONING. 1. Slip the honing back upon the knife and screw the handle into place. Make a scratch upon the end of the back which comes nearest the handle, so that the back can always be replaced in the same position. Make certain that the knife edge is straight, or practically so. Refer to Straightening the Blade (p. 205) if this is not the case.

2. Wash the surface of the yellow Belgian hone and lay the stone upon the table or sink draining board, with one end toward you. Flood the hone with distilled or filtered water and keep a beaker of such water at hand.

3. Take the handle of the knife in the right hand, holding it between the distal joints of the thumb and fingers. The end of the knife to which the handle is attached is called the "heel," its opposite end the "toe." Hold the toe of the knife very lightly between the tips of the thumb and first finger of the left hand, allowing the back of the knife to rest lightly against the second finger. These points are illustrated by Figure 24.

4. Place the knife upon the hone, at the end nearest to you, with the heel lying at the right-hand side of the stone and the edge away from you. This position is shown at 1 in Figure 25. Proceed to hone the blade *from heel to toe,* following the motions illustrated and described in the figure to which reference has been made. During the process of grinding, the knife should rest upon the hone *by its own weight.* Do not exert pres-

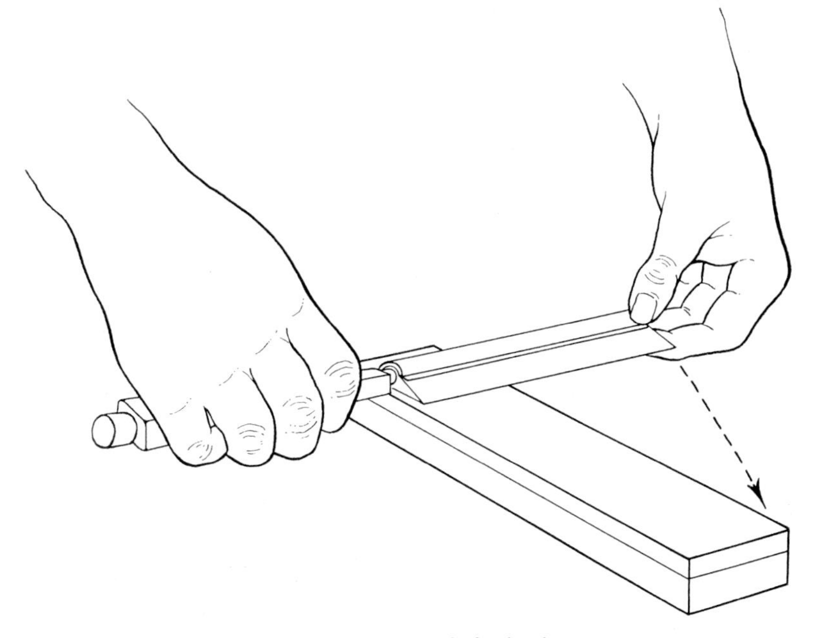

FIG. 24. Method of holding knife for honing movements.

sure upon it, as this will cause the edge to become deeply scratched or blunted. Add more water occasionally to keep the hone well moistened. At intervals of about 5 minutes, rinse off the hone in order to remove from it particles of steel which have been ground from the knife. After a little practice, the movements illustrated in Figure 25 will become blended into a continuous motion following the shape of a broad figure eight.

Grind the blade from heel to toe in the above manner, until all nicks visible to the unaided eye have been removed. Then continue the process, stopping every few minutes to observe the cutting bevels under the compound microscope at a magnification of about 100×, until the entire edge appears nearly smooth. Under higher magnification the edge will show small, evenly distributed irregularities, and the bevels will be seen to bear uniformly spaced, parallel scratches. The length of time required to obtain this result cannot be stated even approximately, as it depends upon the hardness of the steel, the condition of the knife's edge, the size and grit of the hone, and the rapidity of the stroke. A slightly dull knife can be brought to this condition within 10 or 15 minutes, while a badly nicked blade may require an hour or two of grinding. When the result is satisfactory, wash the knife and its bevel.

5. Resume grinding in a similar manner, but *reverse the motions to right and left, so as to grind the blade from toe to heel.* Continue this motion until microscopic examination shows the shallow scratches on the bevels to be *uniformly crossed.* This condition is usually reached within 10 or 15 minutes. While doing this, be careful to put no pressure whatever upon the knife. During the last few dozen strokes, lighten the weight upon the knife edge by exerting a slight upward tension upon the corresponding side of the handle. Then wash the blade thoroughly. This process produces a smoother edge than is obtained by the more usual process of honing in one direction only. If the knife is to be used for cutting sections 8 μ or more in thickness, it is permissible to strop it (step 7) without further honing.

6. To obtain a still smoother edge, which is necessary for cutting perfect sections less than 8 μ thick, continue the grinding upon the blue-green or Arkansas hone. Use the motions previously described, and keep the stone well flooded with grit-free water. Hone the knife from heel to toe until examination under the microscope shows the scratches of the Belgian hone to have been replaced by the finer scratches of the blue-green hone. Then reverse the motion and hone from toe to heel until the remaining fine scratches are uniformly crossed.

Honing on Plate Glass. It is possible to grind a knife quite well on a piece of plate glass to which an abrasive powder has been applied. Some workers prefer this method. The plate glass should be ¼ or ⅜ of an inch thick, about 16 or 18 inches long, and at least an inch or two wider than the length of the knife blade. The edges should be beveled slightly to

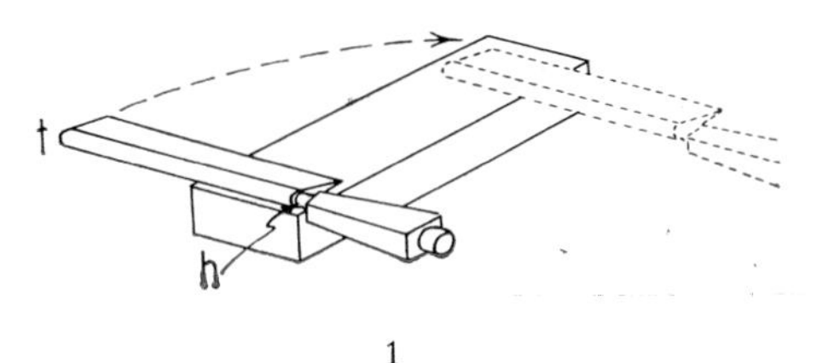

1

1. Push knife full length of hone, in a diagonal direction from heel (*h*) to toe (*t*).

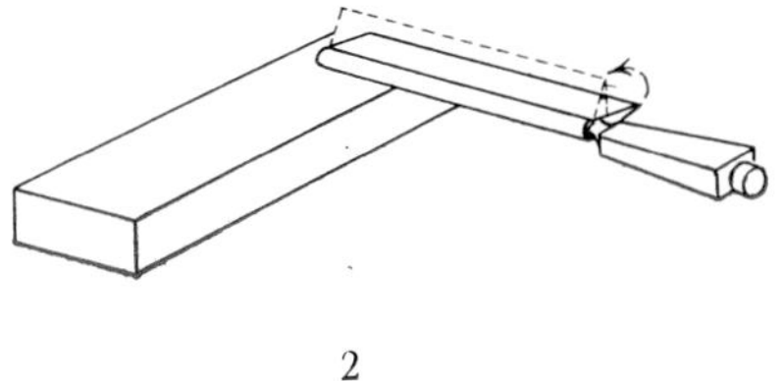

2

2. Rotate knife at least 45°, so as to lift edge from hone while back rests on hone.

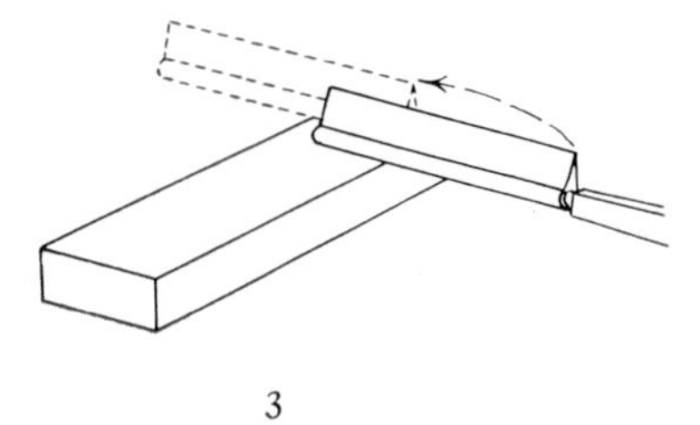

3

3. Swing knife across hone until heel just passes edge of hone and back of knife lies close to end of hone.

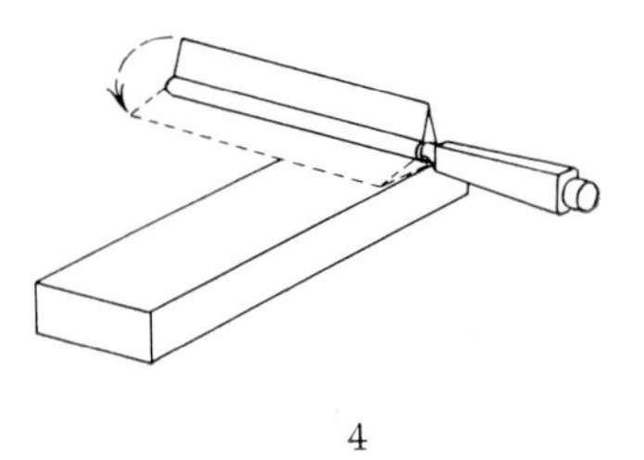

4

4. Continue to rotate knife until blade lies flat on hone, edge toward operator.

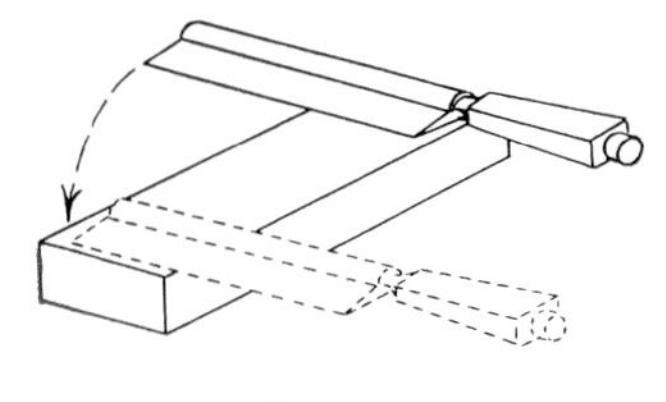

5

5. Draw knife full length of hone in a diagonal direction from heel to toe.

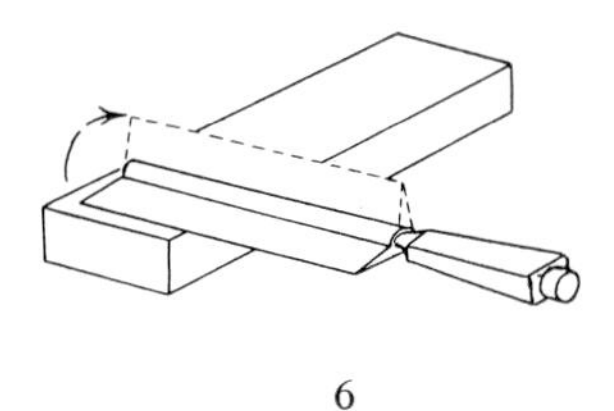

6

6. Rotate knife at least 45°, so as to lift edge from hone while back rests on hone.

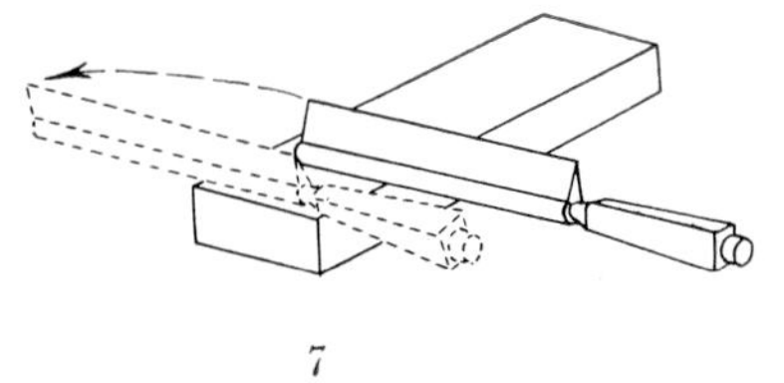

7

7. Swing knife across hone until heel just passes edge of hone and back of knife lies close to edge of hone.

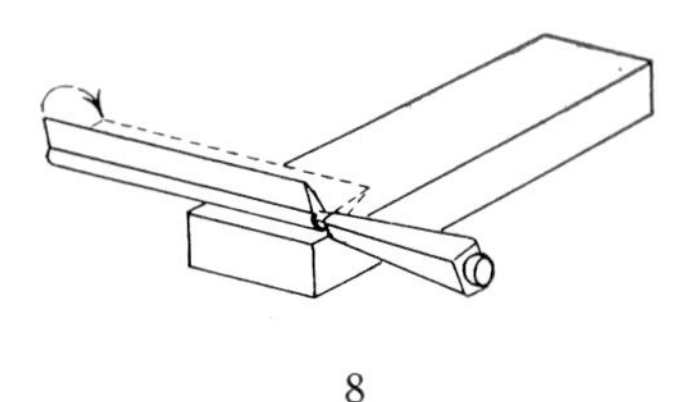

8

8. Continue to rotate knife until it lies flat on hone and is in position to repeat movement 1.

FIG. 25. The series of movements used in honing a microtome knife.

protect the operator from being cut. Sliding of the glass and possible accidents will be prevented if the glass is mounted in a wooden tray; its upper surface should be slightly higher than the sides of the tray.

A new plate of glass will not work properly until it has been prepared by applying a rather fine abrasive as Carborundum (silicon carbide) 1000 mesh and rubbing it vigorously against the plate with a smaller piece of flat glass.

After adding a suspension of the abrasive in a dilute solution (about 0.1 to 0.5%) of a neutral soap, the operator may proceed to sharpen the knife in much the same way as on a standard hone. However, the knife is pushed and pulled straight, for nearly the full length of the plate.

If the knife is neither badly nicked nor dull, a very fine grit (as levigated alumina powder) will be suitable. However, if the knife needs a more drastic reconditioning, use Carborundum 1000 mesh, then follow this with levigated alumina powder. It is important to wash off the plate glass occasionally and then apply a fresh suspension of the abrasive, in order to remove small pieces of steel ground from the knife. Use very little abrasive, especially toward the end of the sharpening process. Do not press the knife too hard against the glass, as this may round the edge instead of sharpening it.

Because of the fact that the knife is pushed and pulled in a more or less straight course over the glass plate, the fine lines formed by abrasive

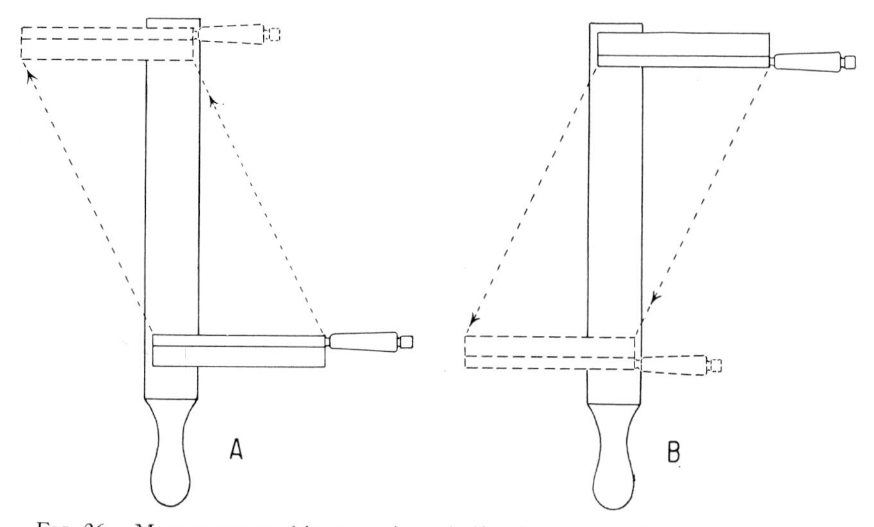

FIG. 26. Movements used in stropping a knife. Inexperienced workers should make the stroke shown in *A*, rotate the knife just enough to lift its cutting edge from the strop, bring it back, let down the edge, and repeat the stroke several times. The knife should then be turned over, placed at the other end of the strop, and several strokes made as shown in *B*. The two phases of the process should be repeated several times. Experienced workers may rotate the knife and swing it across the strop at each end of the stroke, and thus strop it with an easy, continuous movement.

particles on the cutting bevels will of course be nearly at right angles to the edge, instead of diagonal to it.

After sharpening a knife on a glass plate, it is advisable to polish the cutting bevels on a leather strop, in the usual way. However, some workers polish on the glass plate, using a very thin suspension of an exceedingly fine abrasive, as an alumina polishing powder in which the particle size is about 0.1 μ.

METHOD OF STROPPING. 7. Polish the bevels of the knife first upon a clean rouged strop. In stropping, the back of the knife precedes the edge. The diagonal stroke, covering the entire length of the blade, is from toe to heel (Fig. 26*A*). At the end of each stroke, rotate the blade sufficiently to lift its edge from the strop. Pull it back to the starting point, and repeat the stroke (Fig. 26*B*). In this manner, make a dozen or more *light*, rapid strokes on one side of the knife. Then turn the blade over and make about the same number of strokes on the other side, beginning naturally at the opposite end of the strop. Repeat these operations for about 10 minutes. The most important fact about stropping is to make many *light* strokes. A heavy stroke will bend the surface of the strop and blunt the knife edge. A skillful operator can rotate the knife and swing it across the strop at the end of each stroke, as in honing, and so make alternate strokes on both sides of the blade. It is not advisable to try this until you have some practice in stropping, since it may result in damage to the strop.

8. Complete the polishing process by *light* strokes on the plain strop, which must be *scrupulously clean*. About 50 strokes upon each side of the knife will generally be sufficient. Finally, remove the back and handle from the knife and put them in their proper places in the box.

STRAIGHTENING THE BLADE. Remember that the knife edge must be straight. It, like the hone, becomes "sway-backed" after repeated sharpening operations. Whenever curvature of the edge becomes marked, grind the knife, with its back in place, on plate glass which is flooded with water and sprinkled with moderately fine abrasive (as Corundum 303 or Carborundum 1000 mesh). After the edge has been straightened, wash the blade thoroughly and proceed with the usual method of sharpening and polishing.

MICROTOME KNIFE SHARPENERS. Several designs of mechanical knife sharpeners, some of them to a large extent automatic, have been brought to a high level of perfection. Two distinctly different types are illustrated. In one of these (Fig. 27), the knife is stroked by a vibrating ground glass plate impregnated with a rather coarse abrasive for honing and a fine powder for polishing of the cutting facets. The holder turns the knife over after a few strokes, and a timing device (pre-set according to an estimate of the time likely to be required for honing and polishing of a particular knife) stops the operation when the cycle is completed.

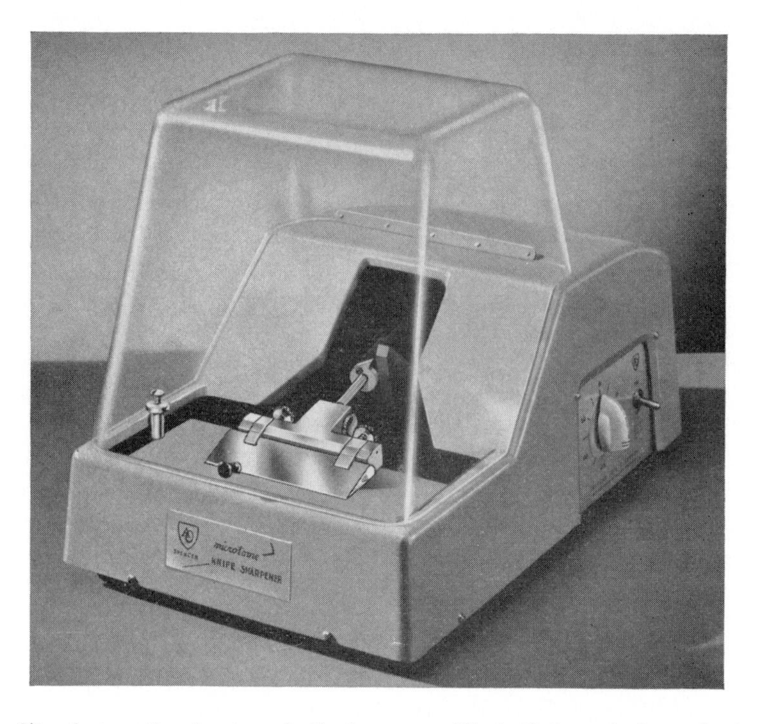

Fig. 27. Automatic microtome knife sharpener. The knife is stroked at a fixed angle against a vibrating honing plate to which an abrasive powder is applied. (Courtesy of American Optical Corp., Buffalo, N.Y.)

In another type (Fig. 28), the action comes a little closer to simulating the process of hand honing. The wet hone moves as the cutting facet of the knife is pressed against it, and at the end of each stroke the knife is turned over. No abrasive is used with the hone. This machine also has a built-in timing device.

Mechanical sharpeners are expensive, but in biological and clinical laboratories in which knives are used a very great deal, they are likely to be a source of economy. However, the operator must still be able to distinguish a knife which has been properly sharpened from one which has merely been processed according to the instructions supplied with the machine.

Professional Microtome Knife Reconditioning. A number of manufacturers and distributors of microtome equipment provide a knife-sharpening service which is very satisfactory. The reconditioning is almost always carried out on automatic grinding and honing machines. The cost is usually proportional to the length of the blade. If, for any reason, satisfactory sharpening cannot be carried out in the laboratory, the use of a professional reconditioning service will, in the long run, be worth the

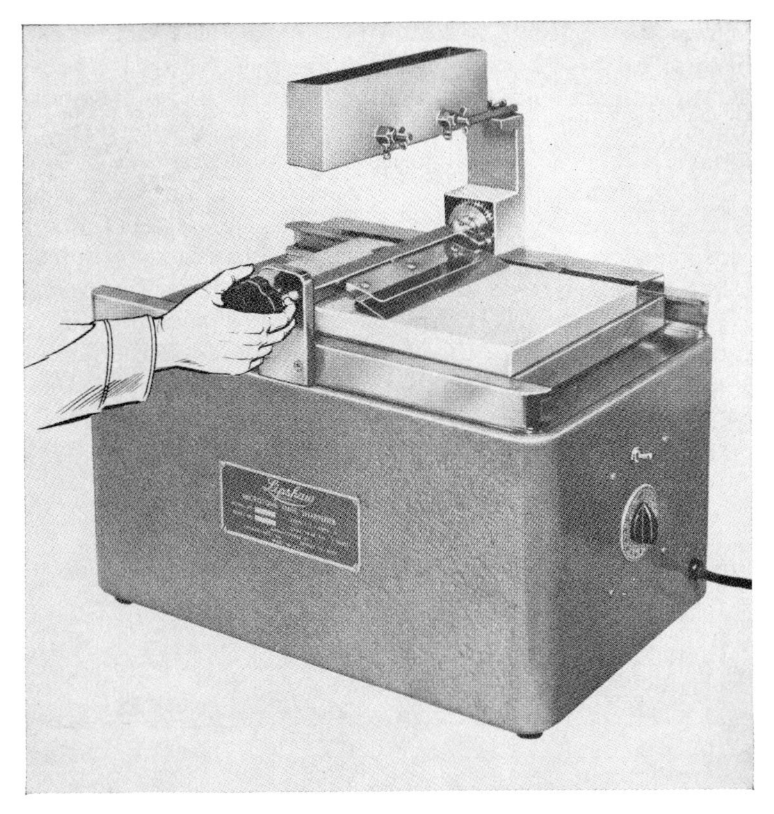

FIG. 28. Automatic microtome knife sharpener. The knife is held with slight pressure against a hone which moves back and forth. (Courtesy of Lipshaw Manufacturing Co., Detroit.)

expense, because routine substitution of razor blades for knives is inadvisable.

CARE OF MICROTOME KNIVES. (1) When a knife is not in use, keep it in the padded box provided by its manufacturer. (2) If a knife is to be stored for a long time or kept in an extremely damp climate, smear it lightly with petroleum jelly or paraffin oil in order to prevent it from rusting. Before using a knife which has been so treated, wash it with xylene or toluene in order to remove all traces of the grease. (3) Handle the knife with great care, to avoid cutting yourself or damaging the delicate knife-edge. (4) Use an old knife, disposable knife, or razor blade in a holder to cut blocks of tissue believed to contain grit or areas of calcification.

ORIENTATION OF SECTIONS

Orientation is concerned with the placing of an object so that its parts have a definite relation to the plane of section. This matter deserves care and consideration. If one wishes to investigate the structure of an organism or a part of it by studying sections, then obviously the sections must be cut with reference to its length, width, thickness, or some part of its surface. Only this makes it possible to deduce from sections the correct form of the structures and the positions which they occupied in the entire specimen.

It is usually necessary to section examples of every object in more than one plane, for a structure may have altogether different appearances when cut in different planes, and can be understood as a three-dimensional object only after it is viewed from several angles. For example, a tubular structure appears in cross section as a little ring; in longitudinal section, as two parallel lines. Its correct form is obvious when both aspects are seen and compared.

Embryos, small organisms, or entire body regions of such objects are

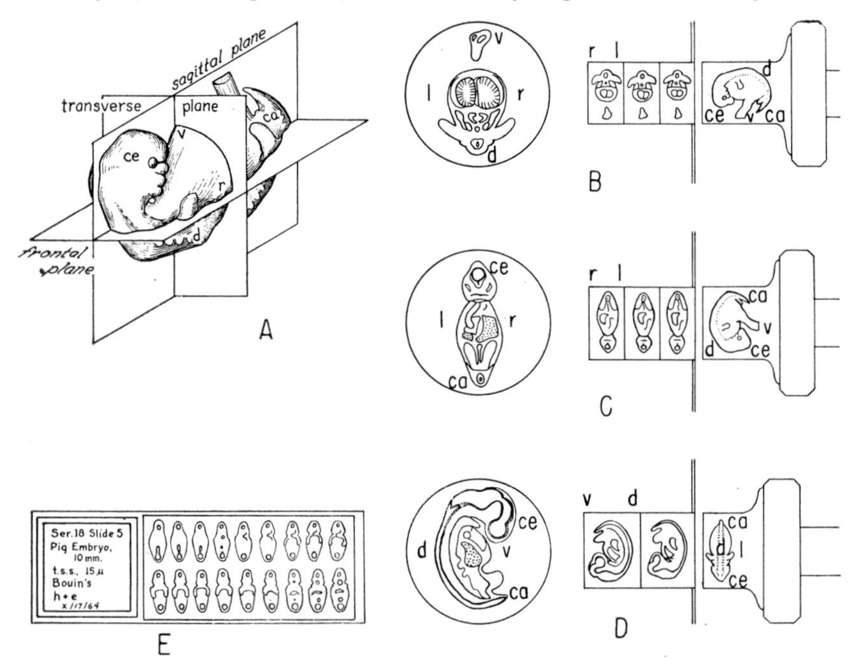

FIG. 29. Planes of sectioning as applied to a 10 mm. pig embryo, and arrangement of sections on a slide. *ca*, Caudal, *ce*, cephalic, *d*, dorsal, *v*, ventral, *l*, left, *r*, right. *A*, Plane of sectioning. Owing to the curvature of the pig embryo, the transverse and frontal planes do not apply in the same way as they would to a more nearly straight animal, as a tadpole.) *B*, Embryo embedded for transverse sectioning, and representative sections as they will come from the knife and as they will appear in the field of view of a compound microscope. Note that the image is inverted. *C*, Frontal sections. *D*, Sagittal sections. *E*, Correct mounting and labeling of serial sections. The paraffin block was closely trimmed so that space has been used economically. The rows of sections are straight and in proper order; the coverglass overlaps the sections a short distance on all sides; the rows of sections end far enough from the right hand end of the slide so that the edge of the coverglass will clear the slot in a slide box.

generally sectioned parallel to all of the three fundamental planes of symmetry. It will be recalled that these planes are *transverse* (across the longitudinal axis), *sagittal* (median longitudinal dorsiventral plane), and *frontal* (longitudinal dextrosinistral plane). These planes, as they apply to a pig embryo, are illustrated in Figure 29*A*. Pieces of tissue removed from large animals are ordinarily cut with reference to the length of the organ they represent. For example, *longitudinal* and *transverse* sections are ordinarily made in the case of nerves, blood vessels, and parts of the digestive and urinary tracts. Skin, other epithelia, and connective tissues are generally sectioned with respect to the surface of the structure, therefore *vertically* or *tangentially*.

It is easier to understand the correct location of the various parts represented in sections if the sections are placed upon the slide in such a position that, when they are viewed *through the microscope*, the various parts appear in realistic positions. This point is illustrated in Figure 29*B*, *C*, and *D*. At *B* we see a pig embryo which is to be cut transversely. It is mounted on the object carrier with its head toward the knife and its right side down. When it is cut, the right side of the embryo will be toward the left of the section and the dorsal side will be away from the observer. Under the microscope these relations appear reversed. Comparable results are obtained in the case of frontal (*C*) and sagittal (*D*) sections by orienting the embryo as shown. These principles can easily be applied to objects of various types.

Serial sections are arranged in the order of lines of printing. This is illustrated in Figure 29*E*, and explained further in connection with the paraffin method. When several representative sections, each from a different region, are to be mounted on a slide, arrange them so as to suggest their natural order.

The first step in orientation is to place the specimen in the embedding medium so that it can easily be mounted as desired in the microtome. When embedding is done in a dish or paper box, that surface of the object which is parallel to the intended plane of sectioning is ordinarily placed next to the bottom of the container. Thin, flat objects which will not stand upon their sides or ends should be allowed to lie flat. When embedding in nitrocellulose is done on an object holder the specimen must at that time be arranged in its final position.

REFERENCES

GIOVACCHINI, R. P., 1958. Affixing Carbowax sections to slides for routine staining. Stain Technology, *33*, 247–248.

JONES, R. M., THOMAS, W. A., and O'NEAL, R. M., 1959. Embedding of tissues in Carbowax. Tech. Bull., Registry of Med. Technol., *29*, 49–52.

RIOPEL, J. L., and SPURR, A. R., 1962. Carbowax for embedding and serial sectioning of botanical material. Stain Technology, *37*, 357–362.

WADE, W. H., 1952. Notes on the Carbowax method of making tissue sections. Stain Technology, *27*, 71–79.

CHAPTER 14

Methods of Embedding in Paraffin

Paraffin was apparently first used by Klebs, in 1869, to form a supporting coat around tissue specimens which were to be sectioned. Two years later, Born and Strickler devised a method for saturating tissues with paraffin, in order to support their structures both externally and internally. This method consisted of six important stages, and these have survived to the present time as the basis of the paraffin method. They are: (1) fixation; (2) washing; (3) treating the tissue with alcohol for a sufficient time to insure the replacement of all water by alcohol (*dehydration*); (4) replacement of the alcohol by an oil capable of dissolving paraffin; (5) saturating the tissue with melted paraffin (*infiltration*); (6) rapidly cooling the tissue and surrounding paraffin to form a solid block (*embedding*). In 1881, Giesbrecht and Bütschli refined the process considerably and suggested that chloroform be used as the solvent for paraffin, in place of an oil. During the next 50 years other solvents were introduced and the paraffin method was modified and improved in various ways, but basically it remained unchanged.

A number of methods have been worked out for dehydrating tissues in a solvent that is miscible with water or ethyl alcohol on one hand, and with paraffin on the other. These methods obviate the need for replacement of the alcohol used for dehydration with another solvent, such as toluene, that is miscible with paraffin. Dioxane and tertiary butyl alcohol are the more popular of the solvents used for both dehydration and infiltration. The fumes of dioxane, unfortunately, are poisonous, at least to certain persons or under certain circumstances. It should be used with great care for this reason, and because it accumulates peroxides which may render old bottles of it dangerously explosive.

Advantages and Disadvantages of Embedding in Paraffin. Paraffin and nitrocellulose are the two embedding media in general use. Each is particularly adapted to certain kinds of tissue or types of work, but the paraffin method offers the following advantages: (1) It is relatively rapid. (2) Paraffin, being nearly inflexible, is easily cut into very thin sections. These can be made at 2 μ when necessary, but for ordinary purposes sections are cut at thicknesses of 5 to 15 μ. (3) Paraffin sections can be made to adhere to one another, forming ribbons of *serial sections*, from which the relationships of various parts may be reconstructed.

Paraffin embedding is, however, unsuitable for some materials or purposes to which nitrocellulose is well adapted. These include: (1) very hard tissues, including bones, teeth, and the lens of the eye; (2) very large objects, such as entire brains; (3) tissues injected with gelatin; (4) very watery tissues, which would be shrunken by hot paraffin; (5) materials to be cut in thick sections (over 20 μ).

Because of the rapidity and convenience of the paraffin method, most workers use it whenever possible, employing nitrocellulose only for objects such as those listed in the preceding paragraph. A thorough understanding of these methods, and considerable skill in their execution, are necessary for efficient microscopical work in many fields of biological science.

THE CONVENTIONAL PARAFFIN METHOD

1. FIXATION. Fix the specimens in a reagent which will suitably preserve the structures which are to be studied. A beginner working with vertebrate tissues would do well to try either Bouin's fluid or Zenker's fluid. Contractile organisms should first be anesthetized in an appropriate manner. Bulky tissues should be cut into thin slices, with proper reference to the desired plane of sectioning, and it may be a good idea to inject them first with the fixing agent. Allow the fixative to act for the proper length of time.

2. WASHING. Wash the material in the manner appropriate to the method of fixation employed. To summarize, transfer tissue from Bouin's fluid, other picric acid mixtures, or mixtures containing mercuric chloride (as Heidenhain's "Susa" fixative) to 50% alcohol. After one to several hours, replace this with 70% alcohol, and change the latter 2 or 3 times, at intervals of 6 hours to several days. Transfer formalin-fixed material directly to the lowest grade in the alcohol series, and proceed with dehydration (step 3). From fixatives which include large amounts of alcohol (Carnoy's fluid, or Carnoy and Lebrun's fluid), place tissues in 95% alcohol. After fixatives containing chromic acid (Flemming's fluids, Karpechenko's fluid), potassium dichromate (Zenker's fluid, Helly's fluid), or osmium tetroxide, wash material in running water for 12 to 24 hours. Obviously, these statements are general and do not apply to certain special methods.

Special Treatments Necessary in Certain Cases. Consider the following possibilities. If one of them applies to the tissue in hand, proceed as instructed; otherwise go on to step 3.

a. If the fixative contained mercuric chloride, treat the material with iodine (p. 93) while it is in 70% alcohol, or note upon the label and in other records that it has not been iodized, so that treatment with iodine will be carried out after sections have been deparaffined.

b. Calcium deposits, if present, must be removed by an acid. In this case, proceed with dehydration (step 3) up to 70% alcohol and then follow one of the methods given on pages 146 and 147. The use of an acid fixing solution (such as Bouin's fluid) ordinarily softens tissues which are only slightly calcified, and they seldom require further decalcification.

c. If it is desired to stain the material before sectioning it, proceed according to the method chosen. A discussion of the advantages of staining tissues before embedding or after sectioning will be found on page 328. Embryos being prepared for routine work are often stained before embedding with alum cochineal (p. 344), alum hematoxylin (p. 357), or Weigert's iron hematoxylin method (p. 367). The latter two stains are also suitable for pieces of tissue in case staining before embedding is an acceptable procedure. Certain methods of metallic impregnation used for tissues of the nervous system and sometimes for other tissues are regularly carried out before embedding.

3. DEHYDRATION. The next step is to *dehydrate* the tissue, by replacing all of the water it contains with alcohol. Ethyl alcohol is generally used, but methyl alcohol or isopropyl alcohol will serve equally well in the paraffin method.

There are two essential facts to remember concerning dehydration. It must be *gradual* and it must be *thorough*. There is a great difference in density between water (specific gravity 1.00) and absolute ethyl alcohol (specific gravity 0.79). If material is transferred from water or weak alcohol directly to strong alcohol, violent diffusion will occur, causing shrinkage and distortion of the tissues. In order to bring about dehydration without damaging the delicate tissue structures, material is carried through a series of alcohols, each member of the series being a little stronger than the preceding one. For all but the most delicate tissues, a series consisting of 15, 30, 50, 70, 80, 95, and 100% alcohols will certainly be satisfactory. Many workers omit 15 and 80% alcohols from the series when processing routine histological materials. If one wishes to study details of cell structure or cell division in delicate tissues such as ovaries, testes, young embryos, or larvae, it is advisable to employ a more closely-spaced series of alcohols and other reagents subsequently used preparatory to embedding. In this connection, refer to the section on Slow Method for Delicate Objects, page 224.

Directions for preparing dilutions of alcohol are given on page 22.

Dehydration must be complete because the presence of water in the tissues will interfere with the penetration of the de-alcoholizing agent and consequently of the paraffin.

Containers for Material. Throughout the processes of dehydration, de-alcoholization, and infiltration, keep the material in wide-mouthed glass receptacles with air-tight covers. Corked bottles, shell vials, or stender dishes will do, but screw-capped jars of ½ to 2 ounce capacity are unequaled for economy, convenience in handling, and effectiveness in preventing the entrance of moisture from the air. However, squat shell vials (without necks) are perhaps more suitable for extremely small objects such as eggs and larvae of invertebrates.

Quantity of Alcohol to Use. Always add 2 or 3 times as much liquid as is needed to cover the material. It is not advisable to use the alcohol over again, except when many specimens are being handled by the same method. The traces of fixing agents absorbed by the alcohol might prove injurious to materials being prepared by different methods. Used alcohol can be purified by re-distillation, if facilities are available. "Waste" 95% and absolute alcohols may profitably be saved for use in alcohol lamps or for cleaning used slides and bottles.

System for Arranging Reagent Bottles and Containers of Material. A system is described in Chapter 3 for arranging bottles of alcohols and other reagents upon a paper which is ruled so as to provide a vacant square in front of each bottle (see Fig. 1, p. 22). Jars or dishes of material are placed upon the square in front of the bottle from which they have been filled, and are moved to the corresponding square each time a different reagent is put into them. If care is taken to avoid misplacing the containers, this system enables one to keep track of numerous materials. If this scheme is not followed it will be necessary to note upon the label each change of reagent.

The Length of Time Which Objects Must Remain in Each Grade of Alcohol. This depends upon their size and permeability. The same is true of subsequent steps, so the following classification of objects is made in order to avoid repetition.

Class A. Small and permeable objects: protozoa; hydra; planaria; chick embryos up to the 48-hour stage. Objects of this class should remain for at least 30 minutes in each grade, up to and including 95% alcohol.

Class B. Objects of moderate size: slices of vertebrate tissue about 4 mm. thick; 10 mm. lengths of earthworms; mammalian embryos up to 12 mm.; 3 to 5 day chick embryos; amphibian eggs and small larvae. Such objects should remain for at least 1 hour in each grade up to and including 95% alcohol.

Class C. Large, dense, or impermeable objects: 5 × 5 × 10 mm. pieces of liver, kidney, prostate gland, or other dense tissue; mammalian embryos of 15 mm. or over; portions of large nematodes (such as *Parascaris*); insect and other arthropod materials. Objects of this class should remain for at least 3 or 4 hours in each grade up to and including 95% alcohol.

The foregoing statements refer to *minimum* periods of time. If there is a large amount of material in the container, allow half again as much time, and change the alcohol once, because diffusion of water from the tissue may appreciably lower the concentration of alcohol. If it is more convenient, material may safely be left for days in 70, 80 or 95% alcohol. It is not advisable to leave specimens for more than 6 hours in alcohol weaker than 50%, since the tissues may become dissociated to some extent.

Procedure. If the material was washed in water,[1] place it first in the lowest grade in the alcohol series (ordinarily 15% alcohol). If it has been washed or preserved in alcohol (usually 70 or 80%), dehydration has already begun, and is simply continued by transferring the material to the next higher grade of alcohol. Observe the foregoing instructions as to containers and quantity of liquid to use.

Replace the alcohol at proper intervals with the next higher grade in the series (in other words, replace 30% alcohol with 50% alcohol, and so on) until the material is finally in absolute alcohol. If the objects are of considerable size, the alcohol may be changed by simply pouring off one grade and pouring in the next. Some histologists tie pieces of tissue in little bags of cotton gauze, attach a tag to each, and move these successively to jars that contain the alcohols, de-alcoholizing agent, and paraffin.

If the specimens are very small and delicate, carefully withdraw the alcohol from them with a pipette and pour the next grade slowly down inside of the jar or vial. Small specimens must be allowed to settle completely before making a change of fluid, and care must be taken not to stir them up while pipetting off the liquid.

It is well to tint small and colorless objects by adding a small amount of eosin or safranin (to make a solution of about 0.1 to 0.2%) to the 95% alcohol. This will stain the specimens and make it easier to orient them later in melted paraffin and on the microtome object holder. The stain does no harm and is easily removed.

Change the absolute alcohol once or twice, depending upon the quantity and nature of the material. *It is essential that the material remain in absolute alcohol until every trace of water has been removed from it.* A few extra hours in absolute alcohol will do no harm, but insufficient exposure may lead to failure. To insure complete dehydration, leave small and permeable objects (class A) in absolute alcohol for at least 1 hour; slightly larger or denser ones (class B) for at least 2 or 3 hours; and those of greater size and density (class C) for 8 or 10 hours. During the course of this time, the absolute alcohol must be changed. If the material is readily permeable, and in small pieces, one change will be sufficient. However, if the objects are large or dense, or if there is a considerable quantity of material in the container, two or three changes should be made so that complete dehydration will be accomplished with certainty.

[1] This may have been done either to remove the fixing agent, or to follow staining.

4. DE-ALCOHOLIZATION ("CLEARING"). All of the free water originally present in the tissue has now been replaced by alcohol. However, it is not yet possible to introduce paraffin into the material, since paraffin and ethyl or methyl alcohol will not mix. Therefore, the next step is to replace the alcohol with a liquid that will dissolve paraffin. The paraffin solvents commonly employed are: (1) hydrocarbons of the benzene series (toluene, xylene, benzene); (2) essential oils (cedarwood oil, wintergreen oil, bergamot oil);[2] (3) other organic solvents (such as chloroform, tertiary butyl alcohol, and isobutyl alcohol). These substances differ in the extent to which they harden tissues, in their effectiveness as paraffin solvents, and in their ability to take up traces of water. *Toluene is recommended as a satisfactory de-alcoholizing agent for general use.* Terpineol-toluene, made by mixing 3 parts of toluene with 1 part of terpineol, is perhaps superior for some purposes. This reagent will absorb any traces of water which may remain in the tissues even after careful dehydration, and it does not harden materials excessively. Tissues which have been permeated by toluene, terpineol-toluene, or similar solvents appear more or less translucent, and for this reason the introduction of such a liquid is commonly referred to as "clearing." This phenomenon occurs because the solvents have high indices of light refraction, as do the tissues. In consequence, light rays pass more nearly straight through solvent and tissue, with little reflection at the surface of the latter. The clearing effect is purely incidental, but it serves as a useful indicator that the essential process of de-alcoholization is proceeding to completion.

The de-alcoholizing agent, like the alcohol, must be introduced *gradually.* However, the steps need not be so numerous, since alcohol has now hardened the tissues sufficiently so that they withstand greater changes of density without suffering damage. By using a mixture of absolute alcohol and a de-alcoholizing agent, any traces of water which might possibly remain in the tissue can be removed at the same time the de-alcoholizing agent is introduced gradually. The following method is sufficiently gentle for all ordinary purposes. In the case of extremely delicate cytological material, follow the method given on page 225.

Tightly covered receptacles must be used for material in mixtures of alcohol and de-alcoholizing agents, as absolute alcohol rapidly absorbs moisture from the air. The mixtures can be re-used several times, if they are kept in tightly stoppered bottles. Upon showing the slightest cloudiness, they must be discarded.

Procedure. Transfer the material from absolute alcohol to a mixture of 1 part toluene (or terpineol-toluene or some other "clearing" agent) and

[2] Essential oils are volatile, odoriferous oily substances, obtained almost exclusively from plants, although a few of them are produced synthetically. They include many types of organic compounds and are chemically unrelated to true oils and fats.

3 parts of absolute alcohol. If the specimens are small and permeable (class A), 30 minutes in this mixture should be sufficient; if they are moderately thick and dense (class B), leave them for an hour or two; if they are quite large and very dense (class C), leave them for 3 hours or more. After an appropriate length of time, replace this mixture with one consisting of equal parts of toluene and absolute alcohol, leaving them in this for the same length of time that they were immersed in the first mixture. Follow this with a mixture of 3 parts of toluene and 1 part of absolute alcohol.

Finally, replace the last mixture with pure toluene or whatever de-alcoholizing agent is being used. After a few minutes, gently agitate the liquid. The penetration of toluene and similar reagents is indicated by the clearing effect which accompanies it. Tissues of some types become very translucent, while those containing much pigment do not. In examining tissues to ascertain the progress of clearing, hold the receptacle in a bright light, but against a dark background. Areas that are not penetrated will then appear definitely white and opaque. If the material is held against the light, these areas will appear dark. After a little experience, one learns to recognize the appearance of partially and fully cleared tissues.

When the material appears to be cleared, or nearly so, draw off the toluene (or other clearing agent) and replace it with new. Allow the tissue to stand in the fresh fluid until no doubt exists that it is completely de-alcoholized. As a general guide, it may be said that objects of class A should remain in toluene or similar clearing agents for 1 or 2 hours, objects of class B ordinarily require 3 or 4 hours, and those of class C may require 6 hours or longer.

If the tissue is not completely de-alcoholized, paraffin will fail to infiltrate the regions which still contain alcohol, and these will be certain to tear out of the sections as they are cut. Should the tissue continue to show definitely white or opaque spots after it has remained for many hours in the clearing agent, it is probable that dehydration was incomplete. In this case, return it to absolute alcohol for several hours, and then repeat the process of de-alcoholization.

5. Infiltration with Paraffin. The tissue is now ready for the introduction of paraffin. Before explaining the way in which this is carried out, it will be necessary to consider some important facts about paraffin and the apparatus employed.

Types of Embedding Paraffins and Their Uses. Paraffins packaged expressly for embedding fall in two general classes: those that are essentially mixtures of pure paraffins, and those adulterated with plastic polymers which confer upon them a more cohesive quality when sectioned. These latter preparations are probably more widely used today than pure paraffins. A few good ones, readily available from dealers in scientific supplies, are: Embeddol (Hartman-Leddon Company, Philadelphia), Histowax (Matheson, Coleman and Bell, Norwood, Ohio), Paraplast (Sherwood Medical Industries, Inc., St. Louis, Mo.), and Tissuemat (Fisher Scientific Co., Fair Lawn, N.J.).

A more or less pure paraffin, and even the paraffin-plastic mixtures just referred to, can often be improved by adding a synthetic resinous mounting medium (p. 452) to the extent of about 5% of the total weight. Piccolyte (CCM: General Biological, Inc., Chicago) is known to be good for this purpose, and can be purchased in the form of dry lumps. Simply grind up some of the lumps with a mortar and pestle, add the powder to the paraffin, and melt the two together in a paraffin oven. Occasional stirring will be necessary to make the mixture homogeneous.

It may be desirable to have on hand paraffins of two different melting points, which may be designated as *soft* or *hard* paraffins. In most laboratories, however, hard paraffin is apt to be used much more extensively, if not exclusively. Most of the paraffin-plastic mixtures which are now popular for embedding are available only as relatively hard paraffins.

Hard paraffin is employed for materials which are to be cut into relatively thin sections (12 μ or less). It affords maximum support to the tissues, and provides sections which hold their shape well. Use of hard paraffin also facilitates the handling of thin sections, as it reduces the likelihood that they will stick to the knife, ribbon tray, or other objects. In moderate and cold climates, a hard paraffin, melting at about 56° C., is generally satisfactory throughout the year. Paraffin of a still higher melting point (58° to 60° C.) is required if hot weather affects the temperature in the laboratory.

Soft paraffin is used for embedding materials which are to be cut into rather thick sections (14 μ and over), especially when complete series of sections are to be made. Thick sections of soft paraffin adhere firmly to each other, and do not roll or crack, whereas thick sections of hard paraffin may separate, curl, and crack. A soft paraffin, melting at about 45° C., is satisfactory the year around in moderate and cool climates. In very hot weather, paraffin of a higher melting point (about 50° C.) may be necessary.

Paraffin Baths and Ovens. Many types of apparatus have been devised for keeping paraffin at a constant temperature just above its melting point. Electric ovens with accurate thermostatic controls are most commonly used. *The all important fact concerning the paraffin bath or oven is that it must maintain a constant temperature slightly above the melting point of the paraffin. Overheating causes tissues to become so hard and brittle that they cannot be sectioned satisfactorily, if at all. An oven set for a temperature of 57° C. can be used for both a soft and a hard paraffin.*

General Remarks on Infiltration. The introduction of paraffin, like that of the alcohol and clearing agent, should be made gradually in order to avoid damaging delicate structures. The common practice of dehydrating tissues slowly, step by step, then throwing them directly into toluene or xylene, and from that into melted paraffin, is inconsistent, to say the least.

The following procedure is sufficiently gradual for ordinary zoological, histological, and pathological work. A more gradual process for cytological material is described on page 225.

It is a good idea to keep containers of melted soft and hard paraffins in the oven, ready for use. If much work is being done, also keep in the oven tightly closed vessels (preferably 1 or 2 ounce jars with molded screw caps) of the following mixtures: toluene 3 parts, melted paraffin 1 part; toluene 1 part, melted paraffin 1 part; toluene 1 part, melted paraffin 3 parts.[3] If the method is not being constantly used, it may be more practical to keep only the second mixture. The first may then be prepared, as needed, by mixing the second with an equal amount of toluene; and the third by mixing it with an equal volume of melted paraffin.

Preliminary Infiltration with Paraffin Mixtures. Remove the clearing agent[4] and cover the material with the warm mixture of 1 part melted paraffin and 3 parts of toluene. The preparation of this and other mixtures used has been mentioned in the preceding paragraph. Put the container, uncovered, into the paraffin oven. Allow it to remain there for 30 minutes if the specimens are small and permeable (class A); leave it for 1 or 2 hours if the material is moderately thick or dense (class B); and for 3 or 4 hours if the specimens are very large or dense (class C).

Then replace the mixture with one consisting of equal parts of paraffin and the clearing agent. Place the material in the oven again, for the same length of time that it remained there in the first mixture. Finally, use a mixture of 3 parts of melted paraffin and 1 part of the clearing agent.

Another way of partially infiltrating the tissue without using the rather high temperature of a paraffin oven for such a long time is to place the specimens into a fresh change of the clearing fluid, and to add to this enough melted paraffin, or shavings of paraffin, to saturate it. The specimens are allowed to stand in the mixture, at room temperature, at least overnight. The container is then placed for a few hours where the mixture will be warmed gently, as on top of a paraffin oven or in an incubator set at about 37° C., and more paraffin is added. After this, it is placed in the oven, and still more melted paraffin is added in installments. When the tissue has remained long enough in a mixture which is approximately three-fourths paraffin, transfer it to pure paraffin.

[3] If some other clearing agent, such as terpineol-toluene, is used, make similar mixtures of it with paraffin.

[4] This may be used again for the first change of clearing fluid, or for making mixtures with alcohol.

Transfer to Pure Paraffin. Pour off the last mixture and replace it with pure melted paraffin. Put the material back into the oven, where it must remain until it has become completely infiltrated with paraffin. During this period the paraffin must be changed several times in order to carry off any clearing fluid which has diffused from the object. Even very small amounts of a clearing agent, if allowed to remain in the tissue and paraffin, will ultimately interfere with section cutting.

The Length of Time Objects Are to Remain in Melted Paraffin. This depends, of course, upon the size and permeability of the objects, and may be gauged roughly by the length of time which was required to clear them. In general, small and permeable objects (class A) are well infiltrated within an hour; thin slices of moderately permeable tissue and similar objects listed in class B require 2 or 3 hours; and larger or denser specimens (class C) require longer periods of time. However, 4 to 6 hours will be adequate for almost any object if the preliminary infiltration with paraffin mixtures has been carried out as recommended.

There is a common belief that exposure to melted paraffin must be curtailed to the very minimum in order to avoid excessive hardening of the specimen. This may be true in the case of certain types of material, but animal and plant tissues of many kinds seem not to be harmed by a few extra hours in the oven, *provided the oven is not too hot.* The real danger which must be avoided is that of overheating. Keep the oven just warm enough to melt the paraffin thoroughly. Temperatures above 58° C. become increasingly harmful.

Changing the Paraffin. Change the paraffin 3 times if the objects are of small or medium size (class A or B), 4 times if they are larger (class C). The intervals at which these changes should be made may vary from 10 or 20 minutes to 1 or 2 hours, and depend on the size of the objects being processed.

Note. The last change of paraffin will probably contain only very small traces of the clearing solvent. It can therefore be used right away for a first infiltrating bath. If the solvent was a highly volatile one, as toluene or xylene, the paraffin can even be used for embedding if it is poured into a metal container and left in the oven for a few days. Relatively non-volatile solvents such as terpineol, cedarwood oil, or wintergreen oil cannot be eliminated in this way.

6. EMBEDDING. The paraffin-saturated objects are now ready to be placed in a small container, with sufficient paraffin to cover them, and quickly arranged as desired. The paraffin is then rapidly cooled to form a solid block. In this process of *embedding*, two especially important requirements must be fulfilled. First, the material must be arranged so that individual objects can easily be cut out and sectioned in the desired plane. (The subject of orientation of specimens for sectioning is discussed in Chapter 13, p. 207.) Second, the paraffin must be cooled as

rapidly as possible, because slow cooling may be accompanied by formation of coarser crystals; this results in poor cutting quality. Rapid cooling is brought about by immersion in cold water.

Embedding Containers. When a number of small objects are to be embedded in one paraffin cake, standard size Syracuse dishes are very convenient containers to use. First make certain that the inside diameter of these dishes is slightly greater at the top than at the bottom, otherwise, the paraffin cakes will not come out of them easily. Small Syracuse dishes, about 1 inch in diameter, are useful for embedding single objects. Stender dishes of various dimensions are suitable, although they are deeper than Syracuse dishes, and this may make manipulation of specimens more difficult.

Boxes of stiff writing paper may be recommended for objects of small to medium size. Figure 30 illustrates a simple method for making these boxes.

A number of companies that manufacture supplies for microtechnique produce expendable or re-usable molds for embedding. Some of these have a place to write pertinent data, or have provisions for inserting a paper tab to be embedded with the specimen. Certain molds permit the operator to cast a thick shank of paraffin which can be inserted directly into the microtome, thus obviating the need for mounting the specimen on a metal object disk or some other holder. Many of the molds have been developed with a view to cutting corners in a clinical laboratory routinely processing a lot of material, especially large specimens. The finished slides may be studied only briefly, with a view to establishing the presence or absence of some pathological condition. Little attention can be paid, under these circumstances, to individualizing the handling of material. Although both the learner and the experienced worker will profit by looking at what manufacturers produce along the lines of embedding molds, they are apt to find that most of them are relatively large and tie up considerable quantities of paraffin. Some of the disposable molds, which function much like boxes of stiff paper, make sense and are very convenient. Those in which a shank of paraffin is cast are of decidedly questionable value for really fine work.

Metal L's, placed together upon a glass or metal plate are sometimes used to form an embedding box. They are not recommended, because they are clumsy to handle and paraffin commonly leaks out between them.

Instruments and Materials. Embedding should be done on a table or other level surface, immediately in front of the paraffin oven or beside it. A small gas burner or alcohol lamp should be close by. A dissecting needle and a pair of fine forceps will be necessary. For handling delicate specimens, such as embryos, perforated section lifters or spoons are convenient. These can be purchased, or they can easily be cut out of thin metal, or even out of Bristol board. When very small objects are to be

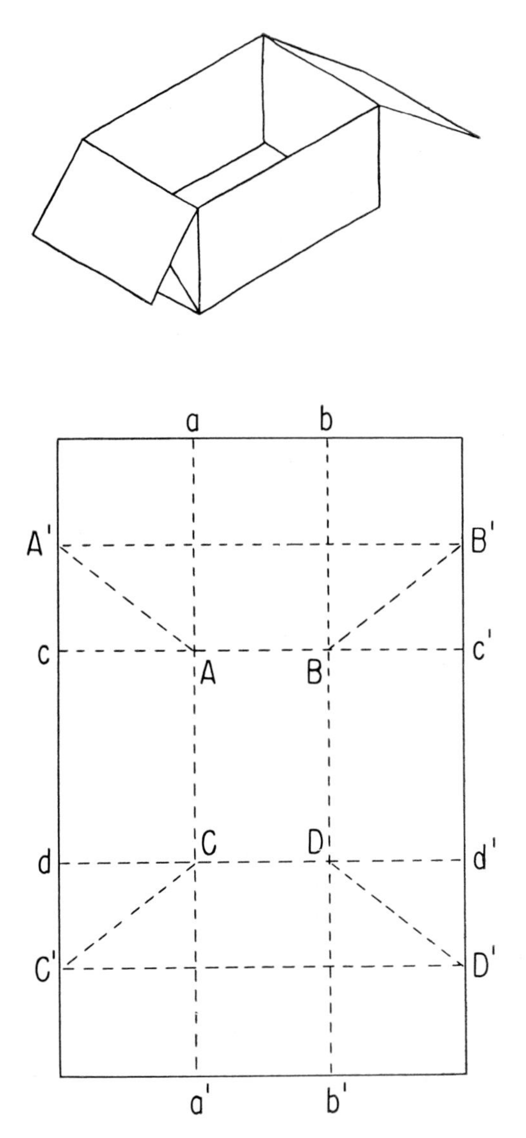

FIG. 30. Method of folding a paper box for embedding tissue in paraffin. Cut a stiff piece of paper of proper size and shape for the box desired. Fold it along lines a–a' and b–b'; then, without unfolding it, add folds c–c' and d–d'. The distances A–a and D–b' should be at least half again as long as A–c and D–d'. Place A–c against A–a, and pinch out the fold A–A'; in the same way make folds B–B', C–C', and D–D'. Now turn the projecting triangular flap AA′c around against the end of the box; do the same with the flaps at the three other corners. Finally, turn down, and over the outside, the rectangular flaps $A'B'$ and $C'D'$, pinching the folds tightly. The operations of folding will be facilitated if a wooden block of appropriate size serves as a mold. It may be convenient to have blocks of several sizes at hand for this purpose.

embedded, a wide-mouthed pipette should be on hand. A large dish or shallow pan, nearly full of very cold water (preferably containing some pieces of ice), should be placed in a convenient place on the table.

Procedure. When all is ready, proceed as follows:

If a glass or porcelain dish is being used for embedding, rub a very thin coat of glycerol on the inside of it. With pencil, write on a slip of stout paper the name or number which will identify the specimens. If a paper box is employed for embedding, write this information directly on it.

Pour into the embedding container a little more paraffin than will be needed to cover the specimens. Immediately remove the material from the oven and quickly transfer it to the embedding dish. Objects of considerable size and toughness, such as most vertebrate tissues, can be handled with forceps previously warmed over the burner. Embryos and other delicate objects must be handled cautiously with a perforated section lifter or spoon. *All objects in melted paraffin should be handled with extreme care*, as they are very likely to break or crumble. If the material consists of many small or fragile objects (such as ova, protozoa, or small larvae), remove the infiltration container from the oven and pour off most of the paraffin from it. Then gently agitate the container and immediately pour its contents into an embedding dish which has been partially filled with melted paraffin.

Working rapidly but carefully, with instruments heated periodically in the flame of an alcohol lamp or Bunsen burner, arrange the objects in the embedding dish. If a number of objects are being embedded together, arrange them systematically and leave sufficient space between them so that each can easily be cut out with several millimeters of paraffin surrounding it. If specimens are arranged somewhat as shown at *A* in Figure 31, any one of them can easily be cut out without injury to the others. Working with a chaotic mess, such as that shown at *B*, can only lead to frustration. It is best, whenever possible, to place each specimen so that the surface which is parallel to the intended plane of section lies next to the bottom of the dish. However, this is not always practical. For example, thin, flattened objects sometimes cannot easily be made to stand on edge, and must lie flat.

Should the paraffin begin to solidify before you have finished arranging the material, heat a large section lifter and touch it to the surface of the paraffin. In this way, the upper layers of paraffin can be kept melted for several minutes. In some cases it may be necessary to warm the entire dish before the arrangement of specimens can be completed. This should be avoided whenever possible, as it causes the objects to sink to the bottom of the dish, leaving no protective layer of paraffin beneath them, and also because a heated dish may retard the cooling process enough to allow formation of coarser crystals.

Dip the prepared label into melted paraffin. Bend it at a distance of

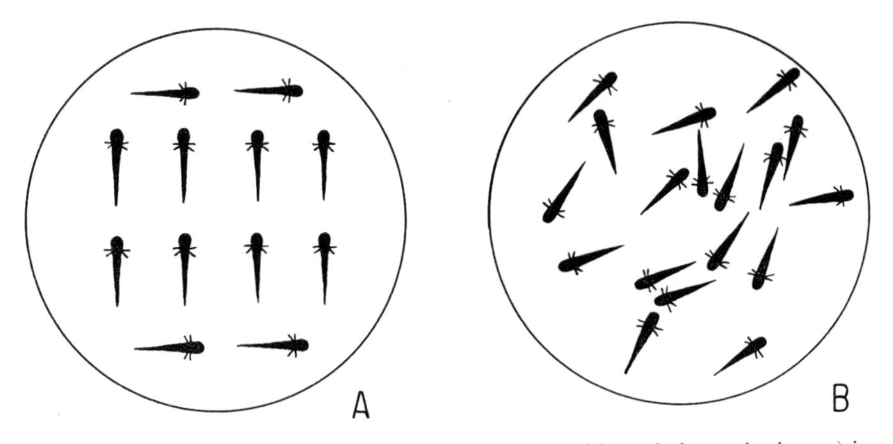

FIG. 31. Correct and incorrect arrangement of small objects (salamander larvae) in a paraffin block. *A*, Specimens arranged in rows with several millimeters of paraffin between them, so that any one of them can be cut out without danger of injuring the others. *B*, Careless, disorderly, and wasteful distribution of objects.

about $\frac{1}{4}$ inch from one end. Hook it over the edge of the embedding dish so that the short, bent-over portion dips down into the paraffin. This is unnecessary if a paper box is being used, as the label will have been written directly on the box.

Hold the embedding dish by its edges and cautiously move it to the pan or dish of cold water. Lower it into the water, almost to its rim, but no further. Blow gently upon the paraffin, to hasten cooling, until a thick film has formed on its surface. Then gently submerge the embedding dish and allow it to drop to the bottom. If the embedding dish is submerged before a thick film of solidified paraffin has formed, the force of liquid running over its edges will cause the paraffin to spurt upwards in a "geyser." If this happens, it may not be serious. However, if re-embedding appears to be desirable, allow the entire mass to cool, dry it in the air, then melt it in the paraffin oven.

If a paper box is used for embedding, solidification of the paraffin is accomplished by simply floating it upon the cold water.

If many cakes are being made at one time, chill the liquid frequently by adding more ice. The paraffin blocks may also be transferred to a second dish of cold water. Leave them in the water until they are thoroughly cooled; 20 minutes should be sufficient for all but very large blocks. Each paraffin cake should float to the surface as soon as it is thoroughly cooled, provided the embedding dish is of proper shape. If a block fails to do so within 20 minutes, loosen it by running a scalpel around its edges. If necessary, cut away some paraffin in order to loosen the block, but do not heat the dish for this purpose.

Inspect the paraffin cake critically. It should have a smooth, homogeneous appearance, and should be entirely free from white spots or bubbles. A granular texture indicates formation of coarse crystals, which may result from use of a poor grade of paraffin or from slow cooling of the block. Milky spots scattered throughout the block may also indicate that the paraffin is of poor quality, but usually they mean that the paraffin contains a considerable amount of the clearing agent. This is almost certainly the case if the milky spots are concentrated around the tissue. If any of these faults is found to a substantial degree, it will be necessary to put the blocks back into the oven. When they have melted, change the paraffin. If poor paraffin was responsible for the trouble, use a paraffin which has been tested and found satisfactory. Leave the specimens in the oven for 30 minutes to 2 hours, depending upon their size and nature, as well as upon the extent of impurities evident. Then embed them again.

STORAGE OF MATERIAL IN PARAFFIN

Material embedded in paraffin does not deteriorate for a great many years. In fact, the very safest way in which to store material for a long period of time is to embed it in paraffin. However, the paraffin blocks must be kept in a *cool place* and protected from dust. Empty cardboard slide boxes make convenient receptacles for large blocks, while coverglass boxes or pill boxes serve well for smaller ones. Label each box carefully, showing not only the name of the tissue, but also details concerning fixation, the de-alcoholizing agent which was employed, and the date, as well as any other pertinent information. Arrange the boxes according to a definite scheme (taxonomic or otherwise) and store them in a cool place. Keep them away from heaters, radiators, and heating pipes, as well as from windows through which direct sunlight passes. Any temperature which is high enough to soften the paraffin may lead to some undesirable crystallization.

SLOW METHOD FOR DELICATE OBJECTS

A more closely spaced series of alcohols and other reagents should be used in cytological work. This is important in the study of chromosomes, mitotic spindles, cytoplasm, cytoplasmic inclusions, and other details of cell structure. The quality of results obtained will more than compensate for the extra time and care required.

1 and 2. FIXATION AND WASHING. Carry out these steps in the usual way.

3. DEHYDRATION. Increases of 10% (10, 20, 30, 40%, and so on) of ethyl or methyl alcohol cause little appreciable damage and seem quite as satisfactory as the still lengthier series recommended by some authors. To dehydrate material with the utmost gentleness, some workers in-

troduce strong alcohol with a dropping apparatus, or allow it to diffuse through a semipermeable membrane into the container of material. These tedious procedures are necessary only in special cases.

If the material was washed in water, place it in 10% alcohol. If it was washed in alcohol, transfer it to alcohol 10% stronger than that used for washing. In either case, change the alcohol at proper intervals, each time introducing an alcohol 10% stronger, until the material is brought into absolute alcohol. Change the absolute alcohol once or twice, depending upon the quantity and nature of the material. Concerning the length of time to leave various kinds of objects in each grade of alcohol, see page 213.

In case the material was fixed in a fluid containing mercuric chloride, iodize it while it is in 70% alcohol. Should calcium deposits be present, remove them by adding 3% of hydrochloric acid to the alcohol.

4. De-Alcoholization. Toluene or terpineol-toluene is satisfactory for de-alcoholizing the majority of delicate objects. For a discussion of these and other de-alcoholizing agents, see page 215.

Cover the tissue successively with four or more graded mixtures of absolute alcohol and the de-alcoholizing agent. In most cases, a series consisting of $\frac{1}{5}$, $\frac{2}{5}$, $\frac{3}{5}$ and $\frac{4}{5}$ toluene, terpineol-toluene, or other clearing agent is satisfactory. Allow the tissue to stand in each of these mixtures for about the same length of time as it remained in each grade of alcohol.

Replace the last mixture with pure toluene or other de-alcoholizing agent. When the material appears to be cleared, or nearly so, pour off the de-alcoholizing fluid and add fresh. Then allow the tissue to stand until there is no doubt that it is completely cleared (see p. 216).

5. Infiltration with Paraffin. *Preliminary Infiltration.* Prepare a small table of wire gauze. This may be done by bending down both ends of a rectangular piece of screen wire and then notching each end so that it forms two tapered legs. Stand this table in the dish or jar of clearing agent which contains the specimens, being careful not to injure the latter. Adjust the level of the liquid to several millimeters above the screen table. This table is to support pieces of paraffin, whose weight would injure the specimen if allowed to rest upon it. Instead of using a table, one may place the paraffin in a bag of cotton gauze and suspend this so that it barely dips into the liquid. Another device which will serve the same purpose is a cylinder of perforated, thin cardboard. The cylinder is placed in the center of the dish, with the specimens outside of it, and is then filled with paraffin chips. Generally speaking, it is advisable to use hard paraffin for delicate objects, because it is likely that these will eventually be cut into thin sections. The total volume of paraffin placed on the screen table, or in the bag or cylinder, should be about half that of the liquid in the container. Cover the container and allow it to stand until a large part of the paraffin has been dissolved. This will require many hours or even days.

If all of the paraffin dissolves, add more. Should the mixture in the dish become inconveniently large in volume, at any stage of infiltration, remove a part of it, leaving just enough to cover the specimen. It is advisable to leave tissues in the slushy saturated mixture overnight, or for 18 to 24 hours. They may safely remain in it for several days or even weeks. In fact, the cutting quality of material seems to be improved by allowing it to stand in this mixture for at least 3 or 4 days.

Next place the container on top of the paraffin oven or in some other place (as in an incubator) where it will receive gentle heat. When all the paraffin has dissolved, refill the screen shelf, cylinder, or bag. Leave the container uncovered, so that the clearing agent may evaporate. After it has stood for several hours, or overnight, the material is ready for the next step.

Transfer to Pure Paraffin. Put the container into a paraffin oven or bath maintaining a temperature just above the melting point of the hard paraffin. When the mixture has completely melted, pour it off and replace it with melted hard paraffin.

Keep the material in the oven long enough to insure thorough infiltration. During this period, change the paraffin three times. The total time may vary from 1 to 6 hours, and depends upon the size and nature of the objects. A classification of objects, with approximate time required for each general type, will be found on page 219.

6. EMBEDDING. Embed in the usual way (step 6, p. 219).

EMBEDDING SEVERAL PROPERLY ORIENTED SMALL SPECIMENS SIMULTANEOUSLY

After a relatively small organism, such as a turbellarian about 0.5 mm. long, is embedded in paraffin, it is generally impossible to see it well enough to be certain that, even after excess paraffin is trimmed away, it will be sectioned in exactly the desired plane. Moreover, after sectioning has begun, it is often difficult to tell which part of the ribbon actually contains slices of the animal. A technique enabling one to arrange several specimens as desired before embedding, so that they may be processed as a unit, makes working with small organisms much more pleasant. The following technique has proved to be remarkably dependable.

1. Fix the specimens in whatever way seems most appropriate. After fixation (or after storage in alcohol), bring the specimens to water.

2. Find some kind of thin and delicate plant tissue, such as a frond of the marine green alga *Ulva*, the leaf of a broad-leaved moss such as *Mnium*, or a flower petal which is not obviously hydrophobic, hairy, or full of air spaces. Put the tissue into fresh water, wipe off any organisms or detritus adhering to it, and cut it into pieces about 1 cm. in diameter; moss leaves may be left intact.

3. Flatten a piece of the tissue, moist but not dripping wet, on a microscope slide. Over the tissue, spread some fresh egg albumen or the mucus from the surface of an aquatic snail or the mantle of a clam. Place the preparation on the stage of a dissecting microscope.

4. With a fine pipette, transfer a few specimens to the slide next to the tissue. Only a little water should be carried over, so that the tissue is not flooded. With the tip of a dissecting needle, push the specimens one at a time onto the tissue and arrange them in a row or in whatever pattern seems appropriate, orienting them with a view to the way they are eventually to be sectioned (Fig. 32). If one works fast, perhaps five or more specimens can be oriented before the water on their surfaces is dangerously close to evaporating.

5. From the tip of a pipette held just above the specimens, let a drop of 70% alcohol fall on them. If the albumen or mucus is of the right quantity and consistency, and if the specimens are in close contact with it, most of them should begin to adhere firmly as soon as the alcohol has started to denature the proteins.

6. When the specimens seem to be well affixed to the carrier tissue, gently flood the whole preparation with 70% alcohol. With a sharp razor blade, trim the sheet of carrier tissue to a neat rectangular shape. If one is consistent in arranging the specimens with respect to the length and breadth of the eventual rectangle, he will know how they are oriented when he starts to trim the paraffin block preparatory to sectioning. It is probably best not to have more than a very little carrier tissue in front of the first specimen and behind the last specimen to be sectioned. Then, if the specimens in the series follow closely upon one another, almost every inch of ribbon showing the carrier tissue will contain sections of the specimens. Careful examination of the ribbon under a dissecting microscope will enable the operator to discard early and late portions of the ribbon which will be of no value. However, when the specimens are very

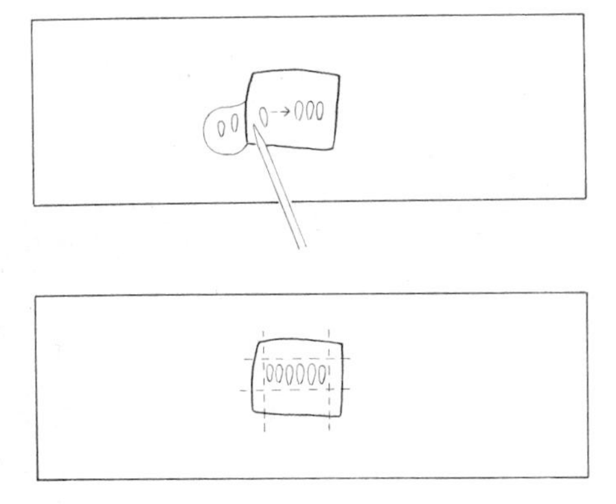

FIG. 32. Arranging specimens on a piece of carrier tissue and trimming the tissue for preparation of serial sagittal sections.

small, it is better to mount a few too many sections than to risk throwing away some useful material.

7. Transfer the piece of carrier tissue to a dish of 70% alcohol, using a pair of very fine forceps and taking care not to bend or pinch the tissue in such a way that the specimens are loosened. Leave the material in 70% alcohol for 30 minutes.

8. Complete dehydration in 95% and 100% alcohols. Replace the 100% alcohol gradually with toluene in the traditional manner, and then partially infiltrate the tissue with paraffin dissolved in toluene. In processing the tissue through alcohol and toluene, avoid handling it any more than necessary. Removing the liquid with a dry pipette and carefully adding the next liquid to the dish is advisable.

9. From the paraffin-toluene mixture, transfer the carrier tissue to pure melted paraffin. If the paraffin-toluene mixture has solidified, place it in the oven until it has melted and the tissue can be withdrawn with heated forceps without knocking off some of the specimens affixed to it. Leave the tissue in the first change of melted paraffin for about 30 minutes, then transfer it to a second bath of paraffin for another 30 minutes, or carefully pour off the first bath and replace it with the second.

10. With heated forceps, transfer the carrier tissue to melted paraffin in the embedding container. Use a heated needle to orient it as desired, keeping in mind the way the specimens are to be sectioned. Allow the paraffin to solidify in a bath of cold water as usual, then proceed with trimming the block and microtoming.

This method does not always give completely satisfying results, for the specimens may not stick to the carrier tissue to begin with, or they may fall off when the tissue is pulled through a surface film of partially solidified paraffin, or if they are accidentally poked. In general, however, if the

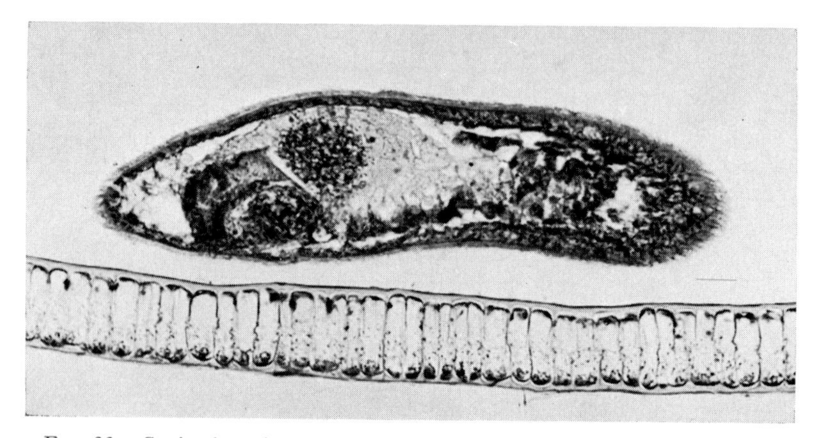

FIG. 33. Sagittal section of a small acoel turbellarian affixed to a piece of carrier tissue.

specimens are rather firmly affixed to the carrier tissue by the time they are in 95% alcohol, most of them will probably stay on.

The plant tissue may not respond well to staining, for no effort was made to fix it properly, but this is of no consequence. The carrier tissue remains on the slides with the specimens (Fig. 33), but this is hardly objectionable when one considers that without it, the problem of getting serial sections of properly oriented small specimens would be a difficult one indeed.

Some plant tissues serve well after preservation in 70% alcohol. However, it is necessary to wash out the alcohol thoroughly before applying the egg albumen or mucus, so that the adhesive substance is not prematurely denatured.

CHAPTER 15

Cutting and Mounting Paraffin Sections

The following directions apply to the sectioning of materials embedded in paraffin and the affixation of sections to slides. The steps are numbered to follow consecutively after step 6 in the conventional method of embedding in paraffin described in Chapter 14.

7. TRIMMING AND MOUNTING THE BLOCK ("BLOCKING"). If the cake of paraffin contains numerous objects which are to be sectioned separately, first cut it into several pieces, each of which includes a row or group of specimens. From one of these pieces cut out a small block which contains a single specimen. To cut a paraffin cake, make a scratch with the point of a sharp scalpel and then deepen the scratch with the blade, until there is a deep groove along which the cake can be broken apart with the fingers. Do not attempt to cut the paraffin by pressing on it heavily with the blade, as some of the paraffin close to a specimen may be mashed, or the cake may crack apart without following a predetermined line.

Make certain of the plane in which the object is to be sectioned. The specimen should have been embedded in a position that makes identification of its various planes possible (see pp. 207 and 226).

With a sharp, straight-edged blade trim the paraffin block to a rectangular shape, allowing several millimeters of paraffin to remain on each side of the embedded object. It is well to leave a considerable thickness of paraffin on the side opposite that which is to be presented to the knife.

Note. The paraffin trimmed off blocks being shaped for microtoming should be accumulated, dried in air (in case any water is adherent to it), and remelted. Its consistency for cutting may actually be better than that of new paraffin. When large amounts of paraffin are being re-used regularly, filtering through coarse filter paper, with the help of a funnel having a jacket for circulation of hot water, may be worth the trouble. However, if paraffin which has been salvaged is allowed to stand in an oven undisturbed for a day or two, the obvious detritus settles to the bottom. Careful decanting should separate clean paraffin from the sediment, which can be discarded. This extent of purification should be sufficient.

Select an object holder of appropriate size. This may be a metal disk (Fig. 34) furnished by a manufacturer of microtome accessories, a piece of hardwood which has been well infiltrated with paraffin, or a piece of vulcanized fiber. Shallow grooves cut into a wood or fiber holder will assist in making the affixation of the paraffin block to the holder more secure. If the face of the holder is not already coated with paraffin, dip it into melted paraffin or pour a little melted paraffin upon it to make a coating about 1 mm. thick. This substrate, after cooling, will provide a firm anchorage for the block containing the object. It will also increase the distance between the base of the object and the holder, lessening the chance that the edge of the knife will ever strike the holder.

Lay the specimen block upon the table, turning upward the surface which will be cut first. Take the object holder in the left hand, or stand it upon the table. (A metal object disk may be made to stand upright by placing it in a narrow-mouth bottle of appropriate size or in a hole drilled in a wooden block.) Heat a section lifter or the handle of a metal scalpel in the flame of an alcohol lamp or Bunsen burner and use it to melt the paraffin on the holder. Remove it, quickly put the object block into the melted paraffin upon the center of the holder, and gently press upon it. Then hold a heated needle or scalpel handle for a few seconds against each side of the joint formed by the paraffin block and the paraffin on the holder. Quickly immerse the holder and block in cold water for at least several minutes.

With a single-edge razor blade, carefully shave down the sides of the block. Trim each side straight and make the opposite sides parallel, so that the face of the block is a perfect rectangle (Fig. 35*A*, *B*). Rounded and wedge-shaped blocks are undesirable. Figure 35*D*, *E* shows the types of ribbons obtained from blocks which have not been trimmed properly. It is customary to trim the sides to about 2 mm. from the widest part of the specimen. However, if one plans to mount the sections in a complete series, it may be best to trim the block to about 1 mm. from

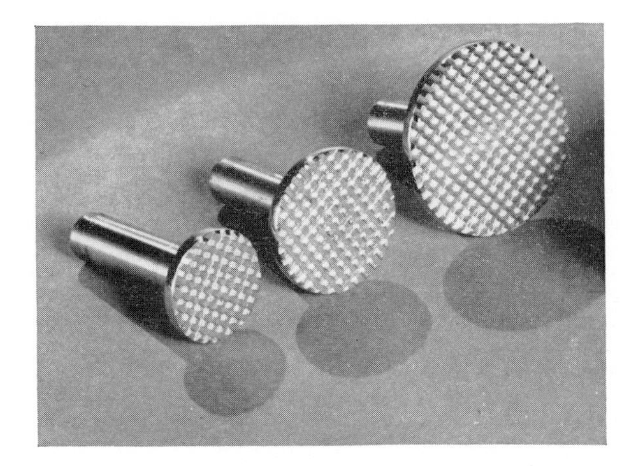

Fig. 34. Microtome object disks. (Courtesy of American Optical Corp., Buffalo, N.Y.)

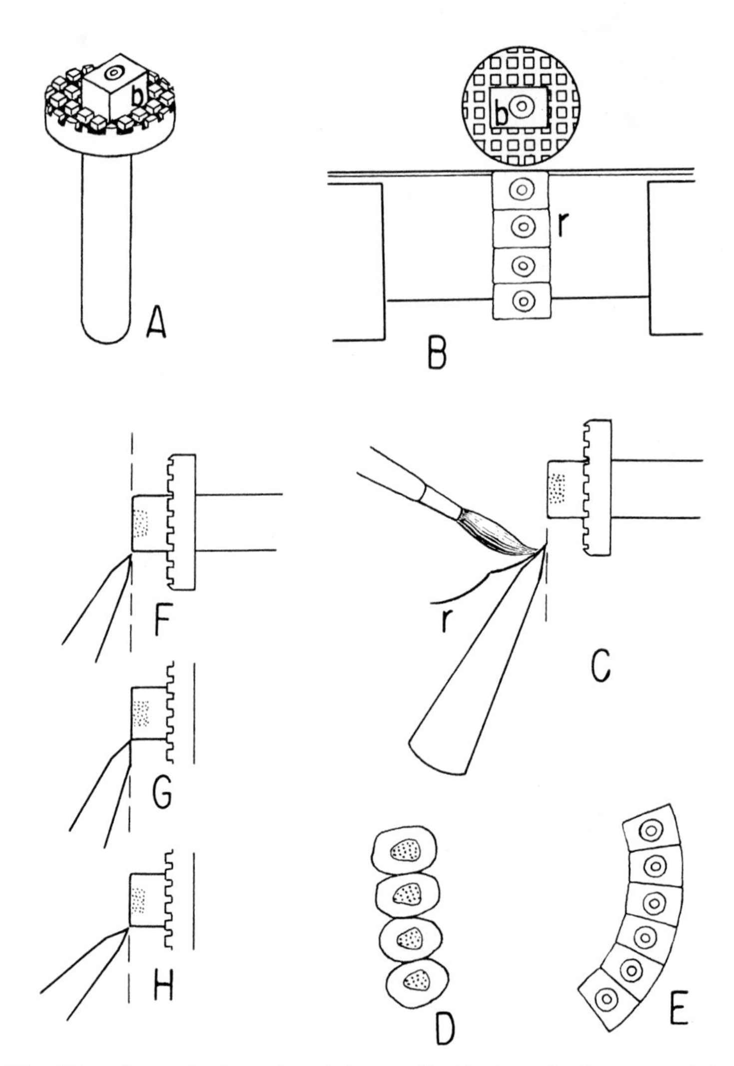

Fig. 35. Trimming and orientation of the paraffin block, and adjustment of the knife in the rotary microtome. *A*, Paraffin block containing object (*b*), properly trimmed and mounted on object holder. *B*, Front view, showing paraffin block containing object (*b*) trimmed to form a perfect rectangle, with its long edge parallel to the knife. The resulting ribbon (*r*) is straight, and will hold together well because of the length of the adhering edges of the sections. *C*, Side view, showing the correctly tilted knife, and use of a camel's hair brush to prevent the sections from curling upward. *D*, Sections which will not adhere to one another because the block was not trimmed to a rectangular shape. *E*, Crooked ribbon resulting from a wedge-shaped block. *F*, Detail of correctly tilted knife; the cutting facet makes an angle of about 5° with the plane of section (dashed line). *G*, Knife insufficiently tilted; the shoulder of the knife strikes the block. *H*, Knife tilted too much acts as a scraper.

the object. The width of the ribbon will be narrower and its total length will be shorter, so that more sections may be affixed to each slide.

8. SECTIONING. *The Microtome.* Rotary microtomes are designed for cutting paraffin sections in the form of ribbons. Before proceeding further, the beginner should familiarize himself with the construction of this instrument, which is explained in Chapter 13 and illustrated by Figure 20, page 193. Examine the clamp which grips the object holder and note how it may be adjusted to turn the object in various directions. Find the feed screw. Turn the handle slowly and observe the operation of the feeding mechanism. Find the small crank by which the object carrier may be moved back and forth. In some models, this should be turned only when the block is at the upper level of its track. Locate and study the mechanism which regulates the thickness of sections. Examine the knife holder, locating the screws which clamp the blade into position and those which control its height and the angle of its tilt. Determine how the knife may be moved laterally.

The microtome is a delicate machine with many moving parts and accurately ground bearings. It cannot function properly unless it is kept *clean* and *well-oiled.* Cover the instrument when it is not in use. Lubricate it frequently, following the manufacturer's instructions.

In the rotary microtome the knife edge is transverse to the line along which the paraffin block moves. As the block strikes the knife edge, the sections are compressed to a slight extent as they are cut. However, the consecutive sections adhere to one another by their edges, so that a continuous ribbon is formed.

In the sliding microtome, the edge of the blade is set at an angle to the block in which the specimen is embedded, so that the knife slices through the block instead of chopping it. The sliding microtome may be used to advantage for many types of material, such as large masses of muscle, connective tissue, cartilage, bone, and entire organisms. If such objects do not cut well on the rotary microtome, use a sliding microtome. Proceed in the same way as for cutting nitrocellulose sections, but leave the block and knife dry.

The Knife. Never lose sight of the fact that a sharp, smooth knife edge is the first requisite of good section cutting. Concerning the selection, sharpening, and care of the microtome knife, see page 196.

The Tilt of the Knife. It is important to tilt the knife as a whole with reference to the angle formed by the cutting facet and the line of sectioning (Fig. 35*C*, *F*). For most materials, this angle should be very acute (about 5°) so that the cutting edge passes freely through the material, without acting as a scraper.

A slightly diminished tilt is advantageous for cutting very hard objects or those which have a decidedly heterogeneous composition. However, if the tilt is reduced too much (as shown at *G* in Fig. 35), the shoulder of the cutting facet will strike the block ahead of the knife edge, mashing the paraffin so that the surface of the block assumes a crumbly, opaque appearance. The impact of the shoulder on the block may even compress the block and displace the knife slightly with each stroke, and when an occasional section does come off, it is thick and battered. If the tilt is only slightly too small, the sections will be wrinkled ("telescoped") or thicker at one edge than at the other.

A slightly increased tilt is sometimes helpful in cutting soft tissues which are embedded in soft paraffin. Excessive tilting, as shown at *H* in Figure 35, causes the knife to act as a scraper and also brings into prominence any slight irregularities in its edge. These irregularities, together with particles scraped from the material and accumulated on the knife, make longitudinal scratches or splits in the ribbon. In extreme cases of over-tilting, the knife may lift portions out of the block. An excessively tilted knife emits a metallic rasping sound as it passes through the block, especially if the embedded tissue is at all hard.

Orienting the Block. It is assumed that the orientation of the object has been considered in trimming and mounting the block. Now clamp the object holder, bearing the paraffin block, into the microtome. Turn it so that the long sides of the block are exactly parallel to the knife edge. Figure 35*B* illustrates this point. The undesirable results likely to be obtained if the block is not trimmed to a rectangular shape are shown in Figure 35*D* and *E*. Adjust the object clamp so that the specimen will be cut in the desired plane. Accuracy in orientation is very important, for the reasons explained on page 207; carelessness in this matter may make interpretation of structure from the sections more difficult. During these manipulations, keep the mechanism of the microtome locked, so that it cannot be turned accidentally and cause injury to the specimen or the knife edge. *When working about a microtome in which the knife is in place, be extremely careful not to cut yourself or to injure the knife blade by striking it with any hard object.*

Operation of the Rotary Microtome. The microtome should be placed upon a sturdy, level table, with the wheel at your right hand and the knife toward your left (Fig. 36). Lay a ribbon tray or shallow cardboard box close to the knife end of the instrument. If the material was stained before embedding, place a piece of smooth white or light-colored paper on the floor of the tray or box; otherwise, a dark surface is best. Have at hand a scalpel with a rounded blade, a pair of forceps with curved tips, and a small camel's hair brush. A dissecting microscope will be useful for examining the ribbons of sections. The room should not be too warm or too cool.

Set the regulator of the microtome at the figure which corresponds to the thickness of sections desired. For general histological and pathological work, sections are usually cut at a thickness of 8 to 12 μ. Sections

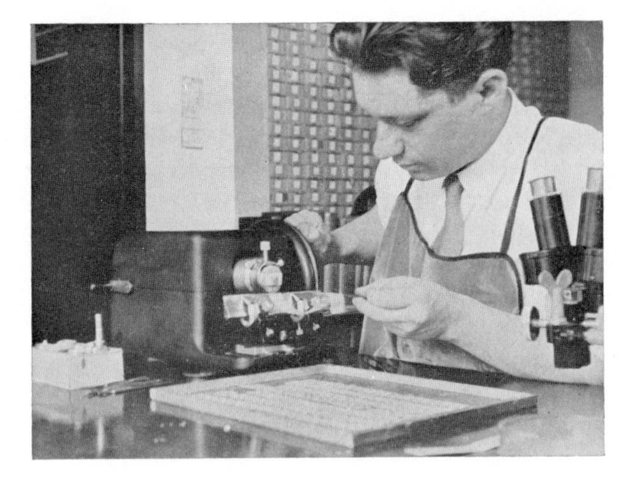

FIG. 36. Operation of the rotary microtome.

5 to 10 μ in thickness are generally used for study of mitosis and cell inclusions; sections less than 5 μ thick are employed only for special cytological purposes. Relatively thick sections (15 to 30 μ) are commonly used in vertebrate embryology and neurology, when general topography, rather than minute detail, is to be studied.

Move the knife holder to bring the edge of the knife reasonably close to the paraffin block, and tighten the clamp which holds it in place. By turning the little crank which operates the feed screw, the paraffin block can be advanced slowly and with precision until it is very close to the edge of the knife. However, as the feed excursion in most rotary microtomes is only about 2 or 3 cm., it is a good idea to begin microtoming with the advance mechanism set near its starting position. Otherwise, the feed mechanism may reach the end of its course while sectioning is in progress, which will necessitate resetting the advance mechanism and re-establishing the relationship of the specimen to the knife.

Now make absolutely certain that every adjustment screw or lever in the knife holder and object clamp is tightened securely. Failure to observe this precaution is frequently the cause of serious injury to the knife, the specimen, or the operator.

Release the lock which holds the drive wheel in place. Take the crank in your right hand and turn the wheel clockwise at a moderate speed (about 2 revolutions per second). Hold the camel's hair brush in your left hand. When the knife first encounters the face of the block, which is likely to be somewhat irregular in spite of careful trimming, only partial sections will be obtained. These first sections may not adhere to one another. However, as soon as complete sections begin to come off, they should form a continuous ribbon.

If the sections show a tendency to curl upward, hold the camel's hair brush close to the knife edge (as shown at *C* in Fig. 35) to prevent them from getting into the path of the block. If the ribbon should curl downward, place the brush under it or carefully hold the first section with a pair of forceps. If the sections do not adhere to one another, or persist in curling, try cutting with a somewhat faster stroke. Also consult the suggestions under Regulation of Temperature (below) and Difficulties Encountered in Sectioning (p. 237). If the ribbon is crooked, it may be necessary to trim the upper and lower surfaces of the block until the edges are more nearly parallel to the knife edge.

Watch the sections carefully as they are cut! If they are scratched or split, discontinue microtoming and wipe the cutting facet of the knife which is nearer to the block, by a gentle upward movement of the little finger. Again cut a few sections. If the scratches persist, refer to paragraph *c*, under Difficulties Encountered in Sectioning (p. 237). Should the sections have wrinkles across them, consult paragraph *e* under the same heading. With some kinds of material, best results are obtained by sectioning with a rapid stroke; with others, a slower motion is preferable. In general, use a fairly rapid stroke (about 2 turns per second) to cut thin sections of objects in hard paraffin, and a somewhat slower stroke for thick sections in soft paraffin. The formation of ribbons depends upon heat produced by impact and friction between the knife and block. Therefore, good ribbons are most likely to be obtained by cutting with a fairly rapid stroke. Yolky eggs, as those of amphibians, commonly show a tendency to crack and tear. The same trouble sometimes occurs in sectioning mammalian embryos through the regions of the brain and liver. This may often be avoided by turning the microtome with extreme slowness while the knife passes through the object. After completing the cut, turn the wheel rapidly until the block strikes the knife and welds the sections together, then slow it down again.

When the ribbon reaches the left side of the tray, or slightly beyond this, stop the microtome and lay the ribbon within the tray near the top. With a rocking motion of the scalpel, separate this length of ribbon. Then lift the ribbon which still adheres to the knife and resume sectioning. In this way continue to cut and lay successive lengths of ribbon upon the tray until there are as many as you desire. In making serial sections, of course, the entire specimen must be cut and special care must be taken to keep the sections in perfect order. Proceed now to Examination and Selection of Sections, page 240.

Regulation of Temperature. To obtain ribbons in which sections adhere tightly to one another, and to avoid compression or curling of sections, it may be necessary to regulate the temperature to a certain extent. The proper temperature will depend upon the melting point of the paraffin and the thickness at which the sections are to be cut. For sections of a given thickness, the higher the melting point of the paraffin, the warmer the room should be. For paraffin having a particular melting point, the temperature should be lower for thin sections than for thick ones. The following table contains specific information on this subject.

Thickness of Sections (in μ)	*Room Temperature (in degrees F.)*	
	For Soft Paraffin (Melting Point 45–50° C.)	*For Hard Paraffin (Melting Point 52–58° C.)*
2–4	Not recommended	60° Cool by ice
5–7	Not recommended	64° in
8–10	Not recommended	70° hot weather
12–14	70°	75°
16–20	74°	Not recommended
22–25	77° Warm knife	Not recommended
25–30	80° if necessary	Not recommended

When it is not possible to control the temperature of the entire room, which is often the case, the air in the vicinity of the microtome should be warmed or cooled. A table lamp or Bunsen burner will provide heat, when necessary. A similar effect may be produced very temporarily by warming the knife and knife block in an incubator or by applying a heated metal instrument momentarily to the upper and lower surfaces of the paraffin block. Cooling may be accomplished by holding a lump of ice against the knife as often as necessary and by soaking the paraffin block in ice water, or by allowing ice water to drip on the block occasionally. Several special devices have been invented for regulating the temperature of the knife and paraffin block.

Difficulties Encountered in Sectioning. *a. The sections curl and do not adhere to one another.* As has been pointed out before, a moderately rapid cutting stroke generally favors the formation of good ribbons. Failure of the sections to adhere is very commonly caused by the knife or block being too cool. Correct this by following suggestions given in the preceding paragraphs. Another remedy is to apply a small amount of softer paraffin or beeswax to the top and bottom of the block with a heated scalpel handle. Make sure that the upper and lower edges of the face of the paraffin block are straight and parallel to the edge of the knife. If the trouble persists, try diminishing the tilt of the knife slightly. A dull knife may also contribute to the problem.

b. The ribbon is crooked. Nearly always this is caused by failure to trim the paraffin block so that its upper and lower surfaces are parallel. Irregularities in the knife edge sometimes produce a crooked ribbon, even when the block is properly trimmed. First check the shape of the block, correcting it if necessary. If this does not remedy the difficulty, try using another portion of the knife edge.

c. The sections are scratched or split. This is caused by nicks in the knife, or hard particles in the paraffin block or on the edge of the knife. If the knife is tilted too much, the difficulties due to any of these causes will be accentuated. First try wiping the cutting facet of the knife which is nearer to the block, by a gentle upward movement of a finger moistened with toluene. Usually this will remove hard particles adhering to the knife edge. If it does correct the difficulty, repeat it whenever scratches appear. Sometimes it is necessary to wipe the facet of the knife which is farther from the block. For this purpose use a soft camel's hair brush moistened with toluene. Always wipe toward the edge. If scratches still appear, move the knife so as to use another part of its edge. Should the trouble persist, examine the knife blade under a microscope.

If the knife is found to be sharp and even, try decreasing its tilt slightly. If none of these procedures eliminates scratches, the paraffin or the material itself is probably at fault. Perhaps the paraffin contains grit. If so, re-embed the material in clean paraffin. The material may contain hard particles of calcium salts, or it may have been hardened by over-heating. Try soaking the block as described under *d.* If this is unsuccessful, prepare another lot of material, taking care to eliminate the causes mentioned.

d. The material crumbles and tears out of the paraffin. Difficulties of this sort are most likely to be noted if the infiltration of the tissue has not been thorough enough, or if the tissue has become very hard or brittle.

If the tissue is soft and mushy, then incomplete infiltration should be suspected. This may be the result of incomplete dehydration or clearing, or of having the material for too short a time in melted paraffin. If the latter possibility appears to be at the root of the trouble, the difficulty can usually be corrected by putting the material back into melted paraffin for an hour or two, and then embedding it again. If this remedy fails, imperfect dehydration or clearing is probably the cause of the trouble, and it is ordinarily advisable to discard the material.

When the embedded tissue is hard or brittle, this may be the result of using an unfavorable de-alcoholizing agent, or of over-heating the material in the paraffin bath.

Toluene and terpineol-toluene are superior de-alcoholizing agents. Certain materials passed through xylene often become very hard and flint-like.

It is commonly believed that exposure to melted paraffin must be as short as possible, in order to avoid extreme hardening. This may be true in the case of certain materials, but a few extra hours in a paraffin oven often seem to do no harm if the oven maintains the correct temperature. The real danger is that of over-heating.

Softening material by soaking. Material which is too hard to cut, or from which the connective tissue tears out, may often be made to acquire a better consistency by soaking in water. Simply remove the object carrier and paraffin block from the microtome and immerse them. A small amount of water will penetrate into the material and soften it. The length of time required for softening depends upon the size and hardness of the specimens. Penetration and softening naturally will proceed faster if a large cut surface is exposed to water. Most objects which are susceptible to this treatment will be softened to an appreciable depth after being soaked for about twelve hours. Refractory objects, however, may require 3 or 4 days. Soaking must also be prolonged if the specimen has not yet been cut into, as in the case of an embryo from which a complete, perfect series of sections is desired.

Soaking in a mixture of 9 parts of 60% ethyl alcohol and 1 part of glycerol may give better results than soaking in water. A mixture of 9 parts of glycerol and 1 part of aniline has been recommended for particularly refractory objects.

e. The sections are compressed and wrinkled. This may be caused by a dull knife, or by cutting sections in an overly warm room. If neither of these accounts for the trouble, it may be that the knife is tilted insufficiently. If the tilt is too small, the impact of the block upon the shoulder of the knife facet produces a dull thud, and the paraffin on the surface of the block becomes white and mushy.

If the knife is tilted properly and the sections appear dry and lusterless, with transverse ridges, the fault may lie in the material. Very often this can be corrected by soaking the paraffin block as directed above under *d*.

f. On its upward stroke the block lifts the sections from the knife. The most common cause of this distressing phenomenon is an accumulation of small particles of paraffin upon the top of the block. This means that the paraffin is too soft, the knife is not tilted enough, or the room is too warm. An accumulation of paraffin on the knife may also be responsible.

g. The sections are badly scratched and the knife emits a ringing sound as the block passes over it on the upward stroke. This may be caused by excessive tilting of the knife, or by hardness of the object. Refer to *c* and *d*.

h. The sections vary in thickness. The variation may be relatively slight, or so great that one or two sections are missed altogether and then a very thick section comes off. The most common situation is one in which thick and thin sections alternate. This trouble may be the result of failure to clamp the knife or block tightly into the microtome. It may also be caused by a microtome which is worn out, or in which the mechanism advancing the object carrier is not working properly. Lack of lubrication may accelerate the functional decline of a microtome.

Insufficient tilting of the knife is another cause of uneven cutting. In this case, the impact of the knife's shoulder with the paraffin block springs the blade back or compresses the paraffin block, or both, so that a partial section or none at all is cut on one or more strokes. When the block has finally advanced far enough for the blade to bite into it, a thick and useless section is delivered.

Very large or hard objects may also cause the knife to spring in the manner just described. Sometimes it is helpful to soak the block in water (see *d*). Embedding such objects in nitrocellulose may be the solution to obtaining better preparations.

i. The ribbon clings to the knife or other metal objects. This is caused by static electricity, formed in sections by the friction of cutting. In extreme cases it is almost impossible to handle the ribbon, which flies to any metal object or to the operator's hand. Under certain atmospheric conditions this phenomenon is so annoying that it may be advisable to postpone sectioning to a more favorable day. When electrification is not extreme, it may be lessened by boiling water in an open receptacle near the microtome. An induction coil is sometimes used to eliminate this difficulty (see Blandau, 1938). A few manufacturers have developed equipment to minimize the effects of static electricity.

Storage of Paraffin Ribbons. Ribbons may be stored for months or years, but they gradually become brittle and more difficult to handle. If it is necessary to keep ribbons for a long time, store them in a dust-proof box and keep them in a cool place so that they will not melt or adhere to the substrate.

9. EXAMINATION AND SELECTION OF SECTIONS. Before proceeding to place the sections on slides, one should determine if they are satisfactory and how they are to be arranged. Some knowledge of the materials involved will be of great help in making intelligent decisions. The time spent in acquiring or refreshing this knowledge will ultimately be repaid many times over.

A wide-field binocular microscope will enable one to examine ribbons very rapidly and thoroughly. If the material was stained before embedding, place the ribbons upon white paper; if not, place them upon black paper or cardboard.

Should any doubt exist as to whether the material is well fixed or shows the structures which the completed preparations are supposed to show, use the following simple method of examination—it may lessen the risk of wasting much time on poor material. Place one or two sections on a slide. Apply several drops of toluene close to the sections and tilt the slide gently, so that the toluene flows over them and dissolves the paraffin. Drain off the toluene and place upon each section a drop of a 0.1% solution of eosin in a mixture of 3 parts of terpineol-toluene (or carbol-toluene) and 1 part of absolute alcohol. As soon as the tissue is stained pink, drain off the stain, clear the section in terpineol-toluene, and place a drop of thin mounting medium or immersion oil upon the sections. Examine them under a compound microscope. The structures can be seen fairly well. If one contemplates making many preparations from a single ribbon, it is a good idea to finish one or two slides by the method which one expects to use in making the permanent mounts.

Representative Sections. In making preparations of most ordinary zoological, histological, and pathological specimens, it is unnecessary to mount entire series of sections. If the organization of the tissues is essentially the same throughout the object (as, for example, in liver or in a particular part of the intestine), select for mounting one or several sections which are complete, typical, and free from scratches or other physical imperfections. When the structure of the specimen differs in various regions, it will be sufficient for most purposes to select one or more characteristic sections from each region.

Serial Sections. Follow each row of sections carefully with the eye, to make certain that all parts of the ribbon have been laid in proper order and with the same side uppermost. Changes in the outline of the specimen should appear gradually. An abrupt change usually indicates that some portion of the ribbon has been misplaced. In this case, a very careful examination should be made and any sections found to be out of order should be put into the proper position in the series. It is equally important to detect any imperfections in the tissues and, in the case of replaceable objects, to discard ribbons of damaged or poorly fixed specimens.

10. STRETCHING AND AFFIXING SECTIONS TO SLIDES. *Preparation of Slides.* The slides must be perfectly clean. Any trace of grease is particularly undesirable. Methods for cleaning and testing slides have been described on page 20.

Two methods for affixing sections to slides will be described. In one of these methods, egg albumen serves as the adhesive substance; in the other, gelatin is denatured by formalin to hold the sections tightly to the glass.

Mayer's Albumen Affixative

Put the whites of several eggs into a bowl. Whip them briefly with a fork or wire eggbeater, until the material is more or less homogeneous. Two or three dozen strokes are usually sufficient; do not beat them until they are white and stiff. After about an hour, skim the foam from the top and pour the remaining liquid into a graduated cylinder. Add an equal quantity of glycerol, and put in 1 gm. of sodium salicylate for each 100 ml. of liquid. Shake the mixture thoroughly. It is a conventional practice to filter the preparation through glass wool, but many workers find the mixture to be uniform enough without this treatment. For handy use, keep a small quantity of the affixative in a vial or bottle equipped with a wooden or glass applicator rod.

Albumen affixative of good quality, ready for use, may be purchased from dealers in supplies for microtechnique.

Haupt's Gelatin Affixative

Gelatin	1 gm.
Distilled water	100 ml.

Dissolve the gelatin at 30° C. in a water bath or oven. Then add:

Glycerol	15 ml.
Phenol, crystals	2 gm.

Stir the mixture well. The solution is often filtered, but this seems not to be really necessary.

Place a small drop of Mayer's albumen affixative or Haupt's gelatin affixative upon a slide. With a clean finger, smear it over the glass and rub vigorously back and forth so that the affixative is uniformly spread over the surface and in tight contact with the glass. Then wipe the slide with the same moist finger, until the film of affixative is extremely thin. If the film is too thick, it may subsequently take up stain and make the preparations untidy.

If Mayer's affixative is being used, place, on the albumen-coated slide, a few drops of distilled water which has been boiled (to free it from dissolved air) and then cooled. If it is necessary to preserve water-soluble substances (such as glycogen or metallic salts) in the sections, 70% alcohol may be used in place of water, or the sections may simply be laid upon the albumenized slide.

If Haupt's affixative is being used, add a few drops of 2% formalin.

Arrangement of Sections Upon the Slides. In deciding how many sections to place on each slide and the manner in which to arrange them, it is necessary to consider (1) the size of coverglass to be used and (2) convenience for study.

Always bear in mind that the coverglass should overlap the sections by at least 2 mm.—preferably a little more than this—on every side. Oxidative action is likely to fade the stain in tissue which is too close to the cover's edge. In cutting off pieces of ribbon, take into account the fact that they will increase 10 to 20% in length when they are heated to flatten the sections.

Sagittal or transverse sections of entire organisms are usually mounted with the ventral side of the specimen nearer the upper edge of the slide. If this is done, the dorsal side of the organism will appear uppermost when the sections are examined under a compound microscope.

Representative sections, if they are small or moderate in size, are generally mounted under 18 or 22 mm. coverglasses, square or circular. If the sections are from an organ which is more or less uniform in structure throughout (as liver or pancreas), two or three will probably suffice for each slide, but if material is plentiful it is well to add as many as the size of the coverglass will permit. When various regions of the specimen differ markedly in structure, select one or more characteristic sections from each region and arrange them on the slide in their natural order. A set of larger sections will usually require a rectangular coverglass of appropriate size.

Serial Sections. Complete series, unless they are very short, are usually mounted under 24 × 50 mm. or 24 × 60 mm. coverglasses. Cut the ribbon into strips about four-fifths the length of the coverglass. Lay the first strip near the upper edge of the slide. If a 24 × 50 mm. coverglass is to be used, the first section should be about 2.5 cm. from the left end of the slide. If a 24 × 40 mm. coverglass is to be used, this distance should be a little greater, but for a 24 × 60 mm. coverglass it must be less. Allow the left end of the strip of ribbon to touch the water first, then lower the remainder of it gradually, being very careful not to trap bubbles of air under the sections. Should air bubbles be formed inadvertently, remove them as completely as possible with the aid of a fine pipette. Lay successive strips of ribbon below the first, in the order of lines of printing. Fill the area which will be occupied by the width of the coverglass, except for a margin of not less than about 2 mm. which must be left between the tissue itself and the edge of the coverglass. If the ribbon is curved, divide each strip into several segments and try to orient these in a straight line. Should the series require more than one slide, prepare successive slides in the same manner.

It is advisable to mount serial sections on slides which have been marked beforehand (preferably at the left end, where a permanent label will be affixed) with a diamond-point pencil. This will prevent mixing one series with another, or getting slides within a series out of order. If serial sections are being prepared from several specimens in one lot of material, it is important to distinguish each by a number or letter. Suppose, for example, that the lot consists of several embryos covered by record card number 50, and serial sections are being made from two of these embryos. The slides of the first series should then be numbered 50-1-1, 50-1-2, and so on; slides of the second series should be numbered 50-2-1, 50-2-2, and so on. If different specimens of the lot are cut in different planes, it may be a good idea to insert a letter before or after the series number to designate the plane of sectioning (T for transverse; S for sagittal; F for frontal). The arrangement and labeling of serial sections is illustrated by Figure 29, page 208. The important points to remember are that (1) the sections must be mounted in exact serial order, (2) the rows must be straight, and (3) the edges of the tissue in the sections should lie no closer than about 2 mm. from the margins of the area eventually to be covered by the coverglass.

Stretching or Spreading Sections. In the process of cutting, sections are unavoidably compressed and curled to some extent. It is therefore essential to warm them until they expand to normal proportions and become perfectly flat and smooth.

Sections should be warmed only to the minimum temperature necessary to accomplish this. Obviously the paraffin must not be melted, as this is likely to distort the sections. A temperature high enough to melt paraffin may also denature the albumen before the sections come into tight contact with it, and as a consequence, the sections may not adhere to the slides after the paraffin is removed.

In working with sections of delicate tissues, best results in spreading them are obtained by accurate control of the temperature. The paraffin and various tissue elements have different coefficients of expansion. Moreover, the dehydrated substances of the tissues absorb water and therefore swell to different extents. When a section floating on water is heated, these differences set up a series of stresses and strains between the paraffin and tissue and between various kinds of structures in the section. The strength of these forces increases with the temperature. In the range of minimum temperatures (37 to 42° C.) at which sections can be spread and flattened, they are not sufficiently strong to tear apart the structures. At higher temperatures, the strains become great enough to tear apart certain structures at the points of least resistance. Except when the temperature has been high enough to melt the paraffin, the effect does not become evident until the paraffin has been removed from the sections. It then appears as a characteristic cracking and separation of masses of cells, often in parallel lines. This distortion is likely to be more conspicuous in the case of soft tissues from organs such as lymph nodes, spleen, testis, gastric mucosa, and cerebral cortex. Sections of the cerebrum often crack so badly that the effect is noticeable to the unaided eye. Cells of columnar epithelium frequently become separated from their basement membrane and appear to be badly crushed. These effects are exaggerated when the sections are finally cleared in toluene, which shrinks the structures to different degrees. The final appearance resembles that caused by cutting with a very dull knife, and it is sometimes erroneously attributed to this.

Warm the slides to a temperature of 37 to 40° C. if the sections were cut in soft paraffin, 39 to 42° C. if in hard paraffin. Very dependable thermostatically controlled warming tables (Fig. 37) are produced by various manufacturers, but if one of these is not available in the laboratory, suitable apparatus may be improvised. A very simple and completely satisfactory device consists of a level sheet of flat metal (such as a cookie sheet) heated from below by an electric lamp. By regulating the

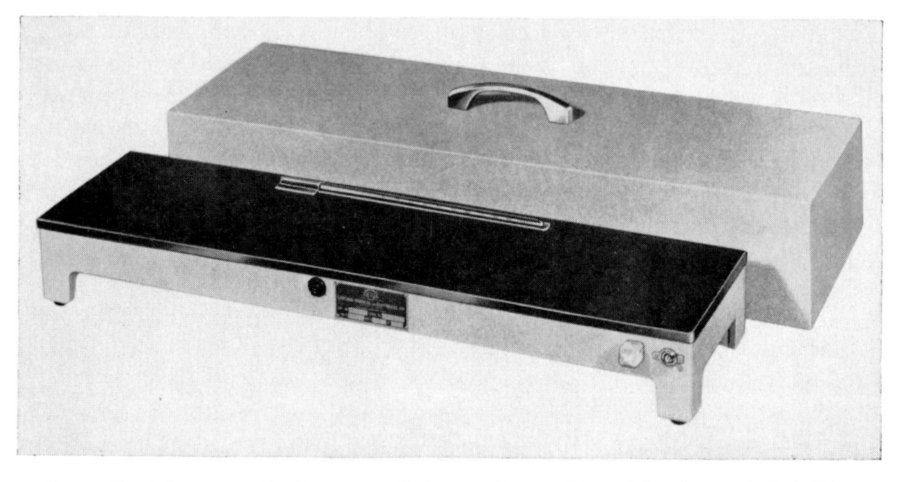

FIG. 37. Thermostatically controlled warming table. (Courtesy of Lab-Line Instruments, Inc., Melrose Park, Illinois.)

distance between the lamp and the sheet of metal, and by using that portion of the sheet where the temperature is optimum for spreading sections, very good results may be obtained. A wooden frame may be built to support the sheet of metal, and a socket for the lamp may be installed permanently within it.

The rapidity with which the sections spread and flatten depends upon their thickness and smoothness, as well as upon the nature of the tissues which they support. At the temperatures recommended, some sections will flatten within 5 minutes, while others may require an hour. Sections of hollow cylindrical objects (such as intestine) are particularly difficult to flatten. It is advisable to slit open and pin out such specimens before fixing them. Sections which contain much connective tissue are also hard to flatten. The process may be aided by gently pulling at opposite ends of the ribbon with needles. If it is necessary to prolong the warming process, add boiled water occasionally to replace that which is lost by evaporation.

Occasionally sections are encountered that cannot be flattened at the above temperatures. In such cases, the temperature of the warming stage should be raised, but not high enough to melt the paraffin. Should the desired result still not be obtained, there is not much that can be done except to follow through with the rest of the procedure and accept the results.

When the sections are flattened, remove the slides from the warming apparatus. Carefully drain off most of the water, leaving the slides just wet enough so that the sections do not adhere too tightly to the glass. With a needle or the point of a scalpel, very carefully move the sections into the exact positions they are to occupy permanently. In doing this, touch only the paraffin which surrounds the actual sections of tissue. If the ribbon is curved, cut apart the sections with a rocking motion of a scalpel blade and straighten the rows. Drain the slide as completely as possible, touching its edge with a piece of blotting paper or cloth.

Many laboratories (especially clinical laboratories) use small thermostatically controlled water baths for flattening sections. Special equipment is manufactured for this purpose by several firms. The water is warmed to the correct temperature, and individual sections or ribbons are simply floated on top of the water. After they have become stretched and flattened, an albumenized slide is dipped into the water under them. With the help of a dissecting needle, the sections are held fast to the slide as it is lifted from the water and drained. The sections are then oriented as desired, and the removal of excess water completed. This method certainly does work well with individual sections or short strips, but it is not suited to handling serial sections when as many as three or four lengths of narrow ribbons are to be mounted parallel on the same slide.

Drying. Keep the slides flat for about half an hour. After this, they may be put into boxes or racks if desired. Dry the slides completely by keeping them in an incubator at about 37° C. for 3 or 4 hours or overnight, or by allowing them to stand at room temperature for at least 24 hours. In a warm, dry climate, preparations will dry in a few hours at room temperature.

Sections affixed to slides can be stored for months or years, but they should be kept in dust-proof boxes. The staining qualities of the material remain substantially unaltered, even after many years.

Reagent and Staining Jars and Slide Holders. Each worker should have on hand 18 or more jars to contain the various liquids in which the slides and attached sections are to be immersed during the course of preparation.

When only a few slides of each tissue are to be handled, Coplin jars will probably be the containers of choice. These are made with ground glass covers, as well as with molded screw caps. The screw caps are effective in preventing loss of liquids by evaporation, and will be preferred when reagents are to be stored in the jars for rather long periods of time. However, Coplin jars with ground glass lids are a little easier to use during the actual operations of transferring slides. It may be advisable to have some of each type available.

Coplin jars are made to take 5 slides. Crowding in more by inserting them in the jar diagonally, or by putting two slides back-to-back in the same groove, is risky if sections are at all close to the edges of the slides.

If more slides than a Coplin jar will hold comfortably are to be treated in the identical manner, it is possible to save much time and labor by

A *B*

Fig. 38. Types of microscope slide dishes in common use. *A*, Coplin jar. *B*, Duplex staining dish. The removable rack is transferred from one reagent to another. (Courtesy of T. C. Wheaton Co., Millville, N. J.)

using racks which can be transferred quickly from one dish to another. Slide racks of glass and metal are available in various sizes and designs, and although they are very convenient for processing large quantities of slides, they have some undesirable features. They are expensive, and the containers into which they fit take large quantities of reagents. The racks carry a considerable amount of liquid from dish to dish, and this necessitates rather frequent renewal of some reagents, especially absolute alcohol and clearing agents. Avoid metal racks having only a thin coating of non-corrosive metal, which rust as soon as this chips off. Slides which are to be coated with Parlodion must be taken individually from the rack and then replaced in it after coating.

A tight spiral of heavy spring wire (preferably stainless steel) is a very useful slide holder. The ends of 6 or 8 or more slides may be inserted between the coils, which should be about $\frac{3}{4}$ inch in diameter. If the spiral is so tight that it is difficult to insert the slides, one end of the spring should be drawn out into a hook which can be placed over a nail driven into the wall or a bench. This enables one to stretch the spiral enough to insert the slides easily. Beakers of appropriate size, or even cheap jelly glasses or jars, may be used for the fluids through which slides on spiral clips are being processed.

Shell vials with an inside diameter of slightly over 1 inch and a depth of about 3 or $3\frac{1}{2}$ inches will hold two slides, placed back-to-back, and require minimum quantities of reagents.

Label and arrange the staining jars as shown in Figure 39. Fill each jar with the proper liquid, to a depth slightly greater than will be necessary to cover the sections. In the diagram, arrows indicate the order in which slides are put into the different reagents. The top line of jars is used for slides which are being moved down through the alcohol series toward water, and the bottom line is for slides being moved up the series to-

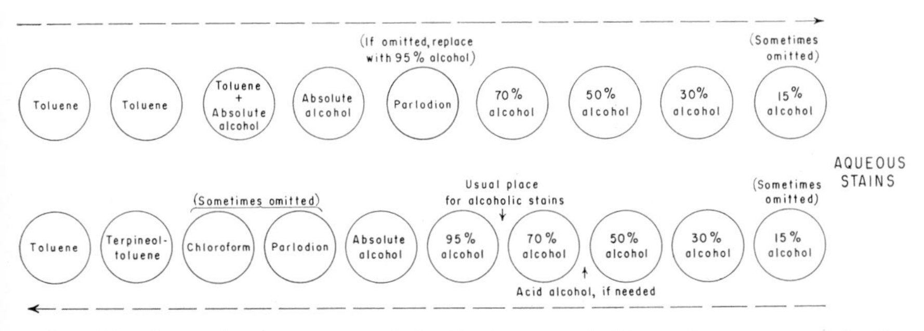

FIG. 39. General arrangement of Coplin jars or substitute glassware used in the treatment of paraffin sections. The arrows indicate the order in which the slides bearing the sections are placed in the various reagents. The jars in the upper row are used for removing the paraffin from sections, coating them with Parlodion, and passing them into water or any desired concentration of alcohol. The jars in the lower row are used for dehydration and clearing the stained sections.

ward toluene. The reason for this is that the toluene employed for removing paraffin will soon contain considerable paraffin; the absolute alcohol following toluene takes up much of that liquid; and all of the alcohols are likely to absorb traces of fixing agents from the sections. These contaminated reagents may prove more or less injurious to stained sections which are being prepared for mounting.

Lay narrow strips of absorbent paper, several layers thick, in front of each line of jars. These are used to absorb liquids which drain from the slides. Paper towels, folded lengthwise, are convenient.

11. Removal of Paraffin. Place the slides in the first jar of toluene of the upper row, and leave them for 3 minutes or longer. Lift them out, hold them above the jar for a few seconds so that most of the toluene on the slides may drip back into the jar, blot their lower edges on the absorbent paper in front of the jar, and immediately place them in the next jar of toluene. Allow them to remain in this for at least 3 minutes. If the toluene is very cold or the sections are more than 20 μ in thickness, leave them in each jar of toluene for 5 minutes. Removal of paraffin can be hastened by warming the toluene very slightly on an electric plate or warming table (never over a flame!). It should be warm, not hot. Sections may remain safely in toluene for an hour or more.

Note. Repeat the procedure of draining and blotting each time you transfer slides to another reagent, except when definite instructions are given to the contrary. *Do this quickly, never allowing the sections to become dry.*

12. Transfer to Alcohol. Place the slides in a mixture of equal parts of toluene and absolute alcohol for at least 1 or 2 minutes. Transfer them quickly, without draining or blotting, to absolute alcohol. Leave them in this for 1 or 2 minutes or longer. If many preparations are being made, it is well to pass them into a second jar of absolute alcohol, in order to assure complete removal of toluene.

Concerning Failure of Sections to Adhere to the Slides. If sections fall off or become loosened when the slides are placed in alcohol, or at any later stage in preparation, the most likely cause is presence of traces of grease on the slides. (Refer to page 20 for methods of cleaning slides.) However, the difficulty may result also from the following causes: failure to flatten the sections perfectly; applying too thin a film of albumen or gelatin on the slides; taking so long to flatten the sections that much of the albumen is washed into the water and later removed when the slides are drained; overheating the slides so much during the process of spreading sections that the adhesive properties of the albumen are weakened; exposing the slides to alkaline solutions which alter the albumen.

13. Further Treatment, Depending upon Method of Staining. What is to be done next varies in different cases. Select a proper method from the following alternatives:

Material Stained Before Embedding. This section applies to stains which are to remain permanently in the tissues. It does not apply to stains used only to make specimens visible, for the purpose of orientation during embedding and sectioning, and which are to be removed.

a. If destaining or counterstaining is not required, proceed directly to step 14. This applies ordinarily to thick sections of embryos which have been stained, before embedding, with one of the carmine mixtures, and to nerve tissues impregnated by Ramón y Cajal's silver methods.

b. If a counterstain is desirable, there are two possibilities. Some counterstains are dissolved in strong alcohol (70 to 95%). These include eosin and orange G. Eosin is widely used as a counterstain for alum hematoxylin; orange G is not really suitable after alum hematoxylin, but it may be advantageously employed after carmine stains. In these cases, transfer the slides directly to the counterstain if this is made up in 90 or 95% alcohol, or pass it gradually to the counterstain, then continue as instructed in connection with the particular stain. Other counterstains are dissolved in water. Among the more common of these are aniline blue and Mallory's aniline blue-orange G mixture, which are sometimes effective as counterstains for carmine. If an aqueous staining mixture is to be used, coat the sections with Parlodion (step 14) and pass them down through the alcohols to water.

c. If the sections are overstained, they should likewise be coated with Parlodion (step 14) and brought into the descending series of alcohols. Destaining is then carried out as described in connection with the stain employed.

Material Not Stained Before Embedding. In the great majority of methods, tissues are embedded and sectioned before they are stained. When this is the case, sections are brought down through the series to 70% alcohol. However, it is advisable first to coat them with Parlodion, except when certain stains are to be used.

14. COATING WITH PARLODION. The purpose of this step is to insure that the sections do not come loose from the slides. This treatment is advised for all sections which are to be stained with hematoxylin or one of the acid coal-tar dyes, or which are to be impregnated with salts of silver and gold. It is also indicated for sections of tissue which was overstained before embedding, and which therefore should be destained.

Of various nitrocellulose products which may be used for coating sections before staining, Parlodion (Mallinckrodt Chemical Works) is especially good. Use a 1% solution in a mixture of equal parts of ether (anhydrous) and absolute alcohol.

Transfer the slides, without draining or blotting them, from absolute alcohol into the solution of Parlodion. After a minute or more, remove the slides individually from the Parlodion. Let the excess from each slide drain off onto a piece of blotting paper, and as soon as the coating congeals to form a very thin, soft film, lower the slide into 70% alcohol. Do this with a steady and even motion. A too rapid, too slow, or unsteady motion may cause ridges or creases in the film. Leave the slides in 70% alcohol for at least 2 minutes. If the sections have been coated, they may be left in 70% alcohol for several hours.

If coating with Parlodion is to be omitted, simply transfer the slides from absolute alcohol to 95% alcohol for 2 minutes, then to 70% alcohol.

15. ELIMINATION OF REMAINING FIXATIVES. In some instances the material has already been stained before embedding, and we may then proceed to destain the sections, if that is necessary, or to counterstain them (step 16). Sections of unstained material are in many cases ready to be taken down through the alcohols and into the stain. However, if certain fixatives were used, the tissue may still contain substances which will interfere with staining or make the preparations unsightly. Removal of such substances should be accomplished before taking further steps.

a. Mercuric chloride is the worst offender. Even though the material was iodized before being embedded, black specks of a mercurial compound are likely to be scattered through the tissues. To eliminate this, place the slides in 70% alcohol containing a sufficient amount of an iodine solution (p. 93) to give it a light wine color. Examine the sections frequently. When mercury deposits are no longer observed, transfer the sections to 70% alcohol and leave them until they are free from the color of iodine.

b. Picric acid is removed simply by leaving sections in 70 or 50% alcohol until they no longer show a yellow color. Extraction of picric acid may be hastened by warming the alcohol. The use of an alkali for this purpose is not recommended.

c. Chromic acid or potassium dichromate, if not completely washed out, may give the tissue a yellow, brown, or greenish color. To remove the foreign color, pass sections down the alcohol series to water, then place them in a 0.25% aqueous solution of potassium permanganate for 5 to 10 minutes. Rinse the sections in distilled water, and place them in a 5% aqueous solution of oxalic acid for 5 to 10 minutes. Finally, wash them in running water for 20 to 30 minutes. This method is also useful for removing unwanted dyes.

16. STAINING, COUNTERSTAINING, DEHYDRATION. Ordinarily, the sections are stained first with a nuclear dye (such as alum hematoxylin, iron hematoxylin, or safranin) and then with one or more dyes (such as eosin or fast green) to bring out cytoplasm, connective tissue, muscle, and other tissue elements. Following this, the dehydration of the sections is completed, and they are cleared and mounted.

If the reader is not already familiar with the various classes of stains, their uses, and effective combinations, he should refer to Chapters 21 to 24. Explicit directions for the methods most widely used in staining paraffin sections are given in Chapters 22, 23, 24, and 25. In each case, the directions begin with the sections in 70% alcohol and cover all of the steps in staining, destaining, counterstaining, and dehydration. *Note that the steps in most staining methods are lettered in place of being numbered.* This system is intended to avoid the confusion which would result from use of disconnected or repeated numbers. Consider the steps in the staining method as 16*a*, 16*b*, and so on. After the last lettered step, the sections will be in absolute alcohol, and ready for step 17.

In choosing a stain or combination, bear in mind not only the structures which are to be demonstrated, but also the reagent which was used for fixation. Some stains give good results after many different fixatives, while others may be used with success only after fixation in a certain type of reagent. Beginners should first try staining sections of amphibian tissue (fixed in Bouin's fluid or Zenker's fluid) with alum hematoxylin and eosin. Similar sections should then be stained according to Mallory's triple stain method (p. 398) or Heidenhain's "Azan" modification of Mallory's method. Heidenhain's iron hematoxylin method (p. 360) should also be practiced, with and without counterstaining some of the sections in orange G. For the iron hematoxylin method, thin (6 to 8 μ) sections of testis from a crayfish, grasshopper, salamander, or mammal (fixed in Bouin's fluid or one of its modifications) will provide good material. With this background, one should be able to apply successfully any of the other standard or special methods described in this book.

17. CLEARING. In order to render them more transparent, and to preserve them as permanent preparations, the sections are mounted in balsam or some other natural or synthetic resin. Because alcohol will not mix with these resins, it must first be replaced with a solvent which will. Toluene is very good for this purpose. It is sufficiently volatile not to interfere with rapid drying of the mounts, and yet not so volatile that the sections are likely to dry out before the mounting process can be completed. Xylene is somewhat less volatile than toluene, and for this reason may be preferred by some workers. Benzene is too volatile to be used routinely. Toluene is suggested here, in part because this hydrocarbon is more generally useful in the microtechnique laboratory, especially in the method of embedding in paraffin.

Toluene and other solvents of resinous media have high indexes of refraction and, for this reason, clear the tissue. Failure of the sections to clear indicates that they have not been dehydrated thoroughly. Toluene, and also xylene, have some tendency to shrink tissues. On this account it is advisable, except after certain stains, to give the sections a protective coat of Parlodion before clearing them.

Re-Coating With Parlodion. Dehydration in absolute alcohol will have removed the film of Parlodion ordinarily applied to the sections before they were stained. This is desirable, because the Parlodion film generally takes up at least a little of the stain. Before placing the sections in toluene, it may be wise to give them a new coating of Parlodion which will infiltrate the delicate tissue elements with a resistant matrix and protect them from the shrinking action of toluene. Moreover, such a coating will protect the sections from injury while they are being mounted and throughout their useful life as permanent preparations. Pressure on the coverglass will be less likely to harm them and, if a coverglass is broken by accident, less danger will be involved in soaking it off and replacing it than would be the case if the sections were not coated.

Unfortunately, some stains are extracted very rapidly by absolute alcohol and, in addition, impart their color to nitrocellulose. For this reason, the coating should be omitted when such stains are used. The more important of these stains are methylene blue, basic fuchsin, safranin, crystal violet, and fast green FCF. If one of these has been used, simply transfer the slides after a minimum time in absolute alcohol to terpineol-toluene.

Unless the sections have been stained with one of the dyes listed above, transfer the slides to 1% Parlodion in ether-alcohol. After 1 or 2 minutes, drain them, touching one end to blotting paper or filter paper. Observe them constantly. As soon as the coating has congealed to form a soft film, lower the slides into chloroform for 1 or 2 minutes. Do not allow them to stand longer in this reagent, as it slowly extracts some dyes.

Transfer to Terpineol-Toluene. Place the slides in terpineol-toluene until the sections are cleared. No traces of cloudiness should remain in either the sections or the nitrocellulose film. Terpineol-toluene is recommended as the initial clearer because it will absorb moderate quantities of water, whereas pure toluene will become foggy and fail to clear the sections if they contain more than minute traces of water.

Transfer to Toluene. Place the slides in the final jar of toluene used for clearing. (Should the sections or the coating of Parlodion become cloudy and fail to clear again within a minute or two, return the slides to terpineol-toluene for several minutes and then put them again into toluene.) After the sections are cleared, leave them in toluene for another 2 or 3 minutes, or until you are ready to mount them. If the sections contain keratin or chitin, it is advisable to mount them as soon as possible, because toluene causes these substances to shrink and curl. As a general rule, it is best not to leave any sections in toluene more than half an hour, in order to minimize shrinkage. If many preparations are being made, it is advisable to pass them into a second jar of toluene before mounting; the same procedure should be followed if initial clearing was accomplished in toluene instead of terpineol-toluene.

18. MOUNTING. Paraffin sections are generally mounted in Canada balsam or a synthetic resin. When delicate and lightly stained cellular structures are to be studied, some workers mount sections in euparal or

some other sandarac medium, in which case the sections are mounted from absolute alcohol (step 12) without being cleared. The properties, preparation, and testing of various resinous media are discussed in Chapter 27.

Place several clean coverglasses of the proper size in a clean Syracuse dish or on a smooth piece of paper. Stand a balsam bottle of rather thin mounting medium in a convenient place and remove its cover. Light an alcohol lamp or a small Bunsen burner and adjust it to maintain a very small flame.

Take a slide from the jar of toluene, drain it, and identify the side which bears the sections. If the record number or other mark has been scratched on this side, as suggested previously, this will be simple. Otherwise, observe the slide while a strong light falls upon it. The back of the slide, including the area back of the sections, will reflect light evenly. If the slide is turned over, the sections will not reflect light as strongly as the glass.

Quickly wipe the back of the slide with a soft cloth, removing the film of Parlodion if one has been applied. If the sections do not cover most of the face of the slide, also wipe the unoccupied parts of that surface, but be careful not to wipe so close to the sections as to endanger them. *Perform this and the subsequent steps rapidly and never allow the sections to become dry.* If drying appears imminent, dip the slide into toluene, then drain it again.

Most persons prefer to apply the coverglass while the slide is lying on a smooth piece of paper or on a sheet of glass. However, some prefer to hold the slide horizontally in the left hand.

To Mount With a Square or Circular Coverglass. With a glass rod place a small drop of thin mounting medium upon the sections, in the center of the area to be covered. Pick up the coverglass with the tips of a pair of forceps and warm it over a small flame. Holding the coverglass at an angle of about 30° (Fig. 14, p. 161) slowly lower it until it contacts the mounting medium and the side opposite to that held by the forceps touches (or nearly touches) the slide. Then carefully lower the edge which is being held, and simultaneously withdraw the forceps. If the whole operation is performed smoothly, the medium will spread evenly and bubbles of air will not be enclosed in the mount. Should a few small bubbles be trapped in it, they will probably be expelled as the mount dries. They can often be removed by warming the slide gently. In case large or numerous bubbles occur, replace the slide in toluene until the coverglass falls off, then remount it. If the proper amount of medium has been used, it will form a very narrow rim around the coverglass.

To Mount With a Rectangular Coverglass. Place upon the slide 2 to 4 drops of thin mounting medium, arranging them in a row about $\frac{1}{4}$ inch from the edge of the slide which is nearest to you. The exact amount of medium that is to be used depends upon the size of the coverglass. Lift the rectangular coverglass with forceps, holding its upper edge or upper right-hand corner, and warm it over the flame. Place the long edge of the coverglass upon the slide, close to the edge which is nearest to you, making certain that the coverglass is centered with respect to the rows of sections. Slowly and steadily lower the opposite edge to the slide. Observe the remarks in the preceding paragraph concerning exclusion of air bubbles.

19. Drying, Labeling, and Storage. These final steps deserve careful attention. They determine to a very large extent the permanence and usefulness of the preparations which have required so much patient and careful work. These matters are discussed in Chapter 28, page 463, which should be consulted at this point.

Precautions to be Observed in Processing Paraffin Sections. *a.* Alcohols which have become colored by a stain should thereafter be used only for sections which bear the same stain. This is true also of the solutions of Parlodion, chloroform, and terpineol-toluene, which may absorb some stains to a certain extent. It is desirable to keep one series of these reagents for sections which are to be counterstained with eosin or orange G, and another series for sections which are not to be counterstained. In the case of stains which are not regularly employed, use minimum quantities of reagents and discard them when the work is completed.

b. Keep the reagent jars covered tightly at all times when they are not in use. This is especially important in the case of absolute alcohol and the solution of Parlodion. Filter the reagents frequently, in order to avoid the accumulation of debris which might cling to the preparations.

c. As soon as you notice that the toluene in the first jar fails to dissolve paraffin quickly from sections, throw it away. Pour the toluene from the second jar into the first and fill the second with new toluene. One hundred milliliters of toluene should serve to remove the paraffin from several hundred average preparations.

d. Replace the absolute alcohol used for dehydration whenever you notice that a section is not cleared quickly when placed in terpineol-toluene, or when the alcohol has become deeply colored by a stain.

e. The solution of Parlodion used for coating slides should be kept in glass-stoppered bottles, and poured into the jar only when it is needed. Otherwise, it will soon become milky, or thicken to a jelly-like consistency.

f. Should any reagent become cloudy or form a precipitate, read the paragraphs in which it is discussed in an effort to find the reason. An instructor with experience may quickly locate the source of difficulty. If the reagent proves to be fit for further use, filter it; otherwise, discard it before it leads to trouble.

REFERENCE

Blandau, R. J., 1938. A method of eliminating the electrification of paraffin ribbons. Stain Technology, *13*, 139–141.

CHAPTER 16

Nitrocellulose Method

In 1879, about 10 years after Klebs began to use a coating of paraffin to support specimens to be sectioned, Duval and Latteaux introduced the practice of dipping specimens in a solution of nitrocellulose to achieve the same purpose. In time, the technique of dipping evolved into a method of saturating the tissue completely with nitrocellulose after dehydration in alcohol. This is essentially the method still in use, although it has been considerably refined.

"CELLOIDIN" AND OTHER FORMS OF NITROCELLULOSE. Nitrocellulose ("gun cotton") is a term which applies to a variety of cellulose nitrates produced by treating cotton with mixtures of nitric acid and sulfuric acid in various concentrations. The less explosive of these nitrates are quite soluble in alcohol and ether, which makes them suitable for infiltrating dehydrated tissues. A product of this type is referred to as *soluble gun cotton*, or *pyroxylin*, and a solution of it is called *collodion*. A particular type of nitrocellulose prepared by the firm of E. Schering for embedding tissues was patented under the name of "Celloidin," and in time this trade-name came to be used as a sort of generic term for various nitrocelluloses used for embedding.

After World War I, the firm of E. I. du Pont de Nemours & Company developed an excellent product called "Parlodion." This is now being manufactured by the Mallinckrodt Chemical Works. The fact that Parlodion is non-inflammable is a characteristic favoring its use in laboratory work, and makes possible its shipment in completely dry form.

A *low-viscosity nitrocellulose* (nitrocellulose, R. S., $\frac{1}{2}$ second), made by the Hercules Powder Co., though very inflammable, has a good cutting consistency. Because of its low viscosity, it is possible to begin infiltration with a relatively strong solution and to shorten the time required for infiltration. The highly concentrated nitrocellulose, when it has hardened sufficiently, assures a very firm consistency and permits the cutting of sections which are thinner than those generally obtainable with other forms of nitrocellulose.

Advantages and Disadvantages of the Nitrocellulose Method. Embedding in nitrocellulose does not require the use of heat and provides a matrix which is more elastic and more cohesive than paraffin. Certain structures which cannot be successfully sectioned if they are embedded in paraffin can be cut remarkably well after being embedded in nitrocellulose. These include the following: (1) very hard tissues, such as bones, teeth, scales, and the lens of the eye; (2) large objects, such as advanced embryos of vertebrates or entire brains; (3) tissues in which vessels or cavities have been injected with gelatin; (4) watery tissues, which shrink excessively if subjected to hot paraffin. Because of its elasticity, nitrocellulose is also more suitable when it is necessary to cut very thick sections (over 20 or 25 μ).

Infiltration with nitrocellulose takes a considerably longer time than infiltration with paraffin, unless heat is applied; but the use of heat offsets one of the real advantages of the nitrocellulose method. The cutting and handling of nitrocellulose sections seem to require greater manual dexterity, and the mounting of nitrocellulose sections in serial order is at best very laborious. However, the popular belief that thin sections cannot be cut in nitrocellulose is not true. If a suitable product is used and the block in which the tissue is embedded is hardened to just the right consistency, sections can be cut as thin as 5 μ, and even thinner.

The old saying "Use paraffin when you *can*, nitrocellulose when you *must*" is based upon certain obvious disadvantages of the nitrocellulose method. However, there are some workers who would like to use nitrocellulose for all materials, and who yield to the paraffin method only when the efficient preparation of serial sections is required.

CONVENTIONAL NITROCELLULOSE METHOD

This procedure is satisfactory for most purposes. It will take several weeks to complete the process of infiltration and to harden the blocks to a suitable consistency for cutting. The only way to speed up infiltration is to use heat, as directed on page 269.

1 and 2. Fixation and Washing. Carry out these processes exactly as in preparing tissues for the paraffin method (steps 1, 2). Take care to prepare the tissues properly for fixation, cutting them into small pieces if necessary, and to use a fixative which is suitable for the particular tissue at hand. Wash them in the manner which is appropriate. If washing or storage of tissue in alcohol is desirable, use only ethyl or methyl alcohol. Note that tissues fixed in fluids containing mercuric chloride should be treated with iodine, and that bones, teeth, and similar structures must be decalcified. If staining or impregnation with metals is desirable before infiltration is begun, then either of these procedures may be introduced into the schedule as soon as washing has been completed.

3. Dehydration. Dehydrate the tissues by means of a graded series of alcohols, exactly as in the paraffin method (see step 3, p. 212). *Use ethyl alcohol or methyl alcohol.* Isopropyl alcohol is not satisfactory for this

purpose. Leave material in absolute alcohol for at least the length of time suggested on page 214. Longer exposure will do no harm, but *insufficient dehydration will result in failure of nitrocellulose to infiltrate the tissues.* During dehydration and infiltration, keep material in wide-mouth glass bottles with air-tight covers. Screw-cap jars of 1 or 2 ounce capacity are efficient and inexpensive. Once the material has reached absolute alcohol, it is particularly important that the container be closed tightly, because absorption of moisture from the atmosphere may have disastrous consequences.

4. TRANSFER TO ETHER-ALCOHOL. After the material has been thoroughly dehydrated, transfer it to the following mixture:

Ether-Alcohol

Ether, anhydrous	1 part
Ethyl or methyl alcohol, absolute	1 part

Leave small objects in this for at least 12 hours. Pieces of vertebrate tissue 5 or 6 mm. thick should be left in it for 24 hours; larger objects, depending upon their size and density, should remain in ether-alcohol for a longer period of time. The purpose of treatment with ether-alcohol is to saturate the tissue with ether, which is more effective than alcohol as a solvent for nitrocellulose.

5. INFILTRATION WITH NITROCELLULOSE. Nitrocellulose dissolves rather slowly, so that the thin and thick solutions of nitrocellulose for infiltration should be made up a few days before they are to be used. Formulas are given below for preparation of these solutions from either Parlodion or low-viscosity nitrocellulose.

Thin Nitrocellulose

Parlodion (pyroxylin, purified, in strip form)	4 gm.
Ether-alcohol	100 ml.

Rinse the Parlodion with distilled water and dry it before weighing. Place the ingredients in a glass-stoppered wide-mouth bottle which is to be the stock bottle of the solution. Stir or agitate the mixture every day until the Parlodion is dissolved.

Nitrocellulose, low-viscosity	20 gm.
Ether-alcohol	100 ml.

This type of nitrocellulose, being very inflammable, is shipped in water or weak alcohol. Squeeze the water or alcohol out of an amount sufficient to make as much of the solution as is desired, rinse it briefly in 95% or absolute alcohol, and dry it on clean paper toweling or filter paper before weighing it. Use a glass-stoppered wide-mouth bottle for making up the solution, and stir or agitate the mixture at least once a day until the solution is uniform.

Thick Nitrocellulose

Parlodion	8 gm.
Ether-alcohol	100 ml.
or	
Nitrocellulose, low-viscosity	40 gm.
Ether-alcohol	100 ml.

Procedure for Infiltration. Before placing the specimens in thin nitrocellulose, they should be left for a few days in a still more dilute solution prepared by mixing the stock solution of thin nitrocellulose with an equal amount of ether-alcohol. A simple way to begin infiltration is to pour off about half the ether-alcohol in which the specimens have been standing and replace it with about the same volume of thin nitrocellulose.

The more slowly infiltration is carried out, the more thoroughly the tissues will be permeated by nitrocellulose, and the better the results are likely to be. The length of time which the material should remain in each solution of nitrocellulose depends upon the size and permeability of the specimens. Material is conventionally left in thick solutions for a longer time than in thinner solutions, because penetration by thick solutions is slower.

Slices of moderately permeable tissue (such as stomach and brain) about 4 mm. thick should remain in the thinner solutions of nitrocellulose for at least 3 or 4 days, and in thick nitrocellulose for at least 10 days or 2 weeks. It is best to leave them for a week in each of the thinner solutions, and for 3 weeks or more in thick nitrocellulose.

Slices of similar tissues ranging from 5 to 8 mm. in thickness should remain in each of the first two solutions for at least 10 days, and preferably for as long as 4 weeks.

Impermeable tissues such as bones and teeth require a much longer time for infiltration. Very thin slices should remain in each of the thinner solutions for at least 2 or 3 weeks, and in thick nitrocellulose for a month. It is highly desirable to leave such objects in thick nitrocellulose for several months.

On the basis of the above information, one may estimate approximately the minimum time required for large objects of various types. For example, a 40-mm. embryo should remain in each of the first two grades for 3 or 4 weeks, and in thick nitrocellulose for 2 or 3 months. More than a year may be required for the complete infiltration of an entire human cerebral hemisphere.

When an appropriate length of time has elapsed, pour off the dilute solution and replace it with thin nitrocellulose. Use about twice as much nitrocellulose as is necessary to cover the material. After the material has remained in thin nitrocellulose for about three-quarters of the required time, loosen the lid so that some of the ether-alcohol may gradually evaporate. This must be done very cautiously, because too much evaporation will cause the nitrocellulose to solidify, and also because moisture may be absorbed from the atmosphere.

When the material has remained long enough in thin nitrocellulose, and the solution has been somewhat thickened by evaporation, pour it off and replace it with thick nitrocellulose.

When the material has been in thick nitrocellulose for about three-quarters of the necessary time, loosen the lid slightly so that some of the ether-alcohol may evaporate. Should the nitrocellulose become the least bit milky because of absorption of water from the atmosphere, immediately replace it with a fresh solution. Should it solidify, add to it a small quantity of ether-alcohol.

If the material is to be stored indefinitely in thick nitrocellulose, omit the foregoing step, and carefully seal the bottles by dipping their lids and necks in melted paraffin. When it is desired to prepare stored material for embedding, simply loosen the lids a little.

6. Embedding and Hardening. The specimens are next oriented in the desired positions, and the nitrocellulose is hardened by immersion in chloroform, and then in glycerol-alcohol.

Select a grooved fiber object holder of appropriate size. Cut a strip of stiff white paper long enough to overlap considerably when it is wrapped

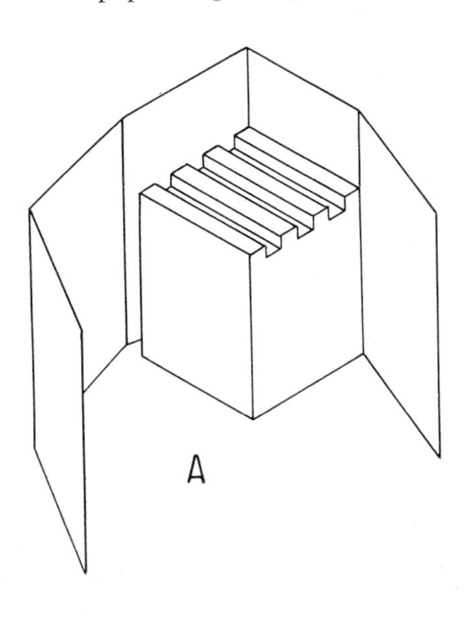

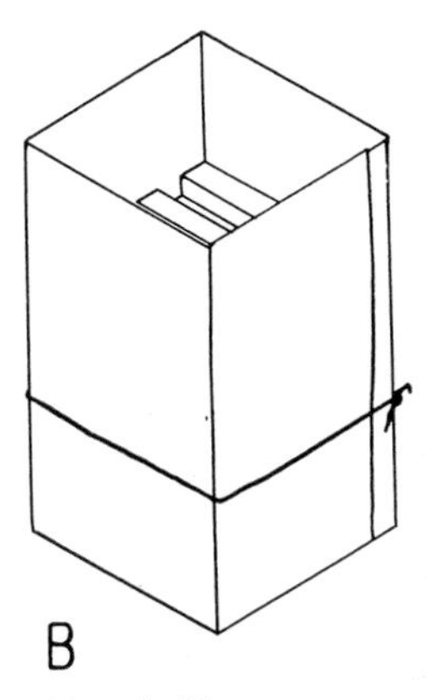

Fig. 40. Method of folding stiff paper around a fiber object holder preparatory to embedding tissue in nitrocellulose. The paper is tied tightly to the object holder, and the thick nitrocellulose in which the specimen is to be embedded is poured into the collar raised above the slotted upper surface.

around the holder, and wide enough to project above the holder, forming a collar which will extend somewhat beyond the object which is to be embedded (Fig. 40*A*, *B*). Write the name or number of the material on this paper, and then wrap it around the holder and tie it in place with a thread.

Immerse the holder and paper in ether-alcohol for a minute or two. Then pour into the cup formed by the paper collar a small amount of thick nitrocellulose from the bottle containing the material. Pick up the desired piece of tissue with a pair of forceps, and place it in the paper cup. Carefully orient it so that the proper face will be presented to the knife. In this connection, it may be advisable to read once again the directions for orienting specimens which are to be sectioned (p. 207). Pour in more nitrocellulose from the bottle which contained the material, until the object is more than covered. Air bubbles may be formed in the mass. If not very numerous, they may be removed by puncturing them with a needle moistened in ether-alcohol or by sucking them out with a pipette drawn out to a fine point. In rare cases where bubbles are very numerous, it is best to expose the block to ether vapor for an hour or two, by placing it in a covered vial or dish containing a very little ether.

Preliminary Hardening by Evaporation. Put the fiber holder bearing the embedded material in a wide-mouthed bottle. In doing so, be careful not to disturb the material. Leave the lid or cork stopper somewhat loose, so that some of the ether-alcohol may slowly evaporate from the nitrocellulose. Allow it to stand thus until the nitrocellulose thickens to form a rather firm gelatinous mass. Should the upper surface of the tissue appear to be in danger of exposure, add more thick nitrocellulose. The rate of evaporation may be controlled by loosening or tightening the lid or stopper. It should be regulated so that small blocks will reach the desired consistency in 18 to 24 hours, and large ones in a longer period, in proportion to their size. Do not allow evaporation to proceed too rapidly or too long, because the nitrocellulose may become so hard and shrunken as to ruin the material. After some experience, one will be able to judge whether the nitrocellulose is firm enough.

This preliminary hardening is important if sections less than 20 to 25 μ thick are to be cut. With material which is to be cut into relatively thick sections, the nitrocellulose can be hardened immediately with chloroform, as directed below.

It is quite possible to embed specimens in boxes of stiff paper or various other molds, including some of the deeper glass dishes (as stender dishes). After the block has become firm by slow evaporation of ether-alcohol (if a glass container has been used, the nitrocellulose will have drawn away from the sides), it is trimmed with a razor blade to the shape and size desired. One end of it is soaked in a shallow layer of ether-alcohol for a few minutes, and this end is then affixed to a fiber object holder with a sufficient amount of thick nitrocellulose. After the cementing nitrocellulose has been allowed to harden gradually in a closed container for at least a few minutes, and preferably longer, the block and object holder are dropped into chloroform and processed further as directed below.

Hardening in Chloroform. Pour enough chloroform into the bottle to fill it to a depth of about 1 cm. and tighten the lid or press the stopper firmly into place. Allow it to stand for several hours or overnight. Material may be left for many days in chloroform vapor without risk of injury.

Pour in more chloroform until the nitrocellulose is submerged. Replace the stopper or lid and allow the nitrocellulose block to remain immersed in chloroform for several hours. Small blocks will be sufficiently hardened within 2 or 3 hours, while very large ones may require one to several days. For all but very large blocks it is a safe and convenient practice to leave them overnight.

Hardening in Glycerol-Alcohol. Remove the block from the chloroform, let it drain a few seconds, and place it in a bottle or dish containing enough glycerol-alcohol to cover it. Glycerol-alcohol is made by mixing equal parts of glycerol and 95% alcohol. The nitrocellulose may at first become slightly milky, but it will soon regain its transparency. The block should remain in the mixture until the nitrocellulose becomes quite hard and inelastic. Small blocks are sufficiently hardened in a few days. Large ones may require several weeks. Material may be stored indefinitely in this mixture.

7. SECTIONING. *The Sliding Microtome.* Material embedded in nitrocellulose is always sectioned with a sliding microtome. The essential feature of this type of microtome is that the blade of the knife is set at an angle to the path along which it moves, so that it is drawn through the material with a slicing stroke. The block is advanced a known distance after each section is cut. The knife blade moves in a nearly horizontal plane, rather than in a vertical plane as in a rotary microtome. Attempts to adapt rotary microtomes to cutting nitrocellulose have been made without much success.

Examine a sliding microtome. Locate the screws which clamp the knife into the holder, and those used to regulate its tilt and slant. Examine the object clamp and learn how to orient the block in any desired position. Find the lever or screw which regulates the thickness of sections. Push the knife block back and forth on its track and observe the action of the feeding mechanism. In case the microtome is not provided with an automatic feeding mechanism, operate the lever provided for hand feeding.

Slant of the Knife. This refers to the angle existing between the knife edge (*k* in Fig. 41*A*) and the line of motion followed by the knife (*d*). In cutting most types of material it is desirable to set the knife very obliquely, so that the knife edge is at an angle of approximately 10

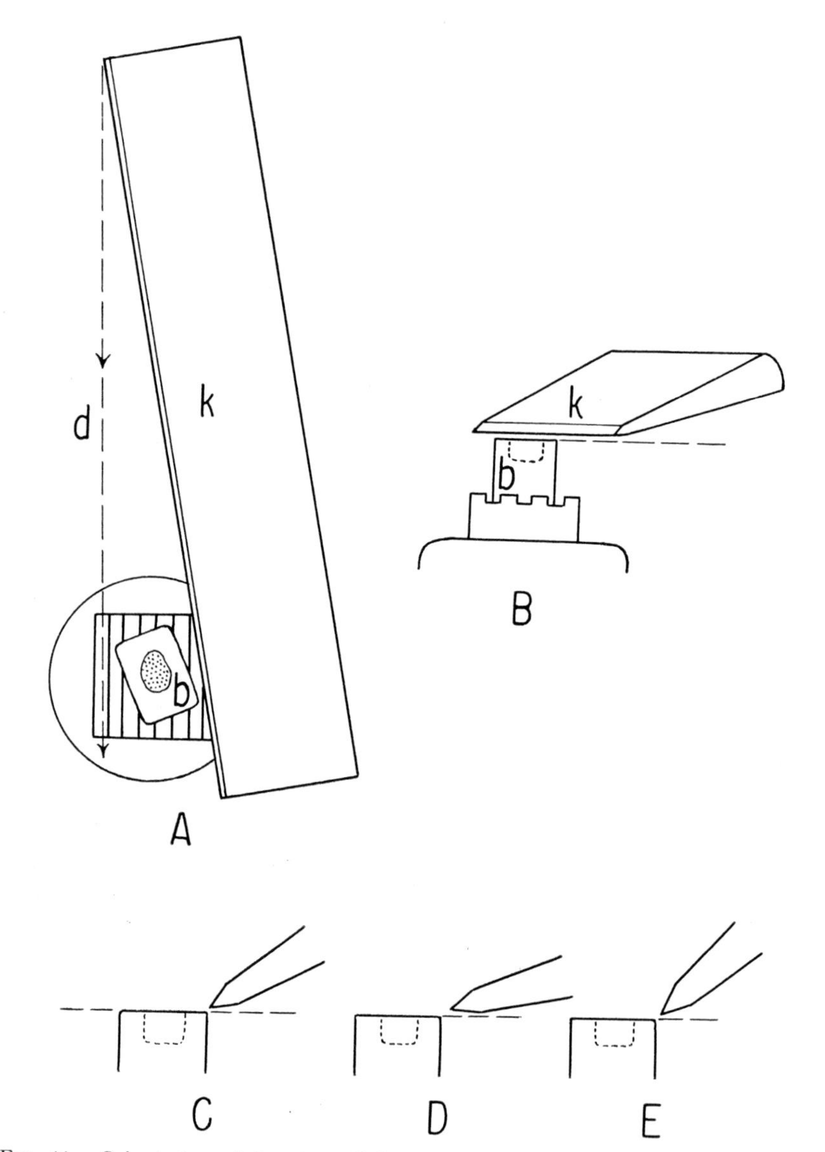

FIG. 41. Orientation of the nitrocellulose block and adjustment of the knife in the sliding microtome. *A*, View from above, showing the knife (*k*) slanted to an angle of about 10° to its line of motion (dashed line *d*), and the nitrocellulose block containing the object (*b*) turned with its longest side at a slight angle to the knife edge. *B*, End view, showing the knife correctly tilted. *C*, Detailed view of correctly tilted knife. The lower cutting facet makes an angle of about 7° with the plane of section (dashed line). *D*, Knife insufficiently tilted; the knife's shoulder strikes the block. *E*, Knife tilted too much, so that it acts as a scraper.

degrees to the line of motion. In case the material is very soft, or the nitrocellulose is not well hardened, it is desirable to increase the angle by 2 or 3 degrees (no more).

Tilt of the Knife. This refers to the angle formed between the plane of section and a line which passes from the edge to the back of the knife. As in cutting sections on a rotary microtome, the tilt is adjusted with reference to the angle between the cutting facet of the knife and the plane of section. The tilt of the blade as a whole is not a reliable index until its relation to the angle of the cutting facet has been determined for the particular knife being used. A discussion of this matter has been given in Chapter 15, page 233, and should be reviewed at this time.

Because nitrocellulose is more elastic than paraffin, the knife should be tilted slightly more than in cutting paraffin sections. For sectioning the majority of materials, the lower cutting facet of the knife should form an angle of 6 or 7 degrees with the plane of section (Fig. 41*B*, *C*). The knife should be set in this position in a manner similar to that described for the rotary microtome. A rectangular piece of cardboard of the proper width may be held squarely on the top of the microtome, and a line drawn across it at the proper angle used as a guide for setting the knife. When the cutting facet has been adjusted to the proper angle, the general tilt of the knife may be recorded by making another line on the cardboard, along the end of the blade. The knife may then always be re-set by placing the piece of cardboard in the same place and tilting the knife so that it coincides with this line.

Sharpening the Microtome Knife. A sharp, smooth, and durable cutting edge is a primary requisite of good sectioning. The process of sharpening microtome knives has been fully described and illustrated in Chapter 13.

Orienting the Block. The approximate orientation of the specimen was given consideration at the time it was embedded. Remove the paper collar, trim away excess nitrocellulose with a sharp scalpel or razor blade, and clamp the object holder into the microtome, turning it so that the longest side of the specimen is nearly but not quite parallel to the edge of the knife (see Fig. 41*A*). Adjust the object clamp so as to bring the specimen into exactly the desired position with reference to the plane in which the knife is to move.

Operation of the Sliding Microtome. The microtome should be placed in front of the operator in such a way that he may pull the knife directly toward him. A small dish of 70% alcohol should stand beside the microtome, and the operator should be provided with a camel's hair brush (No. 5 or slightly larger) and a medicine dropper.

Thickness of Sections. Set the regulator of the microtome to give sections of the desired thickness. For general topographic study, the sections may be 20 μ or slightly thicker. When it is desired to follow nerve fibers or blood vessels for some distance, or to include entire small parasites (such as *Trichinella* encysted in muscle) or other structures in a single section, they should be considerably thicker. For instance, sections of tissue to show injected capillaries ought to be cut as thick as 70 or 80 μ, and infected muscle tissues for the study of *Trichinella* should be cut at about 40 μ. For the study of certain histological details, however, the sections should be as thin as it is possible to cut them.

Cutting. Turn the little crank which moves the object carrier up and down, so as to bring the face of the nitrocellulose block almost to the knife edge. Make absolutely certain that every screw in the knife holder and object holder is well tightened. Failure to observe this precaution may result in serious damage to the knife or the material. Flood the knife and nitrocellulose block with 70% alcohol. This may conveniently be done with a medicine dropper, which can be filled as often as necessary from the dish standing beside the microtome. Some workers prefer to use the camel's hair brush to make liberal applications of alcohol to the knife and block containing the tissue.

Slide the knife back and forth on its track with a slow and steady motion of the right arm, being careful not to push down on it or to lift it. The first sections cut will be incomplete and probably will not contain tissue, so they may be discarded. However, keep the knife and block well moistened with alcohol, adding more alcohol before each cutting stroke. If the sections curl during the cutting process, hold the brush very lightly on the surface of the block, and let its tip slide up on the blade with the section. As soon as acceptable sections begin to come off, gently slide them along the knife a short distance from the cutting edge toward the tip of the knife. After several sections have accumulated, they may be washed off into the dish by a gentle stream of alcohol, or they may be transferred with the brush. Some workers prefer to use a paper spatula for handling the sections and keeping them flat.

Difficulties Encountered in Sectioning. a. The sections are scratched or split. This may be caused by nicks in the knife or by very hard particles in the material, and the difficulty will be accentuated if the knife is tilted too much (Fig. 41*E*). Try another portion of the knife edge. If the scratches persist, examine the blade under a microscope; if it shows nicks or other imperfections, sharpen it. If the knife is in good condition, then the tissue itself is probably causing the trouble. Hard particles of salts of calcium or silicon may be present, and the only way to solve the problem may be to prepare another lot of material, taking care to decalcify or desilicify it thoroughly.

b. The material is soft and mushy, and crumbles or falls out of the sections. This means that the tissue is not thoroughly infiltrated, because it was not completely dehydrated or because it was left for too short a time in nitrocellulose. The latter fault may be corrected by placing the block for several days in ether-alcohol and then re-infiltrating it. It is hardly worthwhile to try to salvage incompletely dehydrated material, as its structures are likely to have been injured by infiltration.

c. The sections vary in thickness. Such variations may be relatively slight, or they may be so great that one or two sections will be missed altogether and then a very thick one cut. Most commonly, the difficulty is manifested by the production of alternate thick and thin sections. This trouble may be caused by a microtome which is not in good working order, but it may also be caused by the operator pressing down excessively on the knife block, or pulling it upward during some cutting strokes.

Insufficient hardening of the nitrocellulose is another common cause of uneven sections. The elastic mass is compressed by the impact of the knife, and when the block has been finally advanced far enough for the blade to bite into it, a thick section is chopped off. Sections of uneven thickness may also be due to not tilting the knife enough (Fig. 41*D*). In this case, the shoulder of the knife strikes the nitrocellulose block and compresses it, so that at best only a partial section is cut. When the block has advanced further, the knife finally bites into it and cuts a thick section.

If the sections are of uneven thickness, first make sure that there is no looseness or other imperfection in the mechanism of the microtome. If the microtome itself appears to be functioning properly, then make sure that the knife is sharp, and try increasing the tilt of the knife slightly. If the trouble persists, harden the block for a longer time in glycerol-alcohol.

8. Staining, Dehydration, and Counterstaining. Basically, procedures for staining and dehydrating nitrocellulose sections are similar to those employed for paraffin sections. However, they are ordinarily kept in shallow glass dishes during the treatment with various reagents, rather than being affixed to slides, and dehydration is carried out only as far as 95% alcohol. Absolute ethyl or methyl alcohol must be avoided because it dissolves nitrocellulose. (Absolute isopropyl alcohol may be used. However, it has been pointed out elsewhere that staining or destaining in higher concentrations of isopropyl alcohol may not be desirable.) When a good many sections of one kind are being handled, the various reagents are poured into the dish containing the sections, and then poured off or removed with a medicine dropper when they have acted for the desired length of time. The reagents may be used several times unless they have taken up stains or other chemicals which might prove injurious to subsequent groups of sections. When a very few sections of one kind are to be stained, the various reagents may be placed in shallow dishes arranged in the same order as the staining jars used in the paraffin method, and the sections transferred, a few at a time, through the series.

Turn to the section dealing with the nuclear stain to be used, and follow the directions given there. Alum hematoxylin (Chapter 22) is a good stain for histological and pathological material in general. See also the summaries of methods recommended for various materials given in Chapter 29.

Directions for washing and destaining are described under the various staining methods. After completing these processes, pass the sections up through the alcohol series to 70% alcohol, unless other specific directions are given.

If the sections are not to be counterstained, or if the counterstain is to be employed in a clearing agent, transfer the sections directly to 95% alcohol. They should remain in this for 3 to 10 minutes, depending upon the thickness of the sections and the number of them in the dish. They should not be left in 95% alcohol for much longer than the time specified, because this reagent will gradually soften the nitrocellulose. Then proceed to step 9.

If the sections are to be counterstained in an alcoholic solution of eosin or of some other dye, transfer them from 70% alcohol to the staining solution. Such stains are commonly dissolved in 90% alcohol. Specific directions for staining with eosin are given on page 383. Avoid crystal violet and other dyes which stain nitrocellulose deeply. Pour off the stain, rinse the sections in one or two changes of 95% alcohol, and allow them to stand until the stain is sufficiently differentiated.

9. Clearing Sections. Nitrocellulose sections should not be cleared in pure toluene or xylene, because these substances will not mix with 95% alcohol, and also because nitrocellulose sections tend to shrink and become brittle in these clearing agents. Terpineol, phenol, and beechwood creosote are good clearing agents miscible with 95% alcohol as well as with toluene, so that they remove alcohol and water as they clear. Used alone, they may cause the sections to swell and pucker; this seems to be especially true of beechwood creosote. A mixture of 3 parts of toluene and 1 part of any one of these agents makes a satisfactory solution for clearing nitrocellulose sections. Terpineol-toluene is recommended, because terpineol is less pungent and does not endanger the skin as much as either phenol or beechwood creosote.

Fill a small glass dish with terpineol-toluene, to a depth of about 5 mm. to 1 cm. Then lift one section at a time with a section lifter, forceps, or paper spatula, and float it on the clearing mixture. Be careful to place the section on the surface of the liquid in such a manner that bubbles of air are not caught beneath it. As the sections clear, they will sink to the bottom of the dish.

If a good many sections have been placed in one dish of terpineol-toluene, it is advisable to renew this reagent when they are all cleared, as it will have absorbed a considerable amount of water. This will not be necessary if only a few sections have been cleared.

The sections may remain for several days in terpineol-toluene, but it is not advisable to leave them for a longer time. The dish should be kept tightly covered.

10. Mounting Sections. Carefully examine the sections under low magnification, as with a binocular dissecting microscope, and discard those which do not pass through the desired structures, or which are torn, too thick, or otherwise imperfect.

Place, in a clean Syracuse dish, a number of cleaned coverglasses of sufficient size to cover a section and extend at least 2 or 3 mm. beyond it on every side. Stains will soon fade from the edges of the sections if they are barely covered.

Place a small drop of rather thick mounting medium on a clean slide. If the section is to be mounted under a square or circular coverglass, the mounting medium should be placed in the middle of the slide. If a rectangular coverglass is to be used, the mounting medium should be placed about half way between the middle and one end of the slide.

Now pick up a section by one corner, being careful to hold it only by the nitrocellulose which surrounds the material. Touch one edge of the section to a piece of filter paper to absorb the excess clearing agent that runs from it. Then lay the section on the drop of mounting medium which has been placed on the slide, being careful to put it down in such a way that bubbles of air are not caught beneath it. If the section is folded or wrinkled, carefully straighten it out. In cases where much excess nitrocellulose surrounds the tissue, it may be advisable to trim this off with a sharp scalpel. Quickly place a drop of mounting medium on the section, warm a coverglass over a small flame, and slowly let it down on the section. If the coverglass is lowered carefully, as described under the paraffin method (p. 253), the medium will spread evenly and no air bubbles will be trapped in it. The spreading of the medium may be facilitated by holding the slide high above a small flame. Then *very gently* press down on the coverglass with a blunt instrument, so as to flatten out irregularities in the section and force out any small air bubbles which may have been caught near the edge of the coverglass. Finally, wipe off any of the mounting medium which may have exuded and put the slide away in a horizontal position to dry. Drying may be accelerated by placing the slides in an incubator at about 37° C. or slightly higher, but temperatures above 40° C. are to be avoided.

Affixing Nitrocellulose Sections to Slides

There are occasions when it is advantageous to handle nitrocellulose sections in much the same way as paraffin sections. Nitrocellulose is stained so deeply by some aniline dyes that cleaner preparations can be obtained if sections are affixed to slides and the nitrocellulose is dissolved out of them before they are stained. If serial sections must be prepared and the material has not been stained before embedding, the only practical way to keep them in order while they are being passed through various reagents is to mount them on slides a few at a time, as they are cut.

A number of ways have been devised to affix nitrocellulose sections. The albumen-clove oil method, outlined below, is one of the better techniques.

PROCEDURE. 1. Smear Mayer's albumen affixative (p. 241) evenly in a thin film on portions of the slide where sections are to be mounted.

2. Drain off some of the 70% alcohol from the sections, then lay them on the albumenized slides in the positions they are to occupy.

3. Place a sheet of cigarette paper or a piece of very smooth filter paper over the sections. While holding the paper down firmly with a finger of the left hand, rub a finger of the right hand back and forth across the paper, to press the sections into firm contact with the albumen. The position of the paper must not shift during this process.

4. Peel the paper away very carefully. Using a medicine dropper, quickly apply clove oil to those portions of the slide where sections have been affixed. The slide may be left in a horizontal position on the table, or it may be immersed in a staining jar of clove oil. Leave the sections in clove oil for at least 10 minutes. A longer time will do no harm.

5. Pass the slides through at least three changes (5 minutes each) of 95% alcohol, to remove all of the clove oil.

6. If the sections are to be stained right away, transfer them from 95% alcohol into ether-alcohol, and leave them in this until all of the nitrocellulose has been removed. Twenty minutes will probably be long enough.

If the sections are not to be stained immediately, they may be passed gradually to 70% alcohol and stored in this. Whenever desired they may be passed up the alcohol series to ether-alcohol, and then processed further as directed in step 7.

It is not necessary to remove the nitrocellulose from the sections before staining them by methods which are not likely to color the nitrocellulose deeply. However, as has been pointed out before, certain aniline dyes stain nitrocellulose very strongly. Differentiation of the dye can be controlled more carefully if one does not have to contend with an excess of dye in the nitrocellulose.

7. Transfer the slide bearing the sections to 95% alcohol for about 5 minutes; then pass it down the graded series of alcohols to water or whatever concentration of alcohol serves as a solvent for the stain. From this point on, handle them as if they were paraffin sections. After staining, counterstaining, and dehydration in 100% alcohol have been completed, it is a good idea to coat the slides with 1% Parlodion before clearing them, unless there is danger that certain of the dyes used for staining will be

removed excessively in the time it takes to coat the sections. The coating with Parlodion and subsequent clearing of the sections is carried out in exactly the same way as directed for paraffin sections (see p. 251).

RAPID NITROCELLULOSE METHOD, REQUIRING HEAT

It has been pointed out previously that one of the advantages of embedding tissues in nitrocellulose is that infiltration may be accomplished at room temperature. However, there may be occasions when the nature of the material is such that nitrocellulose is the medium of choice (as, for instance, when relatively thick sections must be cut), but the time for infiltration must be shortened as much as possible. Under these circumstances, dehydration and infiltration may be carried out in a paraffin oven at a temperature of 56 or 57° C. *Warning: Do not use an oven which has a type of thermostat which may cause ignition of vapors of ether.* The possibility that an explosion may take place should be very carefully considered before the method is attempted. In any case, use tight-fitting screw-caps or corks which can be wired down tightly to the thick lip of the bottle. The high pressure which builds up within the bottles speeds up the rate of infiltration.

Procedure for Final Dehydration, Infiltration, and Embedding. The steps are numbered as in the conventional nitrocellulose method, and apply to pieces of tissue about 4 or 6 mm. thick. Larger pieces should remain for a longer period of time in each change of alcohol, ether-alcohol, and the solutions of nitrocellulose. Steps 3 to 5 are carried out at a temperature of about 56° C.

1 and 2. Fix and wash the material as in preparation for embedding by the conventional nitrocellulose method.

3. Complete dehydration in 70, 80, 95%, and two changes of absolute alcohol. Leave the tissue in each for at least 30 minutes.

4. Place the material into ether-alcohol for 1 hour.

5. Transfer the tissue to thin nitrocellulose for about 2 hours. Complete infiltration in thick nitrocellulose, leaving the material in this for from 12 to 24 hours.

6. Remove the bottle of thick nitrocellulose from the oven and allow it to cool slowly to room temperature. Do not accelerate the cooling; otherwise, bubbles may form in the tissue.

Embed the tissue and harden the blocks of nitrocellulose as in the conventional method.

DOUBLE EMBEDDING IN NITROCELLULOSE AND PARAFFIN

This method makes it possible to cut thin serial sections of some types of material which cannot easily be sectioned if embedded in paraffin alone.

These include yolky eggs, insects and arachnids, and tendons, which contain a great deal of dense connective tissue. The method is too complicated and tedious, as well as unnecessary, for use on other types of material. It may be summarized as follows:

1. The objects to be embedded must be small. It is very difficult to section by this method specimens which are more than 4 or 5 mm. in diameter, and better results can be obtained if still smaller pieces of tissue are used. Organisms such as insects and arachnids, which are apt to present problems in infiltration, may have to be subdivided with a razor blade. In most cases, subdivision will be more easily carried out after fixation.

2. Dehydrate the material, infiltrate it with nitrocellulose, and embed it, as described in this chapter. Do not mount the material on a fiber holder, but embed it in a small dish or paper box. Omit preliminary hardening by evaporation. Harden the block in chloroform vapor for 18 to 24 hours. Then quickly trim it to within 1 or 2 mm. of the specimen on every side, and immerse it in chloroform.

3. After the nitrocellulose block has remained for 12 to 18 hours (or longer) in chloroform, transfer it to terpineol-toluene. It should remain in this mixture until it is thoroughly cleared. The liquid should be replaced with new when the clearing is complete or nearly so. If it does not clear within a reasonable length of time (allow 3 or 4 days for a block about 5 × 5 × 7 mm.), transfer it to a mixture of equal parts of terpineol and toluene until it is thoroughly cleared, and then back to the standard mixture.

4. Pour off most of the clearing fluid, leaving just enough to cover the block. Then pour in an equal amount of melted paraffin and place the container in the paraffin oven.

5. At the end of 1 or 2 hours, depending upon the size of the block, pour off the mixture and replace it with pure melted paraffin, or pick up the block carefully and transfer it to melted paraffin. Leave the block in melted paraffin for 2 or 3 hours, changing the paraffin three times during this period. Then embed and section it in the same way as any specimen which has been infiltrated with paraffin.

6. Affix the sections to slides in the usual way.

7. Remove the paraffin in toluene or xylene, pass the slides into alcohol, and stain them as directed for paraffin sections. If the sections are very friable, pass them from toluene into terpineol-toluene, and then into 95% alcohol, in order to avoid dissolving the nitrocellulose. The removal of nitrocellulose, which will take place if the sections are passed through absolute alcohol, is not serious if they are subsequently coated with Parlodion as directed in step 14 in Chapter 15, page 249.

8. Stain, dehydrate, clear, and mount the sections as if they were ordinary paraffin sections.

CHAPTER 17

Freezing Method

Freezing is the most direct process by which tissues may be hardened sufficiently to permit the cutting of thin sections for microscopic study. Because of the rapidity with which finished slides may be prepared when the freezing method is employed, it is of special value to the pathologist. Sections of frozen material are also useful when tissues are to be subjected to microchemical tests or impregnated by certain metals, because the technique does not involve the use of reagents which dissolve or combine chemically with substances in the tissues.

MICROTOMES AND FREEZING APPARATUS

The method most commonly used to freeze tissue for sectioning utilizes compressed (liquid) carbon dioxide, which is allowed to escape slowly into a hollow object carrier on which the tissue is mounted. A number of companies produce devices suited to this purpose (Fig. 42). Compressed

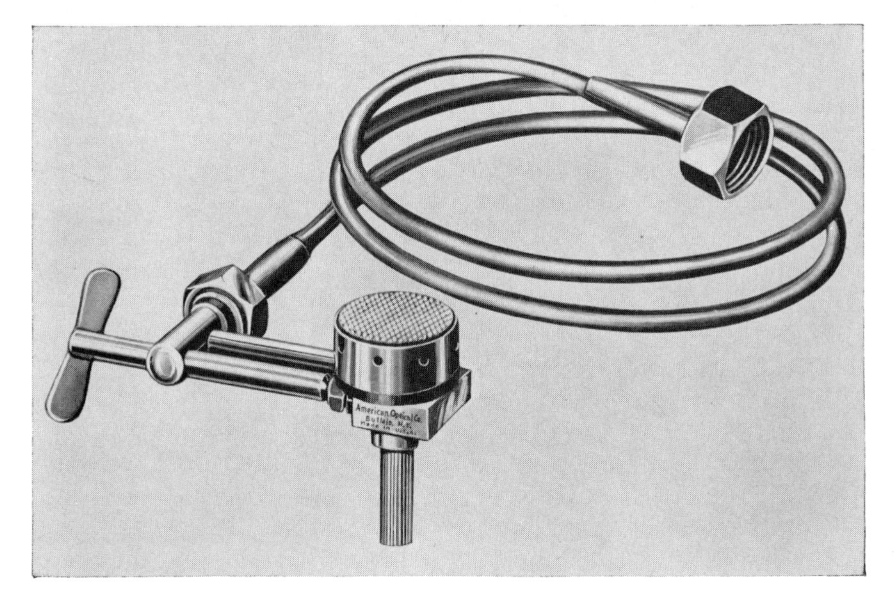

FIG. 42. Microtome freezing attachment for use with liquid carbon dioxide. (Courtesy of American Optical Corp., Buffalo, N.Y.)

FIG. 43. A cryostat. (Courtesy of American Optical Corp., Buffalo, N. Y.)

carbon dioxide, in cylinders, may be purchased from manufacturers of refrigerating materials or carbonated beverages. To set up the apparatus, fasten the carbon dioxide tank in a convenient location near the microtome, with its escape tube at about the same level as the freezing box. Clamp the freezing box into the microtome, carefully bend the attached copper tube so that it will not be in your way, and fit its opposite end to the escape pipe of the tank. Make sure that the gasket is within the coupling and then tighten the connection.

Frozen tissues should preferably be cut on a well-made sliding microtome, or on a so-called clinical microtome (Fig. 21, p. 194). A table microtome and hand-held knife can be used, but this equipment does not enable one to cut thin or uniform sections. Most rotary microtomes are not designed for the accommodation of a freezing apparatus.

In recent years, the routine preparation of frozen sections has been simplified by the development of microtomes mounted in refrigerated boxes. These assemblies are called cryostats or cryotomes. In some of them, the microtome is of the clinical type and is operated from the top of the well in which it is mounted; in others, it is of the rotary type, and a handwheel linked to the mechanism that advances the specimen is outside the box (Fig. 43). Cryostats are now regularly used in clinical laboratories which process a considerable volume of frozen sections and also in research laboratories oriented toward histochemistry. In view of the fact that they are expensive, and because detailed instructions applicable to operation of some models are not applicable to others, it seems best to expect prospective users to rely upon instruction booklets issued by manufacturers. However, basic principles followed in cutting sections with cryostats are those used in cutting frozen sections with the aid of freezing attachments.

PREPARATION OF MATERIAL

Fresh Tissues

If sections must be obtained immediately, which is usually the case in operating room work and often the case in histochemistry, cut a slice of tissue about 5 mm. thick. Place it at once upon the freezing disk, with a few drops of gum mucilage (p. 274) or a proprietary embedding medium (p. 275) beneath it. Then proceed to freeze and cut the tissue as described on page 275. The sections will ordinarily have to be fixed in formalin (step 7, *a*) before they can be stained. This method is employed only when speed is the first consideration, because fresh tissues have a poor cutting consistency and their delicate structures are likely to be torn by crystals of ice.

Fixed Tissues

When time is available, first fix and wash the tissues, which will make possible much better results. It is further desirable to embed the fixed and washed tissues in a suitable matrix before freezing them. These processes are carried out as follows:

1. FIXATION. Place slices of tissue up to about 5 mm. thick in 10% neutral formalin or buffered formalin (pp. 80–81). Allow the fixative to act for at least 12 hours at room temperature.[1] Tissues may safely remain in it for weeks. Formalin possesses important advantages as a fixative in the freezing method. It does not necessitate prolonged washing of the tissues. It is favorable to the stains in ordinary use and may precede a wide variety of microchemical tests. Other fixing fluids may be employed, and the fixed material may be stored in alcohol (unless alcohol interferes with tests or stains which are to be applied to the sections after they are cut).

2. WASHING. Wash formalin-fixed material in running water for 2 or 3 hours. Tissue which has been stored in alcohol should be brought gradually to water, then washed in running water for about 2 hours. Material may be frozen, without further preparation, but better results will be obtained if it is next embedded by one of the following methods.

3. EMBEDDING METHODS. *Embedding in Gum and Syrup.* This medium supports the structures and gives them a good cutting consistency, which makes it possible to obtain sections of high qualiy. It has the further advantage of freezing without the formation of injurious crystals. The following solutions are required:

Gum Mucilage

Gum acacia (best grade)	40 gm.
Distilled water	60 ml.
Phenol, crystals (to serve as a preservative)	0.5 gm.

Gum and Syrup

Gum mucilage, prepared as directed above	50 ml.
Syrup (saturated solution of cane sugar, made by adding 52 parts, by volume, of granulated sugar to 30 parts of water; stir occasionally until the sugar is dissolved, but do not apply heat)	30 ml.

Transfer the washed tissue from water to the gum and syrup mixture. Allow it to stand until it is thoroughly saturated. (A piece of tissue 5 mm. thick should be left in the mixture for about 24 hours.) Then proceed to step 4.

[1] If it is not necessary to preserve lipids, fixation may be hastened by means of heat. Fix in 10% formalin for 1 or 2 hours at about 58° C.

Embedding in Gelatin. This method should be substituted for the preceding one when the material consists of isolated fragments (for example, uterine scrapings) or of parts not firmly bound together. It is used also in connection with Del Río Hortega's silver and gold methods for neuroglia. This method should not be used when it is necessary to preserve lipid masses whose distribution may be altered by heating.

After fixing and washing the material, transfer it from water to a 20% aqueous solution of gelatin, warmed to about 40° C. Allow it to stand in the incubator at this temperature for 1 to 6 hours, depending on the size of the objects. Next arrange the objects in one or more groups and allow the gelatin to cool and solidify. Trim out blocks containing the desired material and harden them in 4% formalin for 18 to 24 hours. Before mounting a block on the freezing disc, rinse it with water. Proceed to step 4. Fresh material may also be treated by this useful method. Leave the tissue in melted gelatin for only 15 to 20 minutes, cool it, cut out blocks, and either freeze them at once or harden them in formalin.

Proprietary Embedding Media. Several companies package mixtures of water-soluble resins and glycol that are intended to provide external support for tissues during sectioning. A few drops of one of these mixtures is applied to a rather cold object disk, and the specimen is pressed down on the viscous fluid and positioned. A matrix 3 or 4 mm. thick is built up around the tissue, and a little of the medium is applied to the upper surface as well. The whole preparation is then frozen.

FREEZING AND SECTIONING

Set up the freezing apparatus and microtome as described on page 273. If a sliding microtome is used, the knife should be tilted to about the same degree as for cutting nitrocellulose sections, but should be set *more obliquely*. An angle of about 30 degrees to its line of motion will probably prove optimum. The blade must be heavy enough so that it will not spring upon impact with the frozen object. Obviously, it must be well sharpened.

4. Mounting the Block. Put a few drops of gum mucilage on the freezing disk. Place the tissue in the mucilage and press it very gently down to the disk. In case the tissue has been soaked in gum and syrup or gelatin, first wipe most of the adhering mixture from it. If it has been embedded in gelatin and hardened in formalin, rinse it with water before placing it on the disk.

5. Freezing With Liquid Carbon Dioxide. Close the valve at the freezing chamber and partially open the outlet valve on the cylinder. Cautiously open the valve at the freezing chamber until the gas escapes slowly and evenly. When the tissue (and gum coating, if added) is frozen hard, close the valve. By allowing the carbon dioxide to escape

in short, interrupted blasts, the process of freezing may be more carefully regulated.

If the tissue has been infiltrated with gum and syrup, paint it with gum as soon as it begins to freeze and add one or two more coats as freezing progresses, until the tissue is surrounded by a layer of frozen gum about 2 mm. thick. Tissue not infiltrated with gum and syrup should not be coated with this medium.

Cooling the Knife. This is necessary in order to prevent sections from sticking to the blade. While freezing the tissue, rest a piece of ice or solid carbon dioxide upon the knife, at the angle formed by its side and back. The refrigerant, in melting, will soon shape itself to the knife and remain in place without being held. If you are using a table microtome and hand-held knife (standard microtome knife with handle attached, or a straight-edge razor), cool the blade by dipping it into ice water before beginning to cut, and at frequent intervals thereafter.

6. SECTIONING. Set the regulator of the microtome at 15 μ, unless the work being done requires thicker or thinner sections. Soft tissues can usually be cut as thin as 10 μ, but such thin sections are difficult to handle and are seldom needed for purposes to which the freezing method is applied.

Cut sections with an even and rather rapid stroke. In using a sliding microtome, be careful not to press down upon the knife block or to lift it. If a table microtome is employed, hold the handle of the knife in the right hand, flex the wrist and brace it against the chest, and plane off sections by swinging the body forward and backward from the waist.

Usually the tissue is frozen so hard, at first, that sections break over the edge of the knife, fly off, or curl tightly. If this happens, make sure that the flow of carbon dioxide has been shut off. After a minute or two, cut sample sections at intervals until the specimen has attained the proper consistency to yield perfect sections. Then rapidly cut as many sections as are needed. Should the tissue become too soft before this is accomplished, cautiously operate the freezing device until it regains the desired consistency.

In case the tissue has been frozen without being embedded in gum or gelatin, the sections generally adhere to the knife and several of them may be allowed to collect upon the blade before it is necessary to pause and remove them. If the knife is clamped to the microtome, as it is in all microtomes which advance the object carrier automatically, wash the sections from the blade with a gentle stream of cold water from a pipette. If the knife is held in the hand, simply dip it into a dish of water and the sections will float off.

In case the tissue has been embedded in gum or gelatin, the sections will curl a little as they are cut and will tend to fall from the block rather than to lie upon the knife. Hold a spatula of smooth, stiff paper or thin cardboard in readiness and catch each section the very instant it is cut. Quickly transfer it to a shallow dish of water. If this is not done very rapidly, melting gum will cause the section to adhere to the knife or block, from which there will be little chance of removing it intact.

Failure to obtain good sections of 15 μ or less in thickness is usually due to a dull knife or to careless work. It is sometimes necessary, however, to increase the thickness to 18 or 20 μ, in order to obtain smooth sections of very hard tissue (such as calcified cartilage) or dense fibrous tissue (such as tendon).

STAINING AND MOUNTING

7. Treatment Depending Upon Method of Fixation. *a. Sections of Unfixed Material.* These must now be fixed in order to preserve their structure and to improve their staining properties. Withdraw most of the water from them and replace it with 10% neutral formalin or buffered formalin. If there is reason to hurry, leave them for 3 or 4 minutes in the formalin solution, previously warmed to about 40° C. Otherwise, allow the solution to act at room temperature for at least 10 minutes.

b. Sections of Material Embedded in Gum or a Proprietary Medium. Wash the sections free of the medium, by changing the water several times.

c. Sections of Material Fixed but not Embedded, or Sections of Gelatin-Embedded Material. Proceed to the next step.

8. Affixing Sections to Slides. It is advisable to affix the sections to slides, as loose frozen sections can withstand very little handling without becoming badly twisted and broken. For this purpose, Wright's method is generally quite good, except in procedures in which the presence of a nitrocellulose film upon the section will interfere with the stain which is to be used. Select a good section. Dip a clean slide for about two-thirds of its length into the water, holding it at a small angle to the surface of the water, and pass it under the section. Carefully lift the slide and section out of the water, meanwhile holding the section in place with the tip of a fine camel's hair brush. If properly manipulated, the section will lie perfectly flat upon the slide. Drain superfluous water from the slide. Then touch a piece of filter paper to the slide, close to the edge of the section, in order to absorb remaining water. Allow a few drops of absolute alcohol to flow very gently from a pipette over the section, until this and adjacent parts of the slide are well covered. Take care not to disturb the section, but if it does float loose from the slide, make certain that it remains flat, straightening it if necessary. After a few seconds drain off the alcohol. Then place one or two drops of 1% Parlodion[1] upon the section and gently

[1] Parlodion is the trade-name of a form of nitrocellulose. (See p. 255.) To dissolve it, use equal parts of ether and absolute alcohol.

tilt the slide back and forth until the Parlodion spreads evenly over the section and adjacent surface of the slide. Turn the slide to a vertical position and touch a piece of filter paper to its surface, about 2 mm. below the section. This will absorb excess Parlodion, leaving a thin and even coating over the section. Hold the slide above a jar of 70% alcohol until the coating solidifies to form a film, then lower it into the alcohol.

9. STAINING, DEHYDRATION, CLEARING, AND MOUNTING. *a. For Routine Work.* Pass the sections (preferably affixed to slides) through the alcohol series to water. Stain them progressively with alum hematoxylin (p. 353) and wash them in running water for about 5 minutes. Then pass them up the alcohol series to 70% alcohol. Next counterstain them with eosin Y in 90% alcohol and place them in 95% alcohol until the counterstain is properly differentiated (steps *c* and *d*, p. 385). Transfer them to terpineol-toluene and leave them until every trace of cloudiness has disappeared. Put them into pure toluene for a minute or longer. If cloudiness develops, return the slides to terpineol-toluene for a few minutes and then bring them back into toluene. After they have been in toluene for at least a minute, and are perfectly clear, mount them in balsam or a synthetic resin.

Frozen sections may be stained by other methods, in the same manner as paraffin sections. However, basic fuchsin and other dyes which color nitrocellulose deeply should be avoided, because the nitrocellulose film must be left to hold the sections upon the slide. After staining (and perhaps counterstaining) the sections, bring them gradually to 95% alcohol. Allow them to stand in this for 1 minute, or until the counterstain is properly differentiated. Clear them in terpineol-toluene followed by toluene, and mount them in balsam or some other resinous medium.

b. Microchemical Tests. Proceed according to the method to be used.

c. To Preserve and Stain Lipids in the Sections. Do not expose the sections to alcohol stronger than 70%. Stain the lipids in Sudan III, Sudan IV, or Sudan black B; then stain the nuclei progressively with alum hematoxylin, if desired. Wash the sections in running water and mount them in glycerol jelly or Farrant's medium.

CHAPTER 18

Epon Method

Although techniques of electron microscopy are outside the scope of this text, certain methods used in the preparation of biological materials for electron microscopy can be used to good advantage in light microscopy. Moreover, if these methods are mastered, a good start will have been made toward competence in electron microscopy techniques.

The usefulness of these techniques to both the light microscopist and the electron microscopist depends primarily on three considerations: (1) fixation in a reagent (or combination of reagents) that admirably preserves cells and their constituents; (2) embedding in epoxy resins that, although they are quite hard when "set," are of low molecular weight and penetrate cells and tissues without causing much distortion; (3) cutting thin and relatively uncompressed sections on very sharp knives of broken glass.

One may wonder why the methods of embedding in epoxy resins and sectioning with glass knives, if they produce sections of high quality, are not regularly utilized for light microscopy. As must be appreciated by now, many techniques seeming to be perfect for one purpose are not suitable for another. The disadvantages of the epoxy resin method are directly related to the advantages: (1) the methods of fixation and the nature of the sections themselves (they are permanently infiltrated by the embedding medium) are not suitable for many staining methods; (2) the sections must be thin—around 1 or 2 μ—so that the depth of field is limited, and the perspective is correspondingly restricted. Besides these disadvantages, the procedure has some rather tedious and precarious aspects, and even an experienced worker cannot confidently expect to cut and to transfer successfully to slides a complete set of serial sections of a small organism or small piece of tissue. In spite of these detractions, the technique offers some remarkable compensations, such as beautiful preservation of cellular form, thick, easily observed cilia, recognizable mitochondria, and many other cytological details. Besides, as mentioned above, certain of the more important techniques required for electron microscopy are learned. Perhaps the most important next step toward electron microscopy is perfection of the technique of sectioning to the point that sections about 0.1 μ or even 0.05 μ are obtained with reasonable regularity. However, it is important to point out that sections 1 μ thick can be cut much more easily than sections a tenth of a micron in thickness. The quality and sharpness of the knife, the hardness of the epoxy block, the perfection with which the block is trimmed, as well as various matters of judgment involved in operating the microtome—all of these make a tremendous difference as progress in the preparation of really thin sections is made.

At the present time, the epoxy resin most widely used for embedding is a mixture of Epon 812 (Shell Chemical Company) and certain other substances, two of which serve as hardeners and one as an accelerator of polymerization. Accordingly, detailed instructions will be given here for using this mixture, hereafter to be referred to simply as Epon. Anyone who learns to use Epon for either light microscopy or electron microscopy should be able to adapt readily to techniques for Araldite, Vestopal, and other embedding media.

PRECAUTIONS TO BE OBSERVED IN HANDLING MATERIALS

Several of the substances regularly used in the Epon method are dangerous because of their toxic properties or combustibility, or both. Persons who contemplate working with this technique must be fully aware of the dangers; others who share the same laboratory facilities must also be protected. The hazardous substances will be commented upon briefly in advance of a description of the method as a whole.

Osmium Tetroxide

Being expensive and having certain disadvantages, osmium tetroxide is not routinely used for specimens to be embedded in paraffin or in nitrocellulose, or to be mounted entire. However, it is regularly used in fixation of tissue for electron microscopy and for preparation of 1 μ sections. The methods for washing and breaking ampoules of osmium tetroxide and for preparing and storing solutions of this substance are given in some detail on page 107. Crystals, solutions, and vapors of osmium tetroxide are highly toxic and may cause severe damage to the eyes and mucous membranes of the respiratory system. Use extreme care when making a fixative from the stock solution of osmium tetroxide and when opening the container of fixative in order to put in or take out specimens. It is advisable to conduct these operations under a ventilating hood. After use, the fixative should be carefully and completely discarded in the sinkhole and washed away.

For transferring osmium tetroxide to the container in which the specimen is to be fixed, it is recommended to use a 5 ml. hypodermic syringe with a needle of about No. 18 or 20 gauge. Plastic syringes are inexpensive and can be used several times in close succession if necessary;

they should then be rinsed out and thrown away. Glass syringes should be rinsed thoroughly with tap water, then with distilled water, and allowed to dry before being used again. In any case, it is advisable to discard the needles.

Propylene Oxide

This substance is highly volatile, inflammable, and explosive; moreover, it is toxic if inhaled, and its extreme volatility increases the risk of inhalation. It is used as a bridge between 100% alcohol and the embedding material. When pouring off propylene oxide, put it directly into the sinkhole and flush it away with plenty of water. If this work cannot be done under a ventilating hood, it is a good idea to have the windows open to encourage rapid dissipation of the vapors in the room. The high volatility of propylene oxide makes it necessary to work quickly when pouring it off the tissue and replacing it with a fresh application, for otherwise the tissue is apt to suffer some desiccation.

Epon

Epon 812 and the two hardeners (nadic methyl anhydride and dodecenyl succinic anhydride) which are mixed with it may, individually or collectively, be irritating to the skin of some persons. If the somewhat sticky fluid Epon gets on the skin, it should be washed off immediately with soap and water. The fine dust produced when Epon blocks are sawed must not be inhaled or allowed to contact the mucous membranes. Suggestions for removing dust as it forms are given on page 292.

DMP-30

This rather viscous substance (2,4,6-tri[dimethylaminomethyl]phenol), which smells like fish or rancid fat, is extremely dangerous to at least some persons. It is known to cause dermatitis and is believed to have carcinogenic properties. In any case, it should be handled with extreme care. Since DMP-30 serves as the accelerator for polymerization of the Epon mixture used in embedding, it must be reckoned with not only in its pure form but also in the mixture. If the skin comes in contact with either DMP-30 or the Epon mixture containing it, wash immediately with soap and water.

DMP-30 is used in very small quantities. The best way to transfer it from the stock bottle to the container in which the Epon mixture is being prepared is with a 1 mm. tuberculin syringe equipped with a No. 18 gauge needle. Because of the viscosity of DMP-30, a glass syringe is apt to function better than a plastic syringe; the plungers in plastic syringes often do not fit tightly enough. However, plastic syringes are cheap and can be disposed of after one use. If a glass syringe is used, store it in a container such as a stoppered test tube or jar until it is needed again. Be sure to force out as much of the DMP-30 as possible, and wipe off the needle with a laboratory tissue. (If the syringe is not likely to be used again very soon, it is best to discard the needle.)

CLEANING GLASSWARE AND DISPOSING OF WASTES

After using a piece of glassware for mixing Epon for infiltration or embedding, empty it as completely as possible into a bottle reserved for this purpose. A wooden applicator stick may help with the more viscous mixtures. The stick should then be discarded into the waste bottle. If considerable work with Epon is regularly being done, it is advisable to have one waste bottle for any embedding mixture which is left over, and another for that which has been diluted with propylene oxide. The glassware to be cleaned should be immersed completely in a container of 95% ethyl alcohol and periodically agitated, over a period of one or two days, with tongs such as are used for picking up hot potatoes. It should then be transferred to a second container of 95% alcohol for at least several hours. By this time, the glassware should be nearly clean. Polyethylene pails with snap-on covers are better than glass jars for washing glassware, because they lessen the risk of breakage. Change the 95% alcohol when the volume of glassware processed suggests that the second bath may contain considerable Epon. The alcohol may be disposed of in a sink drain, but it should be flushed down with plenty of water.

After being cleaned in the two baths of alcohol, the glassware should be removed with tongs and placed in hot soapy water. Wear rubber or plastic gloves to give the glassware a final scrubbing, then wash it in clean hot water, rinse it in distilled water, and let it dry where it will not be contaminated by dust or grease.

Remember, when handling the glassware while it is being cleaned in alcohol that, although the Epon may now be greatly diluted, the alcohol may serve as a very penetrating vehicle. Avoid getting it on the skin.

The final Epon mixture, unless diluted with propylene oxide, will gradually harden in the waste bottle. It is best not to dispose of such waste bottles until the Epon is hard, because of the risk of bringing those who handle garbage into contact with the ingredients, especially DMP-30. In some large institutions, there are arrangements for periodically collecting wastes and disposing of them safely.

The mixtures of Epon and propylene oxide used for infiltration are a more vexing problem than the embedding mixture, because the Epon will not harden unless the propylene oxide evaporates. If a ventilating hood is available, and if there is no danger that the evaporating fumes of propylene oxide will ignite, a practical solution is to leave the waste bottle, with the lid off, under the hood. Eventually, if no more Epon diluted by propylene oxide is added, the Epon should harden and the bottle may be disposed of. If evaporation under a hood is impractical, then perhaps it will be best to take the nearly full bottle outdoors, pour the Epon into cans to allow the propylene oxide to evaporate quickly under supervision, and then cover the cans and let them stand in an out-of-the-way place in the laboratory until the Epon has hardened.

Glassware in which Epon has inadvertently been allowed to harden can be cleaned by soaking it for several days in strong sodium hydroxide or potassium hydroxide. If, after washing out the alkali thoroughly, some residue of Epon still remains, and if this does not come off with a little scrubbing, again treat the piece of glassware in the strong alkali.

FIXATION

Fixation for electron microscopy is almost a science in itself. Proper preservation of mitochondrial cristae, centrioles, ciliary rootlets, pinocytotic vesicles, and other details in specific organisms may require considerable experimentation with respect to pH, substances used for buffering, osmolality, and temperature, as well as the active ingredients of fixatives. However, there are several more or less standard procedures of fixation for electron microscopy, and these are generally quite good for preparation of material to be studied with the light microscope. This is not to say that fixation which is good for a particular problem to be solved by electron microscopy will always give equally good results for light microscopy. For instance, if one wishes to obtain what he believes to be fine preservation of mitochondria or cell membranes for electron microscopy, he may use a fixative which does not preserve other organelles, or even the cell as a whole, especially well. Therefore, it sometimes happens that 1 μ sections prepared for light microscopy may seem better or worse than might be predicted on the basis of examination of very thin sections with the electron microscope.

As a rule, osmium tetroxide will be used in any procedure of fixation for material to be embedded in Epon, whether for light or electron microscopy, or both. The osmium tetroxide either will be applied as the first and only fixative or will follow a primary fixative of a totally different type. Whether fixation is accomplished in one step or two steps, an appropriate pH should be maintained by buffers.

Only a few combinations will be recommended here. For 1 μ sections, these will probably give useful results. The first two, although they are decidedly hyposmotic to sea water, are quite good for marine invertebrates, as well as for fresh-water and terrestrial invertebrates and for vertebrate tissues.

Osmium Tetroxide Buffered by s-Collidine

Osmium tetroxide, 4% aqueous solution	2 parts
s-Collidine (2,4,6-trimethylpyridine), 0.2 M aqueous solution, pH 7.5 .	1 part

s-Collidine is a somewhat viscous fluid with a pungent aroma. Preparations of acceptable purity, packaged in relatively small quantities for electron microscopists, are available from Eastman Kodak Co. (Rochester, New York) and Polysciences, Inc. (Warrington, Pennsylvania). (Decidedly impure samples will, when mixed with water, form a milky suspension; purifying them by taking advantage of the boiling point of *s*-collidine [168° C.] is a lot of trouble.)
To prepare the buffer, mix 2.67 ml. of *s*-collidine with about 70 ml. of distilled water in a 100 ml. volumetric flask. Add 8 ml. of 1 N hydrochloric acid (p. 433), and swirl the contents. Then bring the total volume to 100 ml. with distilled water and agitate the mixture until it is uniform. Solutions of *s*-collidine keep well in a refrigerator, or even at room temperature, if stored in a tightly stoppered screw-cap bottle.

The mixture should be made up immediately before use. Three milliliters will be sufficient for several small pieces of tissue. Draw the solutions from the stock bottles with separate 5 ml. hypodermic syringes, and make the mixture in a glass container which can be stoppered while fixation is in progress. Among the more serviceable containers are weighing bottles, with ground stoppers, having a height of about 40 mm. and an inside diameter of about 25 mm. Small vials, which are not much taller than they are wide, and which have polyethylene snap-on caps, are also good. Stand the container in crushed ice, preferably in an insulated ice bucket. Be sure that the container cannot tip over. In a few minutes, the fixing reagent will be as cold as the ice (i.e., around 0° C.), and the tissue may be dropped into it. Fix the tissue for 1 hour; during fixation, it will probably blacken considerably.

Osmium Tetroxide Buffered by Sodium Bicarbonate

Osmium tetroxide, 4%	1 part
Sodium bicarbonate buffer, pH 7.2	1 part

To prepare the buffer, dissolve 2.5 gm. of sodium bicarbonate in 100 ml. of distilled water. Bring the pH to 7.2 with 1 N hydrochloric acid, checking the pH with an accurate meter. Once made up, the buffer solution may be stored for some time, but the pH is apt to change. Therefore, it will be necessary to check the pH and perhaps to add more hydrochloric acid.

Mix the solution of osmium tetroxide with the buffer immediately before use. Fix the tissue for 1 hour. The container should be in an ice bath throughout the period of fixation.

Double Fixation in Glutaraldehyde and Osmium Tetroxide

Primary Fixation

2.5% Glutaraldehyde . . . . 1 part

Prepare this by diluting 1 part of a 50% stock solution of glutaraldehyde, biological grade (such as is supplied by Fisher Scientific Company), with 19 parts of distilled water.

0.133 M Phosphate buffer, pH 7.4 . . . . 1 part

For preparing this, it is convenient to have on hand 0.2 M solutions of sodium phosphate, monobasic ($NaH_2PO_4 \cdot H_2O$) (27.6 gm. in distilled water to make a total volume of 1000 ml.) and sodium phosphate, dibasic, anhydrous (Na_2HPO_4) (28.4 gm. in distilled water to make a total volume of 1000 ml.). Mixing 1.9 ml. of the former with 8.1 ml. of the latter yields an 0.2 M buffer solution of pH 7.4. Addition of 5 ml. of distilled water lowers the molarity to 0.133.

Fix the tissue for 1 hour at room temperature. Then pour off or withdraw the fixative and rinse the tissue for a minute or two in 0.133 M phosphate buffer.

Post-Fixation in Osmium Tetroxide

Osmium tetroxide, 4% . . . . 1 part

0.133 M Phosphate buffer, pH 7.4 . . . . 1 part

Fix for 1 hour, cooling the container in crushed ice.

The solutions of glutaraldehyde and phosphate buffer suggested above are approximately isotonic with tissues and embryos of birds and mammals. The proportions of these two ingredients of the primary fixative may therefore be varied to give stronger or weaker concentrations of glutaraldehyde. For cold-blooded vertebrates and fresh-water or terrestrial invertebrates, slightly weaker solutions of glutaraldehyde (1.75%) and buffer (0.1 M) should be tried. For marine invertebrates, the following method of double fixation has given rather good results.

Double Fixation in Glutaraldehyde and Osmium Tetroxide for Marine Invertebrates

Primary Fixation

5% Glutaraldehyde in 0.27 M sodium chloride . . . . 1 part

Prepare this by diluting 1 part of a 50% stock solution of glutaraldehyde, biological grade (such as is supplied by Fisher Scientific Company), with 9 parts of 0.3 M sodium chloride.

0.4 M phosphate buffer, pH 7.4 . . . . 1 part

To prepare, dissolve 11.04 gm. of sodium phosphate, monobasic ($NaH_2PO_4 \cdot H_2O$) in 180 ml. of distilled water. Then, using a pH meter, bring the pH to 7.4 with a 10 N solution of sodium hydroxide (40 gm. of sodium hydroxide in 100 ml. of distilled water). About 6.7 ml. of this solution will be required. Finally, add water to bring the total volume of the sodium phosphate-sodium hydroxide mixture to 200 ml.

Fix the tissue for 1 hour at room temperature. To minimize the formation of precipitates, introduce as little sea water as possible, and then either remove the tissue as soon as it has settled and transfer it to a second container, or remove the fixative with a pipette and replace it with fresh fixative.

After fixation, pour off or withdraw the fixative and replace it with a mixture of equal parts of the 0.4 M phosphate buffer and 0.6 M sodium chloride. Let the tissue stand in this for 15 minutes, but do not agitate the container.

Post-Fixation in Osmium Tetroxide

Osmium tetroxide, 4%	1 part
0.4 M Phosphate buffer	1 part
0.75 M Sodium chloride	2 parts

Fix in this mixture for 1 hour, cooling the container in crushed ice.

WASHING AND DEHYDRATION

Pour off the fixative (or second fixative, if two have been used), or withdraw it with a clean pipette. Rinse the tissue briefly in one or two changes of distilled water, then dehydrate it gradually in ethyl alcohol. About 10 minutes in each of the following concentrations of alcohol is recommended: 15%, 30%, 50%, 70%, 95%.

Now replace the 95% alcohol with a considerable amount of 100% alcohol, rotating and tipping the container to wash down the walls of the container and thereby remove traces of water. After 5 minutes, replace the first bath of 100% alcohol with a second, and leave the tissue in this for 5 minutes.

Replace the 100% alcohol with propylene oxide and agitate the container gently. After 10 minutes, change the propylene oxide, and after another 10 minutes, change it again. It may be a good idea to change the third bath if the amount of tissue or the size of the pieces suggests that replacement of the ethyl alcohol by propylene oxide may not have been complete. Normally, however, one may proceed to infiltrate the tissue with Epon while it is in the third bath of propylene oxide.

INFILTRATION WITH EPON

It is conventional to have two mixtures of Epon, A and B, on hand. These will keep for several months, especially if they are refrigerated in tightly stoppered screw-cap bottles. Polyethylene bottles or disposable glass reagent bottles, such as are used to package many laboratory solvents, serve well for storing Epon mixtures and are simply discarded when empty or when there is reason to suspect that the Epon is no longer suitable for embedding. Just before infiltration is begun, an appropriate amount of A and B are mixed together, and a small amount of DMP-30 is added; the latter accelerates the polymerization of the final mixture.

The ingredients of mixtures A and B, and information concerning the ultimate industrial sources of all four ingredients required for the Epon method, are given below. The manufacturers package these substances in amounts larger than most microtechnique laboratories can conveniently use. However, they can be purchased in practical quantities from dealers who supply electron microscope laboratories and fabricators of plastic products.

Mixture A

Epon 812 (Shell Chemical Company, San Francisco) . . . . . . 62 ml.
Dodecenyl succinic anhydride (DDSA) (National Aniline Division of Allied Chemical and Dye Corporation, New York) . . . . . . 100 ml.

Mixture B

Epon 812 . . . . 100 ml.
Nadic methyl anhydride (NMA) (National Aniline Division of Allied Chemical and Dye Corporation, New York) . . . . 89 ml.

Accelerator

2,4,6-tri(dimethylaminomethyl)phenol (DMP-30) (Rohm and Haas Company, Philadelphia)

In preparing mixtures A and B, use either a graduate cylinder or a pharmaceutical graduate; the latter, because of its shape, will be more convenient for mixing and also more easily handled through the process of cleaning in alcohol. The ingredients of mixtures A and B, and the mixtures themselves, are rather viscous. In measuring them, it is generally best to add the second ingredient of each mixture carefully to the first in the same graduate, and then to stir the contents of the graduate thoroughly before pouring them into the stock bottle. Avoid getting either of the ingredients on the inside wall of the graduate above the line marking the volume to be reached by each.

A good way to prepare the mixtures with adequate accuracy, and one which obviates the necessity of cleaning the graduates afterward, is to measure out quantities of water equal to those of the ingredients of mixtures A and B, and to pour these into rather narrow screw-cap bottles which are suitable for storing the mixtures. Two lines marked with a diamond pencil on the side of each bottle will indicate the level to which the first and second ingredients are to be brought. The bottles are then emptied and dried thoroughly before being used for the Epon mixtures. Immediately after adding the two ingredients, stir them together thoroughly, so that there is no risk that later, when some of the contents are poured out, the mixture will not be uniform.

If the mixtures of Epon (A and B) have been stored in a refrigerator, they should be removed several hours beforehand so that they may gradually come to room temperature. To open the bottles and to pour from them while they are still cold is to court disaster, for the mixtures are likely to pick up moisture from the air.

Into a small pharmaceutical graduate (10 or 15 ml. capacity) slowly and carefully pour 7 ml. of mixture A. Then add 3 ml. of mixture B. In the interest of measuring accurately, avoid getting the viscous Epon mixture on the upper part of the graduate. With a 1 ml. tuberculin syringe equipped with a rather coarse needle (No. 18 is recommended), deliver 0.15 ml. of DMP-30. Now, with a glass rod having a diameter of about 8 mm., stir and churn the contents of the graduate thoroughly for several minutes. *The final mixture must be homogeneous, or it may not harden properly.*

If the final bath of propylene oxide in which the tissue has been standing is deeper than about 0.5 cm., the excess may be poured off into a sinkhole, or withdrawn with a medicine dropper and then disposed of. To the remaining propylene oxide, add, little by little, an equal quantity of the Epon mixture, rotating or shaking the container gently to dilute the Epon uniformly. Let the tissue remain in this mixture for 2 or 3 hours.

Now discard about half of the diluted mixture into a waste bottle reserved for this purpose, and to the rest slowly work in an equal quantity of Epon. The proportion of Epon to propylene oxide should now be about 3 to 1. Let the tissue remain in this for 3 to 5 hours, or overnight if this is more convenient.

As a small amount of Epon is likely to be needed for infiltration, the rest can be covered with a piece of smooth paper or aluminum foil and stored in a dry place until one is ready to embed. However, if more than what is left will be required, a new batch can be made just before embedding. To assist in calculating approximately how much Epon will be needed, about 1 ml. will fill the unit of an ice cube tray to a sufficient depth, and 0.6 ml. will nearly fill a No. 00 BEEM capsule or gelatin capsule (see below).

EMBEDDING

The tissue is now ready to be embedded. A variety of receptacles can be used for embedding; some are better than others if the orientation of the object during sectioning is important. If orientation is no particular problem, polyethylene BEEM capsules (Better Equipment for Electron Microscopy, Bronx, N.Y.), sold by dealers in supplies for electron microscopy, are favored by many workers because the lower end has a four-sided taper which may simplify the earlier stages of trimming if the specimen settles, as expected, into the apical portion. (BEEM capsules with simply a conical tip are also available.) No. 00 BEEM capsules (these have an inside diameter of about $\frac{5}{16}$ inch [8 mm.]) are standard;

No. 3 capsules are considerably smaller but may be preferable for certain types of material.

The deeper halves of gelatin pharmaceutical capsules also are convenient and inexpensive. They come in a wide range of sizes, including No. 00 (conforming to No. 00 BEEM capsules) and No. 3. However, because the hard Epon cylinders formed in BEEM or gelatin capsules will fit into the object holder of the microtome, certain types of specimens requiring no special orientation will only need to be trimmed, instead of having to be sawed out and remounted. Therefore, if both BEEM and gelatin capsules are to be used, it is advisable to decide on the one size adaptable for most purposes and which will fit the object holder with which the microtome is equipped. (A $\frac{5}{16}$ inch collet-type object holder will accept cylinders of Epon which have hardened in No. 00 BEEM or gelatin capsules. See page 291 for a further discussion of object holders.)

If orientation must be made carefully after the Epon has hardened, the most convenient receptacles are portions of polyethylene trays used for making small ice cubes in home refrigerators. The individual units in these trays are usually about 1.5 cm. in diameter, and the trays can be cut up with a razor blade or sharp knife into convenient pieces consisting of several units. Various types of polyethylene lids for vials and the polyethylene plugs used in making shotgun shells are also good, although they must be handled individually, and are in some ways less convenient than portions of ice cube trays consisting of several units.

Pour some of the mixture of Epon A, B, and DMP-30 into each container to be used for embedding. If gelatin capsules or BEEM capsules are to be used, and if the hardened cylinder is to be inserted directly into the microtome after the end containing the specimen has been properly trimmed, it will be well to fill them nearly to the top. Even if the specimens are to be sawed out and remounted, it will be advisable to put in considerably more Epon than is needed to cover the specimens, so there will be something to grip during the sawing procedure. If the units of an ice-cube tray are used for embedding, filling them to a depth of about 4 or 5 mm. should be sufficient. Lids for vials and plugs for shotgun shells are usually so shallow that they should be filled to the brim in most cases.

Now transfer the specimens to the embedding containers. As a rule, the best instrument for transferring specimens is a Pasteur pipette or medicinal pipette in which the attenuated portion has a bore of about 1 mm. or slightly larger, and which has been equipped with a rubber bulb. However, the transfer may be done with the flat end of a toothpick which has been smoothed down, or with a piece of wire flattened to form a sort of spatula. Be careful not to damage the specimens while picking them up or to allow them to dry while they are being transferred, but at the same time avoid carrying over more than a very small amount of the infiltrating mixture.

If gelatin or polyethylene capsules are used for embedding, put only one specimen in each capsule. When flat-bottomed containers are used, several specimens may be embedded at one time. In fact, it is customary to do this, because fewer containers need be kept track of, and if there is at least 3 mm. between specimens and these are neatly arranged, it will be easy enough to cut them out individually after the block has hardened. It is a good idea to make sure they are well separated from the beginning. Space them with the aid of a clean needle or the instrument used for transferring them.

The embedding mixture is hardened in an oven set at 60° C. This oven should preferably be one reserved for this purpose. An oven used for paraffin is decidedly unacceptable, but ovens used for other purposes may be suitable if they are free of moisture, oils, or fumes of solvents. Set the embedding containers in the oven and leave them there for at least 18 to 24 hours. This is likely to be long enough for material that is to be sectioned at 1 μ. A still longer period—about 30 hours—may make the blocks slightly harder, and therefore more suitable for preparing the extremely thin sections required for electron microscopy. Unless the specimens being embedded are very small or very light, they are likely to sink to the bottoms of the embedding containers before the hardening of the Epon has progressed far enough to impede their settling. If there should be an advantage to keeping them off the bottom, as would be the case if one had to have a cushion of Epon on all sides of the specimens in order to trim the blocks with a particular orientation in mind, the Epon mixture should be allowed to polymerize slightly before putting in the specimens; this can be done by letting it stand for several hours at 37° C. The quality of embedding is apt to be perfectly acceptable, at least for 1 μ sections. However, if this procedure is to be followed, it will probably be best to transfer the specimens to a full strength Epon mixture, at room temperature, for at least an hour before placing them in the embedding containers; otherwise, infiltration of specimens still containing a small amount of propylene oxide may not be completed in Epon that has already begun to polymerize.

After the embedding mixture has been in the oven for 24 hours or longer, it will be so hard that it cannot be scratched by a fingernail. It can be scratched by a needle but not readily cut by the needle. Blocks that are too soft after 24 hours can often be hardened a little more if they are returned to the oven for several hours. However, if they are not hard enough after 36 hours, something is almost certainly wrong with the embedding mixture. The ingredients (especially Epon A and B) could be too old, or moisture has gotten into them or into the final mixture, or the final mixture was not homogeneous enough because it was not sufficiently stirred. In any case, final judgment should be reserved until after cutting some sections. Many blocks from which very thin sections cannot be cut are perfectly acceptable for preparation of 1 μ sections.

To remove a cylinder of Epon from a gelatin capsule, soak the capsule in warm water for a few minutes and work off the softened gelatin with the fingers. BEEM capsules are carefully cut down one side with a razor blade or sharp knife and peeled away. Embedments in ice-cube trays and other shallow polyethylene containers are simply turned out by a combination of bending and pushing from below with the thumb; these containers are re-usable.

PREPARING EMBEDDED TISSUE FOR SECTIONING

Epoxy resins, after hardening, cannot be trimmed with the same ease as paraffin or nitrocellulose. If several specimens are embedded in the same container, they will have to be separated by sawing; in fact, removal of any great excess of Epon from specimens embedded singly is most easily accomplished by sawing. Razor blades are used for final trimming, so that the face and four sides are flat and smooth.

If a BEEM capsule or gelatin capsule was used for embedding, and if the capsule was filled nearly to the brim with Epon, it may be possible to insert the hard cylinder of Epon directly into the microtome after proper trimming of the portion containing the specimen. This should certainly be possible if no particular orientation of the specimen is required, as would be the case with something like a small piece of liver or kidney. However, if the orientation of the specimen is not acceptable, or if something other than a capsule of appropriate size has been used, it is necessary to saw out the specimen and mount it on a metal chuck which will fit the object holder of the microtome. For the Sorvall MT-1 microtome, which is the model most likely to be available to the student, there are collet-type object holders which accept chucks with diameters of about $\frac{5}{16}$ inch (capsule sizes 00, 0), $\frac{7}{32}$ inch (capsule sizes 1, 2), and $\frac{3}{16}$ inch (capsule sizes 3, 4, 5). Thus, if size 00 BEEM or gelatin capsules are being used routinely, and if it is expected that at least some embedments are to be used without being remounted, the $\frac{5}{16}$ inch holder should be available, together with a number of metal chucks that will fit this holder. The chucks are easily made by sawing a $\frac{5}{16}$ inch aluminum rod into pieces $\frac{1}{2}$ inch or $\frac{3}{4}$ inch long. With a little extra trouble, they can be smoothed and beveled on the end on which the specimen is to be mounted (Fig. 44). A machine shop can produce a hundred or so rather cheaply. They are re-usable, but somehow a large number of them are always tied up with specimens waiting to be sectioned, so it is recommended that an ample supply be available.

Three general types of instruments are used for sawing Epon blocks. The simplest tool is a hacksaw (or hacksaw blade to which a handle has

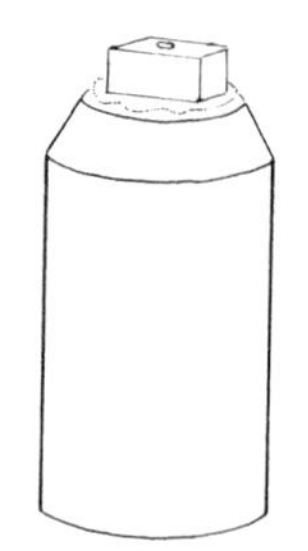

FIG. 44. Metal chuck with small piece of Epon containing the specimen cemented to it.

been affixed) which follows a track in a wooden miter box. See Fig. 8*B* (p. 124) for a practical design. However, it is recommended that the bed of the box be wide enough so that there is room for the fingers to hold the block tightly while it is being sawed. A second type is a small electric tool, such as the Moto Tool (Dremel Manufacturing Co., Racine, Wisconsin), used by makers of models for drilling, filing, and sawing. Any with a mandrel that will accept a dentist's circular saw blade ($\frac{7}{8}$ inch) should serve very well. These tools are cumbersome to hold, because the handle, containing the motor, is rather large. It may be best either to clamp the block in a vise, so that both hands may be used to hold the tool, or to immobilize the tool on a bench and bring the block up to it with pliers. The most convenient instrument for sawing Epon is a dental engine. As the shaft to which the mandrel and saw are joined is turned by a pulley driven by a motor some distance away, the handpiece is small and light; the other hand can be used to hold the pliers gripping the block of Epon. Moreover, the speed at which the shaft revolves can generally be controlled by a foot pedal. If a lot of sawing is to be done in the laboratory, this type of arrangement will be indispensable.

Two precautions must be followed if motorized sawing tools are used. First, because the saws are very sharp, they can cause a serious injury in an instant. They should be in good mechanical order, and the operator must have his wits about him. *No daydreaming.*

Second, these saws produce and scatter a lot of fine dust. If inhaled, or if allowed to accumulate on mucous membranes of the eyes and nose, this dust may be irritating or even dangerous. It is best, therefore, to have the end of a vacuum cleaner hose within a few inches of the restricted area of the table where the sawing is actually done. The hose can be held in position by clamps affixed to a ring stand or some equally strong support. Then it is simple enough to turn on the vacuum cleaner just before sawing. A dust mask may also be worn. In any case, the operator should not only protect himself, but he should prevent the dispersal of dust which may be hazardous to others who may not be aware that the danger exists.

Even when using the hacksaw, some dust will be produced. This can be picked up by a vacuum cleaner, or with moist cloths or paper towels.

If the specimens are so small or so weakly blackened by osmium that they cannot be seen clearly enough with the unaided eye, the Epon blocks should be examined under a dissecting microscope so that the lines to be followed in sawing can be scratched into them with a needle.

When sawing out the specimens or removing excess Epon, be sure to do this with an understanding of how you wish to orient the specimen with respect to the knife. Leave plenty of room on the side to be affixed to a chuck.

A number of proprietary epoxy cements, packaged by manufacturers of adhesives in two separate tubes, are suitable for binding the Epon to a metal chuck. Equal quantities from each tube are mixed thoroughly on a piece of glazed paper, and a small amount of the mixture is applied to the end of the chuck. The block containing the specimen is then pressed down on this, and if the surface facing the chuck is flat, it should stay in position while the cement hardens. Most of the cements will be as hard as the Epon itself if the object holders with blocks affixed to them are placed in the oven used for hardening Epon (60° C.) for about 2 hours.

To remove a specimen from the metal chuck, simply drop the chuck into a screw-cap jar of acetone overnight. When pulled or scratched away, the Epon and epoxy cement will generally separate cleanly from the metal. If a chuck must be pressed into service immediately, break off the specimen with pliers, taking care not to scar the metal any more than is inevitable.

A vise-type holder manufactured for use with the model MT-1 and MT-2 microtomes is convenient for specimens embedded in containers such as ice-cube trays or lids of vials, providing there is enough Epon to one side of the specimen for the block to be gripped tightly. The side of the block to be presented to the knife first will have to be prepared with the aid of a saw and razor blades in the same way as a specimen in a cylinder of Epon or one mounted on a metal chuck.

FINAL TRIMMING OF THE BLOCK

This requires care and considerable patience, and must be completed under a dissecting microscope. A special device (Fig. 45) manufactured for the purpose of holding the specimen during trimming is a tremendous asset, although any relatively small but heavy piece of metal equipped with a collet-type holder or a holder with a setscrew, and which can be placed on the stage of the dissecting microscope, will serve. Some workers hold the Epon cylinder or metal chuck on which the specimen is mounted with one hand and trim with a razor blade held in the other. However, there are enough hazards and frustrations in the Epon tech-

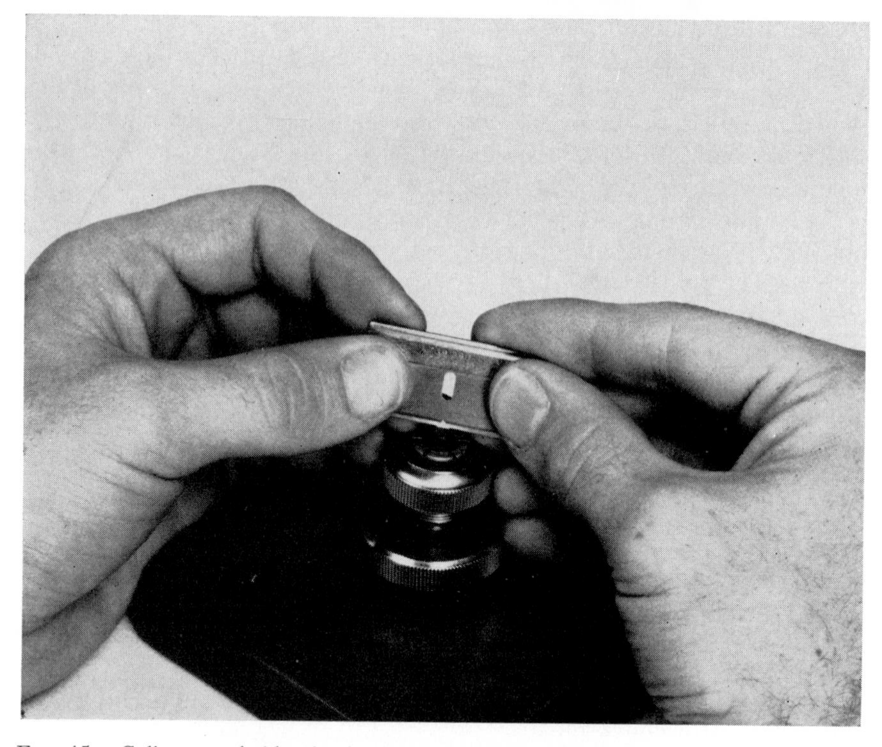

FIG. 45. Collet-type holder for immobilizing metal chucks or cylinders of Epon for trimming with razor blades. (Courtesy of Ivan Sorvall, Inc., Norwalk, Connecticut.)

nique that it seems extravagant to encourage more. Before clamping the chuck or Epon cylinder into position, examine it with a view to understanding how the specimen will be cut and to see how much more Epon will have to be trimmed away.

Remove the grease or oil from several new single-edge or double-edge razor blades of high quality by passing them through two baths of acetone, then wipe them dry with a soft, grease-free tissue (as Kimwipes). If there is a considerable excess of Epon in front of the specimen (i.e., the uppermost side), trim this away with a few shallow transverse cuts (Fig. 46*A*, *B*). (Do not try to speed up the job by cutting off thick slices, because if the razor blade is forced through the block, the cut may be very rough and the tissue distorted.) Removal of excess Epon on the face of the block makes possible a better view of the specimen and makes trimming the sides easier to control.

Establish which side of the face of the block will be nearest the knife edge, and make shallow downward cuts, forming a slope of about 30° with respect to the long axis of the metal chuck or the cylinder of Epon. The last of these cuts should come close to the tissue. (If the piece of

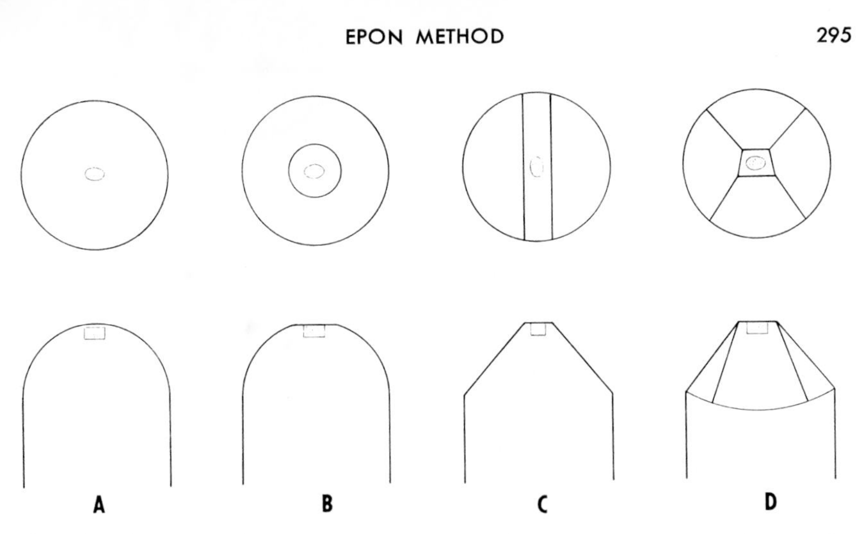

FIG. 46. Steps in trimming a specimen embedded in a cylinder of Epon. (See explanation in text.) The same general procedure is followed if the specimen has been cemented to a metal chuck.

tissue is more than about 1 mm. in diameter, it may eventually be necessary to actually cut into it.) Now rotate the holder or other clamping device 180°, so that similar cuts may be made on the opposite side. When these cuts have been completed, the opposite edges should be parallel (Fig. 46*C*). Rotate the holder about 90°, and make similar cuts. These may be at right angles to the others, but ordinarily it seems best to make the final cuts at an angle of about 80° with respect to the lower edge, and about 100° with respect to the upper edge. Then make comparable cuts on the opposite side. The end result should be a nearly pyramidal block with a smooth face and smooth sloping sides and with the lower edge (i.e., the edge that will first contact the knife) wider than the upper edge (Fig. 46*D*). It is extremely important that the block be trimmed with considerable perfection. Dull corners, irregular margins, rough spots on the sides, and other defects all interfere with cutting, separation of sections from the knife as cut, and formation of ribbons. Ideally, all faces of the block should be glassy and the corners sharp, and no foreign matter should be adhering to it. This is easier to talk about than to do, for the diameter of the face of the block, when it is ready to section, should not ordinarily exceed about 1 mm. Use new razor blades with reasonable liberality. About two or three are likely to be required for each specimen. Most workers do rough trimming with one or two blades, then make final cuts with a new blade.

MAKING KNIVES

In order to cut perfect or nearly perfect sections of tissue embedded in Epon, it is necessary to use very sharp knives made by breaking plate glass. The model MT-1 and MT-2 microtomes currently manufactured by Ivan Sorvall, Inc. are equipped with holders for knives 1 inch (25 mm.) high; the holders supplied with older versions of the model MT-1 were made for knives 1½ inches (38 mm.) high.

The ultimate goal of the knife-making technique is to produce, from a sheet or strip of ¼ inch plate glass, a triangular piece in which at least a portion of the edge at one of the 45° corners is extremely sharp and free of imperfections that might cause the sections to be scratched as they are cut. As Epon and other embedding materials are quite hard, even the best portion of the edge will cut relatively few sections before it begins to get dull or show other signs of deterioration. An adjacent portion of the edge may then be used if it appears suitable. A particular knife may have only about 1 mm. of really good cutting edge, so it is unlikely that one will progress very far with cutting sections of a particular specimen before having to replace the knife.

As glass flows to some extent, knives should be used before their cutting edges deteriorate. For preparation of 1 μ sections, they should not be saved beyond two or three days. Most electron microscopists probably would not keep them even that long.

There are a number of techniques for breaking plate glass into knives. Several of these will be described. Any one of them will yield a certain percentage of knives of high quality. The procedures are directed in all cases toward production of knives 1 inch high; if knives 1½ inches high are required, the initial breaks will have to be spaced 1½ inches apart.

Making Knives from Pre-cut Strips of Glass. The most convenient raw material for making knives is ¼ inch plate glass in strips about 1 inch (25 mm.) or 1½ inches (38 mm.) wide and 14 inches long. These strips are packaged by LKB Instruments, Inc. (Rockville, Maryland) for use in their KnifeMaker (see p. 300). They are expensive but save much time.

Remove only one strip at a time from the paper packaging, and try to avoid handling it by its narrow sides. The strips may be washed in a warm solution of a detergent, rinsed thoroughly in warm tap water followed by distilled water, and then stood on end in a dust-free place until dry. However, the strips are ordinarily clean enough when unwrapped that washing and rinsing may be unnecessary, or even undesirable if there is any chance that residues may persist.

On a flat and clean tabletop, lay the strip so that its long dimension extends from left to right in front of you. The surface showing signs of having been scored before the glass was broken into strips should be lowermost. Pulling a glasscutter toward you, score the glass for about ¼ inch up to the edge nearer to you, at intervals of approximately 1 inch (Fig. 47*A*). The next operation requires a pair of glass knife pliers (sold by Ivan Sorvall, Inc., Norwalk, Connecticut), which are essentially

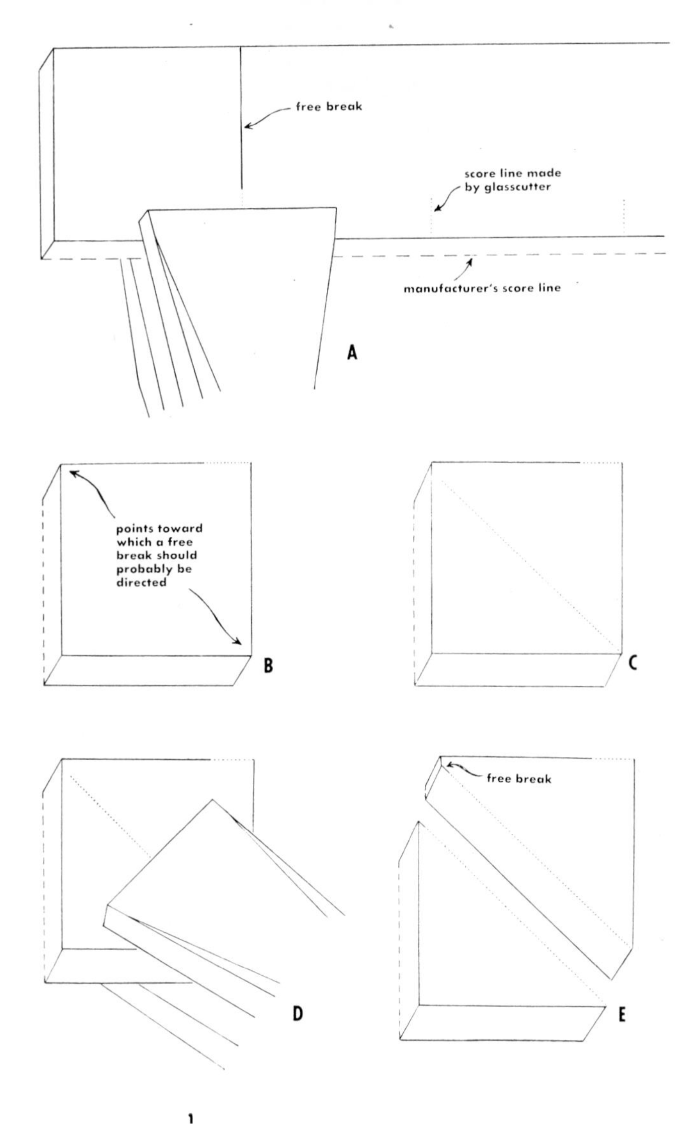

FIG. 47. Steps in preparation of knives from strips of plate glass. (See explanation in text.)

glazier's pliers, with jaws about 1 inch wide, modified to provide pressure points at the center of the lower jaw and both sides of the upper jaw. (Ordinary glazier's pliers without these pressure points can be made by cementing or taping down a half-inch piece of a paper matchstick to the center of the lower jaw, in a direction parallel to the handles of the pliers, and then fixing similar pieces just inside the outer edges of the upper jaw.) At one of the transverse score marks, grasp the strip of glass in such a way that the lower pressure point is directly under the score, and the tips of the pliers overlap the glass about $\frac{1}{8}$ inch (Fig. 47*A*). When pressure is applied, the glass should break along the score mark, and the break should continue away from you until it reaches the opposite edge. The break should lengthen slowly, and this is achieved by applying pressure steadily and slowly. It takes practice. Proceed to break the original strip into squares in this manner, so that a total of 14 squares having nearly equal sides of about 1 inch accumulates. Put the pieces in a clean box, with the original machine score marks facing downward.

Now it is necessary to break the squares of glass diagonally to obtain a cutting edge having a bevel of approximately 45°. At this point, each square must be examined carefully in order to decide where to focus the break.

As the square is viewed with its original machine score lines lowermost, and with the square oriented so that these lines are on the left and right sides, the better points of focus are apt to be just below the corner on the upper left and just above the corner on the lower right (Fig. 47*B*). This is not necessarily true, but as the narrow sides of the strips of glass cut by the manufacturer tend to be quite flat and free of imperfections, good cutting edges will generally be formed against these sides. If the system of knife-making described here is followed, the best part of the edge of each knife, as this is viewed with the side forming the hypotenuse sloping downward toward you, will be near the left margin. Obviously, this margin should not have either a manufacturer's score mark or one made for the purpose of breaking the glass into squares. The important thing, in any case, is to focus the break close to a corner, but actually toward a side that is smooth and flat, and which does not have a score mark on the edge that will be on the left side of the upright face.

Having decided where to focus the break, place the square of glass on the table so that this point will be farthest from you on the left side. Imagine a diagonal line originating about 1 mm. below the upper left corner and extending to the opposite corner. Beginning about 2 or 3 mm. from the left edge, but otherwise following the entire course of this imaginary line, make a score with the glasscutter (Fig. 47*C*). Then grasp the square with the tips of the pliers reaching to an imaginary line between the other two corners, and with the pressure point of the lower jaws directly under the score (Fig. 47*D*). Apply pressure steadily and slowly, until the glass breaks slowly along the score line and finally through the remaining 2 or 3 mm. If the score was angled correctly, the break should not go to the very corner but should be completed on the best edge (Fig. 47*E*). Carefully separate the two triangles, discarding the one on your right and making every effort to protect the prospective knife on your left. As has been pointed out previously, the best part of the cutting edge will probably be near the left margin, when the knife is viewed with the side forming the hypotenuse sloping downward toward you.

As knives are made, they should be stored in some sort of wooden tray with dividers high enough to keep them in an upright position. If they fall or are allowed to lie on their sides, there is a risk that they may contact one another or the sides of the box in such a way that the cutting edges may be damaged.

MAKING KNIVES FROM SHEETS OF GLASS. If strips of glass are not available, sheets of glass cut by a glazier to a size of 4 by 8 inches (or

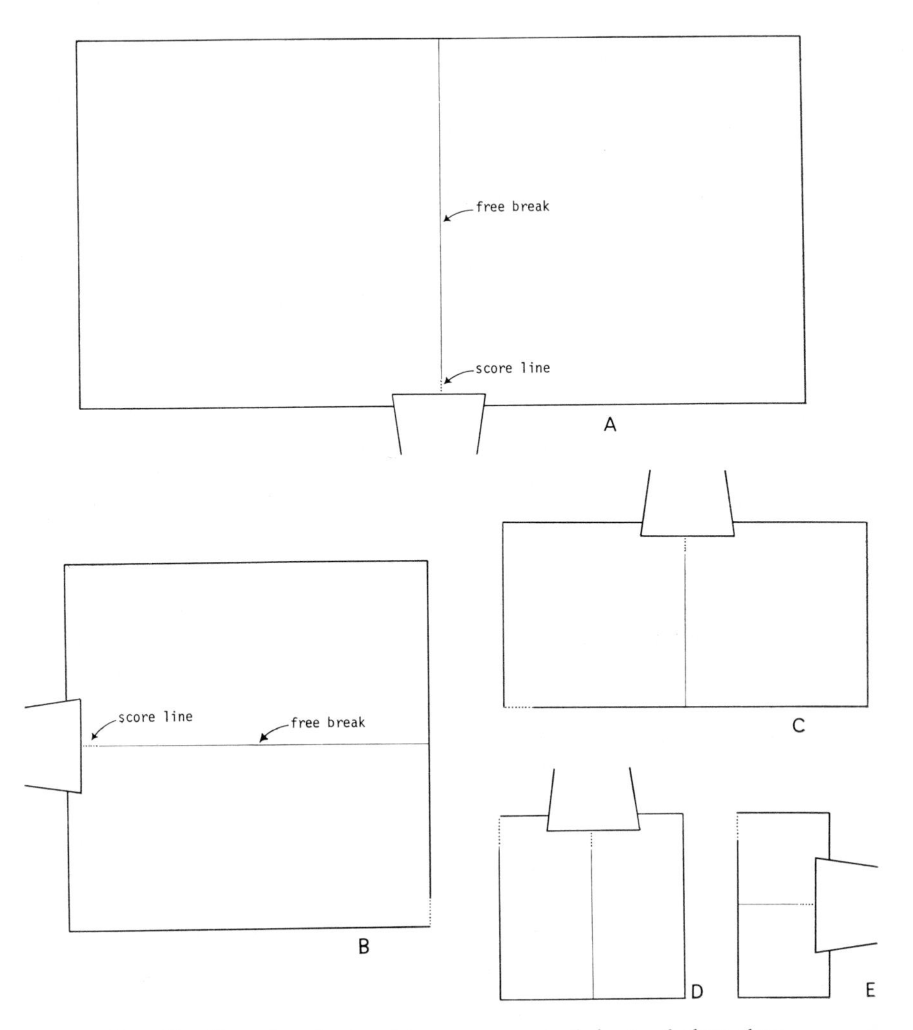

FIG. 48. Procedure for breaking squares out of sheets of plate glass. (See explanation in text.)

6 by 12 inches, if knives are to be made from 1½ inch squares) may be used. Pulling a glasscutter toward you, make a score line about ¼ inch long near the middle of the long edge nearer to you. With the pliers gripping the sheet about half the length of the line away from the edge, apply pressure slowly and steadily until the break along the score line reaches the opposite edge. The first score line and break should separate the original sheet into equal halves (Fig. 48*A*). Now rotate one of the halves so that the smooth edge formed by the free break you have just made is away from you. Score this piece near the middle of the side nearer to you and break it (Fig. 48*B*). Continue subdividing along these lines until the original sheet has been reduced to 1 inch squares (Fig. 48*C–E*). At least one corner of each square should be formed by two relatively flat edges unblemished by score lines. It will be close to this corner of each square that the final diagonal break, as accomplished with squares broken out of strips, should be focused.

Breaking Glass with a Heated Pyrex Rod. Another good way to make knives takes advantage of the fact that if plate glass is scored and then touched along the score line with a white-hot rod of Pyrex glass it will generally crack. The procedure for making knives this way is as follows: Score a 1-inch square diagonally as you would if you were to break it with pliers. Then, *while wearing dark glasses or dark protective goggles*, heat the tip of a Pyrex rod with a diameter of 3 or 4 mm. in the flame of a mixture of laboratory gas and compressed oxygen. When the tip forms a small bead, touch the bead to the score line near the middle of its course, and press it firmly against the glass. In a moment it may crack. If it does not, it may be refractory to being broken this way, but try again after the glass has cooled down. To break off an old bead, immerse the Pyrex rod, while it is still hot, in a jar of water.

If, after cracking, the glass does not fall apart into two triangles, one of which can be used as a knife, it is necessary to pull the pieces apart. For this purpose two pairs of glazier's pliers (without pressure points) are required. The two prospective triangles are grasped firmly and pulled in opposite directions. Do not attempt to bend the glass; just pull.

The principal inconvenience of this method is having to set up a gas-oxygen burner and oxygen tank. This may have to be done in a room remote from the laboratory, or in a shop where considerable dust is generated. Moreover, there are precautions to follow in handling the gas-oxygen burner and the oxygen tank, and a heated glass rod can cause a painful injury if it is picked up accidentally. Never look at the white-hot tip of the Pyrex rod without protection from very dark glasses, and even while wearing these glasses avoid looking at it any longer than is necessary.

The KnifeMaker. A very pleasant way to prepare knives is to use the LKB KnifeMaker (LKB Instruments, Inc.) (Fig. 49). This is an

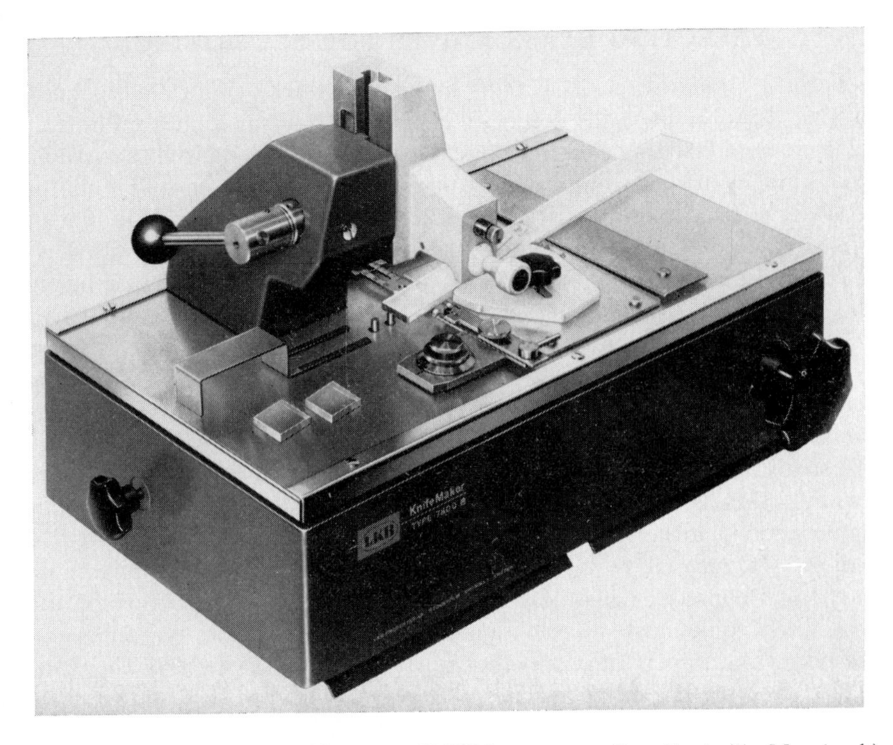

FIG. 49. The KnifeMaker. (Courtesy of LKB Instruments, Inc., Rockville, Maryland.)

expensive machine, however, and if several persons are to use it in the same laboratory, each should receive careful training. The KnifeMaker produces knives of consistently high quality and generally with a nearly straight cutting edge. A fairly large portion of the edge of almost every knife is suitable for 1 μ sections, and at least a small portion of it is generally good enough for cutting ultra-thin sections. Directions supplied with the instrument are detailed, and there is no need to present a watered-down version here.

DIAMOND KNIVES. Most experienced electron microscopists use knives made of polished synthetic diamonds. As diamond is so much harder than glass, knives made of this substance will last a long time if they are not abused, and they are more suitable than glass knives for cutting certain materials. Diamond knives are very expensive (about $400 for a 3 mm. cutting edge), and the cost of periodic reconditioning is about half the purchase price. They are obviously not recommended for beginners, who are apt to make errors of judgment when setting up knives and specimens in the microtome. Even expert microscopists generally use glass knives when they need only 1 μ sections.

SELECTING GLASS KNIVES FOR SECTIONING

For the preparation of sections about 1 μ thick, almost every knife produced by the methods described above will have at least a millimeter or two of good cutting edge. At first, the beginner will do well to examine each knife he makes, using a dissecting microscope with a spot illuminator to observe the characteristics of the cutting edge and to enable him to predict the portion apt to be suitable and the portions to be avoided.

As the knife is examined with the side forming the hypotenuse facing the observer and the cutting edge directed upward (Fig. 50), about half of the edge on the right side will probably show fine serrations. These are linked to diagonal marks (called "checks" or "whiskers") extending away from the edge, which make this portion of the knife unsuitable for sectioning. Moreover, the extreme right hand portion of the edge is usually characterized by a rather steep slope which forms a sort of "spur." Much of the left half of the edge should be free of obvious imperfections, although it is not necessarily really sharp throughout this region. It may curve upward near the left margin, but this is of no particular consequence unless the slope is steep. The portion of the edge likely to be most suitable for sectioning, because of its sharpness and freedom from imperfections, is that portion just to the right of the point where the fracture ridge reaches the edge. The breadth of this good portion will vary from knife to knife. After some experience, the microtomist will be able to judge how much of this he should reserve for actual sectioning, and how much, especially farther to the right, he can use for preliminary facing of the block.

There are various schools of thought with respect to whether a particular knife should or should not be used. Some workers insist that if a knife does not have a considerable portion of the edge (when viewed with a microscope) that appears to be promising, it should be rejected. They then throw away perhaps four out of five knives. At the more permissive end of the scale, there are those who attempt to use almost every knife they make. They figure there are enough things that will go wrong even when the best knife is in the microtome, and that if even a few good sections can be obtained with knives that are marginal, they will be

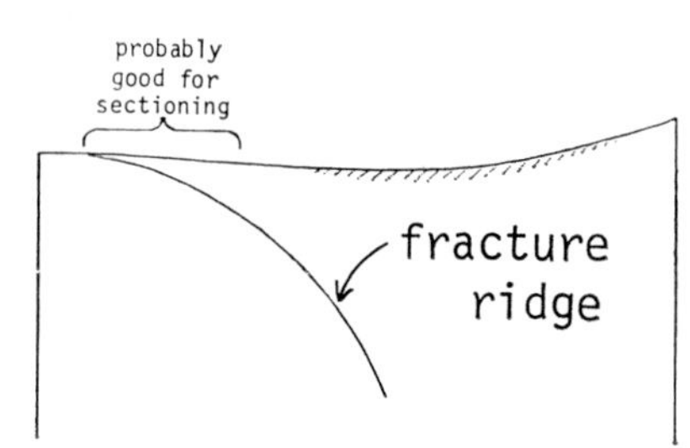

FIG. 50. Characteristics of the cutting edge of a typical glass knife. (See discussion in text.)

ahead in the long run. Perhaps for the beginner, a moderate position is best. He is likely to ruin some good knives before he starts getting perfect sections. On the other hand, he should know good from bad, and if he accepts every knife as adequate he will not know whether his difficulties are related strictly to a poor knife or to other circumstances. Therefore, the beginner is encouraged to be reasonably critical from the start, but not to become obsessed with the idea of making knives with a wide expanse of perfect cutting edge. After he learns the technique of sectioning, he will be better able to make use of really good knives. The moderate approach restrains one from throwing away almost every knife, yet spares one the trouble of mounting section troughs on many obviously unfit knives.

SECTION TROUGHS

Two types of section troughs may be recommended. The advantages and disadvantages of each will be outlined briefly below.

METAL TROUGHS. These can be purchased from suppliers of equipment for electron microscopists or can be made out of thin brass stock by cutting according to the pattern shown (Fig. 51). They are roomy, and therefore collection of sections from them is comparatively easy. However, they must be affixed to knives with care, and must be thoroughly cleaned before they can be used again.

To affix a metal trough to the knife, follow the steps diagrammed in Fig. 51*B*,*C*. Note that the most forward tips of the trough, which eventually will lie on either side of the cutting edge, should first be applied to the sides of the knife some distance below the edge; otherwise, the best part of the cutting edge, which is usually close to the left corner, may be damaged. The tips can be pushed slowly up to their permanent

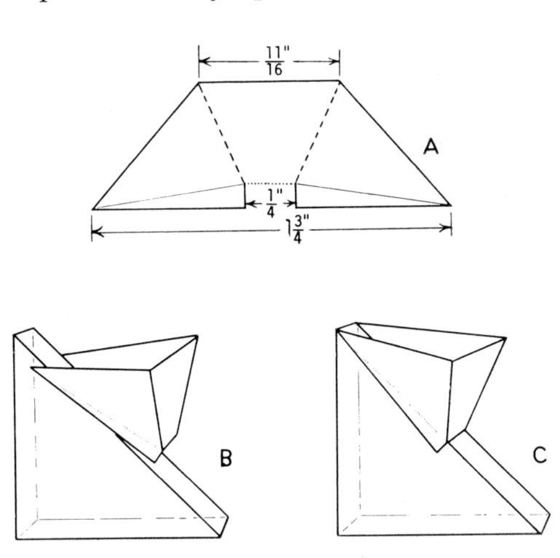

FIG. 51. *A*, Pattern for metal section trough. Use brass sheeting about $\frac{1}{64}$ inch thick. With tinsnips, cut all edges indicated by heavy lines, then break off the small piece remaining in the lower center by bending it with pliers along the dotted line. Bend along the dashed lines to an angle of about 90°, then turn the surfaces below the thin solid line slightly outward so that they will be parallel to the glass and fit tightly against it. A certain percentage of troughs made in this way will not fit well and should be discarded; they generally are not worth the time it takes to file them down or otherwise adjust them. *B* and *C*, Steps in affixing trough to knife.

position at the same time the lower part of the trough is seated on the sloping edge of the knife.

The trough should fit perfectly and not slip away from its permanent position. It may be necessary, before applying the trough, to press on its sides a little to make it hold better.

Before the trough will hold water, it must be cemented to the glass with wax. For this purpose, use a dental wax with a low melting point and which contains no flavorings or oily substances. A very suitable product is Kerr Utility Wax (Kerr Manufacturing Company, Detroit, Michigan). It is colored red—and this is an advantage—but contains no objectionable additives. Heat the upper edge of the deepest portion of the trough in the flame of an alcohol lamp, and more or less simultaneously melt the tip of a narrow strip of wax. Let a drop of the wax roll down the inside of the wall of the trough where it has been heated. If sufficient heat has been applied and not too much cooling has taken place, the wax will fill the crack between the trough and the glass, and by capillarity will reach the point on either side of the knife where the tips of the trough abut the cutting edge. If necessary, apply a little more heat to the top of the deepest portion of the trough. Obviously, if too much heat is applied, the glass at the cutting edge will be encouraged to flow, with the result that the edge will be ruined.

After a knife has been used, the trough is detached, and most of the wax scraped away with a heavy razor blade or old scalpel. It is then placed, along with other troughs being cleaned, into a jar of toluene. Toluene will dissolve the remaining wax. If the toluene is warmed on an electric hot plate to about 40 or 50° C., the removal of wax will be accelerated. Rinse the troughs in a second bath of toluene, then in acetone. Pour off the acetone, and dry the troughs on a clean paper towel. If they are not to be used again right away, they should be stored in a clean, dust-free box.

TROUGHS MADE OF PLASTIC TAPE. A second method of making a section trough uses $\frac{3}{8}$ inch black electrical tape. One end of the tape, cut square across, is pressed to the left side of the knife (as the knife is viewed with its cutting edge farthest from you), so that the upper corner of the tape is at the left side of the cutting edge. The tape is then drawn around the sloping edge of the knife and pressed to the right side of the knife. The excess on the right side is cautiously trimmed off with a sharp razor blade. As the most serviceable part of the blade will probably be near the left side, it is unwise to apply an excess of tape on that side and then have to trim it off.

Once the tape has been affixed, dip a small watercolor brush (No. 1 or No. 2) into melted embedding paraffin or dental wax. Keeping the wax melted while on the brush by judicious heating in the flame of an alcohol lamp, apply it where the tape crosses the sloping edge of the knife and at least a short distance toward the vertical face of the knife. However, do not attempt to apply wax to those edges of the tape parallel to the face of the knife.

Section troughs made of tape have certain disadvantages: (1) they do not have the working room provided by a metal trough; (2) they sometimes leak; (3) the tape may give off oily materials which rise to the surface and contaminate the sections. ("Scotch" electrical tape, produced by Minnesota Mining and Manufacturing Company, is relatively free of visible oils.) However, troughs made of a good tape have two advantages: (1) they do not require cleaning, because they are discarded after use; (2) no great amount of heat is applied to them when the wax is brushed on, so that there is little chance for the best part of the knife edge to be degraded because of heat-induced flow of the glass.

SECTIONING

Of the several ultramicrotomes in wide use, the Model MT-1 ("Porter-Blum") manufactured by Ivan Sorvall, Inc. (Norwalk, Connecticut) is perhaps the best known. In this microtome, the mechanism advancing the specimen block toward the knife consists of a rotating screw which is engaged by a tooth located at the bottom of a vertical forked bar. The fork is suspended by pivots fixed in the rigid casting of the microtome chassis and holds a block by another pair of pivots. The cantilevered arm on which the specimen is mounted is joined to this block. As the screw turns, the top of the fork swings forward slightly, and this motion is translated into a forward movement of the cantilevered arm. The longer the turn of the screw, the greater the advance of the specimen. The extent to which the screw turns is controlled by a ratchet-wheel linked by a pawl to a dial which may be set as desired. Other details of the construction of the MT-1 microtome, and directions for its maintenance, are given in literature published by the manufacturer.

The Model MT-2 microtome is superficially similar to the MT-1. The advance mechanism in this instrument also employs a rotating screw. However, the linkage of the screw to the cantilevered arm is rather different from that in the MT-1. In addition, the MT-2 is supplied with a motorized drive, which can be set for various cutting speeds, but it can also be driven by hand if desired.

There are a number of other ultramicrotomes in which the advance mechanism is of a mechanical type. However, some excellent ultramicrotomes, as that manufactured by LKB Instruments, Inc., take advantage of the principle of thermal expansion. These are decidedly more expensive. Because the MT-1 is the model most likely to be available for instructional purposes, all directions for sectioning will be based on this instrument.

The ultramicrotome is supplied with a relatively cold light source (a fluorescent lamp) and with a stereoscopic microscope enabling observation of the relationship of the specimen to the knife and the progress of sectioning. The microscope also facilitates the transfer of sections from the trough to the slide (or to a grid, for electron microscopy). The lamp can be adjusted in order to illuminate a particular combination of specimen, knife, and trough favorably. Similarly, magnifications most suitable for viewing various operations involved in preparation for sectioning, actual sectioning, and transfer of sections will have to be decided individually.

The illuminator should be left on throughout the course of setting up the knife and specimen and also during the period of sectioning. It is well to leave it on even if sectioning must be interrupted for a few minutes. Although the fluorescent lamp does not produce much heat, it generates enough so that turning it on and off will increase fluctuations in the temperature in the working area and in the temperature of the metal of the microtome itself.

The following tools and other materials should be at hand:

A polyethylene washing bottle ("squeeze bottle") of about 100 or 250 ml. capacity, filled with distilled water. This is convenient for filling the section trough. The dispensing tube should be directed laterally and have a rather small bore, at least at the tip.

A clean pipette with a relatively fine tip, for withdrawing excess water from the trough.

A supply of fine filter paper, cut into strips about $\frac{1}{4}$ inch wide and 2 inches long, for wiping water which happens to get on the vertical face of the knife or on the specimen block.

A supply of laboratory tissues (as Kimwipes) or facial tissues, for removing slight excesses of water from the trough, from around sections after they have been stained and washed, and for mopping up water which has spilled from the section trough onto the knife holder, knife stage assembly, and the track on which the knife stage moves.

A stiff eyelash neatly cemented into a slender wooden applicator stick about 3 inches long. This is used for manipulating sections.

Some new, pre-cleaned microscope slides in a jar of 95% alcohol, and a clean, lint-free cloth for wiping each slide just before it is used.

A bottle or flask of distilled water (preferably double-distilled) and a pipette for transferring small drops of water to a slide.

A device for transferring sections from the trough to the slide (see p. 315 for suggestions).

A warming stage for drying sections and for heating slides during staining (see p. 316).

The total extent to which the specimen block advances during sectioning is relatively short. Therefore, before beginning to cut sections or even to install the knife and the specimen block, it is best to reset the advance mechanism. This is done by pressing down the reset lever near the rear of the microtome base on the left side (Fig. 52). The advance will then automatically jump back to the beginning of its course. Now be sure that the coarse and fine stage advance knobs have been turned counterclockwise through most of their course. Install the knife in the holder but avoid letting the knife be too close to the specimen. Beginners often accidentally allow the knife to contact the block prematurely, thus perhaps ruining the best part of the knife and perhaps also cutting so deeply into the block that it will be severely damaged, or at least require retrimming.

Before inserting the specimen into the holder, immobilize the cantilever arm with the hook-like lock. This precaution should always be followed when manipulating the specimen holder itself or the locking ring that tightens the ball-bearing joint behind it. Insert the specimen, leaving about $\frac{1}{4}$ inch of it projecting, then tighten the jaws. Final orientation of the specimen with respect to the knife edge will probably have to be

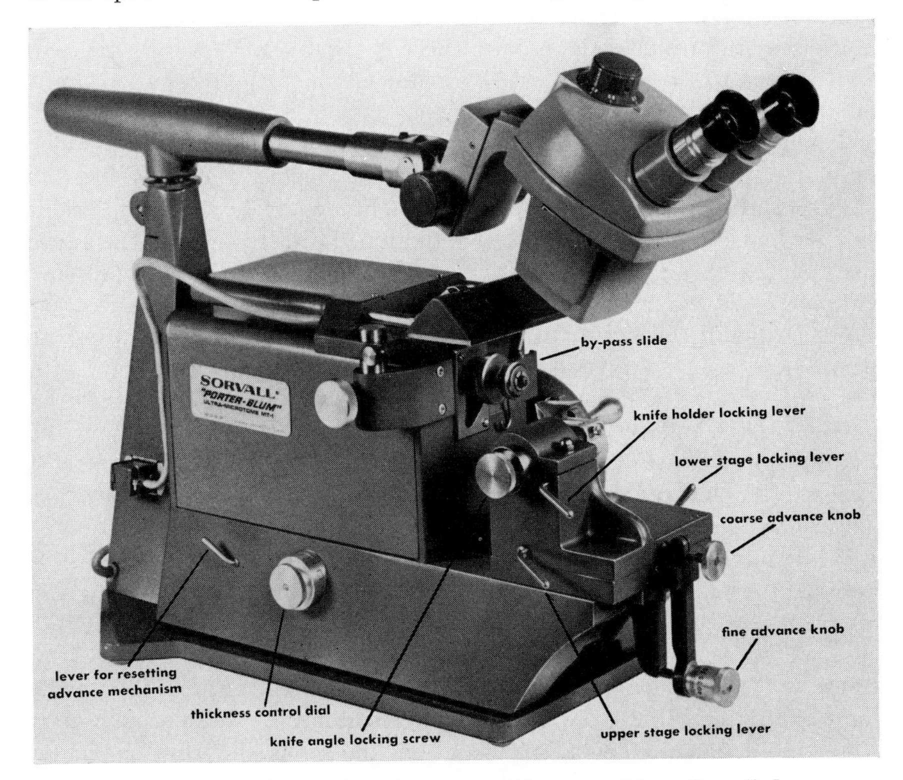

FIG. 52. Model MT-1 ultramicrotome. (Courtesy of Ivan Sorvall, Inc., Norwalk, Connecticut.)

done after the knife is in place and has been advanced close to the specimen. However, a rough alignment may be made at this time with the aid of the ball-bearing joint. Assume, for the moment, that the specimen should project more or less straight toward you, unless it is obvious that compromises made in trimming have left the trimmed portion of the specimen on a tilt with respect to the cylinder of Epon or metal clamped in the holder.

Two rather different types of holders are in use with the MT-1. An older version is essentially a circular drum with a slot for the knife. It has a pressure plate that can be tightened by a screw against the knife, and a screw at the top that is tightened against the sloping surface of the knife. This knife holder was developed for knives $1\frac{1}{2}$ inches high; if used with smaller knives, perhaps an insufficient proportion of the knife, in relation to that projecting, will be firmly clamped. If a 1 inch knife is seated deeply enough to be properly clamped, the drum will have to be rotated counterclockwise at least slightly toward the operator, and this may introduce a new difficulty. The specimen, after being cut by the knife, continues on its downward course briefly, then swings to the operator's left before traveling upward. Unless the vertical face of the knife is well in front of the holder, there is a risk that it will be mashed, as it swings to the left, against the metal in which the rotating drum is mounted.

The newer holder for the MT-1 (Fig. 53) is made expressly for 1 inch knives. The knife is set as far forward as possible, and then the two obvious screws that hold it in place are tightened. The one on the right side (knife-clamping thumb screw) is tightened directly against the knife; the one nearer the observer (rear clamp thumb screw) applies pressure indirectly, via a small triangle of metal (rear clamp) that fits against the sloping edge of the knife. Another screw that supports the knife from below need not be adjusted very often, because once set properly, with the small tool provided with the holder, it should serve with almost all knives broken out of 1 inch squares. The extent to which the knife should project upward out of the holder is checked by raising the height gauge to its elevated position. The cutting edge of the knife and upper edge of the gauge should be at the same level.

While installing the knife, do not work too close to the specimen. If necessary, pull the knife stage assembly toward you after loosening the lower stage locking lever on the right side. Insert the knife and tighten the two screws that hold it in place. Loosen the knife angle locking screw at the left side of the rotating drum and set the vertical face of the knife at an angle of $2°$ to $5°$, to provide clearance for the specimen after it passes the cutting edge. Then tighten the screw. With a particular knife and a particular specimen, the angle may have to be slightly different from that which will be best for another knife and another specimen,

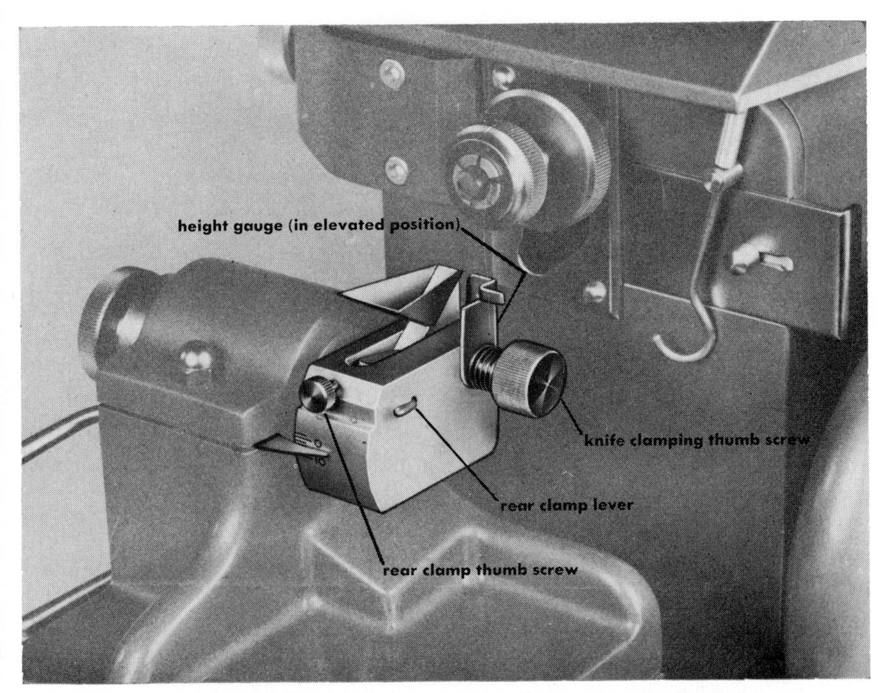

FIG. 53. Knife holder for the Model MT-1 ultramicrotome. (Courtesy of Ivan Sorvall, Inc., Norwalk, Connecticut.)

but this is apt to be of more concern to the electron microscopist than to one who is hoping to cut good 1 μ sections. Newer knife holders for the Model MT-1 are provided with scales making it easy to set the knife for the angle desired. However, it is not difficult to adjust the angle properly even with holders without this scale. The angle may be estimated by sighting the vertical face of the knife with respect to some vertical reference line such as a door, room corner, or window.

Release the cantilever arm and cycle the microtome through one to several complete rotations by turning the handwheel clockwise. On the last turn, stop so that the face of the specimen block is at about the same height as the knife edge. Arrange the lamp and microscope so that the specimen can be seen clearly. Then move the knife stage assembly forward until the knife is about 0.5 mm. away from the face of the specimen. If the knife is a considerable distance away from the specimen, simply slide the entire stage forward, then lock the knife stage into position by firmly depressing the lower stage locking lever. It is risky to push the stage forward until the knife is very close to the specimen. The last bit of advancing should be done with clockwise movements of the coarse knife stage advance knob (the upper knob). Before using either the coarse or the fine advance knobs, however, lift the upper stage locking lever on the lower left side of this assembly, then gently let it be pulled down again. This lever is already under tension and should never be pressed down with a view to locking it. In this respect it is different from the lower stage locking lever on the right side.

Adjust the knife laterally until the specimen is opposite the part of the knife to be used for final facing of the block. Lateral movements of the knife are possible if the knife holder locking lever, which is on the left side of the knife stage assembly near the top, is loosened. Then the knife holder drum is easily moved to the left or the right. After this, the locking lever is tightened again. For "facing" the specimen block (i.e., making the face of it flat and absolutely smooth, so that thin sections can be cut with regularity), it is generally customary to use a part of the knife that is fairly good but not quite perfect enough for thin sectioning. It is also possible to use a knife that is not likely to be particularly good for sectioning but that is adequate for the purpose of facing.

Carefully inspect the relationship of the specimen to the knife, to make sure that the longer of the two parallel sides is parallel to the knife edge, at least if the latter is straight (Fig. 54). If the knife edge is slightly curved in the portion likely to cut good sections, set the lower edge as if the knife were straight. Make these final adjustments while the cantilever arm is locked.

The next step must be done with great care. If the knife is about 0.5 mm. away from the specimen, it must be advanced carefully to a position very close to the specimen but still not touching it. For this purpose use the fine advance knob (or at first the coarse advance knob, then the fine advance knob). Check carefully that the lower edge of the specimen block is parallel to the cutting edge of the knife. If necessary, adjust the ball-and-socket joint until these edges are parallel. If this is to be done, however, it is almost certainly best to back the knife away from the

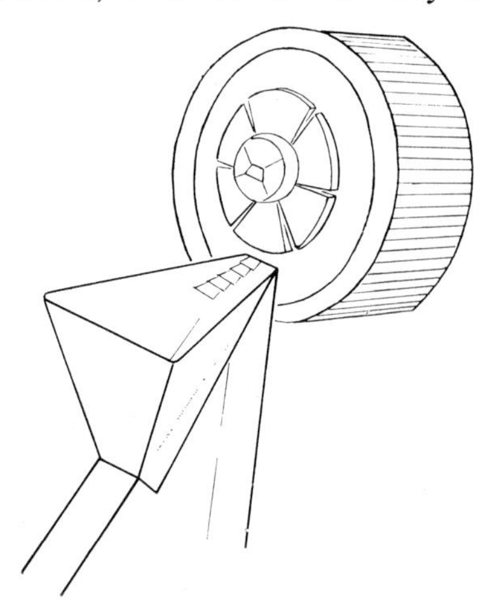

FIG. 54. Correctly trimmed block properly oriented with respect to the knife. The sections should cohere in a ribbon as they are cut.

specimen for at least a short distance. Continue advancing the knife toward the specimen with the fine adjustment screw, *in very small increments*, and rotate the microtome through its cycle after each advance. Eventually a portion of the block will be cut by the knife.

Although the rotating pin in the handwheel is convenient for moderately rapid cycling of the advance mechanism, it is never used for that portion of the cycle in which actual cutting is accomplished. Much better control is achieved if the entire hand is applied to the wheel (Fig. 55). Once the knife has passed through the specimen, the motion of cycling can be speeded up. The beginner will soon learn how to alternate the slow, steady motion required for cutting with a more rapid motion which quickly returns the specimen to the knife.

Once sectioning is started, there is no reason for the left hand to be near the microtome. It is a good idea, in fact, to keep it away in order to minimize fluctuations in temperature.

Continue the alternation of advancing and cutting until the entire face of the block is being cut. Be careful to cut very thin sections; if

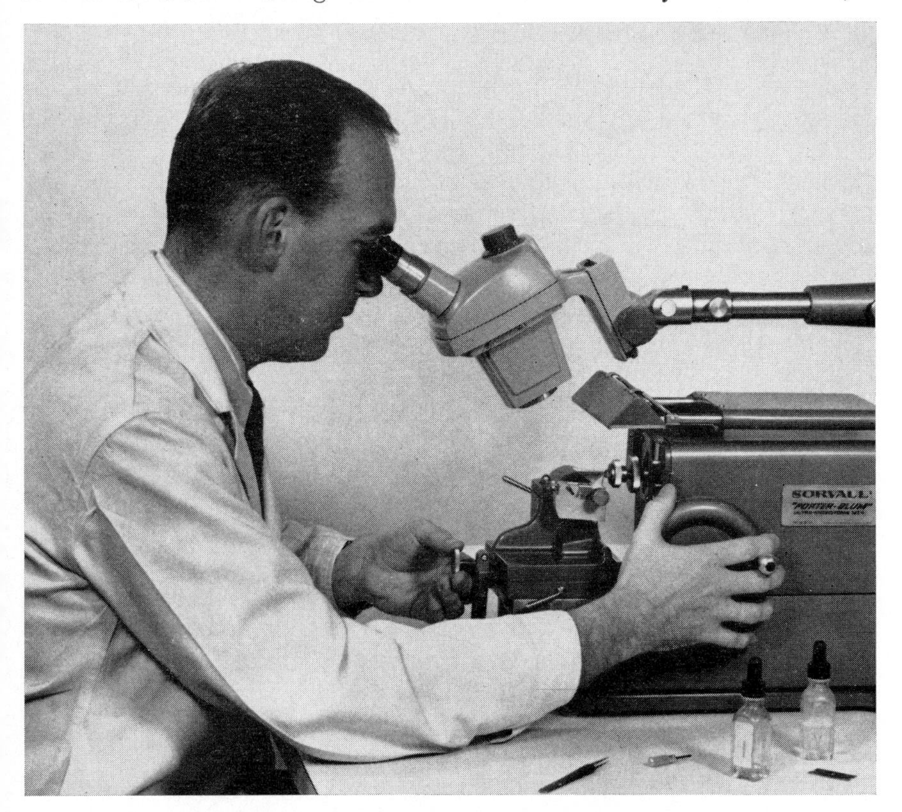

FIG. 55. Correct position of operator's right hand during sectioning. (Courtesy of Ivan Sorvall, Inc., Norwalk, Connecticut.)

thick sections are cut, the knife may be damaged so much that it will cause the face of the block to be roughened, and thus make it difficult to get good sections without again going through the operation of facing the block.

When you are ready to cut thin sections, move the knife laterally so that a good part of its cutting edge will be at the level of the block. If no part of the knife used for facing the block is apt to be suitable, back the knife stage assembly away from the specimen, replace the knife, and then advance it slowly to the point that sections are cut again. Set the thickness control dial to its highest position, i.e., at 19. This setting will yield sections of approximately 0.5 μ, which are usually too thin for light microscopy. Sections 1 or 1.5 μ thick are generally more useful. To obtain 1 μ sections it is necessary, while cycling the wheel, to deflect the specimen holder to the left as the specimen approaches the knife. This is easily accomplished by pulling it gently with the index finger or middle finger of the left hand; using the by-pass slide seems to be less convenient, but some prefer it. Although the specimen has been deflected, it will already have been advanced 0.5 μ, so that when the specimen is cut on the next cycle, the section will come off at about 1 μ. If the specimen holder is deflected twice before the specimen is cut, the section will be approximately 1.5 μ in thickness. This thickness may be favorable for some types of material or for certain purposes, but sections appreciably thicker than this will probably be of limited use. In preparing sections for electron microscopy, continue to change the setting of the thickness control dial until it is set at about 2 or even lower, thus yielding sections of about 0.1 μ or thinner. However, electron microscopists do not rely on the dial setting; they can tell from the interference colors of the sections, as these are viewed with reflected light, approximately how thick they are. Gold sections, for instance, are a little over 0.1 μ thick; silver sections, the most generally useful for electron microscopy, are slightly thinner than this.

Facing the block with a portion of a knife that is unsatisfactory for cutting thin sections may be done without any water in the trough. If this technique is used for facing and the knife is one with a good edge for thin sectioning, then it is necessary to carefully remove the unwanted sections from the knife by means of a very fine brush or an eyelash mounted on the end of a wooden applicator stick. Take care not to damage any part of the knife that is apt to be good for thin sectioning. Ordinarily, if a knife appearing to have a good edge for thin sectioning is installed, the trough may as well be filled with water from the beginning. It is difficult to explain without an actual demonstration just how much water to put into the trough. The edge of the water should definitely come to the cutting edge of the knife, but the meniscus should be slightly concave, never convex. If the water has a convex surface, it will be unsatisfactory for sectioning because, as the face of the specimen block passes the knife, it will pull water down with it onto the vertical face of the knife. This will leave a deposit of water on the face of the block which will cause problems as the block comes down on the knife for the succeeding section. A small polyethylene squeeze bottle (capacity about 100 ml.) is good for delivering distilled water to the trough. A rather fine-tipped pipette with rubber bulb can be used for withdrawing excess water, but it should be clean; strips of filter paper are also helpful.

Once good sections are obtained with reasonable regularity, it is best to continue cutting without interruption until there are enough sections to justify stopping long enough to transfer them to slides. If the sections form a ribbon as they come off the knife, this is a help, as it will facilitate transferring groups of related sections. The sections may separate when they are picked up, but at least they will all end up in a relatively small area of the slide on which they are mounted.

Manipulation of the sections, to guide them away from the edges of the section trough or to break up ribbons longer than can be picked up, is best done with the tip of the eyelash.

As a rule, if sectioning is stopped long enough to pick up the sections, it is found on beginning to cut again, that at least the first one or two sections may be of quite a different thickness than those just collected, and perhaps may be unusable for other reasons as well. This is why, when sectioning appears to be going smoothly, it is advisable to keep cutting until the section trough has so many sections that it would be risky to collect more before picking them up. Of course, when trying to come reasonably close to getting serial sections, sectioning will have to be interrupted frequently in order to transfer groups of only a few sections before they are hopelessly mixed up with others.

How long the particular portion of a knife cutting good sections will last depends on its own characteristics as well as the hardness of the embedded tissue and the skill of the operator. Sometimes one should feel lucky to get 10 good sections; other times, perhaps 30 or 40 may be cut with no hint of trouble.

Among the difficulties encountered in sectioning with a good new knife, a knife beginning to deteriorate, or a knife worthless to start with, the following are more or less obvious at the time sections are being cut or after they move onto the surface of the water in the trough:

1. *Variation in thickness, or skipping, or both.* This may be due to the Epon being too soft, and therefore tending to be compressed on some strokes and cut thick on others. However, it may be due to something being loose, especially in the knife stage assembly or object carrier assembly. Check to see that all screws have been tightened. Unevenness in temperature around the microtome also may cause variation in thickness.

2. *A washboard-like appearance of the sections*, with the corrugations running parallel to the knife edge. Such sections tend to have a dull appearance, instead of being metallic and shiny. The knife may not be sharp enough, the specimen too large, or the speed at which the sections are cut too rapid and/or unsteady.

3. *Scratches in the sections at right angles to the cutting edge of the knife.* These are due to imperfections, generally nicks, in the knife edge. Frequently, they develop after a number of good sections have been cut. If the sections in the areas between scratches appear to be good, it may be best to leave well enough alone and cut some more sections, for it is possible that the final stained preparations will show much useful detail between the scratches. However, sometimes the scratches become so pronounced that the sections tend to fragment, with the result that they become difficult to pick up and transfer to slides.

4. *Each section, as it is cut, is at least partially pulled back over the edge of the knife and down onto the vertical face of the knife.* This usually happens when there is too much water in the section trough, so that some of it tends to be drawn out, wetting both the vertical face of the knife and the specimen. Remove a little of the water, and dry the vertical face of the knife and the specimen with a strip of filter paper. Avoid coming close to the knife edge.

Whenever the knife seems to be at fault, there are two choices: the knife holder drum can be shifted laterally one way or the other to find a better portion of the blade, or the knife can be replaced. If the decision is to try another part of the blade, remember that it may be a little closer to the specimen than the part used previously. Therefore, there is a risk that the first section will be cut so thick that it will put undue stress on the knife. After some practice, it can be estimated if this is likely to happen, and the fine advance screw of the knife stage assembly can be used to draw the knife back a little.

If the decision is to change the knife, then the lever at the right of the knife stage assembly should be loosened, and the assembly pulled away from the specimen. As long as the knife is being changed, it will be well to depress the reset lever; otherwise, at a time when sectioning is going smoothly, the advance mechanism may come to the end of its course and the relationship of the specimen to the knife will have to be worked out all over again. After installing the new knife, bring it reasonably close to the specimen by sliding the knife stage assembly forward, tightening it, then slowly advancing the knife with the coarse and fine advance screws until sections are cut again. During the operation of changing the knife, the object carrier should be immobilized by the hook provided; this protects the mechanism of the microtome as well as the specimen.

AFFIXING SECTIONS TO SLIDES

This aspect of the Epon method is often troublesome and uncertain. None of the three techniques suggested here for transferring sections to slides is foolproof. The uncertainties are largely related to surface tension and to the tendency of the sections to adhere to the instrument used for transferring them. The simplest instrument is the flat end of a regular toothpick. If it is made thinner by careful shaving with a razor blade, and if it has no fibers sticking out of it, it may work well. Use it as a spatula to lift the sections from below the surface. Quickly carry the sections to a drop of water on a slide and then turn the toothpick over. Most of the sections will probably remain in the drop.

Another instrument which works well is a loop of very fine platinum wire set in a handle of metal or wood. The loop should preferably be more or less rectangular, with an opening about 1 × 2 mm. To use the loop, simply immerse it in a part of the trough where there are no sections, then bring it up around a group of them. As the loop is lifted out of the water, surface tension should hold a drop of water supporting the sections. The loop is then touched to a small drop of distilled water on the slide. Periodically, the loop should be rubbed between the fingers; this adds a little oil, thus encouraging formation of a rather discrete drop which will tend to separate completely from the loop when it contacts the water on the slide.

Many workers transfer sections with perforated copper specimen grids on which electron microscopists mount sections for viewing. These are made in two sizes, 2.3 mm. and 3 mm., and in a wide variety of mesh patterns. Grids with a diameter of 3 mm. and 200 mesh, in which about half the area is open, work well for transferring sections, but probably almost any kind of grid of this general type will do. They are held tightly in a pair of fine watchmaker's forceps and used as spatulas. Immerse the grid in the deeper part of the trough, bring it up under a group of sections, and lift it carefully out of the water. Quickly turn it over on a drop of water on a slide and then withdraw it. Hopefully, at least most of the sections will come off. Rubbing the grid between the fingers occasionally will help to prevent the sections from sticking to the grid, but will also make it slightly more difficult to pick up the sections in the first place.

If desired, place a number of drops of water in a row on one slide, and then transfer groups of sections to successive drops. This is obviously the most practical way of trying to collect nearly complete series without involving a large number of slides. However, when nearly complete serial sections are of no particular interest, having several slides with only a few sections on a slide makes possible the use of varied staining techniques, or at least variation in the time that staining is allowed to proceed.

As soon as the sections have been transferred to a slide, they should be placed on a warming stage to evaporate the water and at the same time to remove wrinkles in the sections. A temperature of about 60° C. is recommended. An electrically heated warming stage, such as is used for flattening paraffin sections, is apt to be too cumbersome to have close to the microtome and other accessories, and the temperature of a small electric hot plate may fluctuate so much that there is danger of momentary over-heating. A very good arrangement for drying slides is a smooth sheet of aluminum about 3 mm. thick, supported by a ring stand or some other contrivance over the flame of an alcohol lamp. By raising or lowering the sheet of aluminum, the temperature of the metal can be adjusted to about 60° C.

After the sections are dry, they may be stained immediately, unless there is some reason to delay the staining, in which case the slides should be stored in a dust-free box. As staining of Epon sections is normally carried out at 60° C., the apparatus used for flattening and drying sections will be used again. As a matter of fact, it is not necessary to remove the sections from the warming stage before applying the stain.

STAINING

Because the tissue in Epon sections is permeated by the resin, and because the methods of fixation used before embedding in Epon do not favor staining by most techniques in common use for paraffin or celloidin sections, there is a rather restricted choice of staining methods available. A few methods which have proved to be particularly useful are described below.

Methylene Blue-Azure II

This valuable method, which utilizes a mixture of methylene blue and azure II, is metachromatic in the same way as the methylene blue derivatives in Romanovski-type blood stains. Although originally described by Richardson, Jarett, and Finke, it is commonly referred to as Richardson's stain.

Azure II, 1% solution in distilled water	1 part
Methylene blue, 1% solution in 1% sodium borate	1 part

Place the slide to which the sections have been affixed on the warming plate. Cover the sections with several drops of the staining mixture. After about 1 minute, rinse off the stain in distilled water, or first in tap water and then in distilled water. Shake the slide and stand it on end to dry. Touching a laboratory tissue (such as Kimwipes) or a facial tissue to the slide near the sections minimizes the amount of water left, and therefore the time available for dye to leach out of the sections and leave a slight deposit around them.

Thicker sections (1.5 to 2 μ) ordinarily are more intensively stained than thinner sections (0.5 to 1 μ). If plenty of sections are available and experimentation seems to be in order, the length of the staining period may profitably be varied from about 30 seconds to 2 minutes or longer. The variation in thickness of the sections on a particular slide may be such that some may be better for certain purposes than for others. Variety often brings out one good quality at the expense of another.

Some workers have found that treatment of sections with 1% periodic acid just before staining them may produce preparations which are better in one way or another than those stained by the conventional method. The periodic acid is applied at room temperature, in a Coplin jar or shell vial, for about 5 minutes, then rinsed off in distilled water before the sections are stained on the warming stage.

Crystal Violet and Safranin O

This method of double staining may yield attractive and useful preparations in which the nuclei are stained bright red and most other structures

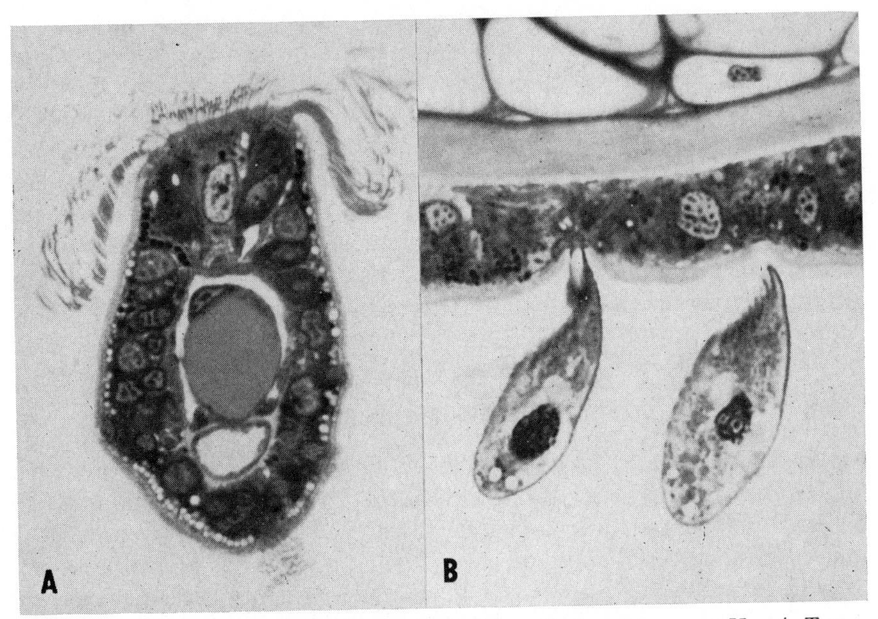

FIG. 56. Epon sections, 1 μ thick, stained with methylene blue-azure II. *A*, Transverse section of a pinnule of a prostomial cirrus of a sabellid polychaete, *Schizobranchia*. Shrinkage is minimal, and this is especially evident in the case of the cilia, which appear much thicker than they would in paraffin sections of comparable material. *B*, A small ciliate, *Ignotocoma*, feeding suctorially on the cirrus of *Schizobranchia*. The delicate suctorial canal and the attachment knob embedded in the epithelium are well preserved.

some shade of violet. Flood the sections with an ample amount of a 1% solution of safranin O, and hold the preparation at 60° C. for 5 to 10 minutes. Rinse the sections in distilled water, and apply a 1% aqueous solution of crystal violet for 1 to 5 minutes, at 60° C. Wash off the crystal violet in distilled water and let the sections dry.

Basic Fuchsin

A 1% aqueous solution of basic fuchsin, applied at 60° C. for about one minute, generally stains 1 μ sections quite heavily and not very selectively. However, lipid inclusions, which after fixation in osmium tetroxide often have a highly refractile and greenish appearance, may stand out very sharply against an otherwise bright red background.

Impregnation with Silver Nitrate

When cell boundaries are not easily recognizable in stained preparations, this technique can be tried. Place slides on which sections have been mounted into a 2% aqueous solution of silver nitrate, in a Coplin jar, for about 10 or 20 minutes. Reduce the silver nitrate, which has been selectively adsorbed by certain structures, by placing the Coplin jar in bright sunlight. When the sections appear to have turned brown (perhaps 10 minutes), wash them thoroughly in distilled water and dry them. If the method succeeds, cell boundaries will be dark brown. Some other interesting results may also be obtained, depending on the material. If the impregnation is too unselective, try placing the slides into distilled water before placing them in the light. The silver nitrate method is apt to require considerable experimentation, whereas any one of the true staining methods outlined above will probably yield useful preparations after the first few attempts. Sections 1.5 or 2 μ thick may be more suitable than 1 μ sections.

MOUNTING IN IMMERSION OIL

The high quality of superior Epon sections cannot really be appreciated unless they are studied with an oil-immersion lens. If the sections are to be used only temporarily and then discarded, the immersion oil can be placed directly on them. However, if the preparations are to be saved, do not remove the oil with a solvent such as toluene or xylene; this will cause the sections to pucker. It is better to apply a coverglass, utilizing the immersion oil as a mounting medium. Even if from the beginning more or less permanent mounts are to be prepared, a synthetic immersion oil is the only really safe medium, especially in the case of sections stained with the azure II-methylene blue mixture and other dyes which tend to be fugitive. When sections as thin as 1 μ or 1.5 μ are mounted in permanent media, a slight fading is much more apparent than would be the case with 10 μ sections.

Synthetic immersion oil, which does not harden, is apt to be a nuisance if it is present in excess. However, if an oil of high viscosity is selected, and if the amount of oil to be applied is estimated correctly, it tends to restrict itself to the area under the coverglass, especially if the slide is kept flat. If oil exudes from around the coverglass, it should be wiped away. The coverglass may be ringed with asphalt cement (p. 462) to minimize the risk that the oil will continue to squeeze out. Although the asphalt cement, being slightly soluble in immersion oil, will slowly diffuse inward and eventually discolor the preparation in the area where the sections are mounted, the discoloration may not be noticeable when the sections are studied with transmitted light.

As the sections will probably be studied with an oil-immersion objective, use a No. 1 or No. 1½ coverglass of appropriate dimensions. Circular coverglasses, which can be neatly ringed on a turntable, will be more suitable than square coverglasses if the preparations are to be sealed with asphalt cement. A rectangular coverglass will probably be necessary if several sets of sections have been mounted on one slide.

CHAPTER 19

Grinding Method for Sections of Bones and Teeth

Thin sections of calcified tissues, prepared by grinding material from which the soft parts have been dissolved, show not only the hard parts but also the form and relations of the cavities which, in life, were occupied by the soft structures.

1. PRELIMINARY CLEANING AND SAWING. Remove teeth or pieces of bone from the body. Free them from adhering muscles and connective tissue. Force out the marrow from long bones. Using a hacksaw with the finest blade obtainable, cut the specimens into pieces according to one or the other of the following methods:

a. To Prepare Bones. For making either transverse or longitudinal sections, place pieces of bone in a miter box (Fig. 8*B*, p. 124) and cut them into lengths of about 1 cm.

b. To Prepare Teeth. A simple method is to clamp the tooth between pieces of cork, in a vise, and saw it either longitudinally or transversely into several sections. These can be cut about 1 mm. thick. This method is adequate for some types of work, but it is laborious as well as inaccurate, and only a few sections can be obtained from a single tooth. When much work is to be done with dental tissues, it is advisable to use special cutting and grinding machines such as those described by Bevelander (1950). Several companies now manufacture equipment for this and related purposes.

2. REMOVAL OF SOFT TISSUES. Soak the pieces in water until the soft parts are completely decomposed. Keep the bad-smelling material in a well-ventilated and unfrequented place. Occasionally shake the container and change the water. The process may require several weeks or months, depending upon the size and density of the object.

If desired, the destruction of soft parts can be accelerated by placing the specimens in a 2% solution of gelatin to which several drops of a culture of *Escherichia coli* have been added. Small pieces of bone treated in this manner will be sufficiently clean within 5 or 6 days.

3. WASHING. Wash the material very thoroughly, by shaking it in several changes of tap water. In preparing bone, go on to step 4. If the material consists of teeth, proceed directly to step 5.

4. DRYING AND SAWING BONE. Allow the pieces of bone to become dry, in which condition they may be stored indefinitely.

Place a piece of dried bone in a miter box and, with a very fine saw blade, cut off transverse or longitudinal sections. Make the sections about 1 mm. thick.

If the material is compact bone, proceed to step 5.

Should the material be spongy bone, place the sections in toluene or xylene overnight. Put them into thin balsam for a day or two, and then into thick balsam for at least 3 or 4 days. After taking them out of the balsam, dry them in a paraffin oven for 5 to 8 days. The matrix of solid balsam will support the thin layers of bone during the process of grinding.

5. GRINDING. There are four ways of doing this. The first three methods involve grinding by hand and are very tedious, although they yield good preparations. Electrically operated hones are almost a necessity when numerous sections are being ground.

a. Grinding with Hones. Lay the section upon a Carborundum hone of moderately fine grit (No. 120) and flood the hone with water. Rub the section upon the hone, in an elliptical path, pressing it gently against the abrasive. After several minutes, turn the section over and grind its other side. Continue in this way, turning the section at short intervals, so as to grind both sides equally. Keep the hone wet. If one part of the section is thicker than another, press upon the thicker portion while grinding, until the inequality is corrected. The proper speed and amount of pressure must be learned by experience. Excessively rapid motion or heavy pressure may crack the section. When the section has become considerably thinner, it may be held upon the hone by means of a rubber stopper, in order to protect the operator's fingers.

When the section is reduced to a thickness of about 0.3 mm., transfer it to a very fine Carborundum hone which has been flooded with water. Cover it with a similar but smaller hone or piece of a hone. Proceed to grind by moving the upper hone in an elliptical path, until the section becomes flexible and somewhat translucent. Then go on to step 6.

b. Grinding with Files. The section can be ground by hand, in a similar manner, upon medium and fine files. The files are not moistened.

c. Grinding by Means of a Glass Plate and Abrasive. Rub the section upon a frosted glass plate which has been moistened with water and sprinkled with pumice or Carborundum powder. Turn the section over at intervals of a few minutes, in order to grind both sides equally. When the section has reached a thickness of about 0.3 mm., lay a second and smaller glass plate upon it and fill the space between the plates with water. Continue the grinding by a rotary motion of the upper plate, until the section appears translucent.

d. Grinding With Electrically Operated Disks. When many sections are being ground, it is advisable to use Carborundum disks, revolved by a small electric motor. Two disks (medium and fine grit) are needed. The abrasive disk is mounted horizontally and its upper surface is used for grinding. The disk is driven at moderate speed by means of a belt extending from a small pulley on the motor to a large pulley on the shaft which bears the disk. Still more convenient is a motor with built-in reducing gears. The section is held upon the disk by means of a rubber stopper and the disk is set in motion. Further details of the grinding process are as described in paragraph *a*, above. Other and more elaborate grinding machines are discussed by Ham and Harris (1950).

6. Polishing. Rinse the section with water and transfer it to a fine yellow Belgian or Arkansas hone. Keep the stone well-moistened. Hold the section under a fingertip and rub it lightly upon the hone. Turn the section occasionally and continue rubbing until it becomes extremely thin and transparent. Inspect it under a microscope, with the iris diaphragm partially closed. When a section has been rubbed down sufficiently, the matrix will appear transparent, with the lacunae and canaliculi of bone (or dentinal tubules of teeth) plainly visible. Continue to rub until this effect is obtained, but no longer. The section will fall to pieces if it is rubbed too long or too hard.

7. Washing. To free the section from debris, put it into a small dish containing water. With a pipette, repeatedly pick up some of the water and discharge it upon the section. Pour off the water and renew it several times during the process.

8. Dehydration. Place the cleaned section in absolute alcohol for 5 minutes or longer. Proceed to the next step, unless the material consists of spongy bone which was infiltrated with balsam before being ground. In the latter case, put the section into toluene or xylene for a day, then return it to absolute alcohol and proceed to dry it.

9. Drying. Lay the section upon a slide and cover it with another slide. Allow it to dry thoroughly; this may require an hour or longer.

10. Mounting. A special mounting technique is necessary to render the lacunae and canaliculi plainly visible. Either of the following methods may be employed.

a. Air Injection. This method is very satisfactory but requires skill in manipulation. Heat a quantity of Canada balsam until a drop of it is found to solidify immediately upon cooling. Prepare several slides and a like number of small circular coverglasses in the following manner: Warm each slide and coverglass and place upon them a very small drop of the heated balsam. If the balsam does not spread out to a slight extent, but forms a globule, warm the glass again. Allow the prepared slides and coverglasses to cool. They may be stored indefinitely if kept free of dust.

Place a dried section of bone or tooth upon the solid balsam on a slide. Lay one of the prepared covers, with the drop of balsam downward, upon the section. Put the slide on a warming table. The instant the balsam

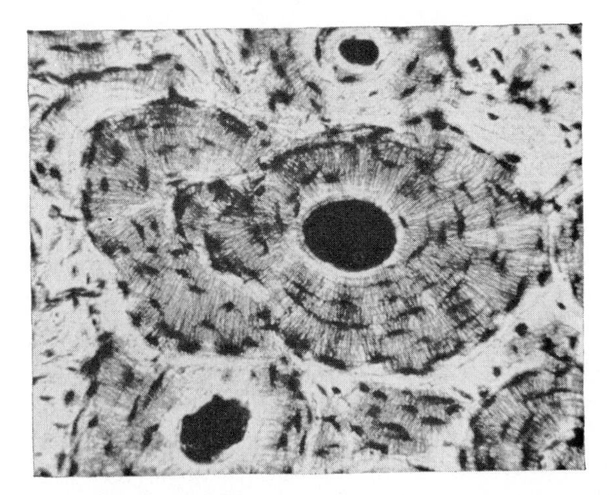

FIG. 57. Ground section of a portion of a human femur. Careful mounting in balsam (warmed after solidification) has cleared the matrix, but left the Haversian canals, lacunae, and canaliculi filled with air. With transmitted light, these cavities appear black.

begins to melt, press gently upon the cover until the balsam spreads to its edge. For this purpose use a small cork, which may conveniently be held by thrusting a pair of forceps into its larger end. If this step is carried out properly, balsam will penetrate and clear the matrix, but the canals and lacunae will remain filled with air. These cavities will therefore appear black, against a practically colorless background (Fig. 57). Should the preparation be heated too much, the cavities will become filled with balsam and therefore will be nearly invisible. Another danger to be avoided is that of breaking the coverglass while pressing it down, which may happen if it is not heated sufficiently, or if it is subjected to too much pressure.

b. Silver Impregnation. Place the section in a 0.75% aqueous solution of silver nitrate for 24 hours. Rinse it with water and put it into 5% sodium thiosulfate for 15 minutes. Wash it thoroughly in water and polish it on a fine hone in order to remove precipitated silver. Rinse it again, place it in absolute alcohol for an hour, and dry it between slides. Mount it in balsam or synthetic resin of ordinary consistency. Lacunae and canaliculi will appear black or brown.

REFERENCES

BEVELANDER, G., 1950. Methods for the study of dental tissues. *In* Jones, R. M. (editor), *McClung's Handbook of Microscopical Technique*, 3rd ed., pp. 285–328. New York: Paul B. Hoeber, Inc. (Reprinted 1961 by Hafner Publishing Co., Inc., New York.)

HAM, A. W. and W. R. HARRIS, 1950. Histological technique for the study of bone and some notes on the staining of cartilage. *Ibid.*, pp. 269–284.

CHAPTER 20

Principles of Biological Staining

One of the purposes of microtechnique is to increase the visibility of structures which are nearly transparent and colorless in their natural state. This is generally done by staining them with dyes or other colored substances. The affinities of intracellular and extracellular structures for particular dyes differ greatly. It is thus possible to produce precise and beautiful color differentiation of nuclei, cytoplasm, musculature, connective tissue, and many other cytological and histological elements.

This chapter will discuss briefly some of the properties of stains used in microtechnique, and will give a simple outline of the process of staining. A general knowledge of this subject is necessary for a proper understanding of the staining methods described in succeeding chapters.

DYES AND OTHER SUBSTANCES USED FOR BIOLOGICAL STAINING

The earliest dyes to be used for biological staining were natural substances derived from plants or insects. Although a few natural products, particularly cochineal and hematoxylin, are still very important as sources of dyes, the majority of dyes now employed in microtechnique are completely synthetic. Certain metallic compounds, which may be selectively reduced or otherwise precipitated on or in specific structures, are also employed as colorants. Each of these groups of substances will be dealt with in greater detail in Chapters 21 to 27, and only a brief résumé of them will be presented here.

Sources of Natural Dyes

Hematoxylin. Hematoxylin, derived from logwood, is not itself a dye, but upon oxidation it is converted into a colored substance called hematein.

When hematein is mixed with certain compounds of aluminum, iron, and some other metals, it forms salts of a deep blue or black color. These colored salts are called "lakes." The metallic substances serve as mordants to bind the dye to constituents of cells and tissues. Certain techniques of using hematoxylin in this way are the most widely employed methods for staining nuclei and chromosomes. They are also used for other purposes in cytology and histology, and for bringing out organs in specimens which are to be mounted entire.

Cochineal. Cochineal consists of the dried bodies of female cochineal insects. The actual dye principle contained in cochineal is carminic acid, and an incompletely purified preparation of it is marketed as carmine. When combined with salts of aluminum or other suitable mordants, carminic acid forms lakes which stain nuclei red or purplish red. Structures colored by carmine are often brilliantly differentiated, yet quite transparent. For this reason, carmine stains are very useful for objects which are to be mounted entire.

Other Sources of Natural Dyes. A number of other dyes derived from plants have been tried for biological staining, but only a few of these have enjoyed wide use. Two or three should perhaps be mentioned here. Orcein, extracted from certain lichens, and now also manufactured synthetically, is commonly employed in a mixture with acetic acid (aceto-orcein) for staining chromosomes in smear preparations. It has, however, also been used for other purposes. Indigo, obtained from various leguminous plants, as well as by synthesis, is converted into a dye known as indigo-carmine, and this is occasionally used as a counterstain for carmine and other dyes. Brazilin, related to hematoxylin in chemical structure and by the fact that it is derived from a similar leguminous plant, has enjoyed some popularity as a substitute for hematoxylin. Like hematoxylin, it is colorless, and must be oxidized before it can function as a dye.

Coal Tar ("Aniline") Dyes

These are brilliant dyes of all colors, produced synthetically from hydrocarbons obtained by the distillation of coal tar. They are commonly called aniline dyes for the reason that aniline, a colorless oily substance, is an intermediate product in the preparation of some of them. Hundreds of these beautiful synthetic dyes have been tried for staining microscopic materials, and many of them are very useful and important. Some of the coal tar dyes are brilliant and powerful nuclear stains. Others are utilized for bringing out cytoplasm, cell boundaries, connective tissues, the matrix of cartilage, and other specific elements of cells and tissues.

Metallic Substances

Salts of silver, gold, osmium, and some other metals are employed to produce color differentiation of certain structures, including nerve cells, nerve fibers, Golgi material, intercellular cement, and flagella. The coloring material is deposited more or less selectively, sometimes in opaque masses, by the reducing action of chemicals or light. It is better to refer to techniques of this sort as methods of impregnation, rather than of staining.

THE CHOICE OF STAINS

In preparing to stain any kind of material, it is important to give careful consideration to the selection of a dye or combination of dyes which will show to best advantage the particular structures which are to be studied. The specific uses of the more important stains will be considered in ensuing chapters. The following general statements, however, will explain the usual manner of approaching the problem.

1. In nearly all cases, a stain which will bring out nuclei will be required. Certain mixtures containing carmine are generally superior for objects which are to be mounted entire, although alum hematoxylin is preferable in some cases. Alum hematoxylin is routinely used for sections in which the general histology is to be demonstrated, but some of the aniline dyes (methylene blue, acid fuchsin, basic fuchsin, and azocarmine G) are also very useful. Iron hematoxylin is the most valuable nuclear stain for cytological preparations; other dyes, however, including basic fuchsin (especially as it is used in the Feulgen method) and carmine (prepared as aceto-carmine), are important for certain purposes.
2. It is generally desirable to stain cytoplasm, connective tissue, and other distinct layers of tissue in one or more contrasting colors. Stains for this purpose must be selected with reference to the structures which are to be studied, and usually should be of colors which contrast favorably with that of the nuclear stain. For example, safranin is a good nuclear stain, and eosin stains cytoplasm and some other tissue elements, but the colors of these dyes are similar and a combination of them is not likely to be used. On the other hand, the pink or reddish color of eosin contrasts effectively with the blue color of alum hematoxylin or methylene blue.
3. It will probably be necessary to use more than one method of staining if a thorough study of an organism or tissue is to be attempted. Histological preparations made with the usual combinations of nuclear and "plasma" stains will show the general organization of the tissue and the types of cells present. It will then be possible to select intelligently special stains for particular structures which are to be examined more critically. For example, impregnation with silver, gold, or mercury is indicated for nerve structures, and certain aniline dyes or osmium tetroxide will demonstrate lipids.

THE PROCESS OF STAINING

The following simple outline is based upon routine methods in general use. It should be understood that the exact order in which the steps are performed differs considerably in various methods, and that altogether different procedures are used in many special techniques.

Specimens which are eventually to be embedded and sectioned can be stained first, but as a rule staining is carried out after embedding and sectioning. The following outline can be applied to either procedure.

PREPARATION FOR STAINING. If the first stain (or mordant) to be used is in an aqueous solution, the sections or specimens are passed gradually through the series of alcohols to water. If an alcoholic solution is to be employed for staining, they are passed down the series of alcohols as far as necessary. As explained in previous chapters, paraffin sections are affixed to slides before they are stained, whereas nitrocellulose sections and entire objects generally are handled in shallow dishes. Frozen sections may be handled in either way, but are likely to survive microtechnical procedures more successfully when affixed to slides.

STAINING NUCLEI. In the majority of methods, the sections or other materials are placed directly into the solution which serves as the nuclear stain. In a few methods, as in the use of hematoxylin with a mordant such as ferric ammonium sulfate, they are first treated with the mordant and then placed in the staining solution.

There are two general methods of applying stains. In progressive staining, the objects are simply immersed in the staining solution (usually a rather dilute solution) until they are stained to the desired intensity. In regressive staining, the objects are left in a strong solution of the stain until they have taken up an excess of the dye; the excess is extracted later. The uses of progressive and regressive staining will be discussed in subsequent chapters. As a general rule, regressive staining produces the more precise differentiation.

WASHING AND DESTAINING. After removal from the staining solution, the objects are washed briefly in water or alcohol to remove superfluous dye. In the regressive method, they are then treated with a dilute acid or some other reagent which removes the stain rapidly from certain structures, but slowly or not at all from others. This step is commonly referred to as destaining, or differentiating. Objects which have been destained are subsequently washed to remove all traces of the chemical used for this purpose.

COUNTERSTAINING. The sections or other material are now passed up or down through the series of alcohols, as necessary, to a dye solution which will bring out the cytoplasm in a contrasting color. This stain is generally referred to as a counterstain. They are left in the counterstain as long as necessary, but usually a very few minutes suffices. Excess stain is then removed in water or in a suitable concentration of alcohol.

DEHYDRATION. This is generally accomplished by passing the objects up through a graded series of alcohols to absolute alcohol (sometimes other dehydrating agents are used) until all water has been replaced. If the counterstain is one which washes out easily, the process must be carried out as rapidly as possible. In some cases, however, the objects may have to be left in strong alcohol for a considerable time to remove the counterstain from certain structures.

Final Steps. The remaining steps in the procedure as a whole depend upon the manner in which material is subsequently to be prepared for study.

1. Specimens which are to be mounted entire, or teased apart, are usually immersed in toluene, xylene, or some other liquid which serves as an intermedium between the alcohol and a resinous mounting medium. This step is called clearing, because the high refractive index of the intermedium makes the specimens more transparent. A resinous mounting medium is then gradually introduced into the clearing agent containing the material, and the specimens are mounted under a coverglass. These steps have been described in earlier chapters.

2. Stained sections are cleared in toluene or xylene (less commonly in other liquids). A resinous mounting medium is then added, and a coverglass applied.

3. Specimens which have been stained entire and are now to be embedded and sectioned are processed according to the paraffin or nitrocellulose methods. Nitrocellulose sections are subsequently dehydrated, cleared, and mounted in a resinous medium. Paraffin sections are affixed to slides in the usual manner, the paraffin removed with toluene, a drop of mounting medium added, and a coverglass applied. Staining of specimens before embedding is usually restricted to nuclear stains or somewhat general stains, and counterstains are usually omitted entirely or applied after the material has been cut into sections. If the latter practice is followed, the sections are counterstained, then dehydrated, cleared, and mounted in the usual manner.

The practice of staining entire specimens before embedding and sectioning them does reduce the number of reagents through which the sections must be processed. However, the variety of staining procedures applicable to entire objects before embedding is rather limited.

The Nature of Staining Reactions

The reasons why the components of cells and tissues exhibit affinities for certain dyes are still not very well understood. Staining reactions appear to be complicated, and most of them probably involve both chemical and physical factors. Some of the ways in which a dye unites with the constituents of cells or tissues, and imparts its color to them, will be discussed briefly.

Chemical Combination of the Dye With Specific Substances. Certain parts of cells are acid, others are alkaline, and some are amphoteric. Similarly, the auxochromes of dyes—those portions which enter into the formation of salts—are acid or basic. One might expect, therefore, that the chromatin of the nucleus, containing acid constituents, would have an affinity for basic dyes, and that these dyes could actually enter into the formation of new compounds with substances in the chromatin. That this actually happens is difficult to prove. Moreover, the fact that acid substances attract basic dyes, and vice versa, may be explained even more satisfactorily by the concept discussed below.

ADSORPTION OF A DYE BY CERTAIN SUBSTANCES. The behavior of many dyes toward various structures can perhaps be explained most satisfactorily on the basis of the phenomenon of adsorption. It is well known that certain ions are more readily adsorbed by some substances than by others, and also that the rate of adsorption is influenced by various factors such as the concentration of hydrogen and hydroxyl ions.

BY SOLUTION OF A DYE IN SUBSTANCES PRESENT IN CELLS OR TISSUES. Staining of this type is largely if not purely a physical process, explained by the fact that a few dyes are very soluble in certain substances existing in tissue. The staining of lipids by dyes such as Sudan IV and Sudan black occurs largely in this manner. It is quite possible that other organic substances may retain dyes in this same general way.

CONVERSION OF A SUBSTANCE INTO A COLORED INSOLUBLE STATE. This is what happens when nerve fibers, Golgi material, kinetosomes of cilia, and various other structures are impregnated by compounds of osmium, silver, gold, and other metals. The metallic substance usually remains as a colloidal deposit within or upon the structures impregnated.

ACID AND BASIC DYES, AND THE EFFECTS OF HYDROGEN ION CONCENTRATION ON STAINING REACTIONS

A dye owes its color largely to certain groupings of atoms, called chromophores. A different part of the compound, known as an auxochrome, may modify the color, but it is also important because it confers upon the molecule the property of electrolytic dissociation, and therefore an acidic, basic, or amphoteric character.

Practically all dyes are supplied as salts of such color acids or color bases. Dyes in which the auxochrome is acidic (due to the presence, for instance, of carboxyl [$-COOH$] groups which furnish hydrogen ions upon dissociation) are salts of sodium, potassium, calcium, or ammonium. Those in which the auxochrome is basic (owing to amino [$-NH_2$] groups and others which furnish hydroxyl ions) are usually chlorides, sulfates, or acetates. In solution, these salts dissociate into ions of positive or negative charge. A solution of an acid dye accordingly contains colored anions and colorless cations, whereas a solution of a basic dye contains colorless anions and colored cations.

THE PROPERTIES OF ACIDS AND BASES. The essential property of an acid is its ability to liberate hydrogen ions, which are positively charged. The effective acidity of a solution is therefore measured by the concentration of hydrogen ions present. A base is characterized by its ability to liberate hydroxyl ions, which are negatively charged. The effective alkalinity of a solution, therefore, is measured by the concentration of hydroxyl ions. True neutrality is represented by the condition of pure water, a liter of which contains 10^{-7} of a gram equivalent of hydrogen ions, and an exactly equal concentration of hydroxyl ions. The concentration of hydrogen ions in a neutral solution is therefore expressed as $H^+ = 10^{-7}$, and that of hydroxyl ions as $OH^- = 10^{-7}$. The product of these is $H^+ \times OH^- = 10^{-14}$, and this product holds good for all solutions, whether neutral, acid, or basic. Because the concentration of hydrogen ions is related to the concentration of hydroxyl ions, the reaction of any solution can be quantitatively expressed by the magnitude of the hydrogen ion concentration. Solutions in which H^+ is greater than 10^{-7} are acid; those in which it is smaller are alkaline. The symbol pH is generally used for the expression of hydrogen ion concentration. $pH = -\log (H^+)$, so that instead of writing $H^+ = 10^{-7}$ we simply write pH 7, and instead of $H^+ = 10^{-5}$ we write pH 5. As pH values are expressed as minus logarithms, an acid solution (which contains more hydrogen ions than a neutral solution) has a pH below 7, and an alkaline solution has a pH above 7.

The acidity and alkalinity of solutions are now routinely measured by the use of electric pH meters, and the effect of addition of an acid or base toward bringing it to the desired pH may be quickly verified by these devices. Because of the deep color of staining solutions, colorimetric determination of pH, by use of indicator dyes, is impractical. In routine biological staining, however, a precise control of pH is usually not necessary. In the case of some staining solutions in common use, a nearly optimum pH (whether it is known or not) may be obtained simply by adding the amount of a standard solution of an acid or base which has been found to achieve the desired result.

THE EFFECTS OF ACIDITY AND ALKALINITY UPON STAINING. The influence of hydrogen ion concentration upon the staining properties of dye solutions may be demonstrated with two of the more common dyes used in biological microtechnique. However, it is not possible to explain fully and with certainty what happens when the relative affinities of certain components of cells and tissues for these dyes are altered by changes in pH.

If sections of tissue are placed in a full-strength solution of alum hematoxylin for a few minutes, then washed in slightly alkaline tap water, it is found that the blue lake formed by the hematein and the mordant is deposited heavily in the nuclei, and to some extent in most of the other structures. By placing the sections in water made more alkaline by addition of a drop of dilute ammonium hydroxide or a little sodium bicarbonate, the rather unselective stain will be intensified. However, if a dilute acid (a 0.5% solution of hydrochloric acid) is applied to the sections, the stain will begin to diffuse out of the tissue. By stopping the process of destaining at the right time, and then neutralizing the acid, a preparation is obtained in which the cytoplasm is nearly colorless, while a sharp stain of nuclear chromatin and some other structures has been preserved.

It must be remembered that during staining with alum hematoxylin, it is not the hematein itself which is adsorbed or otherwise combined with the tissue to any appreciable extent. Hematein is an acid substance having little affinity for any structures in animal tissues, and unless a mordant is used, coloration will be weak. In the case of alum hematoxylin, it appears likely that aluminum hydroxide (formed by hydrolysis of aluminum ammonium sulfate) becomes linked with acidic groups in the tissue, so that the lake as a whole functions as a basic dye. Under acid conditions, the hematein dissociates from the mordant, so that its own color (red) rather than the color of the lake (blue) is seen. This is independent of the effect of increased hydrogen ion concentration on the affinity of the mordant for certain tissue elements.

Similar tests on the influence of hydrogen ion concentration may be conducted with the staining properties of an acid dye. If a 0.5% solution of eosin Y in 90% alcohol is prepared and separate portions of it are adjusted to pH 3.5, 5.4, 7, and 8 by addition of appropriate amounts of hydrochloric acid or sodium hydroxide, the effects of these samples on the sections of tissue properly stained by alum hematoxylin will be quite different.

If the sections are left in the eosin for a few minutes and then placed in alcohol, it will be evident that those which had been in the most acid solution (pH 3.5) are very deeply stained and give up the dye slowly. The sections which had been in the slightly less acid solution (pH 5.4) are less deeply stained, and give up the dye more readily. Those which had been in the neutral solution (pH 7) are stained lightly, and quickly give up the eosin. Finally, the sections from the alkaline solution (pH 8) are scarcely stained at all.

If all of the sections are left in the alcohol until the differentiation of those which had been stained at pH 5.4 is ideal (red blood corpuscles deep pink, and cytoplasm and connective tissue various lighter shades of pink), and they are then cleared and mounted, a comparable gradation is observed. In those stained at pH 3.5, the eosin is still very strong, and may even impair the nuclear stain. In sections stained at pH 7, the red blood corpuscles are less intensely differentiated than those in preparations stained at pH 5.4, and all other extranuclear structures are lighter and more nearly equal shades of pink. Only a weak and diffuse color of eosin, if any is left at all, is seen in sections stained at pH 8.

It has already been pointed out that the action of hydrogen ions and hydroxyl ions in intensifying or weakening the affinity of a dye substance or its mordant for particular structures cannot be explained completely at the present time. However, it does appear likely that components of cells and tissues having acid properties will tend to adsorb or to combine chemically with basic dyes, and that those having basic properties will adsorb or combine with acid dyes. The chromatin of the cell nucleus, which is acid in character, would therefore be expected to have a selective affinity for basic dyes. However, the fact that the affinity of acid material of the cell nucleus for an acid dye such as eosin may be enhanced by increasing the hydrogen ion concentration supports the concept that this particular process of staining is dependent largely upon the phenomenon of adsorption, rather than upon chemical combination.

CHAPTER 21

Cochineal and Carmine Stains

Crude cochineal consists of the dried bodies of females of a scale insect, *Dactylopius cacti*. This species lives upon several kinds of cacti in Mexico and Central America, and especially on the "cochineal fig," *Nopalia coccinellifera*, which is cultivated to encourage propagation of the insect. By grinding and boiling cochineal in water, a deep red solution is obtained. Upon treatment with an alum, this yields a dye known as carmine. Because carmine is the product generally used in biological staining, it has become customary to class all cochineal stains as *carmines*. In the ordinary methods of staining, carmine and other cochineal derivatives stain nuclei a deep red, purple, or blue, but produce relatively little coloration of cytoplasm, connective tissues, and other histological elements.

The active dye principle of carmine and crude cochineal is carminic acid. By itself, carminic acid is a feeble stain, but the salts which it forms with alkali metals, and especially its aluminum salt, act as powerful basic dyes. A simple extract of cochineal does not contain these salts and therefore stains only those tissue constituents which do possess the appropriate metals. For this reason, cochineal is of limited usefulness. If cochineal is dissolved with an alum, it becomes a more effective stain. Carmine includes the necessary metals, and it stains rather strongly without the addition of other salts, although it produces somewhat different effects in a solution of alum.

In 1892, Mayer sought to obtain staining solutions of constant and dependable composition by adding pure carminic acid to salts of aluminum. However, carminic acid is very expensive, and its use seems to result in less brilliant stains than can be achieved with carmines.

Carmine is practically insoluble in water or alcohol which is neutral. However, it dissolves quite readily in water or weak alcohol (up to 35%) which has been rendered alkaline. In strongly acid water or alcohol, its solubility is lower. Both carmine and cochineal dissolve freely in solutions of aluminum ammonium sulfate, forming the salts or "lakes" to which reference has been made. On the basis of these facts, four important classes of carmine and cochineal stains may be recognized:

1. Simple Extract of Cochineal
2. Alkaline Carmine Stains
3. Acid Carmine Stains
4. Alum Cochineal and Alum Carmines.

A category for "Other Carmine Stains" will include picro-carmine, which cannot be classified under any of the headings listed above.

Cochineal was the first dye to be used for staining microscopic material, having been employed as early as 1779 in Sir John Hill's studies of wood fibers. The practice of histological staining first became generally known through the use of carmine by Joseph Gerlach, in 1858. He employed this dye with some success for demonstrating the nuclei and other structures in human tissues. Improved methods of staining with carmine were developed soon after this. Until 1862, when Beneke introduced one of the coal tar dyes, carmine was the only dye employed for microscopic objects.

Although carmines have been supplanted by other dyes in most work with sections, they remain among the best nuclear stains for objects which are to be mounted entire. This is due not only to their great penetrating power, but to the fact that they differentiate structures brilliantly and precisely, yet leave them quite transparent. Another advantage of carmines is that they seldom fade in thick balsam mounts, while other dyes commonly do this. Carmines are used sometimes for staining objects which are subsequently to be sectioned. However, hematoxylins are usually preferable for this purpose. Carmines are employed in both progressive and regressive staining, but the latter method generally gives a more sharply differentiated preparation.

Stains of this group give excellent results after many fixing agents. Following fixation in mixtures containing mercuric chloride (except Zenker's fluid and related mixtures), they produce a very brilliant stain. Fixation in formalin often permits remarkably good carmine staining, in spite of the widespread belief that it is unsuitable for this purpose. As a rule, carmine stains (other than Mayer's carmalum) are effective after fixation in Bouin's fluid and other picric acid mixtures. They seldom give good results on material fixed in potassium dichromate, chromic acid, or osmium tetroxide.

A great number of cochineal, carmine, and carminic acid staining solutions have been formulated, but many of these give similar results. The following selection will cover the requirements for all ordinary work.

SIMPLE EXTRACT OF COCHINEAL

Mayer's Alcoholic Cochineal

Cochineal, powdered	10 gm.
Alcohol, 70%	100 ml.

Allow to stand for 5 days, shaking occasionally, and filter. The stain keeps well.

In general, this is not so useful as borax carmine or alum carmine, but it possesses several unique properties which make it valuable for certain purposes. It is very penetrating and therefore advantageous for some arthropods. Because it contains no free acid, it can be used for specimens in which calcareous structures are to be preserved, such as plutei and metamorphosing larvae of echinoderms. The solution contains no metallic "lakes" and consequently the formation of these active coloring compounds depends upon the presence of certain salts in the cells. As the salt content of cells in various tissues differs in kind and amount, the intensity and color of the stain vary correspondingly. Nuclei are usually stained blue or purple, chitinized elements are stained red, and gland cells become gray-green.

PREPARATION OF MATERIAL. Mercuric chloride and formalin are good fixatives to precede this stain. Bring material to 70% alcohol. If it was fixed in mercuric chloride, iodize it according to the conventional procedure, then wash out excess iodine in clean 70% alcohol.

PROCEDURE. *a.* Transfer material to the stain. Staining requires from several hours to a week, depending upon the size and nature of the specimens.

b. If overstaining occurs, which seldom is the case, reduce the intensity of the stain by treating material with 0.1% hydrochloric acid in 70% alcohol.

c. Wash the material (with or without previous destaining, as the case may be) with repeated changes of 70% alcohol, until it ceases to give up color, and transfer it to 80% alcohol. Then dehydrate, clear, and mount the specimens in the usual manner, beginning at step 6, page 155.

ALKALINE CARMINE STAINS

In alkaline solutions, carmine molecules carry negative charges. Because most cell structures bear positive charges, alkaline carmine solutions stain diffusely. In strongly acid solutions, however, the carmine molecules become positively charged. Thus, when a diffusely stained specimen is transferred to an acid solution, the positively charged carmine leaves the cytoplasm and other substances bearing positive charges, but remains in nuclei, mucus, and other negatively charged constituents. This is the basis for the regressive methods of staining with carmine.

Grenacher's Alcoholic Borax Carmine

Carmine	3 gm.
Sodium tetraborate ("borax"), c.p.	4 gm.
Distilled water	100 ml.

Boil until the carmine is dissolved (½ hour), or—and this is better—allow the mixture to stand at room temperature until this occurs.

Then add:

70% alcohol (methyl or ethyl)	100 ml.

Allow the solution to stand for a day or two and then filter it.

Grenacher's Method of Staining

After being fixed, washed, and hardened in 70% alcohol, the specimens are transferred to 50% alcohol for a time, and then to the staining solution. Ordinarily, they are left in this for 24 hours or longer. After staining, they are passed by way of 50% alcohol to 70% alcohol containing 1 to 3 ml. of hydrochloric acid per 100 ml., and allowed to remain in this until the tissues assume a bright, transparent color. They are then washed in neutral 80% alcohol, counterstained if desired, dehydrated, cleared, and mounted; or they may be embedded in paraffin or nitrocellulose according to the usual procedures. With certain types of material, this method yields a perfectly acceptable stain, but in some cases the coloration is too diffuse. The modification detailed below gives superb results on many types of specimens which are to be mounted entire, and makes borax carmine one of the best stains for such material.

Borax carmine is quite commonly used for staining material which is to be embedded for sectioning. For this purpose, it is inferior to alum cochineal or alum hematoxylin. Both of these penetrate better, and the latter stains nuclei more precisely.

Lynch's Precipitated Borax Carmine Method

This method gives a highly selective stain, much superior to that obtained by Grenacher's original procedure. Precipitated borax carmine is suitable only for objects which are to be mounted entire. Superb results are obtained with hydroids, medusae, ectoprocts, small and medium-sized trematodes, tapeworm scoleces and small proglottids, entire small annelids, cladocerans and other small crustacea, entire small clams and mussels, tunicate larvae, *Branchiostoma* ("amphioxus"), ammocoetes larvae of lampreys, and many other organisms. The stain not only shows the general topography of the organisms, but brings out nuclei and cell boundaries, even in specimens of considerable size.

This method is not recommended for nematodes, insects, or other specimens which possess an impermeable cuticle. It is also unsuitable for organisms such as *Volvox* and echinoderm larvae, which contain large, thin-walled cavities, because it will collapse the cavities, leaving precipitates within them. Other stains are preferable for large tapeworm proglottids, large flukes, or comparable specimens in which there is a considerable amount of parenchyma. It is generally rather impractical to stain protozoa with Lynch's method, because of the difficulty of extricating them from precipitated carmine; but with skill and patience the large species can be handled successfully, and can be manipulated into excellent preparations. The stain is strongly acid, so it should not be used for material in which it is desired to preserve calcareous structures.

PREPARATION OF MATERIAL. Fix in Bouin's fluid, formalin, alcohol, or any mercuric chloride mixture not containing potassium dichromate. After fixation in Bouin's fluid or any other picric acid mixture, treat material with several changes of 70% alcohol and bring it gradually to 30% alcohol. It is not necessary to remove every trace of yellow color. Following fixation in mercuric chloride mixtures, transfer material gradually to 70 or 80% alcohol, iodize it, and pass it down the alcohol series to 30% alcohol. It is best to harden formalin-fixed material by bringing it to 70% alcohol and allowing it to stand in this for 12 hours or longer before passing it down the alcohol series to 30% alcohol.

PROCEDURE. *a.* Transfer the material from 30% alcohol to Grenacher's borax carmine. Allow it to remain in the stain for at least 3 or 4 hours, or preferably overnight. Then add concentrated hydrochloric acid, drop by drop, while gently agitating the dish, until the carmine is precipitated as a flocculent substance of brick-red color. About 6 or 8 drops of acid will be required for each 10 ml. of the staining mixture. Allow the material to stand for 6 to 8 hours or overnight.

b. Add to the above an approximately equal volume of 3% hydrochloric acid in 70% alcohol. Agitate the container, to mix its contents thoroughly. Let it stand until the specimens have settled to the bottom; they will settle within 2 or 3 minutes, unless the objects are extremely small. Most of the precipitated carmine will remain in suspension. Remove it by drawing off the greater part of the liquid with a pipette. If the liquid drawn off contains a considerable number of specimens, place it in a shallow dish; otherwise discard it. Now refill the container with acid alcohol, mix the contents, allow the material to settle, and after several minutes again draw off most of the liquid. By repeating this process several times, practically all of the precipitated carmine may be removed.

If the objects are very small, and a considerable number of them have been removed with the precipitated carmine, further dilute the liquid which has been taken from the original container. Allow the specimens to settle and withdraw most of the liquid. Repeat this process until as many of the specimens as it is possible to salvage have been separated from the precipitated carmine. Return the specimens recovered in this way to the original container.

c. Place the material in a shallow dish, with sufficient 3% hydrochloric acid in 70% alcohol to cover it to a depth of about 5 mm. The acid will slowly extract the stain. Replace this liquid as often as it becomes deeply colored. Examine the specimens under a low-power microscope from time to time, inspecting them more and more frequently as their color becomes lighter. Allow destaining to continue until nearly all color is extracted from cytoplasm and connective tissues. In some cases (for example, gland cells), it may be necessary to allow the cytoplasm to retain

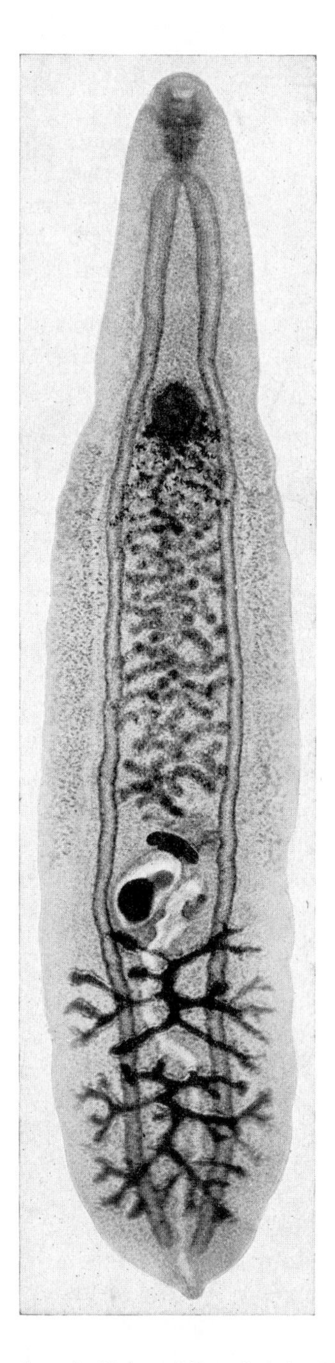

FIG. 58. *Opisthorchis sinensis* (Oriental liver fluke). Fixed in Gilson's fluid; stained by Lynch's precipitated borax carmine method.

considerable stain, in order not to decolorize the nuclei. Nuclei should remain a deep pink color, which will change to a transparent red when the material is cleared. In organisms of considerable size, the parts having a dense cellular structure, such as the digestive tract, gonads, or brain, will be stained quite deeply, while connective tissues and the cytoplasm of epithelia will be practically colorless. If destaining is stopped too soon, all tissues will have a dense red color, without adequate differentiation. If destaining is carried too far, the nuclear stain will not be sharp enough.

The process of destaining is likely to take at least an hour. If the stain is extracted very rapidly and there seems to be danger that it will soon become too weak, transfer the material to 2% or even 1% acid alcohol. Small objects, such as young chick embryos or scoleces of tapeworms, will usually be destained to a sufficient extent within 3 or 4 hours; a 12 mm. amphioxus may require 8 to 16 hours; and chick embryos of more than 60 hours' incubation sometimes require several days. The process may be hastened somewhat by using stronger acid alcohol (4 or even 6%). If night comes before the material is sufficiently destained, pour off the strong acid alcohol and cover the material with neutral 70% alcohol, or with very weak acid alcohol, made by adding 3 or 4 drops of concentrated hydrochloric acid to 100 ml. of 70% alcohol. Return the material to stronger acid alcohol the next morning.

When observation indicates that destaining may have progressed far enough, place one or two specimens in absolute alcohol for a few minutes, clear them in terpineol-toluene, and examine them critically under the microscope. If they appear satisfactory, stop the destaining of the material in the dish. If the trial specimens are too deeply colored, allow the material in the dish to destain until a subsequent trial shows the desired result.

d. When the stain has been properly differentiated, draw off the acid alcohol and replace it with neutral 80% alcohol. Change this 2 or 3 times within a period of a half-hour to an hour, depending on the size of the specimens.

Counterstaining and Completion of Preparations. Begin with step 6, on page 155, and follow the directions for counterstaining, dehydrating, clearing, and mounting.

ACID CARMINE STAINS

In strongly acid solutions, carmine molecules are positively charged. On this account they are adsorbed by nuclei and other negatively charged substances. Cytoplasm and other substances bearing positive charges repel the similarly charged carmine molecules and therefore remain uncolored. For this reason, strongly acid solutions of carmine are always progressive stains. Less strongly acid solutions contain some negatively charged carmine. They differentiate partly in the manner just described and partly by reversal of charge occurring when the stained material is treated with an acid.

Dilute Acidulated Borax Carmine

Alcohol, 70%	100 ml.
Hydrochloric acid	1 ml.
Add, with constant stirring	
Grenacher's borax carmine	12 ml.

Allow the mixture to stand at least 1 hour, then filter it.

This is essentially an acid progressive stain, the employment of an alkaline solution in its preparation being merely incidental. It provides a very fine method for staining small and permeable objects which are to be mounted entire. These include some protozoa, embryos and larvae of echinoderms, trochophores of annelids, and veligers of molluscs. As this stain penetrates rather feebly, it is unsuitable for larger specimens and for any organisms which possess an impermeable cuticle. It is not adapted to staining objects which are subsequently to be sectioned.

PREPARATION OF MATERIAL. Fix in Bouin's fluid, formalin, alcohol, or any mercuric chloride mixture not containing potassium dichromate. Wash and bring it into 70% alcohol. Iodize material which has been fixed in mercuric chloride.

PROCEDURE. *a.* Draw off from the material the 70% alcohol in which it has been preserved, and cover it with 25 or 30 times its volume of the above staining solution. Agitate the container 2 or 3 times in order to insure uniform exposure of the material, and then let it stand overnight.

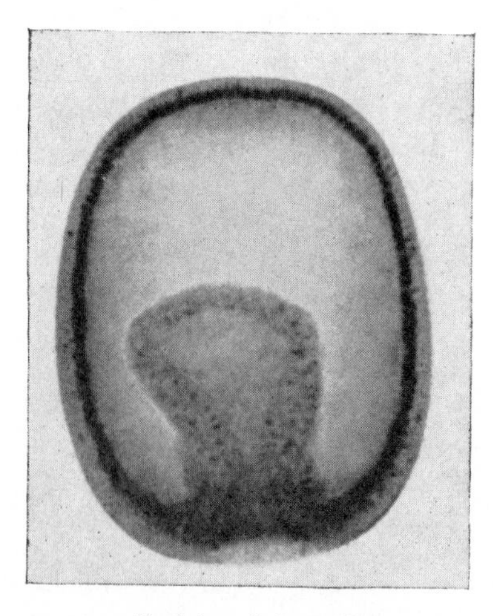

FIG. 59. Gastrula of a sea star, *Patiria miniata*. Fixed in Bouin's fluid; stained by the dilute acidulated borax carmine method.

b. Withdraw a few specimens and examine them under the microscope. The nuclei should be stained a bright rose-red, and the cytoplasm should be colored only very lightly. If this is the case, proceed to step *c.* Should the stain be too faint, make a new staining solution which contains only 0.5% of acid, and repeat the process. Do not leave the objects in the stain for any considerable length of time after they are stained satisfactorily, because they may gradually become decolorized. Should overstaining occur, destain to the proper extent with 1% hydrochloric acid in 70% alcohol.

c. When a satisfactory stain has been obtained, transfer the material to 80% alcohol. Change this once or twice, in order to remove all traces of acid.

d. Counterstain, dehydrate, clear, and mount the material as directed beginning with step 6, page 155.

Material which proves refractory to staining by this method should be passed down through the alcohols to water, and stained with precipitated borax carmine, alum cochineal, or picro-carmine.

Aceto-Carmine

Acetic acid, glacial	45 ml.
Distilled water	55 ml.
Carmine	0.5 gm.

Boil for 10 minutes, allow to cool, and filter.

This solution is extremely valuable as a combined fixative and stain for small organisms, smears, or crushed preparations of fresh tissues. It acts very quickly and gives a sharp chromatin stain. This enables rapid determination of the condition of nuclei and even detailed study of the chromosomes. Suggestions for making preparations of testis are given on page 176, and the technique of demonstrating salivary gland chromosomes of *Drosophila* is outlined on page 500.

BELLING'S IRON ACETO-CARMINE STAIN. The addition of a trace of iron to aceto-carmine causes it to become a stronger and more precise stain. This stain has found great favor among plant cytologists, who have used it chiefly for chromosomes of pollen mother cells. It also has proved superior to ordinary aceto-carmine for animal cells. In animal cytology it is particularly valuable for staining chromosomes in smears of testis and of the salivary glands of *Drosophila.*

The simplest method of preparing the stain is to place a steel dissecting needle, an iron nail, or a small file in a bottle of aceto-carmine (see formula above). After a few minutes the solution will become appreciably darker in color. Remove the steel or iron object. An alternative method is to add to the aceto-carmine, drop by drop, a solution of ferric hydrate in 45% acetic acid. Add just enough ferric hydrate to turn the solution bluish-red; too much will cause precipitation. Dilute the resulting solution with an equal volume of unmodified aceto-carmine.

Iron aceto-carmine is used in the same way as aceto-carmine. The techniques for testis (p. 176) and salivary glands of *Drosophila* (p. 500) may be adapted to other material. The chromosomes appear reddish at first, but acquire a purple color within a few minutes. The acetic acid causes cytoplasm gradually to swell. Gentle pressure upon the coverglass will therefore spread apart chromosomes, so that they can be studied very readily. If the preparations are then blotted with filter paper and sealed with petroleum jelly or vaspar (a mixture of equal parts of petroleum jelly and melted paraffin), they will remain in satisfactory condition for a week or ten days.

Preparations stained with aceto-carmine (with or without iron) can be examined to best advantage by the use of artificial light passed through a green or blue-green filter.

In recent years, aceto-orcein (below) has become very popular for staining chromosomes in smear preparations. It is preferred by many workers. Aceto-orcein may be applied in the same way as either of the aceto-carmine mixtures described above. It stains chromosomes dark purple.

Aceto-Orcein

Acetic acid, glacial	45 ml.
Distilled water	55 ml.
Orcein (natural orcein is recommended[1])	1 gm.

Dissolve the orcein in hot acetic acid, cool the solution, and add the water.

[1] For a long time orcein has been derived from certain lichens containing lecanoric acid When the lichens are boiled, the lecanoric acid is converted into orcinol, one of the resorcinol compounds. Treatment of orcinol with ammonia, in the presence of oxygen, results in the formation of orcein. Orcein is now produced synthetically, although the product varies considerably.

Mayer's Hydrochloric Acid Carmine

Carmine	4 gm.
Alcohol, 80%	100 ml.
Hydrochloric acid	30 drops.

Heat in a water bath for 30 minutes; filter while hot, and cautiously add dilute ammonium hydroxide (1 part of concentrated ammonium hydroxide to 3 parts of water) until the odor of the acid no longer prevails, and the carmine begins to precipitate.

Some zoologists prefer this for invertebrates of many kinds. It is good for tapeworm scoleces and proglottids, and for cysticercus and plerocercoid larvae of tapeworms. On proglottids, it gives a denser and less well-differentiated stain than does picro-carmine. On flukes it gives results which are good, but which generally do not equal those obtained with precipitated borax carmine.

PREPARATION OF MATERIAL. Fix, wash, and harden material as suggested for precipitated borax-carmine.

PROCEDURE. *a.* Transfer material from 70 or 80% alcohol to the stain. In the case of small larvae, leave them in the stain for 2 or 3 hours. Stain cysticerci, adult cestodes, or trematodes for 24 hours to several days, according to size. The specimens should be deeply stained.

b. Replace the stain with 1 or 2% hydrochloric acid in 80% alcohol. Watch the process of destaining, replacing the acid alcohol as often as it becomes deeply colored.

c. When the cytoplasm appears light or nearly colorless, while nuclei remain sharply stained, replace the acid alcohol with neutral 80% alcohol.

d. Change the alcohol several times, at suitable intervals, in order to remove all acid.

Begin at step 6 on page 155, and proceed with dehydration, clearing, and mounting.

ALUM COCHINEAL AND ALUM CARMINES

The colored salts or "lakes" which are formed in solutions containing aluminum compounds and cochineal, carmine, or carminic acid, consist essentially of colored molecules bearing positive charges. Like other positively charged dye substances, these stain nuclei and extranuclear structures which are negatively charged. This does not explain, however, why some combinations of alum with cochineal or carmine produce rather intense coloration of the cytoplasm which is removable by treatment with acid. Possibly some of the color-bearing molecules are negatively charged.

Alum Carmine

Carmine . . . . 1 gm.
Aluminum ammonium sulfate ("ammonia alum") . . . . 5 gm.
Distilled water . . . . 100 ml.
Boil for 15 minutes; add water to replace that lost by evaporation; cool, filter, and add
Thymol . . . . 0.2 gm.

This is a very easy stain to use, as it seldom overstains. It gives acceptable results with many kinds of small and permeable invertebrates. It does not penetrate so well or give such brilliant results as precipitated borax carmine. Alum carmine is a rather weak stain for objects which are to be sectioned.

PREPARATION OF MATERIAL. Fix and wash material as recommended for carmine stains in general. If it has been hardened in 70% alcohol, which is generally desirable, transfer it by way of a graded series of alcohols to distilled water.

PROCEDURE. *a.* Place specimens in the stain and leave them until inspection shows the nuclei to be strongly stained. Should an extensive and undesirable coloration of cytoplasm also occur, treat the overstained material with 0.2% aqueous hydrochloric acid, until only the nuclei retain considerable stain.

b. Wash specimens in running water for 1 hour, or in repeated changes of tap water until no more color is given off.

c. Pass material up the alcohol series to 95% alcohol. If a counterstain is desired, add a little orange G solution to the alcohol. Then treat the material with 2 changes of absolute alcohol over a period of 30 minutes to 6 hours, according to the size and nature of the specimens. Turn to page 156, and follow the directions for clearing and mounting, beginning at step 7.

Partsch's Alum Cochineal

Cochineal, powdered	6 gm.
Aluminum potassium sulfate ("potassium alum")	6 gm.
Distilled water	100 ml.
Boil for ½ hour, add water to replace that lost by evaporation, allow to cool, filter and add	
Thymol or methyl salicylate	0.2 gm.

Alum cochineal is a good stain for invertebrates and embryos which are to be mounted entire. It is not so brilliant or so selective as precipitated borax carmine, and it is decidedly inferior for protozoa, hydroids, ectoprocts, and other relatively permeable organisms. However, it penetrates better than any other carmine stain, and because of this it is very fine for hookworms and other nematodes, acanthocephalans, and arthropods. It is also good for large trematodes (such as *Fasciola* and *Fasciolopsis*), but rather poor for cestodes. Alum cochineal is useful for thin-walled organisms (such as chaetognaths and salps) which would be collapsed by precipitated borax carmine. In chick embryos it brings out the somites very clearly, but gives a rather opaque effect.

Alum cochineal is the best of carmines for staining embryos or other objects which are later to be sectioned for the study of general topography.

PREPARATION of MATERIAL. Fix and wash material as for carmine stains in general. Then pass it down through the alcohols to distilled water.

PROCEDURE. *a.* Cover specimens with several times their volume of alum cochineal. Leave young embryos or other small and permeable forms in the stain for at least 6 hours, or overnight. Specimens as large as a 72-hour chick embryo or a 10 mm. pig embryo should remain in the stain for 24 hours. Relatively impermeable organisms, such as nematodes or crustacea, must be stained for at least 3 or 4 days.

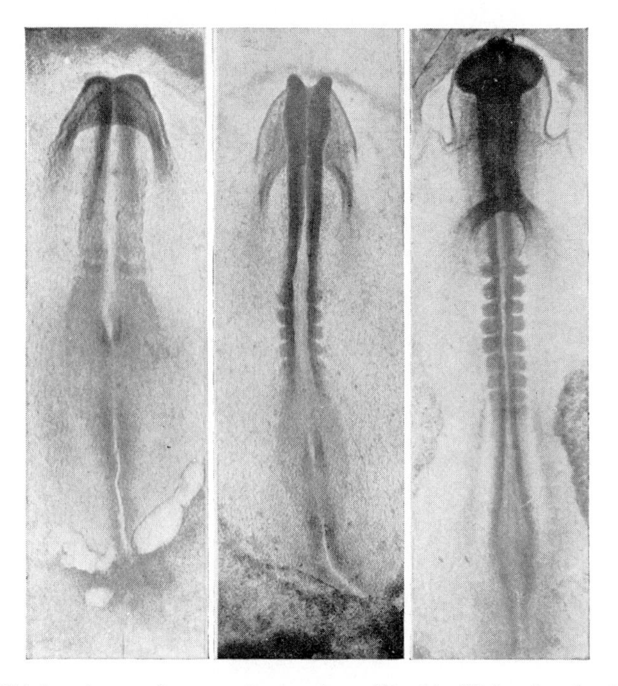

FIG. 60. Chick embryos of one to nine somites. Fixed in Kleinenberg's picro-sulfuric acid mixture; stained with alum cochineal.

b. Draw off the stain and wash the material with distilled water, changing this at intervals of 5 to 10 minutes, until the water no longer becomes tinged with stain.

c. If the specimens are to be sectioned, dehydrate them and embed them in paraffin or nitrocellulose. Sections may be counterstained advantageously with orange G (1 to 3 minutes in a 1% solution in 90% alcohol) or with Mallory's aniline blue-orange G mixture. If the latter is used, bring the sections gradually to water and proceed as described on page 399. After either counterstain, dehydrate, clear, and mount the sections in the usual way.

If the specimens are to be mounted entire, pass them up through the alcohol series to 70% alcohol. Place the material in a shallow dish of 70% alcohol, acidulated with 0.5 to 1% hydrochloric acid. Examine the material under the microscope at frequent intervals, changing the acid alcohol whenever it has become deeply colored, until the specimens are of a transparent pink color. Continue the destaining until microscopic examination shows that only the nuclei and areas of dense cytoplasm remain sharply stained. Other areas should become nearly decolorized. Then place a specimen in absolute alcohol for a few minutes, clear it in terpineol-toluene, and examine it critically.

If the stain proves to be properly differentiated, proceed to the next step. Otherwise, continue the acid treatment until the desired result has been obtained.

d. Replace the acid alcohol with neutral 80% alcohol, and change this 2 or 3 times.

COUNTERSTAINING AND COMPLETING PREPARATIONS. Turn to page 155. Begin at step 6, and follow the instructions for counterstaining, dehydrating, clearing, and mounting the material.

Mayer's Carmalum

Carminic acid	1 gm.
Aluminum ammonium sulfate ("ammonia alum")	10 gm.
Distilled water	200 ml.
Heat until ingredients are dissolved, cool, filter, and add	
Thymol	0.1 gm.

The advantage of this solution is that it will stain material which may be refractory to all other carmines, including material fixed in chromic acid or osmium tetroxide, and also old material. The stain, however, is generally dull and lacks the sharp differentiation obtained with other carmine stains. As a rule, employ other carmines when you *can*, this one when you *must*. Use the stain in the same way as alum carmine (p. 343).

Mayer's Paracarmine

Carminic acid	1 gm.
Aluminum chloride	0.5 gm.
Calcium chloride	4 gm.
Alcohol, 70%	100 ml.

Heat the mixture or allow it to stand until the ingredients are dissolved; filter.

This solution has some value for bulky objects in general, as it penetrates somewhat better than borax carmine. However, it is not so brilliant or so precise as the latter. It does not penetrate so well as alum cochineal, and it appears not to be good for nematodes. Paracarmine is particularly suitable for ascidians and other organisms containing large cavities, because it is unlikely to form precipitates within them. However, the presence of calcium salts in a specimen will cause the stain to precipitate. For this reason, it should not be used for corals, calcareous sponges, or echinoderms. It stains large trematodes (for example, *Fasciola*) very brilliantly and with fine differentiation. It is probably inferior to alum cochineal for objects which are to be sectioned.

PREPARATION OF MATERIAL. Fix and wash material as for carmines in general.

PROCEDURE. *a.* Transfer the material from 70% alcohol to the stain and leave it there for 12 to 48 hours, depending upon the size of the specimens.

b. Place material in 2.5% acetic acid in 70% alcohol, until the stain is properly differentiated. Change the acid alcohol whenever it becomes deeply colored.

c. Transfer material to neutral 80% alcohol.

d. Change the alcohol several times, at intervals of 15 to 30 minutes.

Turn to page 155. Begin with step 6 and follow the directions for dehydrating, clearing, and mounting.

OTHER CARMINE STAINS

Picro-Carmine

Carmine	1 gm.
Ammonium hydroxide	2.5 ml.
Distilled water	7.5 ml.
Stir until the carmine is dissolved, then pour the solution into	
Picric acid, saturated aqueous solution	50 ml.

Add a trace of thymol, to prevent growth of molds, and let the solution stand in an open vessel until it evaporates down to one-quarter of its original volume. Filter it to remove the precipitate which has formed. Allow the filtrate to evaporate to dryness. The crystalline picro-carmine may be kept indefinitely. For staining, dissolve as follows:

Picro-carmine	0.5 gm.
Distilled water	100 ml.
Thymol	0.2 gm.

Picro-carmine was devised as a double stain for sections, and is of special interest as the earliest example of a double stain. In the latter part of the nineteenth century, it was widely used for this purpose. With the advent of better double stains, picro-carmine fell into disuse. However, it is an excellent stain for tapeworm proglottids, large flukes, and some other specimens in which there is a dense parenchyma. In such material it brings out the organs with great clarity, but leaves the parenchyma nearly colorless. Picro-carmine is also a good stain for delicate organisms such as *Volvox*, rotifers, and larvae of echinoderms and tunicates. It stains these more powerfully and more precisely than dilute acidulated borax carmine. Picro-carmine sometimes gives a more precise stain than alum cochineal. In many cases it will give a good stain with material which has remained in alcohol or formalin for a long time and is refractory to all other stains. Picro-carmine destroys calcareous structures. It cannot be recommended for amphioxus and other organisms which contain much muscular tissue, as the picric acid gives this tissue a diffuse and disagreeable color which is difficult to remove. A similar effect occurs in the case of embryos containing considerable yolk.

Preparation of Material. Fix specimens in 10% formalin, alcohol, or any picric acid or mercuric chloride mixture. Avoid chromic acid, potassium dichromate, and osmium tetroxide. After picric acid or mercuric chloride fixation, wash material in 70% alcohol; iodize material fixed in mercuric chloride.

Procedure. *a.* Pass specimens washed or stored in alcohol gradually to water, and place them in the stain. Material fixed or stored in formalin needs only to be washed in water before it is transferred to the stain.

b. Leave small organisms, such as *Volvox* and rotifers, in the stain for 2 or 3 days; tapeworm proglottids or large flukes should be left in it for a week. This stain is reported to have a dissociating effect, and it may be a good idea not to leave material in it without an occasional examination.

c. Wash the material with several changes of distilled water.

d. Pass it up through the alcohols to 70% alcohol. Change this once or twice, at intervals of several hours or a day, in order to remove most of the picric acid.

e. Place material in 70% alcohol containing 0.5 or 1% hydrochloric acid, until the stain is properly differentiated. Change the acid alcohol as often as it becomes deeply colored, and control the destaining as suggested for alum cochineal. When the desired intensity is reached, replace the acid alcohol with neutral 80% alcohol. Change this several times, until no more color comes out of the specimens.

Turn to page 155. Begin at step 6 and follow the instructions for counterstaining, dehydrating, clearing, and mounting the material.

COUNTERSTAINING TO FOLLOW CARMINE STAINS

It is often desirable to counterstain entire specimens with a dye which will bring out structures not colored by carmine. Among these are cilia, the perisarc of hydroids, nematocysts, setae on arthropods, and various other delicate structures. Fast green FCF generally gives rather good results. Light green and indigo-carmine may also be recommended. Methods for staining with these dyes are discussed in more detail on page 390.

In the case of sections of larger embryos and other specimens which were stained with carmine before embedding, Mallory's aniline blue-orange G mixture is sometimes very suitable. Directions for its use are given on page 399. Either aniline blue or orange G by itself may also yield a good counterstain.

CHAPTER 22

Hematoxylin Stains

Hematoxylin (or haematoxylin) is derived from the logwood tree *Haematoxylon campechianum*, a leguminous species native to the east coast of southern Mexico and Central America. This tree, which was utilized for the production of dye long before the arrival of Spanish explorers, was introduced into Jamaica about 1718, and its cultivation spread to other islands of the West Indies as well as to some regions of South America.

The heartwood of the tree is a brownish-red color, and this is the basis for the generic name *Haematoxylon* ("blood wood"). After the bark and sapwood have been cut away, the heartwood is handled commercially as small logs. Chips of the wood are steeped in water to prepare an aqueous extract, which is evaporated. The solids are then dissolved in ether, and after the ether solution is evaporated, the residue is redissolved in water, filtered, and crystallized. The crystalline preparation is not pure hematoxylin, but it is adequate for use in biological staining and in dyeing textiles.

Hematoxylin is not itself a dye. In logwood or an extract of logwood, or in a solution of crystalline hematoxylin, it becomes gradually oxidized to hematein, a reddish-brown colored acid.

Hematein is a rather feeble stain for tissues, and has little use in cytology or histology unless it is combined with a mordant in the form of a salt of aluminum, iron, or some other metal. The mordant confers upon it

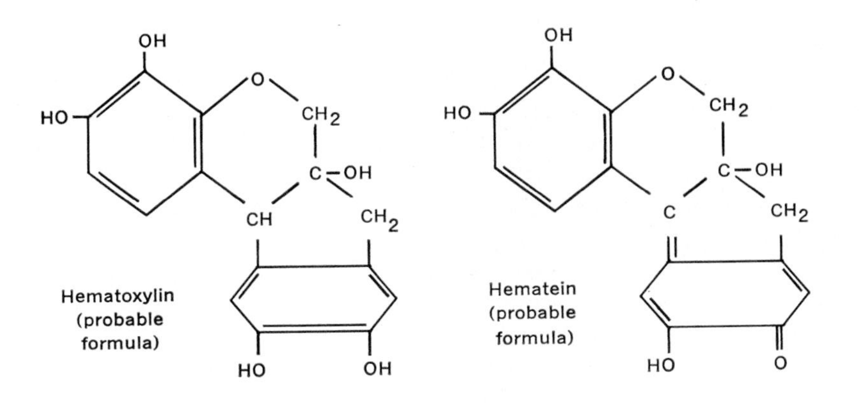

the properties of a color base, and thus contributes to its effectiveness in nuclear staining. The discovery of Böhmer (1865) that a deep blue nuclear stain could be obtained with a mixture of hematoxylin and an alum led to the elaboration of various techniques which are still important today.

From the facts given above, it is apparent that the preparation of useful hematoxylin stains is accomplished in two important steps: (1) the oxidation of hematoxylin to hematein, generally referred to as "ripening"; and (2) the combination of "ripened" solutions with suitable metallic salts which serve as "mordants."

Ripening. A freshly made solution of hematoxylin contains very little of the active color-bearing substance, hematein, and consequently produces a weak and diffuse stain. After a period of ripening, the solution contains sufficient hematein to be capable of giving a strong and well-differentiated stain when it is used in combination with appropriate mordants. As oxidation continues, hematein becomes converted into products which are not suitable for precise staining; these, in turn, are eventually oxidized to form insoluble compounds which possess no staining properties. An old solution, no longer containing sufficient hematein to be satisfactory, is spoken of as "over-ripe," and gives a less selective, brownish stain. Since a hematoxylin solution must be well-ripened, but not over-ripe, in order to yield good results, it is desirable to control the process of ripening so that the solution reaches a hematein-rich stage quickly, and then retains this quality for a long period of time.

Some writers have maintained that staining reagents should be made with hematein itself, in order for them to have enough hematein in solution from the very beginning. In practice, however, this has not always been satisfactory, for some hematein preparations are not of sufficient purity, and solutions of hematein quickly become over-ripe.

Natural ripening of aqueous hematoxylin solutions occurs within 3 to 6 weeks, depending upon the temperature of the room. Solutions containing only hematoxylin and its derivatives then remain in optimum condition for 6 weeks to 3 months. Aqueous solutions which also contain certain alums, such as aluminum ammonium sulfate, remain in good condition for many months.

Hematoxylin dissolved in strong alcohol ripens very slowly, reaching an optimum condition in four to six months, but after that it may keep for years without becoming over-ripe. A stock solution, consisting of 10% hematoxylin in absolute alcohol, may be kept on hand, ready for dilution with water when needed. This procedure is particularly useful in the iron hematoxylin method of staining.

Artificial ripening can be brought about within a few minutes by treating a hematoxylin solution with an oxidizing agent such as hydrogen peroxide, mercuric oxide, or potassium permanganate. But in spite of its convenience, chemical acceleration of ripening is to be avoided in the case of solutions containing only hematoxylin and its derivatives, except in situations where it is desirable to oxidize a small quantity for immediate use. The addition of an energetic oxidizing agent to a solution will induce the formation of undesirable products of advanced oxidation. A solution can be ripened to an appreciable extent, without bad effects, by simply boiling it for a few minutes.

Compounds of hematein and aluminum oxidize less readily than hematein alone, so that alum hematoxylin solutions may be ripened quickly and satisfactorily by the cautious introduction of an oxidizing agent. Mercuric oxide, added to the boiling stain, is the safest of these. Instructions for its use will be given in connection with methods for preparing alum hematoxylin.

MORDANTS. As has been explained, mordants are metallic salts which possess the property of combining with color-bearing substances, such as hematein. The deeply colored compounds thus formed are known as "lakes." Hematein yields its most valuable stains when it is combined with salts of aluminum or iron. The salts of these two metals are employed in different ways and produce stains which differ in a decided and characteristic manner. Separate portions of this chapter, therefore, will be devoted to alum hematoxylins and iron hematoxylins, and a third portion will include the less important mixtures of hematoxylin with other mordants.

ALUM HEMATOXYLINS

Alum hematoxylins consist essentially of "ripe" hematoxylin and a salt of aluminum, the latter ordinarily being aluminum ammonium sulfate ("ammonia alum"), or aluminum potassium sulfate ("potassium alum").

The hematoxylin and mordant are combined in a single solution and form a lake which stains the chromatin of nuclei a deep blue or purplish color. Nucleoli are usually stained less intensely. Cytoplasm which contains certain secretory products may be stained to some extent, but the cytoplasm of most cells is colored very lightly if at all. The matrix of hyaline cartilage and bone is stained light to medium blue, but fibrous connective tissue usually either fails to take up the stain, or gives it up quickly when treated with a dilute acid.

Alum hematoxylins are the most popular nuclear stains for routine histological and pathological work. The formulas generally used are those of Delafield, Ehrlich, Harris, and Mayer. With any one of these mixtures, more or less precise nuclear stains can be obtained. But none of them possesses all the qualities which might be desired of a nuclear stain for general histological purposes. The most important function of such a stain is to bring out vividly the nuclei of the various types of cells present, and to show something of the structure of these nuclei. Generally speaking, its coloration of the cytoplasm (except in the case of some secretory cells), connective tissue, and most other tissue elements should be quite delicate, thereby allowing eosin or some other counterstain to accentuate the differentiation. In addition, a stain for routine use should be dependable and easy to employ. It should be compatible with a wide variety of fixing agents, should retain optimum staining properties for a long time, and should yield a stain which will last for many years if the preparations are mounted properly. Although Delafield's hematoxylin is probably used more widely than any of the other mixtures, it is by no means the most desirable. Occasionally it produces an excellent stain, but generally it fails to bring out nuclear structures well enough, and it may give a muddy, diffuse color to cytoplasm and connective tissues, spoiling these for precise counterstaining. Harris' hematoxylin probably satisfies the requirements of a good stain better than Delafield's mixture.

The solution now to be described as Standard Alum Hematoxylin is a modification of Harris' hematoxylin. It is a thoroughly satisfactory nuclear stain for general histological purposes. It will serve most purposes at least as well as any of the mixtures mentioned in the preceding paragraph, and it is decidedly superior to all of them in some respects.

Standard Alum Hematoxylin

Hematoxylin, "light crystals," certified (if not certified, the worth of the sample will have to be established by trial)	0.5 gm.
Aluminum ammonium sulfate	0.3 gm.
Alcohol (ethyl or methyl), 50%	100 ml.
Mercuric oxide (red)	0.6 gm.

Preparation of the Staining Solution. Add the aluminum ammonium sulfate and hematoxylin to the alcohol in a beaker, and place the beaker in a bath of boiling water. Stir the contents occasionally until the hematoxylin and alum are dissolved. When the solution begins to boil, add the mercuric oxide *slowly* and boil it for 20 minutes. Then add enough 50% alcohol to replace that which was lost by evaporation. Allow the solution to cool and to stand for at least an hour or two, preferably overnight. Filter it through two thicknesses of filter paper and store it in a tightly stoppered bottle. This solution is sufficiently ripe to be used at once, although its staining power will increase slowly for a month or more. It will keep for 6 months to a year, depending upon the temperature.

Instead of dissolving the hematoxylin crystals as directed above, 5 ml. of a 10% stock solution of hematoxylin in absolute alcohol can be substituted. If this is done, use 95 ml. of 50% alcohol for preparing the hematoxylin and alum mixture.

Methods of Staining With Standard Alum Hematoxylin

This stain may be used by either the progressive or regressive method. It may be applied to sections, or to specimens which are subsequently to be sectioned, as well as to small and transparent organisms which are to be mounted entire. If the sections are to be used for critical study of structural details, it is best to stain after sectioning. The practice of staining tissues before sectioning saves time and effort, but the process cannot be controlled with complete certainty and the staining of various portions is often not uniform. The method is, however, of value for embryos and other objects intended for the study of general morphology, rather than of cellular details. It is also useful for eggs of amphibians and other specimens which contain a great deal of yolk, because it eliminates passage of friable sections through a long series of reagents.

Progressive Method of Staining. This method is rapid, easy to control, and in the majority of cases produces good results just as surely as the regressive method. It is applicable to sections as well as to small and readily permeable specimens which are to be mounted entire.

The solution for progressive staining is prepared by diluting standard alum hematoxylin with a considerable excess of a saturated aqueous solution of aluminum ammonium sulfate. The alum acts as a differential decolorizer and prevents overstaining.

To prepare a progressive staining solution from newly made standard alum hematoxylin, mix 20 ml. of the concentrated alum hematoxylin with 80 ml. of a saturated aqueous solution of ammonia alum. The staining power of this solution will increase for 3 weeks or a month, as more hematoxylin is oxidized to hematein. During this time it should be diluted if it overstains average sections in less than 30 minutes. Ordinarily, it is necessary to add 30 to 40 ml. of ammonia alum solution at the end of 1 week, and to add about the same amount within the next 2 or 3 weeks. The staining power of the fully ripened solution will remain almost constant for 6 to 8 months in a moderate climate. Should the concentrated alum hematoxylin be more than a month old, mix 12 ml. of it with 88 ml. of a saturated solution of ammonia alum. The solution will be ready for immediate use and seldom will require further dilution. It should be filtered occasionally, especially after it has been poured out into Coplin jars or other dishes used for staining.

Preparation of Material. Fix pieces of tissue or entire small organisms in Bouin's fluid, formalin, Zenker's fluid, Heidenhain's "Susa" fixative, or almost any fixing agent except chromic acid-osmium tetroxide mixtures. Wash and harden them properly. If sections are to be made, use the paraffin, nitrocellulose, or freezing method. Affix paraffin or frozen sections to slides by the usual methods. Nitrocellulose sections may be handled in dishes or affixed to slides by the albumen-clove oil method (p. 268). Smears are handled in the same manner as sections.

Procedure. *a*. Pass the sections or other material down through the series of alcohols and wash them in water for at least 2 or 3 minutes. Slides bearing sections may be allowed to stand in a pan or dish through which tap water is running. In handling loose nitrocellulose sections, or material for whole mounts, simply change the water once or twice.

b. Place material in the staining solution. After 20 minutes, remove a section or specimen, rinse it in tap water, and quickly examine it under the microscope. If the nuclei are stained a deep blue and other structures are colored very delicately, as described earlier in this chapter, remove the remainder of the slides or material to water and proceed to wash them as directed in step *c*. If the nuclear stain is too faint, replace the slide in the staining solution, wait 10 or 15 minutes, and examine it again. If necessary, repeat the process. If a satisfactory stain is not obtained within 45 minutes, it is probable that the material is unfavorable or the solution is too weak or not suitable for some other reason. Test the solution by staining some other material in it. If necessary, strengthen it by adding a few milliliters of standard alum hematoxylin. If the fault lies in the material, wash it thoroughly in water and stain it by the regressive method.

c. When a satisfactory stain has been obtained, wash the sections or other material *very thoroughly* in tap water. Sections affixed to slides may simply be placed in a dish or pan under the faucet, and washed for 15 minutes or longer in running water.

If overstaining should occur, proceed by the regressive method, beginning at step *b*, page 355.

Loose sections or material for whole mounts should be rinsed with two changes of distilled water, then placed in tap water. This should be changed at least 10 or 12 times, at intervals of several minutes. Such objects may also be washed by tying a piece of fine netting over the dish and allowing water to drip slowly into it.

Thorough washing is necessary in order to remove all traces of alum, which will cause the stain to fade. The weak alkaline reaction of tap water in most localities will turn the stain blue.

d. Dehydrate material in the usual manner, up to 70% alcohol.

COUNTERSTAINING AND COMPLETING PREPARATIONS. In handling sections, the next step is usually to counterstain with eosin, as directed in steps *c* and *d* on page 385. The sections are finally dehydrated, cleared, and mounted in the usual manner. Instructions for the preparation of entire small specimens will be found beginning at step 6 on page 155.

Regressive Method of Staining. This more laborious method must be employed for materials which do not readily take up the stain from a dilute solution. These include specimens fixed in Zenker's fluid and materials which have been stored in alcohol for a long time. By this procedure, various extranuclear structures may be stained more intensely if desired. The method is applicable also to small crustacea and some other invertebrates which are not readily penetrated by a dilute solution, although carmine stains are usually preferable for such objects.

The staining solution is prepared by adding 1 drop of concentrated hydrochloric acid to each 100 ml. of standard alum hematoxylin (full strength). After this, the solution will remain slightly acid due to the presence of sulfuric acid produced by dissociation of the aluminum ammonium sulfate. If the stain is being used regularly, filter it every few days. Otherwise, filter it before use.

PREPARATION OF MATERIAL. Follow the directions given in connection with the progressive method.

PROCEDURE. *a.* Place the sections or other material in acidulated standard alum hematoxylin for 20 minutes, or until they are deeply stained.

b. Destain the tissues in order to extract most of the color from structures other than nuclei. Use one of the following methods:

1. Remove the slides, sections, or other objects from the stain, and rinse them with distilled water. Place them in water containing 0.5% hydrochloric acid. In order to facilitate the removal of stain, move the slides occasionally or, in the case of loose sections or whole objects, gently agitate the dish. Observe the material constantly with the low-power objective of the microscope, until the cytoplasm and connective tissue retain little or no color, while the nuclei are still sharply stained. The stain becomes pink or red in the presence of acid. Bear in mind that it will appear much deeper when, subsequently, it is turned to a blue color. Recognition of the medium or light pink color of sufficiently destained tissues is learned quickly. When the destaining is thought to have progressed far enough, wash the material in tap water, to remove acid and excess

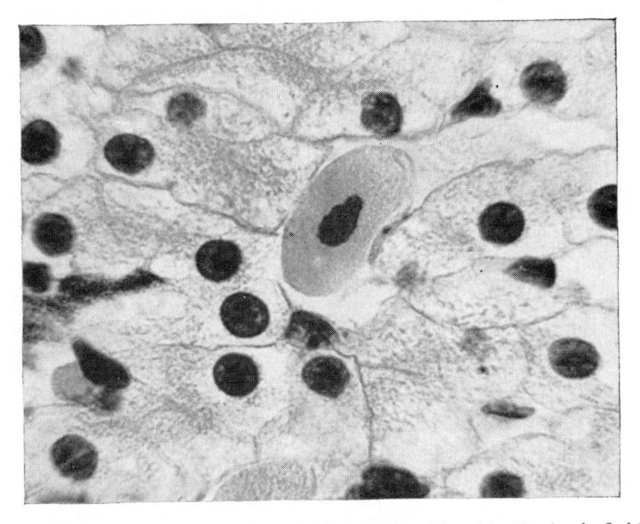

FIG. 61. Liver of a salamander, *Amphiuma means*. Fixed in Zenker's fluid; paraffin section cut at 10 μ; stained with alum hematoxylin and eosin.

alum. This causes the stain to become intensely blue. To wash sections affixed to slides, simply stand the container of slides in a pan under the tap. Wash loose sections or other material with repeated changes of tap water, or tie a piece of gauze over the dish and allow water to trickle into it. As soon as the material is blued, examine a section or specimen under the microscope. Should the stain appear too dark, return the material to acidulated water for a short time, and then wash it again. Should the stain seem too light, repeat steps *a* and *b*.

In destaining numerous sections from the same specimen, the simplest method is to remove the slides from the stain, rinse them in distilled water, and place them in a pan through which tap water is running. Then destain one preparation, noting the number of seconds in which the desired result was obtained. The remaining sections may then be destained satisfactorily by placing them in the dilute acid for the same length of time.

In all cases, the final washing in tap water should continue for at least 15 minutes in running water, or 30 minutes in standing water.

2. A second method is to transfer the sections or other materials from the stain to distilled water or tap water and wash them until they cease to discharge color. Pass them through 15, 30, and 50% alcohol, and then destain them to the proper extent in 70% alcohol containing 0.5 to 1% hydrochloric acid. When the differentiation is thought to be satisfactory, stop the destaining and blue the stain by transferring the material to 70% alcohol which has been rendered alkaline. To prepare the alkaline alcohol, add a few drops of *very dilute* ammonium hydroxide or a solution of lithium carbonate or sodium bicarbonate. (Warning! If too much alkali is added, the sections are likely to drop off the slides, especially if they have not been coated with Parlodion.) Then pass the materials into neutral 70% alcohol. Workers who favor destaining in acidified alcohol claim that materials are slightly more transparent in alcohol than in water, and that the acid alcohol penetrates them with greater speed and uniformity.

3. A third method, useful for materials containing yolk or deep-staining connective tissue, is to wash them in water for a few minutes and then to destain them in a saturated aqueous solution of picric acid. When only the nuclei appear brown, wash the tissues thoroughly in tap water, to blue the stain. Then transfer the sections gradually to 70% alcohol.

Destaining can also be carried out in a saturated aqueous solution of aluminum ammonium sulfate. This method yields a stain which is less restricted to nuclei than is the case with preparations made by the picric acid method, and it gives very poor results with osseous tissue.

Counterstaining and Completing Preparations. For sections, turn to page 385 and follow steps *c* and *d*, which describe the method of counterstaining with eosin. Then clear and mount the sections in the

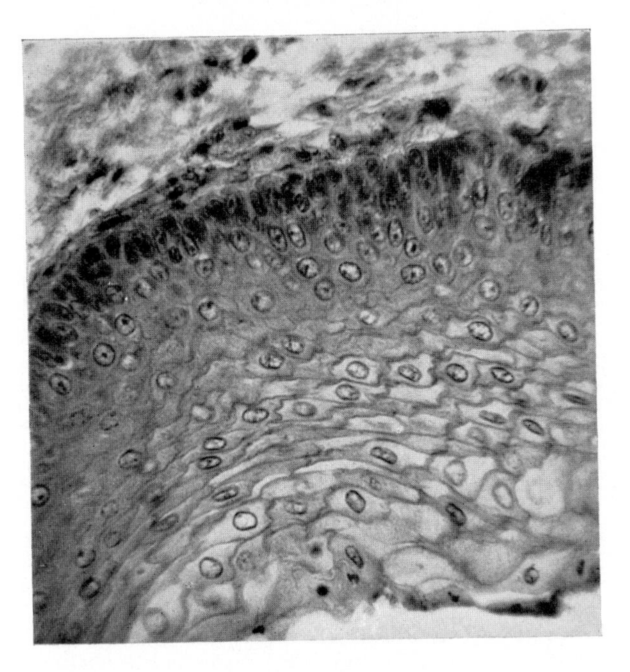

FIG. 62. Mucosa and submucosa of human esophagus. Fixed in Zenker's fluid; paraffin section cut at 8 μ; stained with alum hematoxylin and "triosin."

usual way (steps 17 and 18 in the paraffin method, steps 9 and 10 in the nitrocellulose method).

For objects which are to be mounted entire, turn to page 155, step 6.

Staining Entire Organisms or Pieces of Tissue Before Sectioning. The advantages and disadvantages of this procedure have been discussed on page 353. It is of value chiefly for embryos and other objects in which general structure, rather than cellular detail, is to be studied.

PREPARATION OF MATERIAL. Fix thin slices of tissue, entire embryos, or other organisms in Bouin's fluid or any of the commonly used reagents except chromic acid-osmium tetroxide mixtures. Wash them according to the appropriate method and harden them in 70% alcohol. The thickness of tissue slices should not exceed 4 or 5 mm. They may be larger in the intended plane of section, but if possible should not exceed 1 × 1 cm. Embryos as large as 15 or 16 mm. may be handled successfully without subdivision; if larger, they should be cut into several pieces after being hardened.

PROCEDURE. *a.* Transfer the specimens from 70% alcohol successively into 50, 30, and 15% alcohol, leaving them for 1 to 3 hours in each. Then put them into distilled water. Change the water after an hour and allow the specimens to stand in water for another 2 hours.

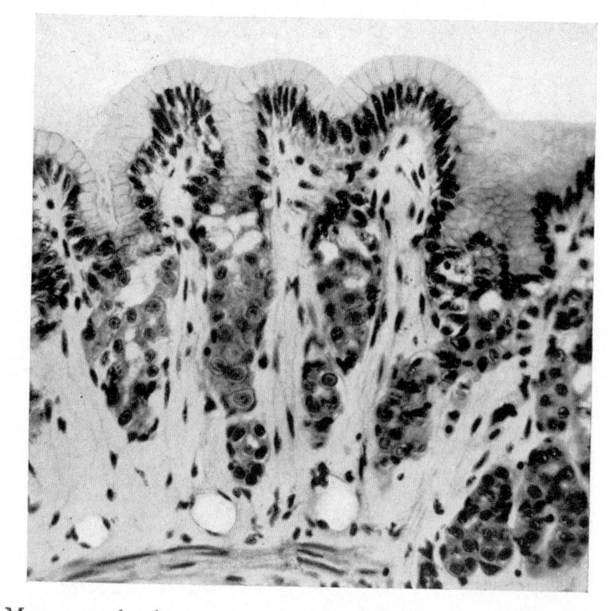

FIG. 63. Mucosa and submucosa of the stomach of a frog. Fixed in Bouin's fluid; stained with alum hematoxylin *before embedding* in paraffin; sectioned at 8 μ; counterstained with eosin.

b. Cover them with several times their volume of alum hematoxylin, of the strength used for progressive staining (see p. 353). Leave them in this for 24 hours.

c. Pour off the stain and rinse the specimens with several changes of distilled water. Next, cover them with 20 or more times their volume of the following mixture: 1 part distilled water; 1 part saturated aqueous solution of aluminum ammonium sulfate. Leave the specimens in this solution, agitating the container occasionally and renewing the liquid when it becomes deeply colored, until the tissue is of a medium blue color.

In most cases it is not necessary to extract a great deal of stain from the specimens, because only the outer layers are likely to be overstained. The amount of stain which must be removed and the rate at which extraction proceeds vary to such an extent that destaining may require from 3 or 4 hours to 48 hours. Only through experience can recognition of the shades of color which various tissues show when they are properly destained be learned. Even then, mistakes are likely to occur. If the removal of a tiny bit of tissue will not interfere with the usefulness of the specimen, snip off a fragment when you think that the material has attained the proper color. Place this tiny piece of tissue in absolute alcohol for a few minutes, clear it in terpineol-toluene, and mount it. Crush it by pressing upon the coverglass. Microscopic inspection of this sample will enable you to determine whether destaining of the outer layers has proceeded far enough.

When the specimens have been destained sufficiently, wash them in running water for at least 6 hours, or preferably overnight.

SECTIONING, COUNTERSTAINING, AND MOUNTING. Embed and section the specimens by the conventional paraffin method or nitrocellulose method. If the paraffin method is used, follow steps 3 to 12, then proceed according to the appropriate alternative listed under step 13.

If the nitrocellulose method is used, and the hematoxylin stain is satisfactory, counterstain the sections with eosin. Then clear and mount according to the procedure regularly employed with nitrocellulose sections (steps 8 to 10). If the stain is too heavy, place the sections in 70% alcohol containing 1% hydrochloric acid, until they are destained to the desired extent. Transfer them to alkaline alcohol (p. 356) until they are blue, and then place them in neutral 70% alcohol. Counterstain them with eosin, and clear and mount them.

In the case of either paraffin or nitrocellulose sections, should the stain prove too faint when the sections are examined at some point before they are mounted permanently, it is possible (even if not always convenient) to pass them down the series of alcohols to water and to restain them by either the progressive or regressive method.

Conklin's Picro-Hematoxylin

This is a good stain for yolky eggs. It has been much used for entire eggs or embryos of gastropods and for blastodiscs of fish eggs. Ordinary hematoxylin mixtures generally color the yolk so deeply as to mask the nuclear structure. Mix 15 ml. of the stock solution of standard alum hematoxylin (p. 352) with 4 ml. of distilled water, and add 10 drops of Kleinenberg's picro-sulfuric acid fixative (p. 100). The acids prevent the dye from acting upon yolk, but do not interfere with staining the chromatin. Transfer specimens from water to the stain. Nuclei are generally well stained within 10 to 15 minutes. Wash the specimens in several changes of 70% alcohol. Should overstaining occur, destain with 1% or 2% hydrochloric acid in 70% alcohol and then return the material to neutral alcohol. Complete the preparations in the usual manner.

IRON HEMATOXYLINS

Benda (1886) was the first biologist to employ a salt of ferric iron as a mordant for hematoxylin. Heidenhain independently worked out a similar procedure, published in 1892, and this survives as the principal cytological staining method in use today. It has been subjected to numerous modifications, one or two of which are sufficiently important to be described here.

Heidenhain's Iron Hematoxylin

This is a very precise chromatin stain of an intense black or blue-black color. Of all the stains in use, this is the most valuable for the demonstration of nuclear phenomena in mitosis, spermatogenesis, fertilization, and development. A differential cytoplasmic stain, in various shades of gray, may also be obtained by this method. Iron hematoxylin brings out fine structure of many types of cells, and it is particularly good for staining the striations in skeletal and cardiac muscle fibers. After an appropriate fixative, iron hematoxylin will differentiate mitochondria, secretion granules, and various other cell inclusions or organelles.

This method is applicable to sections, which should be of uniform thickness (and not over 10 or 12 μ) if good results are to be obtained. It may be applied effectively to smears and entire protozoa of many types. It can be used with success after fixation in the majority of fixing agents. In the study of mitosis and gametogenesis, Bouin's fluid or chromic acid-osmium tetroxide mixtures, such as those of Flemming, are recommended. To differentiate mitochondria with iron hematoxylin, use Regaud's method for fixation and staining.

The procedure described below is based on the method used by Heidenhain. Several modifications have been introduced to improve the quality and reliability of the results.

Reagents for Heidenhain's Iron Hematoxylin Method

MORDANT

Ferric ammonium sulfate ("iron alum"), violet crystals	2 gm.
Distilled water	100 ml.

Always filter the solution immediately before use. It may be of any strength between 1.5% and 4%, but 2% is a favorable concentration.

STAINING SOLUTION

Hematoxylin (light crystals), 0.5% solution in distilled water, ripened to a deep port-wine color. Add 3 drops of a saturated aqueous solution of lithium carbonate to each 100 ml. of the hematoxylin solution. (The lithium carbonate may be omitted, but see remarks following.)

Preparation of the Staining Solution. As previously indicated, use "light crystals." Certified hematoxylins of American manufacture generally are dependable. A lot of time can be spent on trials involving various domestic and foreign samples, certified or uncertified, with a view to finding a few which are especially good.

For reasons explained at the beginning of this chapter, the stain must be ripe, but it should not be over-ripe. An unripe solution gives a weak and diffuse stain. A naturally ripened solution retains optimum staining properties for 6 weeks to 3 months, and during this time may be used over and over again. Eventually it will begin to impart a brownish stain to the cytoplasm. This indicates that it is becoming over-ripe and should be discarded.

A common and very convenient practice is to make a stock solution of 10 gm. of hematoxylin in 100 ml. of absolute alcohol. This will ripen to a dark brown color in the course of 4 to 6 months, and will keep for many years without deterioration. The 0.5% staining solution is prepared, as needed, by adding 5 ml. of the ripe stock solution to 95 ml. of distilled water. Contrary to general belief, the dilute solution thus prepared is not fully ripe, even if it is made from a stock solution 2 or 3 years old. Although it will stain at once, the 0.5% solution does not give the very best results until it has been standing for a week or two.

Some workers dissolve the hematoxylin in distilled water and allow the solution to stand in a loosely covered container until it becomes a deep wine-red color. Hematoxylin dissolves slowly at room temperature, the process requiring several days, but with the aid of heat it may be dissolved within a few minutes. The process of ripening requires 3 to 6 weeks, depending upon the temperature, but it may be hastened somewhat by keeping the solution where the sunlight will fall upon it.

It is probably better not to accelerate the ripening of stock solutions by chemical oxidizers. Solutions containing only hematoxylin and its derivatives are likely to become over-ripe very quickly after such treatment. When no fully ripened hematoxylin is available, enough for immediate use can be artificially ripened. To as much of a 0.5% solution as will be needed for the work contemplated, add a 0.02% solution of potassium permanganate, drop by drop, until the hematoxylin turns a deep port-wine color.

With some materials, the best color and the most precise differentiation are obtained when the solution has been rendered very slightly alkaline by the addition of a carbonate. As the stain is not sensitive to very small differences of pH, a rough measurement of the carbonate will suffice. The addition of 3 drops of a saturated aqueous solution of lithium carbonate to each 100 ml. of 0.5% hematoxylin gives very satisfactory results.

Preparation of Material. This stain may be used after many different fixatives. For nuclear structure and stages of mitosis or meiosis, fix small pieces of tissue in Bouin's fluid, Flemming's fluid, or related fixatives. For secretion granules in cells and for striated muscle, a formalin-containing modification of Zenker's fluid (Helly's fluid) is suggested. To stain mitochondria, fix small pieces of tissue in Regaud's fluid or some other mixture which preserves them well.

Embed and section tissue by the paraffin method (steps 3 to 10) or the nitrocellulose method (steps 3 to 7). For most cytological work, sections having a thickness of 5 to 10 μ are preferred. The sections must be smooth and of uniform thickness. Otherwise, they tend to stain very unevenly. Affix paraffin sections to slides and pass them through the appropriate series of reagents to 70% alcohol. Nitrocellulose sections may be affixed to slides by means of the albumen-clove oil method (p. 268) or handled loose. If a fixative containing osmium tetroxide was used, the sections should be bleached (see p. 143).

Iron hematoxylin has been extensively used for smears of blood or various soft tissues, and for smears of fecal matter or intestinal contents to be used for parasitological examination. Fix the smears while they are still moist. Schaudinn's fluid is generally used for fecal material in which clinically-important amoebae or flagellates are to be searched for, but some picric acid fixatives (Bouin's fluid, Hollande's fluid) are at least as good.

PROCEDURE FOR STAINING. *a. Transfer to Water.* Pass loose sections, or small specimens not affixed to slides, down through the alcohol series to distilled water. In the case of sections affixed to slides, or smears on coverglasses, allow them to stand for several minutes in a pan through which tap water is running, and then rinse them in distilled water.

b. Mordanting. Transfer them from distilled water to the mordant (a freshly filtered 2% aqueous solution of ferric ammonium sulfate). In most localities, it is not advisable to transfer material from tap water to the iron alum, because tap water commonly contains salts which will cause precipitation of iron compounds upon the material. Such precipitates may stain deeply, causing the preparations to appear covered with something like coal dust. Leave the sections or other material in the mordant for at least 45 minutes, prolonging the treatment to 2 or 3 hours in case a very sharp staining of centrosomes or spindle fibers is desired. Mordanting for many hours, as recommended by Heidenhain and his contemporaries, has no noticeable advantages.

The iron alum solution may be used over and over again *for mordanting*, if it is always filtered before use. *Never mordant material in a solution which has been used for destaining.*

c. Washing Out Excess Mordant. Rinse sections or other material for 15 to 30 seconds in distilled water. When handling smears or sections affixed to slides, next wash them in running tap water for about 1 to 3 minutes. (In the case of loose sections or small organisms, simply change the distilled water 2 or 3 times in rapid succession.) Some workers prolong the washing to 10 minutes or more, in order to avoid carrying any of the

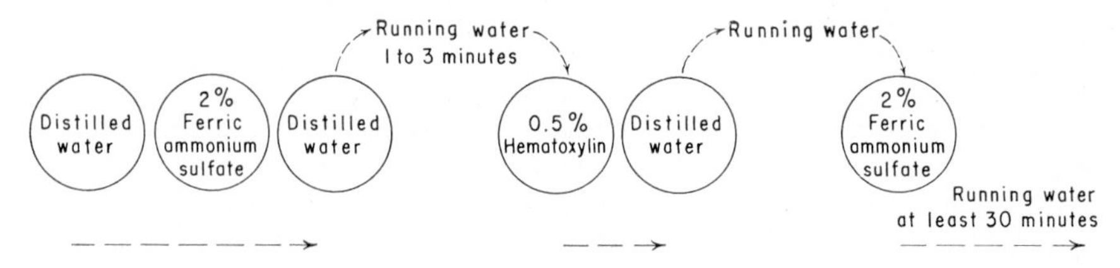

FIG. 64. Arrangement of staining jars for use in Heidenhain's iron hematoxylin method.

iron salt into the stain. However, this may result in slow and inferior staining.

d. Staining. Remove the sections or other materials from water and place them in the staining solution, which has been prepared and ripened by the method described earlier. If you are using the stain regularly, filter it every day. Otherwise, always filter it before use.

The length of time which the material remains in the hematoxylin solution is an important factor in determining the *relative* extent to which various structures will be stained in the finished preparations. After fixation in most reagents, staining at room temperature for about 2 hours, followed by proper destaining, will yield preparations in which chromatin, centrosomes, and red blood corpuscles are stained black or blue-black, and cytoplasm light gray or colorless. Cell membranes, spindle fibers, and striations in muscle fibers will be stained various shades of gray. This variety of intensities of staining is satisfactory for most purposes. Therefore it is a good general practice to stain for 2 or 3 hours.[1] If the period of staining is reduced to 30 or 45 minutes, only the nuclei may be well stained. By prolonging immersion in the stain to 6 or 12 hours, it is possible to make preparations in which centrosomes and spindle fibers are stained more intensely and the ground cytoplasm is well differentiated. Long staining, however, accentuates even the slightest variation in thickness of sections or smears, with the result that some areas may remain black when others have been decolorized sufficiently or even completely. Materials fixed in Flemming's fluid or other mixtures of chromic acid and osmium tetroxide take up the stain slowly, and should be stained overnight at room temperature, or for an hour at 50 to 60°C. The staining of mitochondria will be dealt with in connection with special methods for these structures (Chapter 26).

e. Washing. Remove the sections or other materials from the stain and rinse them with distilled water. Slides to which sections are affixed may then be left for 5 minutes or longer in a pan through which tap water is running. Loose sections or other material should be washed with several changes of distilled water. When stain ceases to come from them, they are placed in tap water.

f. Destaining and Differentiation. After staining, the material is quite black, and must be treated with a reagent which will decolorize the various structures to different degrees. There are two particularly good methods of destaining. Regardless of which method is employed, the process must be controlled with great care. It is the crucial step of the entire procedure.

1. Heidenhain employed ferric ammonium sulfate for destaining, and

[1] If speed is imperative, the same effect may be obtained by staining for 15 minutes in a staining solution heated to 50 to 60° C. The hot stain seems to be more effective than a cold solution in penetrating amoebic cysts, coccidian oöcysts, and similar materials.

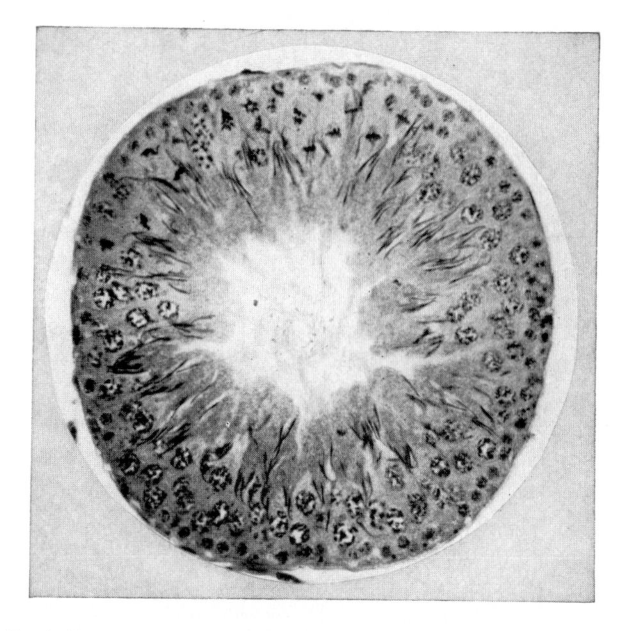

FIG. 65. Seminiferous tubule in the testis of a rat. Testis fixed (in part by perfusion) in Allen's B 15 fluid; paraffin section 6 μ thick; stained with iron hematoxylin. This preparation shows primary spermatocytes in various stages of division, as well as metamorphosing spermatids associated with Sertoli cells.

his method is the one most widely used today, in spite of the fact that it is not the best for all materials. A 2% solution of iron alum is generally acceptable, but a 1% solution acts more slowly, and it may be advantageous to use the weaker solution when very careful control of the process of destaining is desirable.

A 1% or 2% aqueous solution of ferric chloride is also an efficient destaining agent, and with some materials may be less likely to leave a brownish cast in the preparation.

To destain sections or smears affixed to slides, place the slides in the solution of iron alum, and agitate them gently until they no longer give off clouds of stain. When the tissue has been destained to a dark gray color, place a representative slide under the microscope, and flood it with iron alum solution. Observe it constantly. After the cytoplasm and connective tissue have become a very light gray color, but before the nuclear chromatin, centrosomes, and red blood corpuscles lose their sharp black stain, place the slide in running water. If the remainder of the slides are of the same tissue and can be regarded as nearly identical preparations, place all of them in running water at the same time. After the preparations have been washed for 5 to 10 minutes, again examine one or two of them, in order to make certain that they are destained to the proper extent. In most localities, tap water is alkaline enough to cause the stain to assume a blue-gray color. It will then appear somewhat darker than it did in the iron alum solution during destaining. Should the preparations be too dark and opaque, return them to the iron alum solution for a short time, and observe them constantly. When the stain seems to be light enough, return them to running water. Proceed cautiously in this part of the technique. No harm will result if destaining is accomplished in several steps. If destaining is carelessly allowed to go too far, restaining will require extra work and may yield unsatisfactory results.

The exact intensity of staining to be desired differs for various tissues and different purposes. If certain cytoplasmic structures (mitotic spindles, disks in striated muscle) are to be brought out with iron hematoxylin, a sufficient amount of stain must, of course, remain in them. A lighter stain is generally better for counting chromosomes. The cytoplasm should be very light if a counterstain is to be used. Good judgment in these matters comes only through experience.

In order to destain loose sections or small organisms such as protozoa, place them in a shallow dish of iron alum solution. Keep the liquid in motion and renew it if it becomes deeply colored. Frequently examine the material with a microscope. When destaining has progressed as far as desired, quickly draw off the iron alum solution and replace it with distilled water. Change this several times and then proceed to step *g*.

Should destaining inadvertently be allowed to proceed too far, leave sections or other material in the iron alum solution until all stain has left them. Then wash out the excess iron alum (step *c*), and proceed with

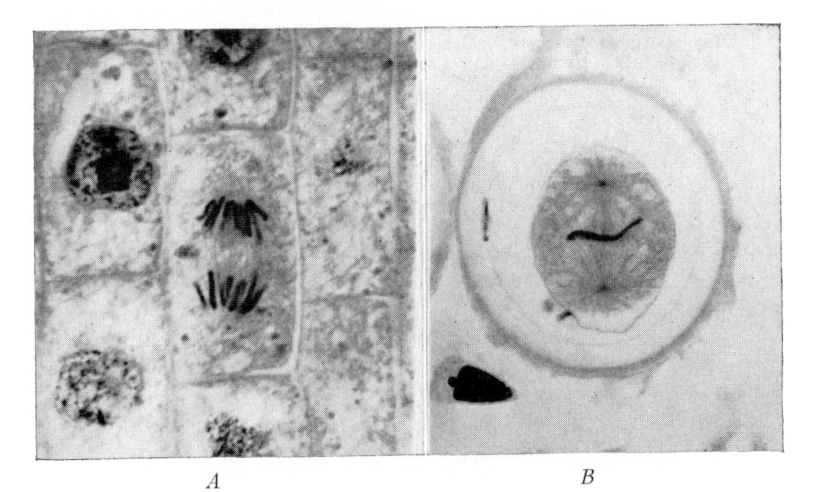

FIG. 66. Cytological preparations stained by Heidenhain's iron hematoxylin method. *A*, Anaphase of mitosis in root-tip of onion, *Allium cepa*. Fixed in Karpechenko's fluid; paraffin section cut at 7 μ. *B*, Metaphase of first cleavage in *Parascaris equorum*. Fixed in Carnoy and Lebrun's fluid; paraffin section cut at 8 μ.

staining, washing, and destaining (steps *d* to *f*). Unfortunately, restained preparations often have a muddy brownish cast.

The iron alum solution in the jar used for destaining smears or sections affixed to slides may be employed only until it becomes considerably discolored. Two hundred milliliters of a 2% iron alum solution will serve for about 50 average preparations. At the end of the day's work, always discard any iron alum solution which has been used for destaining, and discard the solution poured off from nitrocellulose sections or other loose material. These precautions are necessary because an iron alum solution which contains much oxidized stain is likely to leave a diffuse brown color in the tissues. At this point it is advisable to repeat the injunction that *a solution which has been used for destaining should never be used for mordanting*.

2. As a destaining agent, picric acid offers several advantages over iron alum. This substance acts slowly and is easy to control. It often yields a bluer and more transparent stain than is obtained when iron alum is employed. With the use of picric acid, there is less likelihood that staining will be uneven or that there will be unpleasing brown tones in the preparation, especially if it has been left in the stain for many hours.

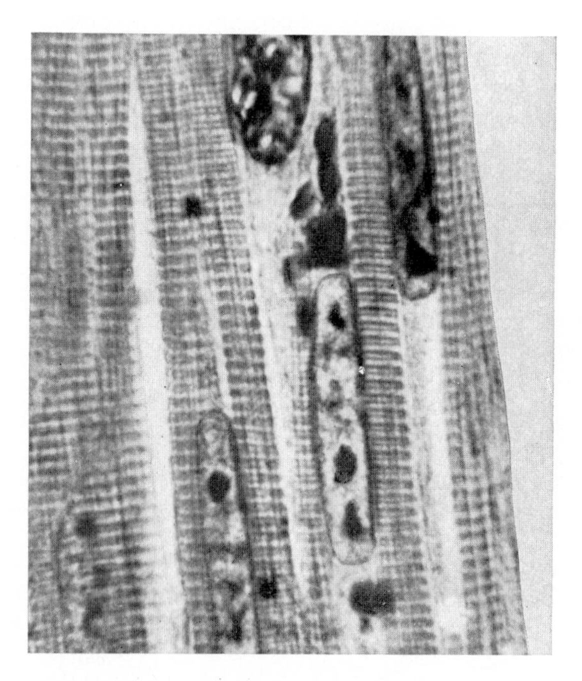

FIG. 67. Striated muscle fibers from the larva of the salamander *Taricha torosa*. Fixed in Bouin's fluid; paraffin section cut at 8 μ; stained by Heidenhain's iron hematoxylin method.

Employ a saturated aqueous solution of picric acid and handle the slides or loose material as directed in 1. The time required for proper destaining may vary from 5 minutes to several hours. Adding 10% glycerol to the destaining solution makes the sections slightly more transparent and slows down differentiation. Until experience has been gained, the yellow color of the solution may make it somewhat difficult to judge the amount of stain which should remain in the preparation. For most purposes, destaining should be continued until only the nuclei and red blood corpuscles appear brown, and all other structures are yellow. Then wash the material in running tap water, or place it in water which has been rendered slightly alkaline (by addition of a few drops of *very dilute* ammonium hydroxide), until the stain is changed to a blue color. Examine several sections or specimens. If they are still stained too heavily, replace them in the picric acid solution until a more nearly optimum degree of differentiation has been attained.

Some workers use a saturated solution of picric acid in 95% alcohol, and blue the stain in slightly alkaline alcohol of the same strength.

g. Final Washing. It is necessary now to wash out of the material every trace of iron alum, picric acid, or other agent used for destaining. If this is not done, the presence of even minute quantities of the destaining agent will cause the stain to fade. In order to wash smears or sections affixed to slides, allow the slides to stand for at least 30 minutes in a pan through which tap water is running. To wash loose sections, tie a piece of netting over the dish containing them, and allow tap water to drip into the dish for an hour or two. Wash loose protozoa or other small organisms with many changes of tap water, over a period of at least 2 hours.

Completing the Preparations. After washing the material, pass it up the alcohol series to 70% alcohol. If counterstaining is desirable, in order to bring out connective tissue, cytoplasm, cell boundaries, or cilia, orange G is a good choice. If it is used, proceed to steps *h* and *i*, page 387. Whether or not the preparations are counterstained, dehydrate, clear, and mount them in the appropriate manner, depending on whether they are paraffin sections (Chapter 15), nitrocellulose sections (Chapter 16), smears (Chapter 12), or entire objects (Chapter 11).

Weigert's Iron Hematoxylin

This is a progressive nuclear stain for histological preparations. It is not good for cytological purposes and should not be confused with the iron hematoxylin methods of Heidenhain or Benda. The mixture gives an almost strictly nuclear stain of a blue-black color, and very seldom shows a tendency to overstain. It may be applied either to sections or to material which is to be sectioned after staining. For staining before sectioning, it is easier to use and perhaps more reliable than any other hematoxylin mixture. The stain, however, is not as vivid as that obtained with alum hematoxylin and is likely to be encroached upon by counterstains.

Reagents for Weigert's Iron Hematoxylin Method

Solution A:

Ferric chloride, 10% aqueous solution	5 ml.
Distilled water	94 ml.
Hydrochloric acid	1 ml.

Solution B:

Hematoxylin 1% in 95% alcohol, well ripened.

To prepare the stain, mix equal parts of *A* and *B*. The mixture is best when freshly prepared, but it remains good for at least 24 hours.

Method for Staining Sections. PREPARATION OF MATERIAL. Prepare material as recommended for the progressive method of staining with alum hematoxylin (p. 353).

PROCEDURE. *a*. Pass the sections down through the alcohol series to water.

b. Either rinse them with several changes of water or wash them for a minute in running water.

c. Place them in the stain for about 20 minutes, or until the nuclei are strongly stained. Some tissues will be stained sufficiently in 20 minutes, while others may require twice as long.

d. Wash them in running water for about 20 minutes.

e. Place them in 15, 30, 50, and 70% alcohols (1 minute or longer in each).

COUNTERSTAINING AND COMPLETING PREPARATIONS. Proceed to counterstain them with eosin, following steps *c* and *d*, page 385. Then dehydrate, clear, and mount them in the customary manner.

Method for Staining Tissues Before Sectioning. PREPARATION OF MATERIAL. Fix whole organisms or slices of tissue (not over 4 or 5 mm. thick) in Bouin's fluid, formalin, or any appropriate fixing agent. Wash them properly, and harden them in 70% alcohol.

PROCEDURE. *a*. Place the slices of tissue or whole organisms into 50% alcohol for one to several hours, depending on their size and permeability.

b. Transfer them to the stain for about 24 hours.

c. Rinse them with distilled water and let them stand in another change of distilled water for about an hour. Agitate the liquid several times.

d. Wash them in running water for at least 4 hours, or preferably overnight.

SECTIONING, COUNTERSTAINING, AND MOUNTING. Embed and section the material in paraffin or nitrocellulose. If the sections were embedded in paraffin, affix them to slides in the usual way; remove the paraffin with toluene, and transfer the sections to absolute alcohol. Place nitrocellulose sections as they are cut into 70% alcohol, and either affix them to slides by the albumen-clove oil method or handle them as loose sections.

Turn to page 385, and follow directions for counterstaining sections lightly with eosin (steps *c* and *d*). Be careful that the sections do not become heavily overstained, as the counterstain may obscure the nuclear stain. Dehydrate and mount the sections in the usual way.

Verhoeff's Method for Elastic Tissue

Elastin, in the form of fibers not occurring in definite bundles, is found in areolar connective tissue, elastic cartilage, walls of blood vessels, air tubes of the lungs, and in a few tendons (the nuchal ligament of the ox). Elastic fibers may be demonstrated by treating fresh connective tissue with 1% acetic acid. The fact that the collagenous fibers become swollen and transparent in acetic acid, whereas elastic fibers do not, makes it possible to distinguish between the two. Various techniques of staining have been devised to bring out elastic fibers, and Verhoeff's method is perhaps the best of these.

PREPARATION OF MATERIAL. Fix the tissue in Zenker's fluid, Helly's fluid, 10% formalin, or Bouin's fluid. If a fixative containing mercuric chloride has been used, it is not necessary to treat the tissue with iodine, because deposits of mercury will subsequently be removed in the course of staining. However, other phases of post-fixation washing should be carried out as required. Embed the tissue in paraffin or nitrocellulose, and cut sections in the usual way. Material may be transferred to the staining mixture from 50 or 70% alcohol.

PROCEDURE FOR STAINING. *a*. Stain for 15 minutes in the following mixture:

Hematoxylin, crystals	1 gm.
Absolute alcohol	20 ml.
Dissolve the hematoxylin in the alcohol on an electric hot plate. Then add:	
Ferric chloride, 10% aqueous solution	8 ml.
Lugol's solution (iodine, 1 gm., potassium iodide, 2 gm, distilled water, 100 ml.)	8 ml.

b. Wash sections for a few minutes in running water. Then place them in a 2% aqueous solution of ferric chloride until the stain is extracted from all structures except the elastic fibers and cell nuclei. The former should be black, the latter gray. Keep the slides moving during the process of destaining. Differentiation usually occurs very quickly. Use of a weaker solution of ferric chloride may be desirable if the progress of destaining is so rapid that it cannot be carefully controlled.

c. Wash the sections for at least 30 minutes in running water, then pass them up the series of alcohols to 70% alcohol. Leave them in this for at least 5 or 10 minutes, so that all traces of iodine will be washed out.

Counterstaining and Completing Preparations. Counterstain the sections lightly with eosin (steps *c* and *d*, p. 385). Then complete dehydration, clearing, and mounting.

Elastic fibers should be jet black, nuclei gray, and other structures various shades of pink.

USE OF HEMATOXYLIN WITH OTHER MORDANTS

Mallory's Phosphotungstic Acid Hematoxylin

This method gives a differential stain in two colors, or sometimes in several shades. It is valuable for a number of purposes. It stains mitotic figures quite well, bringing out centrosomes, spindles, and asters, though not so boldly as does iron hematoxylin. The stain acts rather uniformly, and is therefore quite useful when a long series of stages in spermatogenesis, oögenesis, or cleavage is to be demonstrated on one slide. It is a good method for smooth muscle, because it differentiates muscle fibers from closely associated fibers of collagen, and for bringing out myofibrils. In striated muscle, phosphotungstic acid hematoxylin differentiates the various disks rather sharply. Phosphotungstic acid hematoxylin is as permanent as alum hematoxylin.

Phosphotungstic Acid Hematoxylin

Hematoxylin	0.5 gm.
Phosphotungstic acid	10 gm.
Distilled water	500 ml.

Dissolve the ingredients in separate portions of water, applying gentle heat to dissolve the hematoxylin. When cool, combine the portions. Ripening occurs naturally over a period of several weeks, but it can be brought about immediately by the addition of a few drops of a dilute solution (0.01 or 0.02%) of potassium permanganate. The mixture keeps for years.

Preparation of Material. For mitotic and meiotic figures and muscle fibers, fix material preferably in Zenker's fluid, 10% formalin, Heidenhain's "Susa" mixture, or Carnoy and Lebrun's fluid. The latter is recommended for eggs of nematodes, such as *Parascaris*. Bouin's fluid is less favorable, though not completely useless. For the study of neuroglia fibers, fix thin slices of tissue from the central nervous system in 10% formalin, and wash them in running water for 24 hours.

Wash and harden the material, then embed it in paraffin or nitrocellulose. Pass the sections through the usual reagents to water. If Zenker's fluid or some other fixative containing mercuric chloride was used, and the tissue was not iodized before being embedded, treat the sections as follows: Place them in Gram's iodine solution (iodine, 1 gm.; potassium iodide, 2 gm.; distilled water, 300 ml.) for a few minutes; rinse them in water, and then transfer them to a 0.5% aqueous solution of sodium thiosulfate. After all of the iodine has been removed, wash the sections in running water for 10 minutes or longer.

PROCEDURE. *a.* Place the sections in a 0.25% aqueous solution of potassium permanganate for 5 to 10 minutes.

b. Wash them in water for about 1 minute.

c. Place them in a 5% aqueous solution of oxalic acid for 10 to 20 minutes.

d. Wash them in several changes of distilled water.

e. Stain them in phosphotungstic acid hematoxylin until inspection shows that desired results (described below) have been attained. The time required for staining varies from 2 to 48 hours.

f. Drain the preparations. Dip them several times in 95% alcohol. Dehydration must be performed quickly, because alcohol extracts the color from those elements which are stained red. Place the sections in absolute alcohol for 20 to 30 seconds. Transfer them directly to toluene or terpineol-toluene. Pass them through a second jar of toluene and mount them in balsam or synthetic resin.

RESULTS. Chromatin, centrosomes, smooth muscle fibers, and Q and Z disks of striated muscle fibers are stained dark blue. Neuroglia fibers are also stained blue after fixation in formalin. Mitotic spindles appear light blue, gray, or pink. Collagen, cartilage, and bone matrix are stained various shades of yellowish or brownish red, and cytoplasm is usually gray or pink.

Weigert-Pal Method of Staining with Hematoxylin After Mordanting With Potassium Dichromate

Potassium dichromate alters lipids and renders them less soluble in the usual solvents. Prolonged treatment of lipids with potassium dichromate results also in the deposition of chromium compounds which will form deep blue lakes with hematein. This reaction is the basis for Weigert's differential stain for myelin sheaths of nerve fibers, which has contributed much to our knowledge of the fiber tracts of the central nervous system.

Weigert's original method has been modified many times. The following procedure, a variation of the so-called Weigert-Pal technique, generally gives good results, and is perhaps easier to carry out than some of the others. It demonstrates fiber tracts very well, and brings out gray matter and white matter in sharp contrast.

PROCEDURE. 1. Fix slices of brain or spinal cord in neutral 10% formalin for at least 1 week. If the procedure of staining is not to be carried out immediately, leave them in the formalin.

2. Wash the tissue in running water for several hours or overnight, then place it in a 3% solution of potassium dichromate, which serves as a

mordant. Leave the slices in this solution for 1 to 4 weeks, depending on their size. (One week will be long enough for a slice of tissue about 1 cm. thick.)

3. Wash the tissue thoroughly in running water (several hours or overnight), dehydrate it slowly and thoroughly, and embed it in nitrocellulose.

4. Cut thick sections (about 25 to 50 μ).

5. Bring the sections gradually to water and place them in 100 ml. of 0.5% chromic acid to which 0.5 ml. of 2% osmium tetroxide has been added. This mixture serves as a secondary mordant, and should be allowed to act for one or two hours.

6. Rinse the sections in water, then stain them for 12 to 24 hours in the following mixture:

Hematoxylin, 10% solution (ripened) in absolute alcohol	10 ml.
Acetic acid, glacial	1 ml.
Distilled water	100 ml.

Staining may be carried out in an incubator at about 37° C., or at room temperature. If warmed, the sections may become brittle, but the stain will be quite intense.

7. Wash the sections in water until no more color comes out (an hour or less), then immerse them in 0.5% potassium permanganate for 30 seconds.

8. Rinse them in water and put them into the following mixture:

Sodium sulfite (anhydrous), 1% aqueous solution	100 ml.
Oxalic acid, 1% aqueous solution	100 ml.

The two ingredients should be mixed just before use.

9. Observe the process of differentiation very closely. It may not take more than a few minutes to obtain optimum results. By alternating differentiation in the above mixture with rinsing in water, followed by 30 seconds in 0.5% potassium permanganate, and rinsing in water again, the background of gray matter may be thoroughly bleached, while the white matter will remain dark blue. However, overtreatment with potassium permanganate will remove too much potassium dichromate, and this will impair the staining.

10. When a satisfactory differentiation of gray and white matter has been obtained, wash the sections in running water, then transfer them to a very weak solution of lithium carbonate (distilled water, 100 ml.; saturated solution of lithium carbonate, 0.2 ml.). Leave them in this for 20 minutes to several hours.

11. Dehydrate the sections as far as 95% alcohol, clear them in terpineol-toluene or carbol-toluene, then in toluene, and mount them in a resinous medium.

CHAPTER 23

Synthetic Dyes

Many brilliant dyes are produced synthetically from hydrocarbons obtained by distillation of coal tar. Aniline, a colorless, oily substance having the empirical formula $C_6H_5NH_2$, is an intermediate product in the manufacture of a large number of such dyes. Because the earliest synthetic dyes were prepared from aniline, it became customary to classify all artificial dyes as "aniline dyes." This term is misleading, however, because aniline is not actually involved in the production of many synthetic dyes in common use.

Within a few years after 1856, when the first synthetic dye (mauveine) was prepared by Perkin, several others appeared on the market. In 1862, a violet aniline dye of some sort was used in histological staining by Beneke. As the number of coal tar dyes increased, many of them (including eosin and basic fuchsin) were tried out for biological purposes. In the years 1877 to 1886, Ehrlich made a systematic study of coal tar dyes to test their suitability for biological staining, and he discovered the value of safranin, methylene blue, and various others.

The number of synthetic dyes employed in biological staining is now very large. Relatively few can be discussed in this chapter, but most of those which are in routine use will be included.

Chemistry of Synthetic Dyes

A brief discussion of the chemical composition of artificial dyes will make it easier to explain their classification and uses. These remarks will be made as non-technical as possible. For detailed information concerning dyes and their biological applications, the reader is referred to Baker (1958), Conn (1969), and Gurr (1960).

The chemical basis of a synthetic dye molecule is a benzene ring, or several of these rings linked together. Some of the hydrogen atoms on these rings are replaced by atomic groupings of two kinds, as follows:

CHROMOPHORES. These are groupings which give color to a compound. A benzene compound which carries chromophores is colored, and is known as a chromogen. There are numerous chemically distinct chromophores, and the chemical classification of dyes is based upon the names of these chromophores. Thus we have the nitro, azo, phenyl methane, and other groups of dyes listed on page 375. The basis of color is complex and not fully understood, and it should be noted that the same chromophore may be present in dyes of many different colors.

AUXOCHROMES. The auxochromes are responsible for the affinity of the colored molecule for other substances, including cellular structures. In other words, the replacement of one or more hydrogen atoms of the color-bearing molecule (chromogen) by auxochromes causes the molecule to become an effective dye. Auxochromes may also modify the color.

To illustrate these facts, let us picture the steps in building the molecule of a simple dye, picric acid.

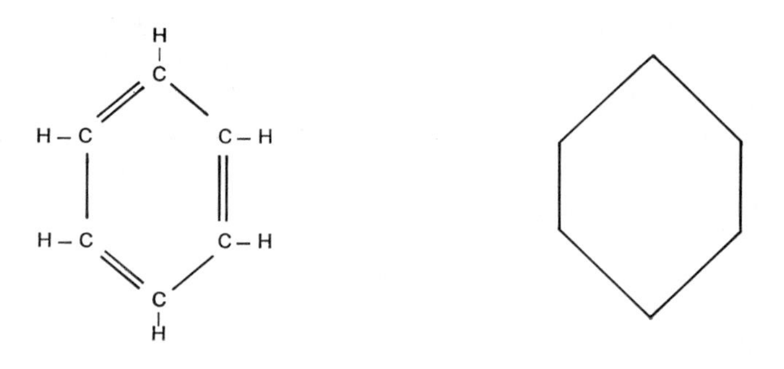

1. Benzene ring and simple hexagon (right) by which it is usually represented.

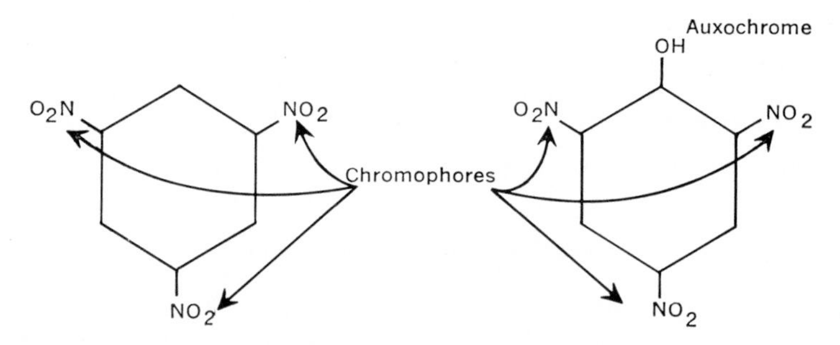

2. Benzene ring with chromophores (NO_2) replacing 3 of the hydrogen atoms. This is trinitrobenzene, a chromogen.

3. Picric acid, a yellow dye, produced by adding an auxochrome (OH) to trinitrobenzene.

PROPERTIES OF BASIC AND ACID DYES. Some auxochromes, such as the amino group ($-NH_2$), have basic properties. Others, such as the carboxyl ($-COOH$), hydroxyl ($-OH$), and sulfonic ($-HSO_3$) groups, are acid. The reaction of a dye is determined chiefly by its auxochromes, although an acid or basic chromophore group may increase the effect of a like auxochrome or decrease that of an opposite auxochrome. Almost any kind of dyestuff (chromogen) may be given basic properties by the introduction of amino groups in suitable positions, or acid properties by the introduction of carboxyl, hydroxyl, or sulfonic groups. For this reason, there are acid and basic members in every chemical group of dyes. As we shall presently see, the dyes used are not pure color acids and bases, but their inorganic salts, which ionize when dissolved.

Ordinarily, basic dyes have an affinity for acid substances (such as chromatin of nuclei), and acid dyes have an affinity for basic constituents (such as collagen of connective tissue). Hence the possession of basic, acid, or neutral properties affords a natural and convenient basis for classifying stains used for biological staining.

Classification of Dyes

In this chapter, the more widely used synthetic dyes (and certain mixtures of dyes) will be grouped according to their basic, acid, or neutral properties. In each category, there will be representatives of several chemical classes. A summary of most of the chemical classes, important examples in common use as biological stains, and the acid or basic nature of each example, is given below.

1. Nitroso dyes
   Example: naphthol green B (acid).
2. Nitro dyes
   Example: picric acid (acid).
3. Azo dyes
   Examples: orange G (acid); Congo red (acid); trypan blue (acid); Janus green B (basic); Sudan III and Sudan IV (weakly basic).
4. Quinone-imine dyes
   Examples: methylene blue (basic); azure A and azure B (basic); thionin (basic); Nile blue sulfate (basic); neutral red (weakly basic); safranin O (basic); azocarmine G (acid); indulins (mixtures; acid and basic forms).
5. Phenyl methane dyes
   Examples: crystal violet (basic); methyl green (basic); aniline blue, water-soluble (acid); fast green FCF (acid); light green SF yellowish (acid); basic fuchsin (basic); acid fuchsin (acid).
6. Xanthene dyes
   Examples: eosins, erythrosins, and phloxine B (acid); phenolphthalein, brom cresol purple, and phenol red (indicators).
7. Anthraquinone dyes
   Example: alizarin red S (acid).

Nomenclature of Dyes

In the past, coal tar dyes were named by their manufacturers with little regard for accuracy or system. A particular dye was sold under various names, and the exigencies of competitive business led to the sale of entirely different dyes under one name. This situation created considerable trouble for microscopists, because biological staining depends upon specific reactions which in most cases are obtainable only if the same compounds are employed consistently. The works of Conn (1969) and Gurr (1960) contain a wealth of information about dyes, and list synonyms for nearly all dyes which have been employed in microtechnique.

When ordering, be sure to specify each dye accurately. Use products which are known to be reliable, or which have been certified by the Commission on the Standardization of Biological Stains.

Permanence of Stains

Synthetic dyes differ greatly with respect to their permanence in microscopic preparations. In general, they do not keep as well as natural dyes, and for this reason they have not replaced carmine and hematoxylin to the extent which might be expected. In view of the obvious advantage of permanent preparations, the account in this text will be limited largely to dyes which are relatively permanent. However, some dyes, in spite of their tendency to fade rapidly, are the only ones known to be suitable for certain cytological purposes and must therefore be considered.

BASIC DYES

Basic coal tar dyes are salts of organic color bases (amino or imino derivatives). For example, methylene blue is the chloride of tetramethyl thionin, or the double chloride of this color base and zinc. As the color bases bear positive electrical charges, they possess, in general, a strong affinity for negatively charged structures such as nuclei, mucin, and the matrix of cartilage.

These dyes are pre-eminently brilliant nuclear stains, suitable for sections or smears of tissue. They may be used successfully after alcohol-formalin mixtures and several other fixatives. Some basic dyes give best results on material which has been fixed in mixtures containing chromic acid, or both chromic acid and osmium tetroxide. Certain basic dyes, crystal violet for instance, are effective cytoplasmic stains; others, such as aniline blue, are valuable for differentiating mucin and some types of connective tissue. Because of this diversity, it is possible to combine basic nuclear dyes with contrasting acid or basic dyes for cytoplasm, connective tissues, and other tissue elements or secretions, thereby producing useful double, triple, or multiple stains.

The majority of basic dyes are employed as strong solutions in water, aniline water, or alcohol. They usually overstain in a short time and ordinarily yield precise stains only upon subsequent differentiation with alcohol, to which some free acid or an acid dye is commonly added. Some exceptions are methyl green, which is employed (in acid solutions) for progressive staining, and methylene blue, which is often used in alkaline solutions. In general, basic dyes are intensified by alkalis and extracted by acids. However, they are not so sensitive as acid dyes to miuute differences of pH. Many of the basic coal tar dyes stain nitrocellulose deeply, so that it is best to limit their application to paraffin sections.

Safranin O

There are many kinds of safranin dyes, but safranin O is the one most commonly used in biological staining. This is a powerful red dye, freely soluble in water and alcohol. A 1% aqueous or alcoholic solution quickly overstains. If the sections are then treated with alcohol, a weak acid, or an acid dye (such as fast green FCF or orange G), the color is extracted from most extranuclear structures, while chromatin retains a brilliant and very precise stain. The importance of safranin as a nuclear stain for sections is enhanced by the fact that it is one of the more permanent of the basic dyes, remaining unaltered for many years if the preparations are free from all traces of acid and are protected from light. Safranin has been used as a differential stain for myelin, elastin, and keratin, but other stains are now preferred for bringing out these elements.

Material fixed in Flemming's fluids or other mixtures containing chromic acid is often stained brilliantly and selectively with safranin, but the dye also gives good results after fixation in formalin or alcohol. Bouin's fluid, other picric acid mixtures, and fixatives containing mercuric chloride are generally unsuitable for material to be stained by safranin.

Safranin, as a nuclear stain, is generally followed by a contrasting plasma stain. Crystal violet (a basic dye), or fast green FCF or orange G (acid dyes), are commonly used counterstains.

Basic Fuchsin

This deep red dye is a mixture of three or four slightly different compounds. It dissolves readily in water and alcohol. Preparations stained with it seldom fade in less than 3 or 4 years, but it is not so permanent as safranin. Basic fuchsin should not be confused with *acid fuchsin*, which has entirely different properties.

General Use in Histology. Basic fuchsin is a good nuclear stain for paraffin sections. Sections should be overstained (5 to 10 minutes) with a 1% aqueous solution of the dye, then differentiated as they are passed up the alcohol series until only the nuclei and certain other structures remain colored. In addition to staining nuclei, it brings out red blood corpuscles, mucin, and so-called "fuchsinophil" granules. A blue, yellow, or green cytoplasmic stain, or a mixture of such stains, is ordinarily applied before the excess fuchsin is removed. Among the better counterstains are Mallory's aniline blue-orange G mixture (p. 399) and Ramón y Cajal's picro-indigo-carmine solution (p. 391). The choice of a fixing fluid depends largely upon the counterstain which is chosen, because basic fuchsin stains well after most fixing reagents.

Bacteriological Staining. Basic fuchsin is one of the more valuable stains for bacteria, either in smears or in tissues. Infected tissues, stained as recommended above, often show the bacteria clearly. The following formula is generally used for staining smear preparations.

Ziehl-Neelsen's Carbol-Fuchsin

Saturated solution of basic fuchsin (about 13%) in 95% alcohol . . .	10 ml.
Phenol, 5% aqueous solution	90 ml.

Acids remove the stain from most bacteria, but fail to do so in the case of a few species, which, therefore, are called "acid-fast." Possession of the acid-fast characteristic makes it possible to identify certain very important pathogenic bacteria, such as *Mycobacterium tuberculosis* and *M. leprae.* Ziehl-Neelsen's method of staining smears with carbol-fuchsin, for identification of these bacilli, is given on page 175.

The Feulgen Method. An important use of basic fuchsin is made in Feulgen's nucleal reaction ("Nuclealreaktion"), a test for deoxyribonucleic acid. Therefore, it serves well as a stain for chromatin. Methods for preparing and applying the necessary reagents are described on page 432.

Applications for Staining Glycogen, Other Polysaccharides, and Mucopolysaccharides. When certain carbohydrates are appropriately oxidized, they react with the Feulgen reagent and thereby are rather intensely stained. Bauer's method for glycogen (p. 446) and the periodic acid-Schiff (PAS) method (p. 447) for a variety of polysaccharides and mucopolysaccharides are important cytochemical and histochemical procedures which take advantage of this reaction. The PAS technique, when combined with a nuclear stain and a general counterstain, produces valuable histological preparations.

Methylene Blue

This beautiful dye serves so many important purposes in microscopy that it requires a rather lengthy discussion. Two principal types of methylene blue are manufactured. One is the chloride of tetramethyl thionin, and is sold as *methylene blue chloride*, or *methylene blue Med. U.S.P.* This type is best for all biological purposes and is quite essential for vital staining, where a relatively non-toxic dye is required. The second type, represented by cheaper commercial grades of methylene blue, is usually a double salt, the chloride of methylene blue and zinc. The toxicity of this compound makes it unfit for vital staining. Its low solubility in alcohol is another disadvantage.

Methylene blue becomes readily oxidized to form azure A and azure B, which are selective stains of a violet color. These oxidation products are probably almost always present, and they accumulate as a solution ages. Their formation may be accelerated by heating the stain with a little sodium carbonate. A solution of methylene blue which contains large proportions of the azures, produced by aging or artificial ripening, is called polychrome methylene blue. The azures are responsible for the differential staining (metachromasy) of certain granules in blood corpuscles and other cells. This matter will be discussed in connection with neutral or compound stains, page 391.

Generally speaking, methylene blue stains do not endure for more than a few years. Use of strictly neutral balsam or synthetic resin for mounting, and avoidance of heat for drying the mounting medium, will prolong the utility of preparations stained by methylene blue.

Supravital Staining. Dilute solutions of methylene blue, in water or an isotonic saline solution, are employed for staining of nerve cells, nerve fibers, and nerve endings. Some suitable procedures are discussed on page 53.

Bacteriological Staining. Methylene blue is widely used for staining bacteria both in smears and in sections of tissue. Sections are stained with a 0.2% aqueous solution of the dye. For staining bacteria in smears, the following mixture is generally used.

Loeffler's Methylene Blue Solution

Saturated solution (about 1.5%) of methylene blue in 95% alcohol	30 ml.
Potassium hydroxide, 0.01% solution in distilled water	100 ml.

Directions for making smears and using the stain will be found on page 174.

Methylene blue is commonly employed as a counterstain for preparations in which acid-fast bacilli (such as *Mycobacterium tuberculosis*) have been stained by Ziehl-Neelsen's carbol-fuchsin method. For instructions, see page 175.

Inclusion in Neutral Stains. Polychrome methylene blue is an essential constituent of the better "neutral stains" for blood and protozoan parasites of blood. The composition and use of these stains is discussed at length on page 391 of this chapter.

Crystal Violet, Methyl Violet, and Gentian Violet

Dye makers, suppliers, and biologists have used these names in a loose and confusing manner. *Methyl violet* is properly defined as a mixture of pararosaniline bases containing respectively 4, 5, and 6 methane groups to the molecule. Methyl violets vary in composition and their color ranges from reddish to deep blue as the proportion of more highly methylated compounds increases. *Crystal violet* is the deepest blue color base in methyl violet and contains 6 methane groups. It is obtainable as a pure substance. *Gentian violet* is a term so deeply rooted in biological literature that it may be difficult to discard. Originally, it was applied to certain deep methyl violets which contained much crystal violet and were diluted with dextrin. In microtechnique, *crystal violet* is called for because its reliability is assured.

Staining Bacteria. Crystal violet has its most important applications in bacteriology, especially for staining smears. The following solution may be used either for staining of all organisms which happen to be in the preparation or for differential staining of certain types by Gram's method (p. 174).

Crystal Violet Solution

Crystal violet	5 gm.
Ethyl alcohol, 95%	10 ml.
Dissolve the crystal violet in the alcohol. Then add a mixture of the following, which has been filtered:	
Aniline	2 ml.
Distilled water	88 ml.

It is a good idea to filter the staining solution after about 24 hours.

Methyl Green

This dye is derived from crystal violet, by the introduction of a seventh methyl group into the molecule. It always contains some unconverted crystal violet or methyl violet. The dye which is available commercially is generally a double chloride of methyl green and zinc. Various preparations differ greatly with respect to their purity and usefulness for biological staining. Methyl green is chemically unstable and therefore should be purchased in small quantities.

Methyl green is a rather specific stain for chromatin in the nuclei of cells in fresh preparations, and this is one of its more important applications. The following solution is recommended:

Acidulated Methyl Green

Methyl green	1 gm.
Distilled water	100 ml.
Acetic acid, glacial	1 ml.

Fresh material is teased, crushed, or smeared, and mounted under a coverglass with a drop of acidulated methyl green. The staining solution may still be effective after considerable dilution. Fixation and staining take place simultaneously and within a short time.

Methyl green has also been used in several methods for staining chromatin in making permanent preparations of fixed material. However, it fades rapidly, and unless there is some special reason why methyl green should be used, some other method (the Feulgen nucleal reaction) is recommended.

ACID DYES

Acid synthetic dyes are salts formed by combination of organic color acids with inorganic bases, usually sodium. For example, acid fuchsin is the disodium salt of rosanilin trisulfonic acid. Color acid molecules bear negative charges, and generally exhibit strong affinities for structures which are positively charged, such as connective tissue fibers. Various acid dyes show to different extents the property of staining some of these substances more deeply than others. Eosins, for instance, stain blood corpuscles, collagen, muscle fibers, and cytoplasm in decidedly different and characteristic shades. Aniline blue can be manipulated in such a way as to color practically nothing but collagenous connective tissue.

An acid stain is usually employed as a *counterstain*, to follow a contrasting nuclear stain of carmine, hematoxylin, or a basic coal tar dye. Addition of a second acid dye, to stain structures not colored by the first, will result in triple stains, some of which are very useful.

Factors Influencing Selectivity

The staining power and selectivity of acid dyes are influenced to a considerable extent by extraneous factors. Failure to recognize and to apply this fact is responsible for the erratic and prevailingly unsatisfactory results obtained with these dyes. Good stains are sometimes obtained by chance occurrence of favorable conditions, but in order to achieve good results consistently it is necessary to give careful consideration to the following matters.

Fixing Agents. The tissue must have been fixed in a reagent which renders it capable of taking up the desired stains. For example, fixation in mercuric chloride or formalin is favorable to eosins and some other acid dyes; objects fixed in chromic acid-osmium tetroxide mixtures are stained weakly and diffusely by eosins, but often prove suitable for staining with orange G or fast green FCF. Definite suggestions concerning the suitability of particular fixing agents preparatory to use of certain stains are given in Chapter 8 and in connection with various special methods.

Preservation. The material either should be stained within a few weeks after it is fixed, or should be stored in paraffin, glycerol, or some other medium which will not affect its staining properties. Long storage in alcohol is to be avoided. For further information on this point, see page 140.

Hydrogen Ion Concentration (pH). The power and selectivity of acid dyes are altered radically by even slight differences in the acidity or alkalinity of the staining solution. All acid dyes stain weakly and diffusely if the solution is at all alkaline. Some of them, however, stain satisfactorily in neutral solutions. The majority, including fast green, the eosins, and some other very important dyes, give the best selective stains only in solutions of a *very slight acidity.* This behavior is explained by the fact that the proteins of the tissues are basic (positively charged), and hence capable of taking up an acid dye only in solutions which are more acid than their isoelectric (neutral) points. An acid dye will stain no protein structures when the solution is less acid than the isoelectric points of any of the proteins present. Conversely, all protein structures will be colored if the solution is more acid than the isoelectric points of all the proteins involved. However, the isoelectric points of the various protein substances in the tissues differ somewhat. Because of this, there lies, between the extremes just mentioned, a range of slight acidity at which some of the proteins react as acids, some are nearly neutral, and others behave as bases. Within this range, the proteins of the various tissue structures, differing in reaction, will take up an acid dye to different extents. The critical concentration of acid in which this differential staining will take place depends also upon the relation between acidity and ionization of the dye, and differs for various dyes.

Very weak acidity and alkalinity, such as that with which we are now concerned, are measured in terms of hydrogen ion concentration, and are expressed as the numerical value of pH. The meaning of this symbol is explained on page 330. Neutrality is represented by pH 7.0, acidity by lower values, and alkalinity by higher values. The hydrogen ion concentration of deeply colored liquids, such as dye solutions, is best measured by an electric pH meter, which gives direct readings rapidly and therefore facilitates quick adjustment of the hydrogen ion concentration with a weak acid or base.

It has been determined by experiment that eosin, erythrosin, and orange G, which are among the more useful acid counterstains for cytoplasm in sections, generally give optimum results in solutions of pH 5.4 to 5.6. If the nuclear stain is weak, which may be the case with material fixed in certain reagents (such as Orth's fluid), the solution should be less acid (pH 6.0). On the other hand, a somewhat more acid solution (pH 5.0) can be recommended if a particularly energetic and selective counterstain is desired, or if the material has been rendered refractory by fixation in a chromic acid-osmium tetroxide mixture or Allen's B 15 modification of Bouin's fluid. These dyes can be employed most conveniently in a 0.5% solution in 90 or 95% alcohol. The pH of the solution can easily be adjusted by adding the proper amount of standard 0.1 normal hydrochloric acid, as shown in the following table.

Quantities of 0.1 Normal Hydrochloric Acid to be Added to 100 ml. of 0.5% Solution of Eosin, Erythrosin, or Orange G in 90% Alcohol, in Order to Obtain Various Hydrogen Ion Concentrations

	ml. of 0.1 N HCl per 100 ml. of 0.5% dye solution:	
pH Desired	*In 90% Ethyl Alcohol*	*In 90% Methyl Alcohol*
5.0–5.2	5.4	4.4
5.4–5.6	4.0	3.2
5.9–6.1	2.5	1.4

The 0.1 N acid is prepared by adding 8.25 ml. of concentrated hydrochloric acid (specific gravity 1.18) to enough distilled water to make 1000 ml. In measuring out the concentrated acid, and in adding the 0.1 N acid to the dye solution, use a serological pipette accurate to 0.1 ml. However, a graduated cylinder is suitable for mixing acid with distilled water, and for measuring the dye solution.

The acidulated staining solution will yield excellent results for about 2 weeks, if it is kept in a stoppered bottle and poured into the staining dish when it is needed. Eventually, however, the pH will rise to the point that it is no longer suitable. Accordingly, the acidulated dye solution should be prepared only when the need for it arises.

Eosins, Erythrosins, and Phloxine

These are red acid dyes, ranging in hue from yellowish to greenish and bluish. Chemically, they are derivatives of fluorescein, containing 3 or 4 atoms of bromine or iodine in the molecule. Several of these dyes are among the finest counterstains for cytoplasm, muscle fibers, blood corpuscles, and connective tissue. They are commonly used *following* nuclear stains of hematoxylin or *preceding* methylene blue, thionin, or other contrasting basic coal tar dyes. When mounted in neutral balsam or synthetic media, the dyes of this group are among the most permanent of coal tar preparations.

Some writers recommend that eosins be used in aqueous solution. However, aqueous solutions stain slowly, and the color is likely to be washed out by the alcohols used for dehydration, so that the finest differentiation is seldom obtained. Removal of the stain can be prevented by acidulating the alcohols, but this is usually out of the question if basic coal tar dyes are used for nuclear stains. Acidulation also may spoil the color and permanence of hematoxylin stains. Alcohol of about 90% is perhaps the best solvent in which to use these and most other acid dyes.

It is very important to employ these dyes in solutions of pH 5.4 to 5.6, as slight differences of acidity alter their action greatly. To obtain a solution of the desired pH, add 0.1 normal hydrochloric acid to the 0.5% solution of any variety of eosin or erythrosin, in the proportions shown in the table on page 383. Detailed directions for using all dyes of this group following hematoxylin are given on page 385.

Exceptionally old or otherwise refractory material may fail to stain in alcoholic eosins. In such cases, using a solution of the dye in a clearing oil may be attempted. Beechwood creosote is frequently used as a solvent, but it tends to swell and pucker sections. A 0.1% solution of the dye in a mixture of 3 parts of terpineol-toluene (or carbol-toluene) and 1 part of absolute alcohol is recommended. Slides bearing sections are transferred to the stain from absolute alcohol or toluene and left for about 3 minutes, or until the sections are slightly overstained. Next they are placed in the same reagent, without dye, until the stain is differentiated. Finally, they are passed through 2 changes of toluene and mounted in a suitable mounting medium.

The several varieties of eosin and its relatives differ considerably in staining properties. The following are generally used in histological work, but each one is adapted to somewhat specific purposes.

Eosin Y (Eosin, Yellowish). This very transparent variety of eosin is probably the best for general use in histology. It is particularly good for staining the eosinophil granules of leukocytes, and is employed in making Wright's stain and Giemsa's stain for blood.

Eosin B (Eosin, Bluish). This variety is favored by some workers, and is usually employed in an aqueous solution. It has, however, a rather dull color, and is less selective and transparent than either eosin Y or erythrosin.

Erythrosin, Bluish. This compound contains iodine in place of the bromine of typical eosins. It has a ruby red cast, and stains red blood corpuscles very deeply. It is less transparent and has a less agreeable optical quality than eosin Y. It is employed in the general manner stated for eosins.

Phloxine B. This is a very bright and powerful dye. It has been used as a substitute for eosin as a cytoplasmic stain preceding methylene blue. It is not recommended as a counterstain following hematoxylin, for the reason that it shows a decided affinity for chromatin.

Counterstaining with Eosin or Erythrosin. PREPARATION OF MATERIAL. Brilliant and selective stains are obtained with material which has been fixed in 10% formalin or Zenker's fluid. However, tissue fixed in Bouin's fluid may show a better preservation of the cells in general, and will stain quite well if it has not been stored in alcohol for a long time. Material fixed in Helly's fluid, Orth's fluid, or in a saturated aqueous solution of mercuric chloride with 5% acetic acid is also favorable.

Chromic acid-osmium tetroxide mixtures are decidedly not suitable fixatives for tissue to be counterstained with eosin or erythrosin, although the finished preparations may be passable.

Cut sections by the paraffin, nitrocellulose, or freezing method. Stain them with alum hematoxylin, and pass them up through the alcohol series to 70% alcohol. Sections which have been stained by the iron hematoxylin method may also be used, but they are generally mounted without counterstaining, or counterstained with orange G.

PROCEDURE. The steps are lettered to follow consecutively after step *b* in the regressive method for staining with alum hematoxylin (p. 354).

c. Transfer the sections from 70% alcohol to a 0.5% solution of eosin Y or erythrosin in 90% alcohol. The staining solution should have been adjusted to a pH of 5.4 to 5.6, as explained on page 383. Leave them in the staining solution until they are overstained.

Sections of tissue which was fixed in Zenker's fluid will probably be overstained in about 45 seconds. Material fixed in other mixtures generally takes a little longer.

d. Rinse paraffin sections, or nitrocellulose sections affixed to slides, in 95% alcohol, and transfer them to absolute alcohol. Loose nitrocellulose sections, or frozen sections affixed to slides by nitrocellulose, should not be dehydrated beyond 95% alcohol; if many sections are being processed at one time, it is best to rinse them for a few seconds in one bath of 95% alcohol and then to transfer them to a second bath. They should remain in the alcohol until the excess stain has been extracted and the various structures are well-differentiated. As a whole, they will then appear light pink, with some areas being slightly darker than others. In dense cellular tissues, the blue hematoxylin stain should be evident to the naked eye. Loose sections, in shallow dishes, may easily be observed under the microscope. By working very rapidly, sections mounted on slides may be examined, but care must be taken not to let them dry out.

When the stain is properly differentiated, striated muscle fibers will be deep pink, bone matrix will be red, and red blood corpuscles will be bright pink. Connective tissue should be medium pink, smooth muscle a little lighter. Cytoplasm in general should be very light pink, except in basophilic cells (such as those of pancreas), in which it is bluish purple, and in acidophilic gland cells (such as parietal cells of the stomach), in which it is bright pink. Cell boundaries are usually a little darker than the cytoplasm.

Coat paraffin sections (or nitrocellulose sections affixed to slides) with Parlodion, and clear and mount them (steps 17 and 18, p. 251). Loose nitrocellulose sections, or frozen sections affixed with nitrocellulose, should be cleared in terpineol-toluene or carbol-toluene, then mounted (steps 9 and 10, p. 266).

Counterstaining With Eosin Dissolved in a Clearing Agent. For most purposes, this method is not so good as staining in alcoholic solutions. However, it is useful for counterstaining sections of some types of material which were stained with a nuclear stain before being embedded, and which may not withstand treatment with alcohol after being sectioned and affixed to slides. Paraffin sections of yolky amphibian eggs and adult specimens of amphioxus are examples. Immerse the sections in toluene to remove the paraffin, then place them in a 0.1% solution of eosin in a mixture of 3 parts of terpineol-toluene (or carbol-toluene) and 1 part of absolute alcohol. Leave them in this for about 3 minutes, or until they are slightly overstained. Then place them in the same mixture without any dye, until the counterstain is properly differentiated. Finally clear them in toluene and mount them.

Counterstaining with a Combination of Eosin, Erythrosin, and Orange G. For a number of years, a combination of eosin Y, erythrosin (evidently erythrosin Y), and orange G was sold under the trade name of "Triosin." A close approximation of the contents of this product can be prepared by thoroughly mixing together 0.6 gm. of eosin Y, 0.3 gm. of orange G, and 0.1 gm. of erythrosin Y, or any multiple of these amounts.

When "triosin" is used as a counterstain for sections stained with alum hematoxylin, the general effect is similar to that obtained with eosin Y. However, cytoplasm, red blood corpuscles, muscle, connective tissue, and other histological elements are usually differentiated in a more varied way and in characteristic shades of red, pink, and orange. The results are often very beautiful as well as useful. The proportions of the ingredient dyes may, of course, be varied at will.

"Triosin" should be used in a 0.5% solution in 90% alcohol, and this should be adjusted to a pH of 5.4 to 5.6, in the manner recommended when any of the constituent dyes are used individually (p. 383). Staining should be carried out according to the schedule given for eosin and erythrosin.

Orange G

This is a valuable stain for cytoplasm, mitotic spindles, collagen, and keratin. In histology and pathology it is widely used in Mallory's triple connective tissue stain. In cytology, it figures as a component of several well-known stain combinations, including Flemming's triple stain.

Counterstaining with Orange G. Orange G is an excellent counterstain following iron hematoxylin in preparations for study of mitosis, meiosis, and spermatogenesis. It brings out cell boundaries and spindles, and tones the cytoplasm. For this purpose, a 1% solution of the dye should be prepared in 90% alcohol, and the solution brought to pH 5.4 to 5.6 by adding, to each 100 ml. of the solution, 4 ml. of 0.1 N HCl if the solvent is ethyl alcohol, 3.2 ml. of the acid if the solvent is methyl alcohol. In neutral solutions the stain is often diffuse and quickly washed out by the alcohols used for dehydration.

Paraffin or nitrocellulose sections to be counterstained are brought gradually, after thorough washing in water, to 70% alcohol.

PROCEDURE. The steps are lettered to follow step *g* in the iron hematoxylin method.

h. Transfer the sections to the solution of orange G and leave them in this for at least 2 or 3 minutes.

i. Rinse paraffin sections (or nitrocellulose sections affixed to slides) in 95% alcohol, and transfer them to absolute alcohol. Leave them in this until it appears that the stain is properly differentiated. Loose nitrocellulose sections should not be dehydrated beyond 95% ethyl or methyl alcohol.

Complete the preparations in the usual way, depending upon the nature of the material (steps 17 and 18 in the paraffin method, steps 9 and 10 in the nitrocellulose method).

Acid Fuchsin

This dye should not be confused with basic fuchsin, which has been discussed in an earlier section of this chapter (p. 377). Acid fuchsin has several important staining properties, and the literature of histology and cytology makes many references to it. It dissolves readily in water and alcohol. Aqueous solutions of 0.2 to 1% are commonly used for staining.

Under certain conditions, acid fuchsin is a useful cytoplasmic stain, often bringing out mitotic spindles and asters clearly. However, for this purpose, the staining solution should be slightly acid (pH about 5.6). In neutral solutions, it stains weakly and diffusely, and in the presence of bases, it will not stain at all.

Acid fuchsin is also used in several methods for staining mitochondria (see Chapter 26, p. 436).

Acid fuchsin exhibits a decided affinity for chromatin and is employed as a nuclear dye in Mallory's triple stain (p. 398). This property renders it unfavorable for use as a counterstain after alum hematoxylin, because it makes it difficult to get a suitable cytoplasmic stain without destroying the vivid blue coloration of the nuclei.

Mixtures of picric acid and acid fuchsin were once popular, and are still used occasionally for differential staining of collagenous connective tissue, muscle, and keratin. When applied to sections of embryos, picrofuchsin often produces a striking differentiation between bone and cartilage. The acid fuchsin stains the matrix of bone deep red, but not the matrix of cartilage.

The proportions of acid fuchsin and picric acid most suitable for various tissues differ. The following will in general prove to be satisfactory. It may be modified in any way which appears advisable.

Van Gieson's Picro-Fuchsin

Acid fuchsin, 1% aqueous solution	10 ml.
Picric acid, saturated solution	100 ml.

PROCEDURE FOR STAINING. *a.* Bring paraffin sections to water and overstain them at least slightly with alum hematoxylin. Overstaining is necessary because of the fact that picric acid extracts the hematoxylin.

b. Wash the sections in running water for at least 5 minutes, or until the nuclei are thoroughly blued.

c. Stain them in picro-fuchsin for 2 or 3 minutes. The time may be lengthened if examination of trial preparations suggests that this is necessary.

d. Dip the sections into 50% alcohol several times. Following this, dip them into 95% alcohol several times, then transfer them to absolute alcohol. Leave them in this about 20 or 30 seconds.

e. Clear the sections in terpineol-toluene (or carbol-toluene). Transfer them to pure toluene, and mount them in synthetic resin or balsam.

In favorable preparations, collagenous connective tissue is stained deep red; bone matrix and the enamel of teeth usually become deep brownish red; ganglion cells are pink; muscle fibers, cytoplasm of most cells, and myelin sheaths are generally some shade of dull yellow, whereas red corpuscles are bright yellow; keratinized and chitinized tissues are deep yellow; nuclei are brown; axis cylinders are reddish brown. Considerable variation in the results actually obtained with various kinds of organs or embryos fixed in different ways may be expected.

Congo Red

This dye resembles acid fuchsin in staining properties, but it fades less rapidly. As a stain for cytoplasm, following hematoxylin or preceding methylene blue, a 0.5% aqueous solution of Congo red (to 100 ml. of which add 0.2 ml. of 0.1 N HCl) may be used. However, in routine histological work eosin is generally preferable.

A distinctive and rather important property of Congo red is that its solutions turn blue in the presence of even the smallest traces of free acids. Its toxicity is low, and it is therefore a useful indicator of free acids in living or fresh tissues.

Picric Acid

This yellow crystalline substance has several uses in staining, although it is better known for its properties as a fixative. It gives a diffuse cytoplasmic stain, but provides a good background or contrast color for selective stains which color connective tissues, certain constituents of cytoplasm, or other structures. For this purpose, picric acid is commonly combined with acid fuchsin, forming van Gieson's connective tissue stain (p. 388). It is sometimes mixed to good advantage with indigo-carmine (p. 391) and other dyes. Picro-carmine (p. 347) is a useful stain for helminths and some other invertebrates.

Picric acid can also be employed to differentiate hematoxylin in both progressive staining (Conklin's picro-hematoxylin method, p. 359) and regressive staining (iron hematoxylin method, p. 366).

Aniline Blue, Water-Soluble

The dyes generally sold under this name are mixtures of at least two compounds, and commercial brands therefore differ much in composition and properties.

Water-soluble aniline blue is an excellent differential stain for collagenous connective tissue. It also brings out, although with a much less intense stain, the matrix of cartilage, mucus, amyloid, and some other hyaline substances. It is employed most commonly, and to best advantage, in connection with orange G and a red nuclear stain. This combination is used in Mallory's triple stain and some of its modifications, described in Chapter 24. Unfortunately, the intensity of this very useful and beautiful stain sometimes becomes perceptibly weakened after a few years.

Sudan III and Sudan IV

Owing to a peculiarity of molecular structure, these dyes are unable to form salts, and hence do not stain protoplasmic structures. However, they are highly soluble in lipids, which they color by simple solution. The value of Sudan III as a specific stain for lipids was described first by Daddi in 1896, and the modern method of using it in alkaline solution was published a few years later. About the same time, Sudan IV was introduced, and owing to its deeper color, it is generally preferable as a stain for lipids. Methods of staining with both Sudan III and Sudan IV are given on page 443. For obvious reasons, materials so stained must be mounted without exposure to solvents of lipids.

Thiazine Red R

This has been used, following hematoxylin, to stain muscle. It is particularly useful for striated fibers. Heidenhain recommended that sections, after being stained with alum hematoxylin or iron hematoxylin, be stained for 30 minutes in a 1% aqueous solution of thiazine red. This brings out striations quite well. A mixture of thiazine red and picric acid has been used to differentiate smooth muscle and collagen.

Light Green SF Yellowish

This is a cytoplasm stain of great precision and beauty, suitable for thin sections, and particularly good for gland cells or cells in mitosis. As a 0.1 or 0.2% solution in 70 or 80% alcohol, it gives admirable results after chromatin staining with iron hematoxylin or safranin, and follows fixatives containing chromic acid or potassium dichromate rather well. It also offers the advantage of differentiating safranin at the same time that it counterstains the extranuclear structures. For these reasons, light green has been widely used, despite the fact that it may fade rapidly. The following dye possesses much the same properties and is generally preferable.

Fast Green FCF

The characteristics of this dye closely resemble those of light green, but it is of a slightly more blue and less transparent color than the latter. It is much more permanent and it will keep well for many years if the preparations are not exposed to strong light.

In a 0.1 or 0.2% solution in 70 or 80% alcohol, fast green is an effective counterstain for various invertebrates (including hydroids and ectoprocts) stained by certain of the carmine methods, especially borax carmine and modifications which employ this mixture. It is also one of the better counterstains for sections or smears stained by the Feulgen nucleal reaction, iron hematoxylin, and safranin. Fast green is employed in modifications of Masson's trichrome stain, replacing light green as originally prescribed.

Indigo-Carmine

This dye is the sodium salt of indigo disulfonic acid. Though formerly made from products of the indigo plant, it is now manufactured synthetically. Indigo-carmine has no chemical relationship or resemblance to carmine. It can be used effectively, as a 0.1% solution in 70 or 80% alcohol, for counterstaining whole mounts of hydroids, ectoprocts, and some other invertebrates stained by certain of the carmine stains. Unfortunately, indigo-carmine sometimes fades within a few years.

Paraffin sections of histological material in which nuclei have been stained by basic fuchsin may be counterstained with beautiful polychromatic effects in picro-indigo-carmine. Leave the sections in 1% aqueous basic fuchsin until they are overstained—five to 10 minutes should be sufficient. Rinse the sections in water and place them in a saturated aqueous solution of picric acid containing 0.4% indigo-carmine. After about 5 minutes, rinse them briefly in water, and transfer them rather quickly up the alcohol series (or omit some of the steps) to 95% alcohol. The basic fuchsin, as well as the counterstain, is removed by the alcohol. When the sections appear to be properly differentiated, complete dehydration in absolute alcohol, and clear and mount the preparations in the usual way.

Azocarmine G

Azocarmine G is neither an "azo" dye nor is it related to carmine. However, it resembles carmine in color. Its properties are exceptional for an acid dye, for it is a powerful and precise chromatin stain. It is also quite permanent. It seems to give best results after fixation in fluids containing mercuric chloride, but it will stain quite well following fixation in formalin or Bouin's fluid.

Azocarmine G is only slightly soluble in cold water. For this reason, staining is carried out at a temperature of about 55° C. The overstained sections are treated with 1% aniline in 95% alcohol to differentiate the stain, and the destaining is terminated by treatment with 1% acetic acid in 95% alcohol. Azocarmine G is used chiefly in Heidenhain's "Azan" modification of Mallory's triple stain, in which it replaces acid fuchsin. Complete directions for this method are given on page 400.

Trypan Blue

The value of this dye as a vital stain was discovered in 1909 by Goldmann, a pupil of Ehrlich. Goldmann injected trypan blue and certain other acid dyes into vertebrates and observed the appearance of colored granules in various kinds of macrophages. If a 0.5% aqueous solution of trypan blue is injected subcutaneously, intraperitoneally, or intravenously into vertebrates (0.5 to 1 ml. for a mouse, proportionately larger amounts for larger animals), the stain is taken up as particles of colloidal dimensions. The entire body becomes colored within a few hours. Colored granules appear not only in macrophages, but also in endothelial cells and in the epithelial cells of the renal tubules. Heavy deposits of dye accumulate in those portions of the bones of young mammals in which osteoblasts are active. The dye can be fixed in the tissues, and Heidenhain's "Susa" fixative has been recommended for this purpose (see p. 61).

NEUTRAL STAINS (COMPOUND STAINS)

Acid and basic dyes, when mixed in solution, interchange ions in exactly the same way as do other salts, and new compounds are thus formed by chemical union of color acids with color bases. These are called *neutral stains*, or *compound stains*. They are insoluble in pure water, but dissolve in water containing an excess of either the acid or the basic dye from which they are formed.

Stains of this type were prepared by Ehrlich before the turn of the century, and he was perhaps the first biologist to understand clearly the difference between acid and basic dyes. Ehrlich's earliest and most important neutral stain mixture (the so-called "tri-acid stain") consisted of acid fuchsin, orange G, and methyl green. He found this to give a differential coloration of the nuclei and various cytoplasmic granules in blood cells. Ehrlich designated as "neutrophil" those granules showing an affinity for the neutral stain, "basophil" to the nucleus and certain cytoplasmic granules taking up free ions of the color base, and "acidophil" to granules exhibiting an affinity for the color acid. This terminology is still in use. Ehrlich's original mixtures and modifications of them have been to a large extent supplanted by mixtures of eosin and methylene blue.

Eosin-Methylene Blue Stains

These are the most familiar neutral stains, having proved to be more satisfactory and convenient than any others for routine staining of blood corpuscles, blood-forming tissues, protozoan parasites of the blood, and other subjects for which neutral stains are indicated. Eosin and methylene blue react as follows: eosin (color acid tetrabromo fluorescein—sodium) + methylene blue (color base tetramethyl thionin—chlorine)→ eosinate of methylene blue, a compound or neutral stain (color acid tetrabromo fluorescein—color base tetramethyl thionin) + sodium chloride.

This mixture was first prepared by Romanovski (Romanowsky) in 1891, and was employed in his classical work on malarial parasites. For this reason, all compound stains of eosin and methylene blue have since been classed as Romanovski-type stains. The characteristic "Romanovski effect" is a reddish-lavender coloration of nuclei in lymphocytes and monocytes, and a red nuclear stain and blue cytoplasmic stain of malarial parasites.

Romanovski noticed that compounds made from old solutions of methylene blue gave the effect described above most conspicuously. At about the same time, Unna pointed out the advantages of a polychrome methylene blue, prepared by heating a solution of the dye with sodium carbonate. This treatment oxidizes a portion of the methylene blue to form an active reddish substance, which Bernthsen (1885) had already described and named methylene azure. In 1898, Nocht recognized that an oxidation product was responsible for the red Romanovski stain of nuclei in protozoan blood parasites, and combined polychrome methylene blue with eosin. Since azure combines with eosin in the same manner as does methylene blue, the resulting solution contained the eosinates of both methylene blue and azure.

As already noted, neutral stains such as the eosinates of methylene blue and azure are insoluble in water, except in the presence of excess acid or basic dye. Ehrlich, Romanovski, and Nocht used aqueous solutions containing an excess of acid dye, but were troubled by precipitates and inconsistent effects. Jenner, in 1899, made an important advance by collecting the precipitate formed by eosin and methylene blue, and dissolving it in methyl alcohol. But he did not prepare a polychrome methylene blue. His stain lacked the essential quality of Romanovski's preparation and eventually fell into disuse.

Jenner's principle of re-dissolving the neutral stain in alcohol was soon combined with Nocht's method of making the compound from polychrome methylene blue. This was accomplished independently by Reuter and by Leishman. The modern method of blood staining employs a solution, in pure methyl alcohol, of a mixture of methylene blue, methylene azure, and the eosinates of both, which is diluted with water at the moment of staining. In 1902, Wright introduced a process of polychroming methylene blue by the action of live steam. Subsequent improvements will be considered following the description of Wright's method of staining.

Another fact of general importance must be touched upon at this point. It is that all neutral stains are greatly influenced by the hydrogen ion concentration of the solution. Results which might be considered optimum for most purposes are obtained only at pH 6.4 to 6.9. Control of pH is necessary for accurate diagnosis and reproducibility, and can easily be accomplished by use of the buffer solution which will be described presently.

Wright's Blood Stain. This stain is popular because of the rapidity and ease with which it can be used. In fact, the majority of physicians and clinical workers think of Wright's stain as *the* blood stain. In terms of quality of results and dependability, however, it is inferior to Giemsa's stain.

PREPARATION OF THE STAIN. The process of making the staining compound is rather complicated. Biologists generally buy it ready-made, either as a powder or as a solution. Certified mixtures of Wright's stain are widely used, but samples may vary considerably, and some may not be entirely satisfactory. To make the stock solution, dissolve 0.1 gm. of dry dye in 60 ml. of absolute methyl alcohol, *free from acetone or acid.* Keep this solution in a tightly stoppered bottle. If a precipitate forms, add sufficient methyl alcohol to re-dissolve it.

PROCEDURE FOR STAINING. *a.* Make blood films, or smears of bone marrow, spleen, or other tissues, in the usual way (p. 169). Allow them to dry. Stain them as soon as possible, preferably at once.

b. Lay the slide on a level surface. With a pipette, quickly flood the smear with the staining solution, noting the number of drops used. Be sure to use a sufficient amount, for otherwise a precipitate may form as the alcohol evaporates. The solution simultaneously fixes and stains the cells. Allow it to act for 1 minute.

c. With another clean pipette, add an equal number of drops of buffered water having a pH of about 6.75 (see formula on p. 395), meanwhile tilting the slide back and forth to insure rapid mixing of water and stain. Allow the diluted stain to act for 2 or 3 minutes. Eosinophil granules are often most sharply differentiated if the time of staining is brief, whereas neutrophil granules tend to show up better if a full 3 minutes is allowed.

d. Wash the preparation in buffered water for 30 seconds, or until the thinner parts of the smear appear pink or yellowish.

e. Dip the slide once into distilled water. Shake off the excess and stand the slide on end, resting it against some convenient object, until the smear is dry. To hasten drying, some workers blot the smear with smooth filter paper. If this practice is followed, it must be accomplished with great care, so that the smear is not damaged.

f. Either cover the preparations or preserve them as dry films. If they are to be covered, use a rather thin solution of synthetic resin or *neutral* balsam, and a No. 1 coverglass. Then put away the slides in a horizontal position to dry at room temperature. To examine an unmounted preparation, place a drop of immersion oil directly upon it and use an oil-immersion objective. The stain keeps best in unmounted preparations, but these are likely to be scratched. In removing oil from an unmounted preparation, do not rub it with lens paper. Instead, add some toluene or xylene to the film, apply lens paper, and draw the soaked paper off the slide. If this process is repeated once or twice with clean lens paper, the oil will be removed without damaging the film.

The characteristic staining which is to be expected with Wright's stain, and the common causes of failure, will be discussed following the paragraphs on Giemsa's stain.

Giemsa's Blood Stain. Owing to unavoidable variations in the polychroming process, Wright's stain and its predecessors are of uncertain composition, and therefore are more or less unreliable. To overcome the disadvantages of such stains, Giemsa developed a more accurately formulated blood stain. The first step in making Giemsa's stain is the preparation of methylene azure, which Giemsa called azure I, and which is a mixture of three compounds (azures A, B, and C). To this is added an equal amount of methylene blue, in order that a blue coloration of cytoplasm may be obtained, and the mixture is designated as azure II. An eosinate of azure II is then prepared. Finally, this eosinate ("azure II-eosin") is mixed with a certain proportion of azure II. Giemsa's stain consists, therefore, of rather definite proportions of methylene blue, methylene azure, and the eosinates of both. It produces consistently good results, provided it is correctly made and is applied in solutions of the proper pH.

Giemsa's stain produces more or less satisfactory results in solutions of pH 6.4 to 6.9, but gives the most striking and characteristic differentiation at pH 6.75. To obtain this optimum pH with certainty and ease, it is best to dilute the stain with a buffer solution. The following is satisfactory:

Clark and Lubs' Phosphate Buffer, pH 6.75

Potassium phosphate, monobasic, crystals, analytical reagent grade	6.8 gm.
Sodium hydroxide, analytical reagent grade	0.9 gm.
Distilled water	1000 ml.

In the absence of facilities for making the above, it is convenient to use buffer tablets or packaged buffer salts, which are calculated to provide the desired pH when mixed according to directions of the manufacturer.

BUFFERED WATER. Mix 1 part of the buffer with 5 parts of distilled water. This does not alter the pH to any great extent.

PREPARATION OF THE STAINING SOLUTION. Giemsa's stain is ordinarily kept on hand as a concentrated stock solution; a small quantity of this is diluted just before use. The dry powder for preparing the stock solution is packaged by a number of firms dealing in biological stains.

Stock Solution of Giemsa's Stain

Giemsa's stain	1 gm.
Glycerol	66 ml.
Methyl alcohol, absolute (acetone-free)	66 ml.

Disperse the powdered dye in the glycerol and place the mixture in an oven at 60° C. for 2 hours. Then add the methyl alcohol. Store the stock solution in a tightly stoppered bottle.

Staining Solution for Giemsa's Method

Giemsa's stain (stock solution)	10 ml.
Buffered water, pH 6.75	90 ml.

Prepare this mixture immediately before use. If only one or two preparations are to be stained, simply mix 1 ml. of buffer solution with 5 ml. of water and add 0.5 ml (8 to 10 drops) of concentrated Giemsa's stain.

METHOD FOR STAINING DRY BLOOD FILMS OR SMEARS. *a.* Make blood films or smears of tissue (spleen, bone marrow) or pathological exudates following the methods given in Chapter 12. Allow them to dry. The preparations should be fixed and stained as soon as possible, because their staining properties begin to decline with age.

b. Fix them in absolute methyl alcohol (acetone-free). Freshly made preparations should be flooded or immersed for 5 minutes; those which are a day old should be fixed for 2 or 3 minutes. Allow them to dry again.

c. Cover them with a freshly prepared staining solution. If only a few preparations are being stained, lay them flat and flood them with the stain. If many are being processed, immersion in staining dishes may be more convenient.

After 20 minutes, remove a slide from the staining solution, or pour the stain off it, and rinse it with *buffered water*. Examine it under the microscope, comparing it with the description given under Appearance of Preparations (p. 397). If the coloration in general is too faint, or if the red corpuscles are not bright pink, return it to the stain for another 10 minutes. Then re-examine it, again replacing it in the stain if necessary. Almost always, the cells in fresh smears will take a fine stain within 30 minutes.

To obtain the best results with old smears, or to bring out very boldly the structure of trypanosomes, it may be necessary to prolong the staining to several hours. Some workers have been troubled by precipitation of the stain after an hour or more. This seems to be a property of certain dye preparations.

d, *e*, and *f*. Wash, dry, and mount under a coverglass, or examine the dried film directly with the oil-immersion lens. (See steps *d*, *e*, and *f* under Wright's Blood Stain.)

METHOD FOR STAINING SECTIONS. This stain is valuable not only for sections of blood-forming organs, but also for bacteria, protozoa, Negri bodies, and various other elements in tissues.

a. Fix small pieces of tissue in Zenker's fluid for 6 to 18 hours, depending upon their thickness. Following the usual method, wash, harden, iodize, and embed them in paraffin. Make sections 5 μ or less in thickness and affix them to slides. Dry them, remove the paraffin with toluene or xylene, and pass the slides down through the alcohols to water. Do not coat them with Parlodion.

b. Prepare a buffered staining solution as directed on page 395. To each 100 ml., add 3 ml. of acetone-free absolute methyl alcohol, in order to retard precipitation. Stain the sections in this for 12 to 24 hours.

c. Wash them in buffered water for 30 seconds.

d. Pass them through the following mixtures, which dehydrate the sections and differentiate the stain: (1) acetone 95 ml., toluene or xylene 5 ml.; (2) acetone 70 ml., toluene or xylene 30 ml.; (3) acetone 30 ml., toluene or xylene 70 ml. The length of time which they should be left in each mixture must be determined by experiment, in order to obtain critical differentiation of the desired structures.

e. Place them in toluene or xylene. When they are fully cleared, transfer them to a second jar of the clearing agent.

f. Mount them in thin synthetic resin or neutral balsam, and allow the preparations to dry at room temperature.

METHOD FOR STAINING WET-FIXED SMEARS. Fix smears in Schaudinn's fixative, transfer them to 70% alcohol, and treat them with iodine to remove precipitates of mercury (p. 93). Pass them down through the alcohol series to water. If they still retain a brownish color, place them in 5% sodium thiosulfate to remove the iodine, and wash them again in running water. Then stain, dehydrate, and mount the smears according to the method given for sections.

APPEARANCE OF PREPARATIONS STAINED WITH GIEMSA'S OR WRIGHT'S STAIN. *Erythrocytes* (red corpuscles) are pink or yellowish pink. Nuclei of those erythrocytes which possess nuclei (lower vertebrates) are deep blue. *Lymphocytes* show dark purplish-blue nuclei and pale blue cytoplasm. *Monocytes* show large nuclei varying from medium blue to reddish lavender, and pale blue cytoplasm. *Neutrophil leukocytes* show dark blue or lilac nuclei. Their cytoplasm should have a faint bluish or lavender tint, with granules of reddish-lilac color. *Eosinophil* (*oxyphil, acidophil*) *leukocytes* show dark, lilac-colored nuclei, and bright pink granules in faint blue or lavender cytoplasm. *Basophil leukocytes* show dark blue or purple nuclei, and pale blue cytoplasm with numerous deep purple granules. *Blood platelets* are light blue with purple or violet granules in their central portions. Parasitic protozoa, such as malarial organisms and trypanosomes, generally stain the reverse of the blood corpuscles. Their nuclei, and also the kinetoplast of trypanosomes, should be red, and the cytoplasm blue. In ideal preparations of trypanosomes, the flagellum is clearly differentiated. Some spirochaetes, including *Treponema pallidum*, are stained red. Most spirochaetes, however, and nearly all bacteria are stained blue.

CAUSES OF UNSATISFACTORY STAINING WITH WRIGHT'S AND GIEMSA'S STAINS. If the red corpuscles stain blue-gray, and the white corpuscles show only blue, the stain is probably too alkaline. If the red corpuscles are bright pink and the white corpuscles stain lightly or not at all, the stain is almost certainly too acid. Use the buffer solution which has been recommended.

REFERENCES

BAKER, J. R., 1958. *Principles of Biological Microtechnique.* New York: Barnes & Noble, Inc.

CONN, H. J. (Editor), 1969. *Biological Stains*, 8th ed., revised by R. D. Lillie. Baltimore: The Williams & Wilkins Co.

GURR, E., 1960. *Encyclopaedia of Microscopic Stains.* Baltimore: The Williams & Wilkins Co.

CHAPTER 24

Combinations of Dyes

To differentiate a variety of histological elements in sections of organs, embryos, or small organisms, two or more dyes of contrasting colors and rather specific action may be combined. In the case of aniline dyes, this is ordinarily accomplished by using a basic dye to stain nuclei, and one or more acid dyes which will counterstain cytoplasm and bring out connective tissue, the matrix of cartilage, mucin, and other constituents. This is not always the case, however. Acid fuchsin is an effective stain for nuclei, and crystal violet and some other basic dyes can be used for staining cytoplasm.

A great many combinations of aniline dyes fulfill the requirements of double or triple staining very satisfactorily. Some aniline dyes unfortunately fade rapidly, and others are useful only for differentiating certain structures (chiefly of a granular nature) which are of interest in special studies. The following well-known combinations will serve the majority of purposes. Intelligent use of these methods should provide the laboratory worker with an understanding of the *rationale* of double and multiple staining, and so enable him to master other combinations or to invent new ones as the need arises.

The uses of hematoxylin and carmine nuclear stains with certain contrasting counterstains are discussed in connection with dyes of these two categories (Chapters 21 and 22). Mixtures of eosin with methylene blue and its derivatives, as used in blood staining, are considered in Chapter 23.

MALLORY'S TRIPLE STAIN FOR CONNECTIVE TISSUE AND OTHER HISTOLOGICAL ELEMENTS

This beautiful combination of brilliant dyes was designed chiefly as a differential stain for collagenous connective tissue and reticulum. Mallory's method is also excellent for demonstrating the general topography of tissues in sections of various organs, advanced embryos, and some small organisms. Fibrous connective tissue, mucin, and matrix of hyaline cartilage are stained deep blue; erythrocytes are stained orange, and myelin sheaths yellow; the cytoplasm of dense cellular tissue (such as liver and kidney) is stained pink, and the nuclei should be red. Muscle, axis cylinders, fibrin, neuroglia, and bone matrix become various shades of red or deep pink. The method is not suitable for the demonstration of finer details of cell structure.

Preparations stained by Mallory's method usually show no sign of fading for a number of years. The acid fuchsin is generally the first of the dyes to weaken. After another few years, the aniline blue may begin to fade. The orange G seems to be the most durable dye in the combination. The substitution of more permanent nuclear stains for acid fuchsin will be discussed later.

Zenker's fluid is known to be very good for fixation of material to be stained by this technique. Mallory regarded the use of Zenker's fluid as essential, but with some types of material good results are obtained after fixation in 10% formalin or Heidenhain's "Susa" fixative. Fair to good results may be obtained after fixation in Bouin's fluid.

For routine histological work, sections may be prepared by either the paraffin method or nitrocellulose method, but as a rule they should be not more than 12 μ thick.

PROCEDURE. *a.* Bring sections gradually to distilled water.

b. Stain for 5 to 30 minutes in a 0.2% aqueous solution of acid fuchsin ("Solution I").

c. Rinse the sections briefly in water, and place them in the following solution:

Mallory's Aniline Blue-Orange G Mixture ("Solution II")

Aniline blue, water-soluble	0.5 gm.
Orange G.	2 gm.
Oxalic acid	2 gm.
Distilled water	100 ml.

Leave the sections in this mixture for 10 to 20 minutes. The shorter period generally yields more transparent preparations.

One of the functions served by oxalic acid is the exclusion of acid fuchsin from connective tissue. At the same time, it restricts the aniline blue to connective tissue, mucin, matrix of hyaline cartilage, and a few other elements.

d. Rinse the sections in water and transfer them to 95% alcohol. Passing them through the series of lower alcohols should be avoided, because too much stain is likely to be extracted before the preparations are finally dehydrated and cleared. Keep the sections in constant motion while they are in 95% alcohol.

e. When the excess dye has ceased to come out of the sections in clouds, transfer paraffin sections to absolute alcohol for at least a minute or two, then clear them in terpineol-toluene or carbol-toluene. No more dye will be extracted after the sections are out of absolute alcohol.

In the case of nitrocellulose sections, differentiation is completed in 95% alcohol, and the sections are then thoroughly cleared in terpineol-toluene or carbol-toluene.

f. Transfer the sections through one or two changes of toluene, and mount them in neutral balsam or a synthetic resin. After some experience, recognition of the color of properly differentiated sections is learned. In case of doubt, they can be examined with the microscope while they are in toluene, provided this is done with speed so that the sections are not in danger of drying out. Generally, the extraction of aniline blue proceeds so rapidly that it may not be advisable to coat the sections with Parlodion between dehydration in absolute alcohol and clearing.

Numerous minor alterations have been made in the method devised by Mallory, usually with the idea of improving the sharpness of contrast between structures having an affinity for acid fuchsin and those having an affinity for aniline blue. One popular modification involves the use of 1 gm. of phosphomolybdic acid or phosphotungstic acid in place of oxalic acid in the aniline blue-orange G mixture. Another requires thorough rinsing in a 1% solution of phosphomolybdic acid or phosphotungstic acid after staining with acid fuchsin and before staining with the aniline blue-orange G mixture. If this procedure is followed, the inclusion of phosphomolybdic acid, phosphotungstic acid, or oxalic acid in the second staining solution is unnecessary.

Heidenhain's "Azan" Modification of Mallory's Triple Stain

The substitution of azocarmine G for acid fuchsin will generally yield a more intense and very permanent nuclear stain. In preparations stained by this modification, the nuclei are a brighter red color than in sections stained with acid fuchsin, and the contrast between the nucleus and cytoplasm is sharper. Erythrocytes tend to be stained red, due to masking of the orange G by azocarmine G.

PROCEDURE. *a.* Bring de-paraffined sections, or nitrocellulose sections, gradually to distilled water.

b. Stain for 1 hour, at 55° C., in a solution of azocarmine G prepared as follows: Dissolve 1 gm. of azocarmine G in 100 ml. of boiling distilled water. Cool to room temperature and add 1 ml. of glacial acetic acid. If the solution is warmed to 55° C. before the sections are put into it, the time required for staining can often be reduced to 30 minutes.

c. Allow the staining solution containing the sections to cool down to room temperature.

d. Differentiate the azocarmine G in a 0.1% solution of aniline in 95% alcohol. (*Caution: Aniline has been demonstrated to be carcinogenic, so it should be handled with care.*) The nuclei should be bright red, and the cytoplasm pale pink. Muscle tissue should be red.

e. Rinse sections in 1% acetic acid in 95% alcohol for about 1 minute.

f. Transfer sections to a 5% aqueous solution of phosphotungstic acid for 1 to 3 hours. The connective tissue should be decolorized by the end of this time.

g. Rinse them briefly in distilled water.

h. Stain sections for 15 minutes to an hour in an aniline blue-orange G mixture prepared according to the following formula:

Aniline blue, water-soluble	0.5 gm.
Orange G	2 gm.
Acetic acid, glacial	8 ml.
Distilled water	100 ml.

i. Rinse them briefly in distilled water, then differentiate the aniline blue in 95% alcohol (and finally 100% alcohol, in the case of paraffin sections). Clear sections in terpineol-toluene or carbol-toluene, then in toluene, and mount them in synthetic resin or neutral balsam.

MASSON'S TRICHROME STAIN (MODIFIED)

Masson's original method of staining actually involved four separate dyes, but two of them (acid fuchsin and ponceau de xylidine) were red. The tissue was first stained with hematoxylin, then with a mixture of the two red dyes, and finally with light green. The exact identity of ponceau de xylidine is not known; perhaps it was similar to what is now sold as ponceau 2R. In any case, Masson's procedure has been extensively modified, partly because of the desirability of finding a substitute for ponceau de xylidine. Ponceau 2R seems not to give completely satisfactory results.

Lillie's modification, described here, is quite different from the original method. Weigert's hematoxylin is used in place of iron hematoxylin, and acid fuchsin and ponceau de xylidine are replaced by Biebrich scarlet.

Tissues to be stained by this technique may be fixed in Zenker's fluid, Bouin's fluid, 10% formalin, or a number of other fixatives. The method is applicable to paraffin or nitrocellulose sections.

PROCEDURE. *a.* Treat sections with a saturated solution (about 7%) of picric acid in 95% alcohol.

b. Pass them down the graded series of alcohols to water, and wash them in running water for 2 or 3 minutes.

c. Stain sections for about 10 minutes in Weigert's iron hematoxylin.

d. Wash them in running water.

e. Stain them for about 5 minutes in 1% Biebrich scarlet in a 1% aqueous solution of acetic acid.

f. Rinse sections in water.

g. Leave the sections for 1 or 2 minutes in the following mixture:

Phosphomolybdic acid	5 gm.
Phosphotungstic acid	5 gm.
Distilled water	200 ml.

h. Transfer sections directly to 2.5% fast green FCF (or aniline blue, water-soluble) in a 2.5% aqueous solution of acetic acid. Stain them for about 5 minutes.

i. Differentiate sections in 1% acetic acid to remove excess fast green (or aniline blue).

j. Dehydrate, clear, and mount them in a synthetic resin or neutral balsam, following procedures suited to paraffin or nitrocellulose sections.

In good preparations made by this method, nuclei are black or blue-black, muscle is red, erythrocytes are scarlet, collagen and mucus are green (or blue, if aniline blue has been substituted for fast green FCF). Myelinated fibers are stained red or pink. The cytoplasm of most cells should show a pleasing pink counterstain to the sharp nuclear stain.

CHAPTER 25

Metallic Impregnation

Salts of silver, gold, mercury, and osmium are employed to produce color differentiation in certain structures, especially elements of the nervous system. Staining is accomplished by precipitation of metallic compounds upon or within cell boundaries, thereby causing them to stand out boldly against a light background. Ordinarily, the process is one of chemical reduction, in which the metal is precipitated in opaque masses. For this reason, coloration by metals is perhaps better designated as *impregnation*, to distinguish it from true staining, in which the coloring matter is retained by the structures in a finely divided state resembling a solution.

Metallic impregnations are of great importance in the study of nerve fibers, nerve endings, and neuroglia. Much of our present-day knowledge of the finer structure of the nervous system has been made possible by methods of silver impregnation, supplemented by methods of impregnation with gold and mercury. Silver impregnation is also important for demonstrating cell outlines, reticular connective tissue, Golgi material, spirochaetes of syphilis, and the structure of some protozoa. Impregnation with osmium is useful for demonstrating lipids and Golgi material. Some techniques utilizing osmium tetroxide will be described in Chapter 26.

IMPREGNATION WITH SILVER

Method for Outlining Cells with Silver Nitrate

The reduction of silver nitrate by intercellular substances, leading to formation of a brown or black deposit, sometimes brings out cell outlines very sharply. This method of demonstrating cells may be applied with special effectiveness to thin membranes, such as the peritoneum, in which the cells of the epithelium as well as those of the endothelia lining the smaller blood vessels and lymphatics are outlined. It is also useful for some other kinds of cells, such as those of the cornea.

Impregnations of this type are classified as "negative," because most of the metal is deposited in intercellular substances. The method is not always successful, but when it is, the results are better than those which can be achieved with other methods.

Impregnation of Cells in Thin Membranes. 1. Stretch the membrane tightly over the end of a shell vial from which the bottom has been removed, or any short tube of glass or plastic, and tie it in place with a thread. Do not allow the membrane to dry out. The mesentery of a frog is very favorable material upon which to practice.

2. Rinse it gently with distilled water to remove blood and other proteinaceous material which is likely to become heavily impregnated.

3. Place it in a 2% aqueous solution of silver nitrate, in darkness, and leave it for 1 minute to 1 hour. The optimum time of exposure varies. In the case of frog mesentery, 10 minutes is usually an appropriate length of time. The concentration of silver nitrate may also be varied from about 0.2 to about 3%.

4. Place the membrane in distilled water, in direct sunlight, until it becomes reddish brown and inspection with a lens shows the cell outlines to be well differentiated. On a bright day this is likely to take place in about 10 to 15 minutes.

5. Rinse it with distilled water, and place it, for about 30 seconds, in a 2% aqueous solution of sodium thiosulfate, to remove all unreduced silver and prevent after-blackening. Wash it in running water for 30 minutes.

6. Place it in dilute alum hematoxylin (p. 353) for 20 minutes, or until the nuclei are well stained. Wash it in running water for 15 minutes.

7. If the result is satisfactory, pass the membrane up the alcohol series to 70 or 95% alcohol. Then remove it from the vial and complete the dehydration. Clear it in toluene or terpineol-toluene and cut it into pieces. Mount these in a resinous medium. The cells should be outlined as described in the introductory paragraph on this method.

Impregnation of Ciliates by Silver Nitrate Methods

Two techniques for impregnating ciliates with silver nitrate are in common use. One of these, developed by Klein, involves drying the ciliates on a slide, placing them in silver nitrate, and reducing the silver in light. The ciliates are distorted to a considerable extent by drying, and that portion of the body surface which adheres to the slide is not impregnated, so that it is not possible to get a composite picture of any one specimen. However, the impregnation of kinetosomes and associated fibrillar elements in those areas of the surface which do show is often remarkably clear.

The other method requires the fixation of ciliates in Da Fano's fluid, although preliminary fixation in Champy's fluid is commonly practiced, because it preserves the form of the ciliates more satisfactorily. Following fixation, the ciliates are enrobed in gelatin made saline by addition of sodium chloride, and then exposed to silver nitrate. This technique, when it works particularly well, enables impregnation of all surfaces uniformly. The results obtained by this method are otherwise similar to those obtained by impregnation of dried ciliates.

For impregnating cilia and demonstrating the arrangement of membranelles or segments of adoral ciliary rows which enter into deep buccal cavities, the Protargol method (p. 419) is generally superior. The use of a silver nitrate method to supplement the Protargol method, or vice versa, is recommended whenever a very complete understanding of the structure of a ciliate is the objective.

Impregnation of Ciliates in Dry Films. This method is applicable to ciliates in fresh water or to symbiotic forms living in fresh-water and land animals. If the ciliates have been grown in a culture medium containing yeast extract, proteose-peptone, or some similar nutrient, almost all of this should be washed out by centrifuging the ciliates and pipetting them into filtered pond water, aged tap water, or distilled water. By centrifuging them again, a concentrate of washed ciliates may be taken from the bottom of the tube with a fine pipette, and a small drop placed on a slide. Symbiotic ciliates should be liberated from a small piece of the organ or tissue in which they live into a drop of aged tap water or distilled water on a slide. Do not use a saline solution.

The drop of concentrated washed ciliates or the drop containing tissue from a host animal should be quickly spread with the tips of very clean forceps until it is so thin that it dries within about a minute or two. The formation of dense aggregates of ciliates, which do not impregnate well, may be discouraged by tilting the slide back and forth as it dries or by stirring those portions which are still wet with a clean needle. If the preparation contains a large number of ciliates, but is too wet to dry quickly, tilt the slide back and forth a few times to distribute them evenly; then, while it is slanted or vertical, touch the edge of a piece of absorbent paper to the excess water which accumulates on the lower side of the smear.

Drying may be accelerated by using an electric fan to keep the air in motion, but heat must not be applied.

After the smears are dry, place the slides directly into a 2% aqueous solution of silver nitrate. They may be exposed to direct sunlight or diffuse daylight immediately after they are put into the silver nitrate, or left in this in the dark for a time and then exposed to light. Sometimes a selective impregnation is achieved most successfully by leaving the smears in the silver nitrate for at least a half-hour, then transferring them to distilled water and exposing them to light. With certain types of ciliates, any of the alternatives outlined gives results which are good for at least some purposes. The smears, as a whole, turn brown during reduction. Examination of the preparations with a low-power objective may enable one to decide whether reduction should be terminated. (If the slide has silver nitrate on it, protect the microscope from contact with it!) Following reduction, the smears are washed thoroughly and allowed to dry, standing on end.

It appears that the concentration of salts and the nature of the proteinaceous material in the smear may be more important in obtaining a good impregnation than the intensity of light used to reduce the silver nitrate. Ordinarily, material in pond water, or from the tissue of an invertebrate which has been comminuted in a drop of water before smearing, impregnates well. Ciliates cultured in rich organic media will yield very poor preparations if they are not washed sufficiently, for they become coated by extraneous material which may impregnate heavily. If, on the other hand, they are separated from nearly all of the salts and organic material in the culture, they may not impregnate at all.

In a good preparation, the kinetosomes and associated fibrillar elements of the surface are black or dark brown, and contrast sharply against a nearly colorless background.

Preparations made by this method are normally studied with an oil-immersion lens, the oil being applied directly to the smear. The oil is removed by dipping the slide in toluene and then drawing a lens paper across it. In smears mounted in a synthetic resin or balsam, the impregnation is likely to deteriorate within a few weeks.

Impregnation by Chatton and Lwoff's Method. 1. Fix the ciliates in Da Fano's fluid (p. 441), or first in Champy's fluid for a few minutes and then in Da Fano's fluid. If the latter procedure is followed, it will be a good idea to change the second fixative once, so that it will be nearly free

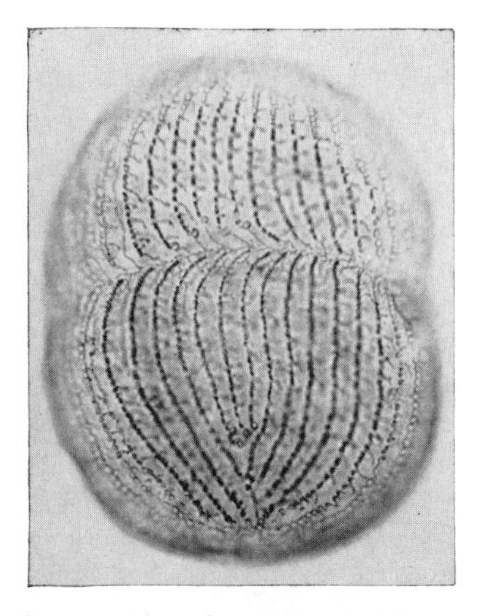

FIG. 68. Silver nitrate impregnation of a ciliate, *Tetrahymena rostrata* (in division). Fixed in Champy's fluid, then in Da Fano's fluid, impregnated by Chatton and Lwoff's method.

of the ingredients of Champy's fluid. Leave the ciliates in Da Fano's fixative for at least an hour.

2. With a fine pipette, transfer a number of the ciliates from Da Fano's fixative to a small drop of the following mixture, warmed on a coverglass to about 35 or 40° C.

Saline Gelatin

Gelatin (powdered)	10 gm.
Sodium chloride	0.05 gm.
Distilled water	100 ml.

Heat and stir the mixture until the gelatin is dissolved. Store the stock in a refrigerator when it is not in use. Even in a refrigerator, however, it will soon deteriorate due to bacterial activity.

Distribute the ciliates through the saline gelatin, and spread this out so that it forms a very thin layer on the coverglass. Allow the gelatin to solidify, preferably by placing the coverglass in a moist chamber (a Petri dish with damp filter paper on the bottom) in a refrigerator.

3. As soon as the gelatin has congealed, place the coverglass in a 2 or 3% aqueous solution of silver nitrate, keeping this cold enough (5 to 10° C.) so that the gelatin has no tendency to soften. It is generally best to keep the preparations in the dark for 10 minutes or longer to insure that no premature reduction of silver will take place. Following this, they should be rinsed in cold distilled water, then placed in a dish of cold distilled water in sunlight or in front of a strong sun lamp. The reduction may be carried out in a Petri dish over a piece of white paper, in a porcelain evaporating dish, or in a Columbia dish. (A white background may be necessary to encourage reduction if the intensity of the light is weak.) If there is any chance that the gelatin may melt because the air temperature is above about 28° C., or because the light has a warming effect, it will be best to keep the dishes containing the preparations in a pan of water cooled by ice.

4. When inspection of the gelatin films or individual ciliates suggests that reduction has progressed to an optimum point, the preparations should be dehydrated gradually and thoroughly, cleared in toluene, and mounted in a synthetic resin or neutral balsam.

The process of enrobing the ciliates in a drop of saline gelatin and subsequent handling may of course be carried out on microscope slides as well as coverglasses.

In the case of certain symbiotic ciliates living in organs which yield proteinaceous material when smears are prepared on slides and coverglasses, fixation may be carried out by dropping the smears face down on the surface of the primary fixative, whether it be Champy's fluid or Da Fano's fluid. After the preparations have remained in Da Fano's fluid for an hour or more, enough saline gelatin may be added to make a very thin layer over the smears. As a rule, the impregnation of individual ciliates in such smears is not as good on one side as on the other.

In the case of some types of material which may be handled as smears on coverglasses, very precise (although not often uniform) reductions of silver may be obtained without the use of the saline gelatin. If gelatin is of no help, it should be omitted, because it makes the preparations thicker and decreases their transparency.

Golgi's Chrome-Silver Methods

In 1875, Golgi described a method of coloring nerve cells by treating pieces of the central nervous system with a solution of potassium dichromate and then placing them in a solution of silver nitrate, in the dark. In the next 20 years, Golgi and others worked out numerous variations of this method, and our knowledge of the nervous system and certain other structures was advanced considerably by the use of these techniques. Chrome-silver methods have been regarded as the classical procedures for demonstrating nerve cells and their processes, neuroglia, bile capillaries, and the finer ducts of various glands.

An important advantage of chrome-silver methods is that they color only certain of the cells present, thus giving very bold and complete images which are not complicated by the presence of other stained structures. However, the reaction involved is so delicate that a different schedule must be worked out for each type of material, and even then the results are more or less uncertain. Another disadvantage is the frequent formation of precipitates which may obscure the structures or constitute misleading artifacts.

A large number of modifications of Golgi's methods have been made in order to adapt them to different types of material. One series of modifications involves hardening the material in formalin before placing it in the solution of potassium dichromate. Some of the more important modifications, known as Golgi's "rapid methods," require hardening the tissue in a mixture of potassium dichromate and osmium tetroxide. The following version of a rapid process has been selected for description because it is easy to carry out and is likely to give good results on a variety of materials. After learning the principles of the method, as outlined in this procedure, anyone who is seriously interested in the study of the nervous system should read carefully the extensive discussion of chrome-silver methods given in specialized works. Experimentation can then be started with the variations which seem most likely to apply to the material to be studied.

Golgi's Rapid Method. 1. With a sharp razor blade, cut slices (2 to 3 mm. thick) from the cerebral cortex, cerebellum, medulla, and spinal cord of a freshly killed animal. Make about five slices from each region. Place them in about 20 times their volume of the following solution, covering the bottom of the container with cotton in order that the fluid may reach the tissue from all sides.

Golgi's Rapid Hardening Fluid

Potassium dichromate, 3% aqueous solution	80 ml.
Osmium tetroxide, 1% aqueous solution	20 ml.

Renew the hardening fluid in case it becomes turbid. The length of time it is allowed to act upon the tissues determines to a large extent which structures will subsequently be impregnated. At the end of 2 days, remove one of the slices taken from each organ and proceed to the following steps. Do likewise at the end of the third, fourth, fifth, and seventh days.

2. Lay the pieces upon filter paper for a few seconds, to drain off the hardening solution. Place them in a glass container of ample size, pour over them a small amount of a 0.75% solution of silver nitrate in distilled water, and move the container back and forth until the production of a brown precipitate has ceased. Pour off the liquid and replace it with fresh silver nitrate solution of the same concentration, in a quantity equal to about 40 times the bulk of tissue. Keep the container of material in darkness for at least 2 days, renewing the silver nitrate solution if it becomes yellow. Although the impregnation is complete within 2 days, tissues may remain safely for weeks or months in the silver nitrate solution, provided this is kept in the dark.

3. Drain the pieces of tissue on filter paper and place them in 80% alcohol, renewing this at the end of 1, 2, and 3 hours.

4. Place them in 95% alcohol for 4 hours, then in two changes of absolute alcohol for 6 hours each.

5. Replace the absolute alcohol with ether-alcohol and leave the material in this for 4 hours.

6. Infiltrate the tissue with nitrocellulose, allowing it to stand in each of the progressively more concentrated solutions for about 12 hours.

7. Embed the tissue in nitrocellulose. Pieces of cerebrum and cerebellum should be oriented in such a way that they may be cut perpendicular to the surfaces of these organs; pieces of spinal cord and medulla should be oriented for transverse sections. Harden the block in chloroform for 2 hours, then transfer it to 70% alcohol, changing the latter 3 times at intervals of 20 or 30 minutes.

8. Cut sections 50 to 100 μ in thickness, in the planes suggested in step 7. While cutting, keep the knife wet with 70% alcohol. Place the sections, as they are cut, into 95% alcohol. When you have finished cutting, renew the alcohol on the sections and allow them to stand for about 5 to 10 minutes.

9. Float the sections, a few at a time, on terpineol-toluene (or carbol-toluene or creosote-toluene). They will sink as they are cleared.

10. Select sections showing a successful impregnation. (Sections from the outer layers of the block are likely to be over-impregnated and to contain precipitates, while those from the center are likely to be under-impregnated.) Immerse each section in toluene for 1 or 2 minutes. Then lay it upon a clean slide, and flatten it if necessary by blotting it gently with filter paper. Immediately place upon it a drop of a *synthetic resin* and apply a coverglass.[1]

Ramón y Cajal's Methods for Neurofibrils

In 1903, Ramón y Cajal published descriptions of some methods for reducing silver nitrate to demonstrate neurofibrils, the finer ramifications of nerve fibers, and nerve endings. Eventually, he described several other methods, each with variants, to meet the conditions presented by different types of material. The following two procedures will serve most purposes satisfactorily.

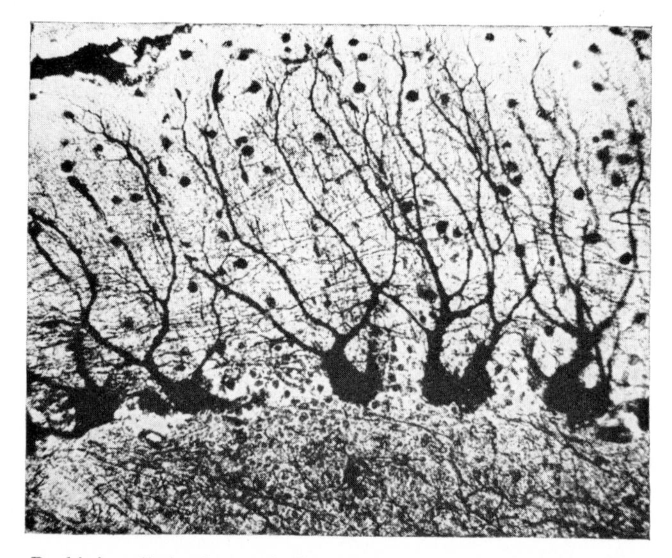

FIG. 69. Purkinje cells in the cerebellar cortex. From a sagittal section of the cerebellum of a cat, prepared by Ramón y Cajal's reduced silver method; paraffin section 16 μ thick.

[1] When balsam is used for mounting the sections, fading will generally occur if a coverglass is applied. For this reason, most workers employing balsam as a mounting medium have simply allowed the balsam to harden in a very thin layer over the sections. If this procedure is followed, it is of course important to keep the slides in a dust-free place during the time the balsam is hardening. The synthetic resins in common use do not seem to have an unfavorable effect on silver deposited by the Golgi method.

Ramón y Cajal's Method for Adult Cerebrum, Cerebellum, Spinal Cord, and Medulla. 1. With a sharp razor blade, cut slices of tissue perpendicular to the surface of the organ. These slices should be about 5 mm. thick. Put them into 70% alcohol for 6 hours.

2. Transfer them to 80% alcohol and leave them for 6 hours.

3. Place them for 24 to 36 hours in ammoniacal alcohol. For cerebrum, cerebellum, spinal cord, or ganglia, prepare this solution by adding 4 drops of concentrated ammonium hydroxide to 50 ml. of 95% alcohol; for the medulla, use 9 drops of ammonium hydroxide.

4. Drain the slices on filter paper, and place them in a relatively large volume of a 1.5% aqueous solution of silver nitrate. Cover the container and place it, for 5 days, in an incubator maintaining a temperature of about 37° C. The material should not be exposed to light during this time.

5. Rinse the slices of tissue with distilled water and place them in the following fluid for 24 hours:

Ramón y Cajal's Reducing Fluid

Pyrogallic acid (or hydroquinone)	1.5 gm.
Sodium sulfite, anhydrous	0.2 gm.
Distilled water	100 ml.
Formalin (neutral)	6 ml.

Dissolve the sodium sulfite completely, then add and dissolve the pyrogallic acid (or hydroquinone). Add the formalin last.

6. Rinse them with several changes of distilled water, over a period of at least an hour. Then dehydrate and embed them in paraffin or nitrocellulose by the usual methods.

7. Make sections perpendicular to the surface of the organ, and about 14 μ in thickness. Affix paraffin sections to slides, dry them, remove the paraffin with toluene, and mount them in a synthetic resin or neutral balsam, under coverglasses. Dehydrate, clear, and mount nitrocellulose sections in the customary manner.

Neurofibrils should appear dark brown or black, and cytoplasm very light brown; nuclei are usually unstained; the background in general is normally pale yellow. If the impregnation is unusually heavy, useful preparations may perhaps be salvaged by cutting the sections thinner. Sometimes the cells are too heavily impregnated to show the neurofibrils, and yet the background is very pale. In this case, do not discard the material summarily, because thick sections (16 to 18 μ) may turn out to be good for showing the general structure of the nerve cells and their processes.

Modifications of Ramón y Cajal's Methods Adapted to Impregnating Sections. The advantages of impregnating sections, rather than blocks of tissue, are the same with silver-reduction methods as with staining techniques. The impregnation process can be controlled so as to give more uniform results, and sections from the same specimen can be impregnated or stained by different and complementary methods. A modification of one of Ramón y Cajal's methods is given here.

1. Fix small pieces of tissue in 10% formalin, or some other fixative which contains formalin but not heavy metals.

2. Wash out the fixative in the appropriate manner. Dehydrate the tissue, infiltrate it with nitrocellulose, and embed and section it in the usual way.

3. Impregnate sections (keeping them separated and in darkness) in the following solution for 4 to 24 hours at about 37° C.

Acidified Alcoholic Silver Nitrate Solution

Silver nitrate	10 gm.
Distilled water	10 ml.
Alcohol (95% or absolute)	90 ml.
Nitric acid, normal (nitric acid, specific gravity 1.42, 6.3 ml.; distilled water, 93.7 ml.)	0.5 ml.

Dissolve the silver nitrate in the distilled water before adding the alcohol and normal nitric acid.

The sections may be transferred to the impregnating solution from 70 or 95% alcohol. They should gradually become brown in color.

4. Rinse them in distilled water.

5. Reduce the silver nitrate in a 2 to 5% solution of pyrogallic acid in 95% alcohol. Add 2 or 3 ml. of formalin to each 100 ml. of the reducing solution.

As reduction proceeds, observe the sections in a shallow dish under the microscope. Keep them in motion to discourage formation of objectionable precipitates. Change the reducing solution several times.

6. Wash the sections in 95% alcohol, clear them in terpineol-toluene, and mount them in a synthetic resin.

Nerve fibers should be light brown to almost black, while the background should be yellowish and transparent. The method is suitable for demonstrating myelinated nerve fibers and tracts in the central nervous system, but is generally not appropriate for bringing out fine unmyelinated fibers or nerve endings.

Ammoniacal Silver Methods

These techniques depend for success upon the following reactions: When silver nitrate and sodium hydroxide are mixed, a precipitate is formed. If ammonium hydroxide is added, the precipitate is converted into a soluble salt of silver, and this may be reduced by formalin. Originally, the ammoniacal silver techniques were worked out by Bielschowsky to demonstrate neurofibrils. However, his methods bring out nerve endings very well, and have also been used for showing the fine collagenous fibers forming the reticulum of various organs.

As a general rule, Bielschowsky's methods for sections are more successful than those intended for impregnation of blocks of tissue before sectioning.

PRECAUTIONS APPLYING TO AMMONIACAL SILVER METHODS. The glassware used must be chemically clean. Keep the silver solution in a glass-stoppered bottle and prepare the ammoniacal silver in a glass-stoppered graduate reserved exclusively for this purpose. The distilled water should be free of suspended organic matter. It can be tested by adding a few drops of silver nitrate to several milliliters of water. If no precipitate or color change is noticeable in 10 or 15 minutes, the water is satisfactory. It is important to use only the purest chemicals obtainable, and to employ a high quality, washed filter paper. In handling loose sections, use glass rods or forceps with paraffin-coated tips.

Bielschowsky's Methods for Neurofibrils and Nerve Endings in Sections. 1. Fix slices of brain, ganglia, peripheral nerves, or tissue containing nerve endings, in 10 or 20% formalin. The slices should be not more than about 5 mm. thick. Leave them in the fixative at least 3 weeks; the time may advantageously be increased to 2 months.

2. Wash them in running water for 18 to 24 hours, then place them in distilled water for one to several hours.

3. Make sections as thin as possible, by the freezing method. Collect them in distilled water. Change this 3 times, at intervals of about 30 minutes.

4. Place the sections in pure pyridine for 24 to 48 hours. This treatment insures a sharp staining of axis cylinders. (However, better impregnation of intracellular fibrils may be obtained by omitting this step.) Wash the sections in repeated changes of distilled water, to remove all traces of pyridine.

5. Place them in a 2 or 3% aqueous solution of silver nitrate, in darkness and at room temperature, for 24 hours.

6. Prepare the following solution immediately before use.

Bielschowsky's Ammoniacal Silver Solution

To 10 ml. of a 10% aqueous solution of silver nitrate in a measuring cylinder, add 5 drops of a 40% aqueous solution of sodium hydroxide. Next add ammonium hydroxide, drop by drop, shaking the cylinder after each addition, until all but a few grains of the precipitate are dissolved. This should not require more than 15 to 20 drops of ammonium hydroxide. Then add distilled water to bring the total volume up to 25 ml. Filter the solution into a clean staining dish.

7. Take the sections, one by one, from the silver nitrate bath. Dip them into distilled water, and place them in the ammoniacal silver solution. Leave them in this until the sections become deep brown in color, which should occur in 10 to 20 minutes.

8. Wash them, for a few seconds, in each of two changes of distilled water.

9. Place them in an ample quantity of 20% neutral formalin for 1 hour, in order to reduce the silver. Then wash them in tap water for 15 minutes.

10. Rinse them in distilled water and tone them for 10 to 20 minutes in a dilute solution of gold chloride (5 drops of a 1% aqueous solution of gold chloride and 10 ml. of distilled water; add 3 drops of glacial acetic acid if a purplish background is desired, or 1 drop of a saturated aqueous solution of lithium carbonate if a grayish-white background is preferred).

11. Rinse the sections with distilled water and leave them in a 5% aqueous solution of sodium thiosulfate for 1 or 2 minutes.

12. Wash them in several changes of tap water, over a period of 1 or 2 hours.

13. Place the sections in 50% alcohol. Float one or more sections onto a slide, being careful to eliminate wrinkles. Flood them with absolute alcohol, then with terpineol-toluene or carbol-toluene. When they have been cleared completely, flood them with toluene. Drain them, add a resinous mounting medium, and apply a coverglass.

In successful preparations, axis cylinders are sharply impregnated, appearing as bundles of fibers. Intracellular neurofibrils are visible. Terminal fibers, end-plates, and muscle-tendon spindles are colored deeply.

Foot's Modification of Bielschowsky's Method for Reticulum. Certain modifications of Bielschowsky's method demonstrate very clearly and completely the reticulum of fine connective tissue fibers in various organs. Foot's modification is widely used because of its convenience for routine work.

1. Fix slices of tissue in Zenker's fluid, Helly's fluid, or 10% formalin. Wash and dehydrate them (but do not treat them with iodine) and embed them in paraffin. Make sections about 10 μ in thickness and affix them to slides. After drying the preparations, pass them through toluene into absolute alcohol.

2. Place them, for 5 minutes, in 0.5% iodine in 95% alcohol.

3. Rinse the sections in water and place them in a 0.5% aqueous solution of sodium thiosulfate until all iodine is removed from them and they become white.

4. Wash them in tap water for 5 minutes, then place them in a 0.25% aqueous solution of potassium permanganate for 5 minutes.

5. Wash out the potassium permanganate in water and transfer the sections to a 5% aqueous solution of oxalic acid for 15 or 20 minutes. The brown color conferred upon the sections by potassium permanganate will be removed.

6. Wash the sections in running tap water for 5 minutes, and rinse them in distilled water.

7. Place them in a 2% aqueous solution of silver nitrate, and leave them in this for 48 hours, in subdued light.

8. Prepare the ammoniacal silver solution as follows, making a fresh lot for each day's work: To 20 ml. of a 10% aqueous solution of silver nitrate, add 20 drops of a 40% aqueous solution of sodium hydroxide. Then cautiously add ammonium hydroxide until all but a few grains of the brown precipitate are dissolved. Add distilled water to bring the solution up to a volume of 80 ml., and filter it.

Remove the slides from the silver nitrate solution, rinse them in distilled water for about 30 seconds, and place them in the ammoniacal silver solution for 30 minutes.

9. Rinse them in distilled water and place them in 5% neutral formalin for 30 minutes, to reduce the ammoniacal silver.

10. Rinse them in tap water and tone them in a 0.1% aqueous solution of gold chloride until the reticulum stands out sharply against a light gray background.

11. Rinse the sections in tap water and place them, for 5 minutes, in a 5% aqueous solution of sodium thiosulfate.

12. Wash them in running water for 1 or 2 hours.

13. Stain the sections progressively in dilute alum hematoxylin (p. 353); this will require approximately 20 minutes. Wash them in running tap water for 20 minutes, or until the hematoxylin stain has been thoroughly blued.

14. Counterstain the preparations in Van Gieson's picro-fuchsin (p. 388) for about 30 seconds. Dip them several times into 95% alcohol, then place them in absolute alcohol for 20 to 30 seconds. Clear them in terpineol-toluene or carbol-toluene, followed by pure toluene, and mount them in a resinous medium.

Finer fibrils of the reticulum should be black; coarser collagenous fibers are ordinarily red or rose; nuclei are stained blue; cytoplasm is usually yellowish gray; elastic fibers and muscle are yellow.

Silver Carbonate Methods

Del Río Hortega devised a number of methods, each with variants, to demonstrate astrocytes, oligodendrocytes, and microglia. Only one of these will be considered here. Because Ramón y Cajal's excellent gold-mercuric chloride method for astrocytes is to be described (p. 426), it is perhaps more appropriate to deal here with a method for bringing out the other two forms of neuroglia. A technique for demonstrating oligodendrocytes and microglia has therefore been chosen. Anyone making a critical study of neuroglia should refer to the methods described by Penfield and Cone (1950).

In carrying out Del Río Hortega's methods, or modifications of them, it is essential to observe all of the precautions concerning cleanliness of glassware and purity of chemicals given in connection with Bielschowsky's ammoniacal silver methods.

Penfield's Modification of Del Río Hortega's Silver Carbonate Method for Oligodendrocytes and Microglia. 1. Fix slices of strictly fresh tissue, 5 to 7 mm. thick, in 10% formalin or in the following mixture:

Ramón y Cajal's Formalin-Ammonium Bromide Fixative

Formalin	15 ml.
Distilled water	85 ml.
Ammonium bromide	2 gm.

Fixation for 5 days or a week is sufficient, but the time may be prolonged to several weeks.

2. Wash the tissue in running water for 1 hour. Make sections, by the freezing method, about 20 μ in thickness, and collect them in distilled water.

3. Remove formalin from the sections by leaving them, for about 12 hours, in a covered dish of distilled water to which ammonium hydroxide has been added in the proportion of 15 drops to 50 ml. of water.

4. Transfer the sections to a mixture of 5 ml. of 40% hydrobromic acid and 95 ml. of distilled water. Place the dish in an incubator, at about 37° C., for 1 hour.

5. Pass them through 3 changes of distilled water.

6. Mordant them in a 5% aqueous solution of sodium carbonate for 1 hour to 6 hours, but no longer than this.

7. Transfer the sections (with or without washing them) to the following solution:

Del Río Hortega's Weak Ammoniacal Silver Carbonate Solution

Silver nitrate, 10% aqueous solution	5 ml.
Sodium carbonate, 5% aqueous solution	20 ml.
Ammonium hydroxide, added cautiously, sufficient to dissolve the precipitate.	
Distilled water, to make the total volume of solution up to	100 ml.

Filter and store the filtrate in a dark bottle. It will keep for a long time without deteriorating.

Impregnate sections in this fluid for 3 to 5 minutes. If the trial sections quickly turn a uniform gray color, proceed to reduce all of them. Otherwise, continue the impregnation, testing a section in the reducing agent every minute or two until the desired coloration is observed. Reduction is usually complete within 1 or 2 minutes.

8. Transfer one or two sections directly to 1% neutral formalin and gently agitate this liquid. The formalin serves as a reducing agent.

9. Wash the sections in distilled water for several minutes.

10. Tone the sections in a 0.2% aqueous solution of gold chloride until they become bluish-gray in color.

11. Rinse them in distilled water and remove excess silver and gold salts by treatment, for about 30 seconds, in a 5% aqueous solution of sodium thiosulfate.

12. Wash them in several changes of distilled water, over a period of 15 to 30 minutes.

13. Dehydrate, clear, and mount the preparations as directed in step 13 of Bielschowsky's method (p. 414).

In successful preparations, both oligodendrocytes and microglia are sharply impregnated, and may easily be differentiated by their structure. The general background should be nearly colorless. Astrocytes may be stained faintly.

Impregnation with Activated Silver Albumose (Protargol)

This method takes advantage of the fact that when silver albumose is activated by metallic copper, silver is selectively accumulated by certain structures. This silver may then be reduced by hydroquinone, which is an ingredient of most photographic developing solutions, and then replaced ("toned") by gold to improve the differentiation of various cellular or histological elements. The technique of impregnation with silver albumose was worked out by Bodian for bringing out nerve fibers and nerve endings in paraffin sections. The method may appear complicated because of the number of reagents required, but it is actually rather simple. In the hands of an experienced worker, it is remarkably predictable.

With some minor modifications, the method of Bodian may be used to impregnate certain types of protozoa to differentiate kinetosomes, flagella, parabasal bodies, and various other extranuclear organelles. The fact that it will usually bring out the nuclei of some protozoa reasonably well may be of some utility in studies in which stages in the mitotic cycle are to be correlated with morphogenetic events involving other structures. The technique may have some applications in work with various invertebrates, especially those in which the organization of ciliary tracts or elements of the nervous system are to be demonstrated.

Excellent preparations of silver albumose for impregnation by Bodian's method are Protargol S (Winthrop Laboratories, New York) and Protéinate d'Argent (Établissements Roques, Paris [available from Roboz Surgical Instrument Co., Washington, D.C.]). (For convenience, the name Protargol will be used hereafter.) They are variable to some extent, and few seem to behave in the same way as the better preparations imported in the years shortly before World War II; however, the variability can be put to good use, because the results obtained with one sample may be complementary to those obtained with another. If impregnation with Protargol is to be done regularly, it is a good idea to test the suitability of new samples before the stocks already in use run out.

Protargol Method for Nerve Fibers and Nerve Endings. 1. The following mixture was found by Bodian to be the best single fixative for pieces of tissue from the nervous system and various organs showing nerve endings:

Formalin	5 ml.
Ethyl alcohol, 80%	90 ml.
Acetic acid, glacial	5 ml.

Fixation of smaller pieces of tissue should be complete in 24 hours. After fixation, wash them in several changes of 70 or 80% alcohol over a 24-hour period, to remove all formalin and acetic acid.

Other fixatives (including Bouin's fluid) may also be used, and for certain purposes some of these give good results. If experimentation with fixatives is contemplated, it would be well to consult a paper by Bodian (1937) which deals with this subject. Fixatives containing chromic acid, potassium dichromate, mercuric chloride, and osmium tetroxide are, for all practical purposes, unsuitable.

2. Following fixation and washing, dehydrate, infiltrate, and embed the tissue in paraffin. After sectioning the material and affixing sections to slides, de-paraffin the sections in the usual manner and transfer them through a graded series of alcohols to water. Do not coat the sections with Parlodion before hydrating them.

3. Prepare the solution of Protargol just before it is to be used. Ordinarily, a 1% solution is employed for impregnation, but some samples seem to work better in a concentration of 1.5%. Dust the powder—a little at a time—on the surface of distilled water in a beaker. Attempting to dissolve the Protargol by dropping it into the water all at once will lead to formation of a gummy lump which will take longer to go into solution. Once the Protargol is dissolved, pour an appropriate amount of the solution into the staining jar.

The activation of Protargol is accomplished by placing a flat coil of bright copper wire (No. 18 or No. 20 gauge) or granular copper on the bottom of the dish in which the impregnation is to be carried out. Five gm. of copper is used for each 100 ml. of Protargol solution. Two and one-half gm. is therefore the right amount to put into a Coplin jar which is filled with 50 ml. of liquid.

Leave the sections in the Protargol solution for 12 to 24 hours, at a temperature of 20 to 37° C. As a rule, 24 hours at 30° C. will give good results, but some samples seem to work decidedly better at room temperature or at 37° C.

4. Wash the sections in distilled water for a few seconds. Then transfer them to the reducing solution, prepared as follows:

Sodium sulfite, anhydrous	5 gm.
Hydroquinone	1 gm.
Distilled water	100 ml.

Dissolve the sodium sulfite completely before adding the hydroquinone.

Leave the slides in the reducing solution for 5 to 10 minutes. It is not likely that over-reduction will take place. Because there is a chance that the impregnation will turn out to be too light if the time is shortened, 10 minutes is suggested as a safe minimum for reduction.

5. Wash out the reducing solution completely. It is preferable to do this in running water and to continue the washing for at least 5 minutes. Then rinse in distilled water. If washing is not thorough enough, the solution of gold chloride (which normally may be re-used a number of times) will be spoiled.

6. Tone the sections in a 1% solution of gold chloride in distilled water. Addition of 3 drops of glacial acetic acid to each 100 ml. of the solution is advised. Leave them in this for at least 5 minutes. During the toning process, the sections turn from brown to an ashen gray color.

7. Rinse the sections briefly in distilled water. This is apparently a rather critical step. If they are washed too little, a reddish precipitate will form on the surface of the glass slides and on the sections themselves when they are exposed to oxalic acid (step 8). If they are washed too long, the intensity of the impregnation will be weakened.

8. Place the sections in a 2% aqueous solution of oxalic acid until they have a purplish or blue color. This generally takes about 5 minutes, but no harm will be done if the sections are left in the oxalic acid for 10 minutes or longer.

9. Wash out the oxalic acid, preferably in running water. Then remove all excess silver and gold by leaving the sections in a 5% aqueous solution of sodium thiosulfate for about 5 or 10 minutes. Following this, wash the sections thoroughly in running water for about 10 minutes. After washing, they may be dehydrated, cleared, and mounted in synthetic resin or neutral balsam.

In a good preparation, neurofibrils, myelinated and unmyelinated fibers, the end-feet of Held, and other nerve elements are sharply defined in a purple, purplish-black, or blue-black color.

Protargol Method for Protozoa and Other Small Invertebrates in Smears. 1. Fix material on coverglasses having a diameter of 22 mm. Drop the smears face down on the surface of the fixative in a shallow dish. They may be left in this dish until fixation is complete (about 30 minutes to 1 hour), but if more than a few smears are to be prepared it is usually more convenient to transfer them, after the films have congealed, to Columbia dishes containing the fixative.

In preparing smears, care should be taken not to make them too thick or too thin. Smears in which the organisms are enrobed in a matrix containing some mucus or other proteinaceous material generally give better results than those in which the organisms appear to be adhering to the coverglass without the help of extraneous material. Ordinarily, good preparations are obtained when symbiotic organisms in intestinal contents, tissue from the ctenidia of molluscs, blood, and material of similar consistency are dispersed on the coverglass. Some suggestions for preparing smears and handling them through various reagents are given in Chapter 12.

Bouin's fluid and Hollande's fluid are known to be good fixatives for ciliates, flagellates, and gregarines which are to be impregnated. Duboscq and Brasil's alcoholic modification of Bouin's fluid may also give satisfactory results with some organisms. Fixatives containing mercuric chloride are almost never suitable; those containing chromic acid, potassium dichromate, or osmium tetroxide are generally thought of as unacceptable, but splendid results have been obtained with impregnation of orthonectids following fixation in Champy's fluid, which contains all three of these ingredients.

2. If the smears were fixed in Hollande's fluid or Bouin's fluid, wash them in water for a few minutes, then transfer them by way of a graded series of alcohols to 70% alcohol. Change the alcohol until it is no longer conspicuously discolored by picric acid or cupric acetate. The alcohol will serve to harden the films as well as complete the washing process. From Duboscq and Brasil's fixative, the smears may be transferred directly to 70% alcohol, which should be changed until the picric acid has been removed. Material fixed in Champy's fluid should be washed in running water for a few hours, then brought gradually to 70% alcohol and left in this for at least an hour or two. Smears fixed by any method, if they are not to be impregnated immediately, may be stored in 70% alcohol.

3. Pass the smears down the graded series of alcohols to water. Then transfer them to a 0.25% aqueous solution of potassium permanganate for 5 to 10 minutes. Wash this out in several changes of distilled water. The smears will now appear dark brown. Place them in 2 or 2.5% oxalic acid for 5 or 10 minutes. This will bleach out the brown color. Then wash the smears well, preferably in running water.

The potassium permanganate-oxalic acid treatment makes possible the impregnation of many organisms which would otherwise be refractory. Most samples of Protargol S and Protéinate d'Argent now available seem not to work well on any protozoa unless these have been bleached. Unfortunately, the bleaching process often alters the smears in such a way that the material is more likely to fall off the coverglasses, especially

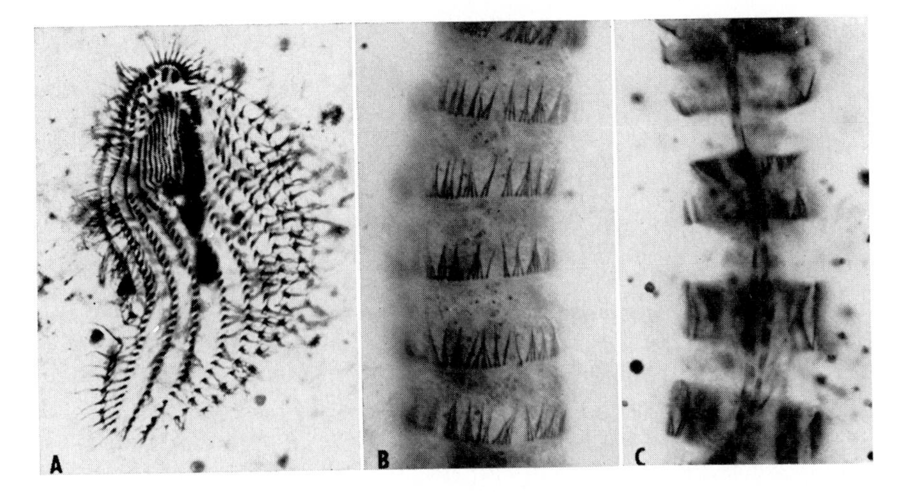

FIG. 70. Small organisms impregnated by the Protargol method. *A*, A marine heterotrich ciliate (fixed in Hollande's fluid), showing kinetosomes and aggregates of kinetosomes; the cilia and two macronuclei characteristic of this species are also evident. *B*, Surface view of an orthonectid, *Ciliocincta* (fixed in Champy's fluid), showing kinetosomes and ciliary rootlets. *C*, Optical section of a portion of *Ciliocincta* (fixed in Champy's fluid), showing longitudinal myocytes.

when the smears are being washed in water after reduction of the silver by hydroquinone.

If tap water has been used for washing out the oxalic acid, transfer the smears to distilled water for a minute or two before proceeding with step 4.

4. Dissolve the Protargol just before use, by dusting it on the surface of distilled water in a beaker. A 1.5% solution is recommended, but a 1% solution of some samples may be suitable. Columbia dishes hold 10 ml. of liquid, so 0.5 gm. of copper (preferably in a flat coil of No. 18 or No. 20 gauge wire) should be at the bottom of each dish. Leave the smears in the activated Protargol for 18 to 24 hours at room temperature. If they are not well impregnated in 24 hours, they are not likely to be impregnated any better in a longer period.

5. Remove the smears from the Protargol solution, rinse them in distilled water, and reduce them in the hydroquinone-sodium sulfite solution (p. 419). Over a period of 5 or 10 minutes, the smears will turn darker brown. It is not likely that they will be over-reduced.

6. Wash the smears thoroughly. Running water is recommended if they will stand even gentle circulation of water. Very commonly they tend to swell and peel off the coverglasses after they have been transferred from the reducing solution to water. Moving the coverglasses individually from one Columbia dish of water to another until they have been thor-

oughly washed is generally much safer than pouring the water out of a dish of coverglasses and filling the dish again.

If the smears do fall off to an excessive extent, when making a new set try to have the consistency of the material more viscous and sticky, even if some tissue or a very little egg albumen must be worked into it.

7. Transfer the smears to 1% gold chloride. It seems to be unnecessary to add acetic acid, as recommended for sections of tissue. Within about 5 minutes, the smears will turn grayish.

A 10 ml. quantity of gold chloride solution may be used for processing about 10 sets of smears, provided it does not become badly contaminated by the ingredients of the reducing solution.

8. Rinse the smears briefly in distilled water. This is an important step. Most of the excess gold chloride should be washed off the smear and coverglass, so that these are not suffused by an objectionable reddish deposit when the preparations are treated with oxalic acid in step 9. However, removal of too much gold chloride from the smears is also undesirable. By dipping each smear twice into a beaker or Columbia dish of distilled water, a satisfactory rinse will usually be achieved.

9. Transfer the smears to a 2% aqueous solution of oxalic acid. In 5 or 10 minutes, they should become purple, reddish-purple, or bluish-purple. When it appears that darkening has proceeded to completion, wash out the oxalic acid, preferably in running water.

10. Place the preparations from either tap water or distilled water into a 5% aqueous solution of sodium thiosulfate. This will remove unreduced salts of silver and gold.

11. Remove the sodium thiosulfate by washing smears (in running water, if possible) for at least 10 or 20 minutes. Then dehydrate the smears gradually; clear them in toluene or some other clearing agent, and mount them in a synthetic resin or neutral balsam.

This method often gives remarkably sharp differentiation of extranuclear organelles. In complex flagellates, as trichomonadids, it brings out flagella, kinetosomes, the parabasal body, costa, and some other extranuclear organelles. When applied to ciliates, it often provides the most complete morphological picture which may be obtained by any single method. In the case of gregarines, it brings out pellicular fibrils and the structure of the epimerite. In orthonectids (a group of so-called "mesozoa"), which are relatively small and ciliate-like although multicellular and quite complex, the method not only brings out nuclei and kinetosomes of cilia, but also the rootlets of cilia and bundles of myofibrils, as well as a number of other details. When impregnation with Protargol is applied to invertebrates such as gastrotrichs, rotifers, and minute turbellarians, some very instructive preparations can be obtained.

As a rule, those preparations which have a purple or reddish-purple cast by the time they are completed are superior to those which are blue.

IMPREGNATION WITH SALTS OF MERCURY

Golgi's Mercuric Chloride Method for Nerve Cells. This method produces a very fine impregnation of individual nerve cells, including their processes and ramifications. It is easy to perform and can be relied upon to give consistently good results if carried out correctly. For these reasons, it is decidedly preferable to Golgi's chrome-silver methods for certain purposes. However, the mercuric chloride method ordinarily impregnates all or most of the cells present, while the silver nitrate method is highly selective. Golgi's silver method is therefore preferable for obtaining clear, unconfused views of certain cells in regions where a great number and variety of cells are present. The mercuric chloride method gives unusually beautiful results on the cerebral cortex, and for routine preparations of this region it is incomparable. It also gives good results on the cerebellum, but cannot be recommended for the spinal cord.

1. With a sharp razor blade, cut out slices 4 to 5 mm. thick from the cerebral or cerebellar cortex. It is advisable not to make the surface of the slices more than 1 cm. square. Larger pieces, or even entire small brains, can be successfully impregnated, but it is difficult to handle and mount large sections.

2. Place the pieces of tissue in at least 20 times their volume of a 1% aqueous solution of potassium dichromate and leave them in this overnight. Then transfer them to the same volume of a 3% aqueous solution of potassium dichromate. They should be left in this for 20 to 30 days,

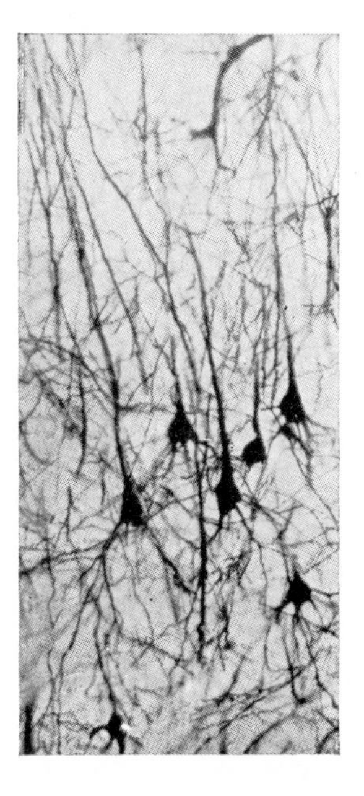

FIG. 71. Pyramidal nerve cells in the cerebral cortex. Nitrocellulose section (50 μ thick) of material impregnated with mercuric chloride by Golgi's method, then toned by Da Fano's gold chloride technique.

and the solution should be changed each day for the first 2 or 3 days. Tissue may remain in 3% potassium dichromate for several months without undergoing deterioration.

3. Rinse the tissue with distilled water and place it in a similar quantity of a 1% aqueous solution of mercuric chloride. Change the solution each day for 3 or 4 days, and later as often as it becomes yellowish. The tissue should remain in this solution for at least 2 weeks if the pieces are small, and a month or more if they are as large as 10 × 10 × 5 mm. They may be left in it for many months.

4. Wash the tissue for several hours in distilled water, dehydrate it with alcohols, and embed it in nitrocellulose by the usual method. During embedding, orient the pieces in such a way that sections may be cut perpendicular to the surface of the cortex. Harden the nitrocellulose in chloroform, and store the blocks in 70% alcohol until they are to be sectioned.

5. Cut sections 40 to 60 μ thick. If the impregnation is very heavy, it may be necessary to make them somewhat thinner, in order to avoid confusing superimpositions. This can be determined by examining a section under the microscope. In order that the fibers may be traced as far as possible, it is advisable to make the sections as thick as they can be made without interfering with critical study of the individual structures.

6. Collect the sections in a shallow dish containing 70% alcohol. Then pass them progressively through lower concentrations of alcohol and wash them in several changes of distilled water.

7. Now tone them by Da Fano's method, as follows: Place them, for 5 to 10 minutes, in a mixture of 5 ml. of ammonium hydroxide and 95 ml. of distilled water. This will convert the white mercury precipitate into a black one. Then wash them with 3 changes of distilled water.

8. Cover them with a 0.2% aqueous solution of gold chloride and leave them in this for 15 minutes.

9. Rinse the sections quickly in distilled water and put them into a 5% aqueous solution of sodium thiosulfate for 5 minutes.

10. Wash them with several changes of distilled water, or in running water for 3 to 5 minutes.

11. Pass them through 30, 50, and 70% alcohol, adding 1 drop of a saturated tincture of iodine in 95% alcohol to every 5 ml. of alcohol. They should remain for 10 or 15 minutes in each grade of alcohol.

12. Transfer the sections to 80% alcohol (without iodine) for 10 minutes, then to 95% alcohol for 15 or 20 minutes.

13. Float the sections, one by one, on terpineol-toluene or carbol-toluene. When they have all sunk to the bottom, place the dish under a low-power microscope and pick out as many perfect sections as are likely to be needed. Place each section on a slide and flood it with a few drops of pure toluene. Drain off the toluene and add a drop of mounting medium

and a coverglass. The preparations may be dried at room temperature, or in an incubator at about 37° C.

As previously stated, nerve cells and all of their processes should be thoroughly impregnated. Preparations made by this method will remain unaltered for many years. The impregnation withstands exposure to light much better than hematoxylin or aniline dyes.

IMPREGNATION WITH GOLD

Gold Chloride Method for Nerve Endings and Plexuses. This method, when it succeeds, produces very beautiful demonstrations of nerve endings and fine plexuses. It is useful also for the study of certain types of connective tissue cells, notably those of tendons and of the cornea. In suitable preparations, the gold is reduced upon the surfaces of these structures, giving them a dark reddish-violet color.

The formation of deposits of gold upon the structures is brought about by an exceedingly delicate reaction, which is not well understood and cannot be produced with certainty. Several trials may have to be made before a satisfactory impregnation is obtained, but the clarity and beauty of successful preparations will justify the effort.

1. Cut out small pieces of tissue (about 5 × 3 × 3 mm.) from a freshly killed animal. For demonstrating motor end-plates, the intercostal muscles of mammals or reptiles provide excellent material.

2. Rinse them with distilled water. Place them in fresh, filtered lemon juice for 30 minutes, or in 25% formic acid for 10 to 15 minutes.

3. Wash the pieces of tissue briefly in distilled water and place them in about 10 times their volume of a 1% solution of gold chloride in distilled water. Allow them to stand, in darkness or diffuse light, for 30 minutes.

4. Rinse them again with distilled water, cover them with about 4 times their volume of 10% formic acid in distilled water, and put them in a dark place. The acid slowly reduces the gold salt; the process usually requires 8 to 24 hours. At intervals of several hours, tease apart fragments of the tissue in glycerol and examine them. When reduction has progressed far enough, the nerve fibers and nerve endings should appear nearly black, and muscle fibers should be reddish or light purple (not dark purple); connective tissue should be nearly colorless.

5. Wash them with many changes of distilled water, over a period of an hour.

6. Place them, for 5 to 10 minutes, in a 5% aqueous solution of sodium thiosulfate, to remove excess unreduced gold.

7. Again wash them thoroughly, as in step 5.

8. *a.* If the specimens are to be teased apart (which is generally the most satisfactory way to mount them, especially in the case of nerve endings in muscle), it is advisable to soften them by treatment with arti-

ficial gastric juice (pepsin, 1 gm.; distilled water, 1,000 ml.; concentrated hydrochloric acid, 3 ml.). Incubate them in this fluid at a temperature of about 37° C., until trial proves that a small piece of the tissue, under a coverglass, can be spread out by slight pressure. The pieces will reach this stage in about 30 minutes to 2 hours. Then wash the tissues in 5 changes of distilled water, leaving them about 10 minutes in each. Transfer them to pure glycerol and renew this twice within an hour. Place a small piece of tissue upon a slide, remove any large excess of glycerol from it with a bit of filter paper, and place upon it a drop of glycerol jelly. Place a coverglass upon the mount, put it under a dissecting microscope, and press upon the coverglass until the tissue is spread enough to show the nerve endings clearly. It is desirable to use a circular coverglass and to seal the mounts, after they have hardened for a day or two, with asphaltum cement. The technique of sealing mounts prepared in nonresinous media is described in Chapter 27, page 461.

b. If the specimens must be sectioned, fix them in 10% formalin for 24 hours, and embed them according to the paraffin method, using alcohols for dehydration. Shorten the steps in dehydration, clearing, and infiltration as much as possible, in order to minimize shrinkage of the tissue. Affix the paraffin sections to slides, place them in toluene to remove the paraffin, and mount them in a resinous medium.

Ramón y Cajal's Gold-Mercuric Chloride Method for Astrocytes of Neuroglia. 1. Cut slices about 5 mm. thick, perpendicular to the surface, from various parts of the brain and spinal cord of a freshly killed mammal (preferably a young cat or dog). Fix them for 2 to 10 days, at room temperature, or for 1 day at about 37° C., in the following mixture:

Ramón y Cajal's Formalin-Ammonium Bromide Solution

Ammonium bromide	2 gm.
Distilled water	85 ml.
Formalin (neutral)	15 ml.

2. Cut sections 15 to 25 μ in thickness, by the freezing method, and place them in a 1% solution of formalin in distilled water.

3. Prepare the following reagent:

Ramón y Cajal's Gold-Mercuric Chloride Solution

Mercuric chloride (crystals)	0.5 gm.
Gold chloride, 1% aqueous solution	10 ml.
Distilled water	60 ml.

Make this solution immediately before use, in a scrupulously clean glass vessel. Dissolve the mercuric chloride by application of gentle heat, add the gold chloride solution, and filter the mixture.

4. Rinse the sections, in lots of 4 to 6, in 2 changes of distilled water, and place each lot in a separate small glass dish, filled with the above solution to a depth of about 1 cm. Flatten the sections and arrange them so that they do not overlap one another. Place the dish in darkness, bringing it into the light for a brief examination at the end of 2 hours and every hour thereafter. After the sections become reddish-purple, microscopic examination should show the astrocytes to be sharply impregnated. The speed of the reaction and quality of the preparations vary with the material and the temperature. Mammalian material is generally well impregnated within 4 to 6 hours at 18 to 20° C., which is about the optimum temperature. Material from lower vertebrates generally impregnates more slowly and gives better results at temperatures of about 25 to 30° C.

5. Rinse the sections with distilled water and place them for 5 to 10 minutes in the following mixture:

Ramón y Cajal's Fixing Bath

Sodium thiosulfate	5 gm.
Distilled water	70 ml.
Alcohol, 95%	30 ml.
Sodium bisulfite	5 gm.

6. Wash them for at least 5 minutes in each of 2 changes of 50% alcohol.

7. Place one or more sections on a slide, by dipping the end of the slide into the alcohol and lifting out the sections upon it. Flood the sections with absolute alcohol and flatten them. After 1 or 2 minutes, pour off the alcohol and place several drops of a 1% solution of Parlodion on the sections. Tilt the slide back and forth to spread the Parlodion. When this has formed a film, place the slide in chloroform for 1 minute. Pass it through terpineol-toluene (or carbol-toluene) until the sections are cleared, and then into pure toluene. Mount them under a coverglass in a resinous medium.

Both protoplasmic and fibrous astrocytes should be black. Nerve cells are generally pale red, but the fibers are uncolored. The background should be unstained or pale brownish purple. Oligodendrocytes and microglia are not brought out by this method.

REFERENCES

Bodian, D., 1937. The staining of paraffin sections of nervous tissues with activated Protargol. The role of fixatives. Anatomical Record, *69*, 153–162.

Penfield, W., and Cone, W. V., 1950. Neuroglia and Microglia (The Metallic Methods). *In* Jones, R. M. (Editor), *McClung's Handbook of Microscopical Technique*, 3rd ed., pp. 399–431. New York: Paul B. Hoeber, Inc. (Reprinted 1961 by Hafner Publishing Co., Inc., New York.)

CHAPTER 26

Some Cytological Methods

Cytology calls into use a wide variety of methods for demonstrating the structure of cells, their chemical constituents, and the changes which take place during mitosis, meiosis, and fertilization. Most of the basic procedures of microtechnique are applicable to cytology, but it is often necessary to carry these out with greater refinement than might be necessary for more routine histological studies.

SELECTION OF MATERIAL. Unless an investigation must be limited to a certain species, it is a matter of importance to select an organism which is likely to show well the cytological details which are to be studied. Methods for preparing some materials commonly employed for demonstration of cell structures and changes taking place in the nucleus are outlined briefly in Chapter 29. Further suggestions may be obtained from the literature of cytology.

EXAMINATION OF LIVING MATERIAL. The importance of studying cells before they have been affected by postmortem changes is obvious. Although it is true that any careful cytological study requires specialized techniques which involve fixation of material, the ultimate goal of the cytologist is to understand the structure of living cells and processes which take place within them. For this reason, extensive observation of living cells is desirable.

Techniques for studying living cells with the aid of isotonic saline solutions, supravital staining, and other methods have been explained in Chapter 6.

KILLING AND FIXATION. Appropriate techniques for anesthetizing and killing certain kinds of organisms are discussed in Chapter 9. In critical work, it may be desirable to compare preparations from individuals treated in different ways, because a particular method of anesthetizing or killing an animal may alter the appearance of cells in a way which might otherwise not be detected.

Remove and fix the desired tissues *immediately* following death of the individual, before postmortem changes occur. It is generally desirable to fix small pieces of tissue, not thicker than about 4 mm. If mixtures of chromic acid and osmium tetroxide are to be used for fixation, the pieces should certainly be no larger than this. A very sharp blade should be used for cutting the tissue apart. If the tissues are dense or soft, it may be advantageous to begin fixation of them by injection through blood vessels (p. 129).

The choice of a fixing fluid is a matter of great importance. The reagent must not only fix the cells with a minimum of distortion, but must also prepare them for the particular stains to be used. It is always advisable to compare the appearances of cells fixed in several fixing agents, in order to minimize the possibility of mistaking artifacts of fixation for normal structures, or overlooking completely certain structures which are preservable by some fixatives but not by others.

The temperature at which the fixing agent is employed is also a matter to which attention should be given. For example, the clumping of chromosomes in orthopteran testes can often be prevented by fixing the tissue (preferably in Allen's PFA 3 fluid) at about 3 to 5° C. In the case of testes of mammals, however, best results with certain fixatives (Allen's B 15 modification of Bouin's fluid) are obtained if fixation is carried out at body temperature.

The length of time the tissue remains in the fixative is frequently an important consideration. Tissues may remain safely in Bouin's fluid, alcohol, or formalin for days. However, fluids containing strong acids (such as nitric or sulfuric acid) should be allowed to act only until fixation is apparently complete. The same precaution applies in the case of chromic acid, which is unstable, and in the case of osmium tetroxide, which has a blackening effect.

Changes of Reagents. These must be made by very gradual steps, in order to avoid shrinkage or swelling of delicate structures. To precede embedding in paraffin, use a closely graded series of dilutions of alcohol. Replacement of absolute alcohol by toluene or some other solvent of paraffin, and infiltration with paraffin, should also be carried out slowly and carefully. The amount of heat used to spread paraffin sections during the process of affixing them to slides should be minimal. It is advisable to coat sections with 1% Parlodion before clearing them preparatory to mounting in a resinous medium. Because examination of cytological material is generally carried out under an oil-immersion objective, the protective coating of the Parlodion will help to hold the cells of the sections together if pressure should accidentally be applied to the coverglass.

The same general precautions concerning gradual dehydration and infiltration apply to cytological material which is to be embedded in nitrocellulose. Any objects which are to be mounted entire should not only be slowly dehydrated and cleared, but also slowly infiltrated with the mounting medium.

Choice of Stains. The conventional iron hematoxylin method, utilizing ferric ammonium sulfate as a mordant, is undoubtedly the most important staining technique employed in cytology. It is used for demonstrating chromosomes, mitotic spindles, the organization of the interphase nucleus, and various extranuclear structures. Even though other methods of staining may be required in a particular cytological study, iron hematoxylin is likely to be indispensable. Results obtained by its use should be carefully compared with results obtained by application of alternative techniques, as well as tests for specific constituents.

Certain staining reactions are regarded as more or less reliable tests for particular substances in cells and tissues. In the case of some of them, the chemistry of the reactions is understood. In the case of others, the repeated correlation of positive reactions with other dependable methods of analysis makes the techniques useful. Generally speaking, however, only tentative conclusions should be drawn from evidence obtained with staining reactions.

Although the selectivity of a certain method of staining or impregnation may be of very great usefulness in a cytological investigation, the application of a variety of methods may be required before a composite picture of a particular type of cell may be realized. When studying the remarkably different results obtained when several unrelated techniques (such as the iron hematoxylin method, the Feulgen nucleal reaction, and impregnation with activated Protargol) are applied to certain material, it becomes obvious that drawing conclusions on the basis of results obtained by the use of a single method is hazardous.

At this point, it will be well to mention once more the importance of using fixatives which will preserve well the structures which are to be differentiated, and which are also compatible with the techniques of staining or impregnation to be employed. A fixative which may be very good preceding iron hematoxylin may be unfit for use before the Feulgen nucleal reaction or special methods for mitochondria and other structures or inclusions.

GENERAL METHODS FOR THE NUCLEUS AND NUCLEAR DIVISION

Staining Methods and Fixatives

The most generally useful technique for staining nuclei and chromosomes in sectioned material and smears is the iron hematoxylin method. Iron hematoxylin may ordinarily be applied with success after fixation in any of the reagents listed below:

Bouin's fluid. (Excellent for many animal tissues and protozoa in which cytological details are to be studied.)

Allen's B 15 modification of Bouin's fluid. (Particularly good for testis tissue of birds and mammals.)

Allen's PFA 3 modification of Bouin's fluid. (Recommended for fixation of testes of orthopterans and some other insects.)

Duboscq and Brasil's fluid. (Penetrates rapidly; good for somewhat resistant reproductive cysts of gregarines and various other protozoa, maturation stages in eggs of *Parascaris*, and similar materials.)

Flemming's fluids. (Excellent for small pieces of permeable tissue.)

Karpechenko's fluid. (Widely used for plant tissues, such as meristems in root-tips.)

Saturated aqueous solution of mercuric chloride with 5% acetic acid.

Carnoy and Lebrun's fluid. (Penetrates very rapidly; useful for fixation of resistant reproductive cysts of protozoa, such as gametocysts of gregarines, and pronuclear and later stages of development of *Parascaris*.)

The Feulgen nucleal reaction is now almost routinely used to confirm results obtained by the iron hematoxylin method. As a rule, it stains only those portions of the nucleus which are rich in deoxyribonucleic acid, and is therefore rather selective for chromatin in both the interphase nucleus and in stages of mitosis and meiosis. In ciliates, it usually brings out micronuclei which sometimes are difficult to differentiate from various other granules stained by iron hematoxylin. The Feulgen method generally works well after fixation in a saturated aqueous solution of mercuric chloride with 5% acetic acid, and also after fixation in Schaudinn's fluid, although the latter is rarely used except for protozoa in smear preparations. It is sometimes successful after mixtures containing picric acid, such as Bouin's fluid and modifications of it.

Aceto-carmine and aceto-orcein are widely used for making temporary stained mounts, especially of chromosomes in smear and squash preparations of animal and plant tissues. The acetic acid in these staining solutions fixes the material at the same time that the dye penetrates. Certain types of cytological material, such as the giant chromosomes of the salivary glands of *Drosophila* (p. 500), are routinely stained by one or the other of these mixtures.

CHROMATIN

Some of the more precise nuclear stains, including iron hematoxylin, also bring out structures which do not contain chromatin. For this reason, methods which are more specific for differentiating chromatin are sometimes required.

Methyl Green. This stains chromatin in the nuclei of fresh cells. In some types of nuclei, however, the coloration is faint, and in the heads of spermatozoa the chromatin generally does not stain at all. Therefore, although the staining of structures in nuclei is a nearly conclusive test for presence of chromatin, the failure of these structures to take the stain does not prove the absence of chromatin. Outside of the nucleus, methyl green sometimes stains cytoplasm and certain glandular secretions such as mucin and silk. Most samples of the dye color extranuclear structures bluish; perhaps this is due to the presence of methyl violet as an impurity.

Methyl green is always used in an acid solution. The staining mixture is generally prepared by dissolving 1 gm. of methyl green in 100 ml. of distilled water, and then adding 1 ml. of glacial acetic acid. This solution may be diluted considerably for use. Fixation and staining of protozoa or of teased, crushed, or smear preparations occur simultaneously, within a very short time. In nearly all nuclei, chromatin is stained bright green.

Methyl green will also stain chromatin in cells which have been fixed in mercuric chloride, but not in those which have been fixed in reagents containing acetic acid or other acids.

As the stain is very difficult to preserve, the use of methyl green is chiefly limited to rather temporary preparations. Alcohol washes it out quickly, but this can be avoided to some extent by dehydrating rapidly, or by using alcohols containing about 1% of the dye. After dehydration, clear the preparations in terpineol-toluene and finally in toluene. The stain generally fades within a few months in resinous media. Likewise, it does not keep well in glycerol, glycerol jelly, or other aqueous media.

Feulgen's Nucleal Reaction. This method is widely used for staining chromatin. A positive Feulgen reaction, yielding a deep magenta color, depends upon the release of an aldehyde from deoxyribonucleic acid (DNA) when this constituent of chromatin is hydrolyzed in weak hydrochloric acid. It is unusual to find any structures outside the nucleus to be stained. Particles of wood ingested by symbiotic flagellates in termites are often stained weakly, and perhaps this reaction also depends upon release of an aldehyde during hydrolysis.

The Feulgen reagent is a solution of basic fuchsin which has been decolorized by sulfurous acid, so that it remains nearly colorless until it becomes recolorized in the presence of free aldehyde groups. It is a modified Schiff's reagent, used by chemists for detecting aldehydes. The process of staining, which may at first appear more complicated than it really is, will be explained as its various steps are described. It is applicable to sections, smears, and to entire small organs or organisms. It yields permanent and beautiful preparations, which justify its routine use in cytological work.

PROCEDURE. 1. *Fixation and preliminary preparation.* The following fixatives, applicable to smears or pieces of tissue, usually give good results with material which is to be stained by the Feulgen method: (1) a saturated aqueous solution of mercuric chloride; (2) a saturated aqueous solution of mercuric chloride with 5% glacial acetic acid; (3) Schaudinn's fluid. A number of other fixatives, including those containing picric acid, may prove suitable for the particular material at hand. Some workers have avoided the use of any fixatives containing formalin, believing that a positive reaction cannot be taken as evidence that an aldehyde has been released during the process of hydrolysis. In general, however, fixation involving formalin is not likely to prove objectionable, because all or nearly all of the formalin can be removed by washing. Moreover, if the purpose of the technique is to bring out chromosomes, micronuclei, and other chromatin-containing elements sharply, rather than to demonstrate DNA quantitatively, the presence of a trace of foreign aldehyde will not affect the results substantially.

If a fixative containing mercuric chloride has been used, it will be advisable to treat the pieces of tissue or smears with tincture of iodine after transfer to alcohol, as directed on page 93. Other fixatives should be washed out by whatever methods are most appropriate.

If the material is to be sectioned, embed it in paraffin. For most purposes, sections 8 or 10 μ thick will be suitable. After affixing sections to slides, and drying them, remove the paraffin with toluene and pass them down the series of alcohols to distilled water. Material in smears on slides or coverglasses, or small organisms which are not to be sectioned, are simply passed down the series of alcohols to distilled water.

2. *Hydrolysis.* Place the sections or smears into normal hydrochloric acid for 1 or 2 minutes at room temperature. The solution is prepared by adding, to 8.25 ml. of concentrated hydrochloric acid (specific gravity 1.17 to 1.185), enough distilled water to bring the volume of the mixture to 100 ml.

Then treat the material for 1 to 5 minutes or longer with normal hydrochloric acid which has been brought to a temperature of 60° C. in an oven or water bath. The time required for hydrolysis (and consequent release of aldehydes) varies, and the optimum must be determined for each type of material. As a rule, however, 5 minutes is long enough.

3. *Staining.* The following solution should have been prepared in advance:

Decolorized Basic Fuchsin (Fuchsin-Sulfurous Acid)

Basic fuchsin (certified for use in Feulgen method) . . . . 1 gm.
Distilled water . . . . 200 ml.

Bring the water to the boiling point and stir in the dye. Cool the solution to 50° C., filter it, and add:

Normal hydrochloric acid . . . . 20 ml.

Cool to 25° C., and add:

Sodium bisulfite, anhydrous . . . . 1 gm.

If basic fuchsin of proper composition has been used, the solution fades within a period of 12 to 24 hours to a pale yellow color, and is then ready for use. If it remains red or dark orange, or if the dye precipitates, the mixture should be discarded and another sample of dye tried. Keep the solution in darkness, in a tightly-stoppered bottle.

Rinse the slides or other materials briefly in distilled water, and place them in the fuchsin-sulfurous acid mixture. The aldehydes liberated by hydrolysis combine with fuchsin, displacing sulfurous acid, and magenta or reddish-purple coloration of chromatin is the result. The time required to obtain a deep stain varies, depending upon the type of material and the completeness of hydrolysis. Generally, if the extent of hydrolysis is sufficient, the reaction proceeds quickly, and a strong stain will be obtained within 30 minutes or an hour. If a satisfactory stain does not result within about 2 hours, either the hydrolysis was allowed to proceed for too short or too long a time, or the material was not fixed in a suitable mixture, or the material is simply not readily stainable by this method ("Feulgen-negative" or "Feulgen-weak").

4. *Washing.* Rinse the preparations with distilled water and treat them with 3 changes of dilute sulfurous acid. This is made up by mixing 200 ml. of distilled water with 10 ml. of normal hydrochloric acid and 1 gm. of sodium bisulfite. Allow the sections or smears to remain 1 or 2 minutes in each change. Then wash them in running water, or in repeated changes of water, for 5 minutes.

5. *Counterstaining.* Recommended counterstains are fast green FCF (a 0.1% solution in 70, 80, or 90% alcohol) and orange G (a 1% solution in 90% alcohol), although other dyes providing suitable contrasts may be used. Before counterstaining, bring the material gradually to the appropriate concentration of alcohol.

6. *Dehydration, Clearing, Mounting.* After counterstaining, rinse the specimens in 90% alcohol to remove excess stain, then complete the dehydration in 100% alcohol (2 changes). Clear the material in two changes of toluene or terpineol-toluene, and mount it in a resinous medium.

MITOCHONDRIA

For many years, cytologists worked at the problem of demonstrating small cytoplasmic bodies called mitochondria, or chondriosomes. (Originally, the term "mitochondrion" was used to indicate chondriosomes of a particular shape. Today, however, it is the favored term.) Mitochondria are not seen in ordinary histological preparations, for they are generally destroyed by acetic acid, which is present in many common fixing agents, and by certain of the solvents used in routine methods. The special techniques needed to demonstrate them and to differentiate them rather conclusively from various other elements of the cytoplasm require patience and skill.

The significance of mitochondria, as centers of activity of enzyme systems involved in cellular respiration, was not appreciated until rather recently. Their physical organization has been so thoroughly explored with the aid of electron microscopy in the last few years that mere recognition of them with the higher magnifications of the light microscope may seem old-fashioned today. In certain situations, however, the identification of mitochondria and correlation of their abundance and distribution with other cytological phenomena may be very effectively done with sections, or with entire cells or small organisms, prepared for light microscopy. Therefore, a few methods for staining mitochondria will be given. Attention is also called to the Epon method (Chapter 18), which combines some advantages of techniques of electron microscopy with those of light microscopy, and which often demonstrates mitochondria well.

Supravital Staining of Mitochondria. Janus green B is a rather specific stain for mitochondria in fresh cells. Not all samples of this dye are equally suitable, however. Staining is generally accomplished by immersing the cells, or small organisms such as protozoa, in a 0.01% solution of the dye made in an appropriate medium, or by coating a microscope slide beforehand with a thin film of Janus green B and then applying the liquid in which the specimens are immersed. These techniques are explained on page 56, in connection with supravital staining. It is also possible to inject a solution of Janus green B into the blood vessels of an animal, after draining the blood (see p. 56).

Regaud's Method for Mitochondria. This is perhaps the simplest technique for making permanent stains of mitochondria. It gives excellent results with tissues (particularly the pancreas) of adult vertebrates, but poor results with embryonic tissues. The stain brings out various intracytoplasmic granules, as well as mitochondria, and therefore is less specific than certain other methods to be described.

1. Fix small pieces of tissue for 4 days in the following mixture, changing the liquid every day:

Regaud's Fluid

Potassium dichromate, 3% aqueous solution	80 ml.
Formalin	20 ml.

2. After fixation, mordant the tissue for 7 days in a 3% aqueous solution of potassium dichromate, changing the solution every other day.
3. Wash it for 24 hours in running water; embed it in paraffin, and cut sections 3 to 5 μ in thickness. Affix the sections to slides, remove the paraffin in toluene, and pass the slides down the graded series of alcohols to *distilled* water. (Do not coat them with Parlodion before passing them from 100% alcohol into lower grades.)
4. Mordant the sections in a 5% aqueous solution of ferric ammonium sulfate, at about 35 to 38° C., for approximately 24 hours.
5. Rinse them in distilled water for about 30 seconds; a longer rinse may remove too much of the mordant.
6. Stain the sections for 24 hours in a well-ripened 0.5 or 1% solution of hematoxylin. It may be desirable to add 3 drops of a saturated aqueous

solution of lithium carbonate to each 100 ml. of the staining solution. The addition of glycerol to the staining mixture, often recommended, is probably superfluous.

7. Rinse them in distilled water for 1 or 2 minutes. Then differentiate the stain in a 5% solution of ferric ammonium sulfate or a 2.5% solution of ferric chloride. The concentration of either may be advantageously reduced to permit more carefully controlled destaining. Observe the progress of destaining under the microscope. In good preparations, the cytoplasm is colorless or nearly so, mitochondria are blue-black, and nuclei are brownish gray. After washing the sections in running water for an hour, dehydrate them, coat them with Parlodion if desired, and clear and mount them.

Modifications of Altmann's Method for Mitochondria. Altmann's original method consisted of fixing material in a mixture of potassium dichromate and osmium tetroxide, making paraffin sections, and staining them with a concentrated solution of acid fuchsin heated to steaming. The stain was then differentiated with picric acid. Altmann's procedure will not be described here, as it has been largely superseded by the following modifications.

Champy-Kull Modification. This gives a brilliant three-color stain. It is applicable to most tissues, but is particularly good for invertebrates and for embryonic tissues of various kinds.

1. Fix small pieces of tissue, for 24 hours, in Champy's fluid (p. 90).
2. Wash them for 30 minutes in distilled water, then place them, for about 24 hours, into a mixture of equal parts of pyroligneous acid and a 1% solution of chromic acid.
3. Wash them for 30 minutes in distilled water. After this, mordant them for 3 days in a 3% aqueous solution of potassium dichromate.
4. Wash the tissue in running water for 24 hours. Embed it in paraffin, and cut sections 3 to 5 μ in thickness. Remove the paraffin from the sections and pass them down the series of alcohols to water. Do not coat them with Parlodion before hydration.
5. Flood the slide with the staining mixture.

Altmann's Aniline-Acid Fuchsin

Acid fuchsin	10 gm.
Aniline water (To prepare this, shake aniline in water to saturation, then filter the mixture.)	100 ml.

(Be careful in handling aniline—it is a carcinogenic substance.)

Hold the slide over a small flame until the liquid steams, then lay it aside to cool for about 6 minutes. Pour off the stain and rinse the slide quickly in distilled water.

6. Counterstain for about 1 or 2 minutes (the exact time should be

determined by experiment) in a 0.5% aqueous solution of toluidine blue O, or in a saturated aqueous solution of thionin. Rinse in distilled water.

7. Place the slide in a 0.5% solution of aurantia in 70% alcohol, keeping it in motion and watching it carefully. As soon as the acid fuchsin ceases to come freely from the sections, which will be within 20 to 40 seconds, dip the slide 5 to 10 times into 95% alcohol.

8. Place the slide in absolute alcohol until excess blue stain is extracted; this will take from a few seconds to about a minute.

9. Clear the sections in toluene and mount them in a resinous medium. If the various processes have been timed properly for the particular material in hand, chromatin will be stained blue, mitochondria (and occasionally Golgi material) will be red, and cytoplasm will be yellow to green. Some experimentation is usually necessary. The stain is not very permanent.

COWDRY'S MODIFICATION. This technique combines the good qualities of Regaud's fixative with Altmann's aniline-acid fuchsin and a methyl green counterstain. It is suitable for the various purposes mentioned in connection with the Champy-Kull modification, and is less difficult to carry out. Unfortunately, the stain usually fades rapidly.

1, 2, and 3. Follow steps 1 to 3 in Regaud's method.

4. Place the slide to which sections are affixed in a 1% aqueous solution of potassium permanganate for 15 to 60 seconds; the optimum time will have to be determined by trial. This is done to remove excess potassium dichromate.

5. Rinse the slide in distilled water and place it in a 5% aqueous solution of oxalic acid for about the same length of time as it remained in potassium permanganate. This removes the brownish discoloration produced by the latter.

6. Rinse the sections for 30 seconds in each of 3 changes of distilled water.

7. Flood the slide with Altmann's aniline-acid fuchsin stain (p. 436). Hold it over a small flame until the liquid steams, then put it aside to cool for 6 minutes. Pour off the stain and wipe it from the slide except for the area occupied by the sections.

8. Hold the slide over a white paper. With a pipette, cover the sections with a small amount of a 1% aqueous solution of methyl green. After 5 seconds (or a longer time, as indicated by experiment), pour this off, plunge the slide into absolute alcohol for a few seconds, and then transfer it to toluene. Pass it through a second change of toluene, and mount it in neutral balsam or a synthetic resin.

In successful preparations, mitochondria are bright red, nuclei are green, Nissl substance is green or blue, and neurofibrils are light brown. If the acid fuchsin does not stain deeply, or is all removed when the methyl green is applied, omit steps 4 and 5. If this fails to correct the trouble, again omit these steps and mordant the sections in a 2% aqueous solution of potassium dichromate before proceeding with step 7. If, on the contrary, the fuchsin stains so intensely that too little of it is removed when the solution of methyl green is applied, increase the time for steps 4 and 5.

Bensley's Cupric Acetate-Hematoxylin-Potassium Chromate Method. Preparations obtained by this technique are less attractive than those made by the Champy-Kull method or other modifications of the Altmann method, but they keep rather well. The stain is more nearly specific than Regaud's technique for mitochondria.

1. Fix small fragments or very thin slices of tissue for 12 to 24 hours in the following mixture:

Bensley's Fluid

Potassium dichromate, 2.5% aqueous solution	8 ml.
Osmium tetroxide, 2% aqueous solution	2 ml.
Acetic acid, glacial	1 drop

2. Wash them in running water for 12 to 15 hours. Dehydrate the pieces of tissue and embed them in paraffin. Cut sections 3 to 5 μ in thickness, and affix these to slides. Remove the paraffin in toluene and pass the slides down the graded series of alcohols to water.

3. Place them in a saturated aqueous solution of cupric acetate for 5 minutes.

4. Rinse them, for a total time of 1 minute, in several changes of distilled water.

5. Place the slides in a well-ripened 0.5% aqueous solution of hematoxylin for 1 minute.

6. Rinse them in distilled water and place them, for 1 minute, in a 5% aqueous solution of neutral potassium chromate. The sections should turn blue-black. If they are a much lighter shade, rinse them in distilled water and repeat steps 3 to 6 as many times as may be necessary to obtain a strong stain.

7. Rinse the sections in distilled water and return them to the cupric acetate solution for 10 seconds. Then wash them, for 1 minute, in each of 3 changes of distilled water.

8. Flood the slide with the following mixture, diluted with 2 parts of distilled water:

Weigert's Sodium Borate-Potassium Ferricyanide Solution

Sodium borate	2 gm.
Potassium ferricyanide	2.5 gm.
Distilled water	200 ml.

Observe the progress of destaining under the microscope. The mitochondria should appear dark blue against a light yellowish background.

9. Wash the sections in running water for 6 hours. Then dehydrate, clear, and mount the preparations. If desired, the sections may be coated with Parlodion before being passed into toluene or some other clearing agent.

GOLGI MATERIAL

As revealed by electron microscopy, Golgi material is a system of vesicles with thin membranes. It is present in cells of most organisms and appears to be involved in the manufacture and packaging of secretory products. In some cells, especially those of glandular tissues, the amount may become more plentiful during periods of increased secretory activity. Methods of demonstrating Golgi material in ordinary histological or cytological preparations generally involve impregnation with silver or osmium.

Mann-Kopsch Method. In 1902, Kopsch discovered that Golgi bodies were preserved and stained by prolonged fixation in osmium tetroxide. A few years later, Weigl improved the technique by fixing tissues in Mann's fluid (devised originally for a different purpose) before osmicating them. The following account of the so-called Mann-Kopsch method is based upon experience with several more recent modifications. The method is applicable to both vertebrate and invertebrate materials.

1. Fix very small pieces of tissue in Mann's fluid (p. 98) for 18 to 24 hours. Tissues of the pancreas and spinal ganglia of a young mammal provide good material for practice.

2. Wash them in distilled water for 30 minutes, changing the water once during this time.

3. Place the pieces of tissue in a small wide-mouthed bottle, provided with a tight-fitting glass stopper, and pour in just enough of a 2% aqueous solution of osmium tetroxide (made as suggested on p. 107) to cover them. Incubate for 3 to 5 days, in darkness, at a temperature of 30 to 37° C.

4. Pour off the osmium tetroxide solution, cover the material with distilled water, and incubate it again for approximately 24 hours.

5. Wash the tissue in running tap water for 3 to 6 hours, dehydrate it, and embed it in paraffin. Make sections 2 to 5 μ in thickness, affix the sections to slides, and allow them to dry overnight.

6. Place one or two slides of the series in toluene to remove the paraffin, mount the sections in a resinous medium, and examine them critically. If the Golgi material is not blackened—and this is unfortunately often the case—start over again with new pieces of tissue. In the second trial, the time of incubation may be prolonged, or carried out at a slightly higher temperature. It may also be advisable to renew the solution of

osmium tetroxide once during the period of incubation. When examination of sample preparations indicates that a good impregnation has been obtained, proceed as follows:

7. *a.* If the Golgi material has been blackened rather selectively, remove the paraffin from the remaining preparations, and pass them into absolute alcohol. The nuclei may then be stained by passing the sections through 95 and 70% alcohol to a 1% solution of safranin O in 50% alcohol. Leave them in this for 6 to 24 hours. Then rinse them in 50% alcohol and complete the process of dehydration. Observe sample preparations under the microscope while the sections are being dehydrated. Only the nuclei should retain the safranin to any great extent. Slow down the passage through the higher alcohols if necessary to achieve optimum results. After dehydration is complete, clear the sections in toluene and mount them.

If it is desirable to bring out mitochondria, pass the sections gradually to water, and proceed with step 8.

b. If the cytoplasm has also been considerably blackened, it is sometimes possible to bleach the preparations to obtain better differentiation. Very often, however, the bleaching process injures the impregnation of the Golgi material.

The following procedure for bleaching is suggested: Treat the sections in a 0.25% aqueous solution of potassium permanganate until the tissue turns brown. Rinse them thoroughly in distilled water, and transfer them to a 2.5% aqueous solution of oxalic acid for a few minutes. This should make the sections nearly colorless. Then wash the preparations in running water for about 20 minutes, and stain the nuclei with safranin, as directed above, or attempt to bring out the mitochondria according to the procedure outlined in step 8.

8. Flood the sections with Altmann's aniline-acid fuchsin (p. 436), and warm the slide over a small flame until the liquid steams. Then set it aside for about 30 minutes. Pour off the stain, rinse the slide with distilled water, and place it in absolute alcohol for a few seconds. Then clear the preparation, preferably in terpineol-toluene, transfer it to pure toluene, and mount it in a resinous medium. The Golgi material should be black, the mitochondria and nucleoli red, and the cytoplasm gray or pinkish. Much experimentation may be necessary, however, before this result is consistently achieved.

Da Fano's Silver Method. In developing methods of silver impregnation for the nervous system, Ramón y Cajal found that the use of uranium nitrate and certain other nitrates favored impregnation of Golgi material. Later, Da Fano found cobalt nitrate to be especially suitable for the purpose, and he also introduced the practice of toning the sections with gold chloride in order to decolorize the cytoplasm. This method often gives very clear views of Golgi material, and is generally reliable.

1. Fix slices of tissue, about 4 or 5 mm. thick, in the following mixture:

Da Fano's Fluid

Distilled water	100 ml.
Formalin	15 ml.
Cobaltous nitrate	1 gm.

This fluid should not be allowed to act any longer than is necessary. Three to 4 hours is long enough for very small objects, such as spinal ganglia or pituitary bodies of mice; 6 hours is about the right length of time for spinal ganglia of cats or thin slices of pancreas; 8 to 12 hours is suggested for pieces of the central nervous system from adult mammals; up to 24 hours, but no longer, should be allowed for fixation of less permeable objects.

2. Wash the pieces of tissue in distilled water for 5 minutes.

3. Cover them with a generous quantity of an aqueous silver nitrate solution. For very small specimens, the concentration of silver nitrate should be 1%; for the spinal cord or tissues surrounded by fat, the concentration should be 2%. Put the container in a dark place, at room temperature, for 1 day if the objects are extremely small, for 2 days if they are larger.

4. Rinse the specimens in distilled water for about 2 minutes. Then, with a very sharp razor blade, cut them down to a thickness of not more than 2 mm., if they are not already as small as this.

5. Place them into the following for 24 hours:

Ramón y Cajal's Reducing Fluid

Hydroquinone	1.5 gm.
Sodium sulfite, anhydrous	0.2 gm.
Formalin, neutral	6 ml.
Distilled water	100 ml.

Prepare the solution immediately before use. Dissolve the sodium sulfite completely, then the hydroquinone. Add the formalin last.

6. Wash the pieces of tissue in 3 or 4 changes of distilled water, for a total of 30 minutes. Then dehydrate them, embed them in paraffin, and cut sections. The sections should be 4 to 7 μ thick in the case of tissues composed of small cells (such as pancreas), and 10 to 15 μ thick in the case of organs consisting of larger cells (such as spinal cord and ganglia). Affix the sections to slides, remove the paraffin in toluene, and pass them into absolute alcohol. Then hydrate them by way of a graded series of alcohols. The Golgi material should appear black against a yellow background.

7. Place one preparation in a 0.2% aqueous solution of gold chloride. Observe it frequently until the yellow background disappears This will

probably occur within a few minutes. The time required should be noted, and the remaining preparations treated accordingly.

8. Rinse the sections quickly in distilled water and place them, for 10 to 30 seconds, in a 5% aqueous solution of sodium thiosulfate. Then wash them in running water for 3 to 5 minutes.

9. Stain them for 15 to 20 minutes in azocarmine G (see formula on p. 400), at 55° C. Rinse them briefly in water, then place them in aniline-alcohol (p. 400) until only the nuclei remain red. Complete dehydration, coat the sections with Parlodion if desired, and clear and mount them. The Golgi material should appear black, the nucleus bright red, and the cytoplasm various shades of pink.

LIPIDS AND GLYCOGEN

In recent years, methods for identifying chemical constituents, secretory products, and reactive sites in intact cells and tissues have increased rapidly. In order to be conclusive, these techniques must be based on known chemical reactions, or specific results obtained by their use must be confirmed by other methods. Morphological staining techniques, on the other hand, utilize dyes and metallic compounds in a more or less empirical way.

The subject of histochemistry, which in the broad sense includes methods for demonstrating specific substances within cells (cytochemistry), is complicated. No really useful treatment of a wide variety of techniques can be given except in a comprehensive work. Only some representative procedures for identifying lipids and certain polysaccharides, particularly glycogen, will be described in this book. For additional methods of this type, and techniques for identifying enzymes, metals, and various other organic and inorganic constituents or products of cells and tissues, the reader is referred to the manuals of Bancroft (1967), Chayen, Bitensky, Butcher, and Poulter (1969), and Pearse (1968).

Lipids

The lipids, which include fats and oils, occur as inclusions in cells and as constituents of Golgi material, mitochondria, membranes, and other structures. Most lipid substances are readily soluble in strong alcohols, ether, dioxane, acetone, chloroform, toluene, xylene, essential oils, and other aromatic hydrocarbons. For this reason, they are not often observed in ordinary histological preparations, having been dissolved by the reagents used in the routine methods of preparation.

The chemistry and identification of various kinds of lipids are beyond the scope of this work. The methods given here are directed primarily at staining or impregnating the true fats, which are esters of fatty acids and glycerol. These are commonly seen as small droplets in the cytoplasm, but may form rather massive inclusions, as in adipose tissue.

There are two particularly useful ways of preserving fatty substances in cells, so that their distribution may be studied in permanent preparations: (1) Fixation of material in formalin or some other aqueous fixative, and mounting it in a medium which does not require passage through fat solvents. If it is necessary to make sections, this should be done by the freezing method. (2) Rendering the fats less soluble in the conventional organic solvents by fixing the material in osmium tetroxide or potassium dichromate. Sections may then be made by either the paraffin or nitrocellulose method, and it will be possible to mount them in a resinous medium.

Staining Lipids with Sudan III and Sudan IV. These dyes stain all true fats, and some other lipids as well. The staining appears largely to be a purely physical process, resulting from the fact that these dyes are soluble in fats, much more so than in the alcohol that is used as a vehicle for their application.

Sudan IV is a somewhat more energetic dye than Sudan III. For this reason, it is generally preferred. The following directions will apply to either dye.

1. Fix material in 10% formalin for 24 hours or longer. If the material is in the form of a thin sheet of tissue, stretch it over the end of a tube, vial, or glass or plastic ring before fixing it. Other types of material may be teased apart with needles or smeared on slides or coverglasses. Large or dense specimens should be sectioned by the freezing method. (Fixation may precede freezing and cutting, or frozen sections may be cut from fresh material, then fixed.)

2. Stain in a saturated solution of Sudan III or Sudan IV in 70% alcohol, until large fat globules appear red and smaller ones various shades of orange. Ordinarily this requires 30 minutes to 1 hour. It is preferable to filter the solution immediately before use, and to carry out the staining in a tightly covered receptacle, because as alcohol evaporates some of the stain will precipitate. A deep stain can be obtained within a very few minutes by using a saturated solution of the dye in 70% alcohol containing 2% sodium hydroxide or potassium hydroxide.

3. Rinse off the excess staining solution with distilled water.

4. If desired, counterstain the preparations in alum hematoxylin to obtain a deep blue coloration of nuclei. Twenty minutes in the dilute solution used for progressive staining (p. 353) is usually about the right length of time.

5. Wash material in running water for 10 to 15 minutes.

6. Mount in glycerol, glycerol jelly, or some other medium which does not contain solvents which remove fats. Place thin glass supports under

the coverglass if there is any danger that the weight of the coverglass will squeeze the fat out of the cells. If it is essential that the mounts be kept for an indefinite period, use circular coverglasses and seal the preparations with a suitable cement (see p. 461).

It is also possible to use Sudan III and Sudan IV for staining lipid inclusions in unfixed material. Protozoa, other small organisms, or cells which have been teased apart are simply immersed in a drop of the staining solution under a coverglass. In a few minutes, the fat droplets will begin to concentrate the dye.

It should be remembered that the color image of a stained structure shows up better when this is brightly illuminated. Small lipid inclusions do not accumulate enough of a Sudan dye to become intensely colored. Their recognition may therefore require widening the aperture of the iris diaphragm of the substage condenser.

Staining with Sudan Black B. This dye, which stains lipids black or blue-black, has become quite popular in recent years. It often brings out fat droplets with considerable intensity, evidently because its solubility in fats is very high. Sudan black B may be applied to fresh or fixed material in exactly the same manner as Sudan III and Sudan IV.

Blackening by Osmium Tetroxide. This substance oxidizes unsaturated lipids, rendering them less soluble. In the process, osmium tetroxide

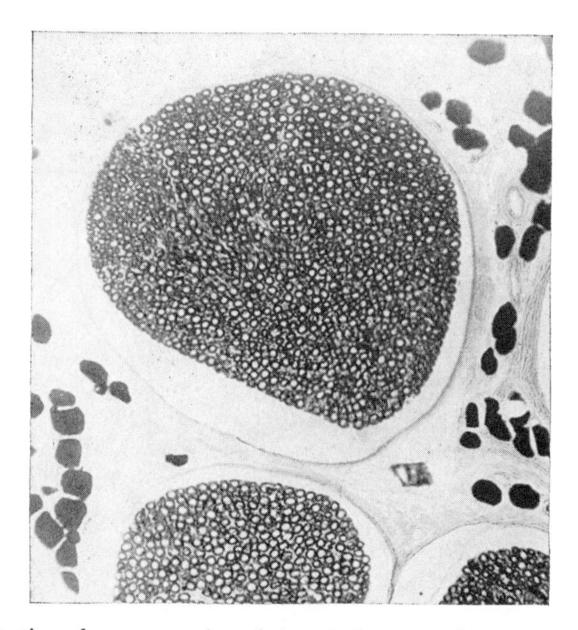

FIG. 72. Portion of a cross-section of the sciatic nerve of a cat. Fat cells and the myelin sheaths of nerve fibers have been impregnated with osmium tetroxide after fixation in formalin; double embedded in nitrocellulose and paraffin; sectioned at 8 μ; lightly counterstained with eosin.

is reduced to a lower oxide, which is black and therefore a useful indicator of the presence of fatty substances. Because osmium tetroxide is soluble in both saturated and unsaturated lipids and might become reduced under the influence of some other substance (alcohol), blackening must not be considered a conclusive test for unsaturated lipids.

A good method for blackening lipids by the use of osmium tetroxide is outlined below.

1. Place small pieces of fresh tissue, or tissues fixed in 10% formalin and rinsed in distilled water, into 2% osmium tetroxide. Use only enough of the solution to cover the tissue. The bottle should have a tight-fitting glass stopper, and should be kept in a dark place.

2. In a week or 10 days, pour off the osmium tetroxide solution (which cannot be used again) and wash the tissue with several changes of distilled water. Then wash it in running water for 2 hours.

3. Dehydrate the material and make thin paraffin or nitrocellulose sections in the usual way. Double embedding in nitrocellulose and paraffin (p. 269)is recommended if the tissue involved is nerve, because it permits thin sections to be cut without tearing out the connective tissue sheaths.

4. *a.* Affix paraffin sections, or sections made by the double embedding method, to slides. Remove the paraffin with toluene and mount the sections in a resinous medium. If a simple counterstain is desired, the sections may be passed, after the paraffin has been removed, into a 0.1% solution of eosin Y in a mixture of 3 parts of terpineol-toluene (or carbol-toluene) and 1 part of absolute alcohol, then cleared and mounted. However, to bring out axis cylinders more sharply, it is advisable to bring the sections gradually to water and to stain them in 0.5% acid fuchsin. The sections may then be rinsed in water and dehydrated and cleared.

b. Dehydrate, clear, and mount nitrocellulose sections in the usual way. They may be counterstained with eosin in 90% alcohol, if desired.

It is also possible to make teased preparations or whole mounts of material in which lipids have been blackened with osmium tetroxide. Protozoa and other small transparent organisms are simply dehydrated, cleared, and mounted. They may be handled conveniently in smears on coverglasses.

The technique for making teased preparations of nerves is explained on page 178.

Glycogen

Polysaccharides to which the term glycogen is applied are found in most animal phyla, as well as in bacteria, fungi, and certain organisms (for example, dinoflagellates) which have affinities with both animals and plants. Glycogen is stained mahogany brown by aqueous solutions of iodine in fresh or properly fixed preparations, is digested by amylases (salivary ptyalin), and may be hydrolyzed to glucose by 0.5 normal hydrochloric acid. The glycogen localized in cells of vertebrates commonly presents a more or less flocculent appearance, and is soluble in cold water. Other types of glycogen ("paraglycogen") occur as very conspicuous granular inclusions which are not soluble in water unless it is heated.

Of the various methods used for demonstrating glycogen in permanent preparations, the technique of Bauer and the periodic acid-Schiff (PAS) procedure may be recommended. In both methods, the Feulgen reagent (i.e., a modified Schiff reagent) is applied to the tissue following treatment with an oxidizing agent. Neither method is absolutely specific for glycogen, because a number of polysaccharides, mucopolysaccharides, and related compounds react, after being oxidized, with the Feulgen reagent. When the PAS method is applied to sections in which there is a wide variety of tissue types, including cartilage and collagen, certain constituents of these will react to varying extents; thus, if the preparations are subsequently stained to bring out nuclei and to provide a general counterstain, rather interesting preparations for general histology may be produced.

Bauer's Method for Glycogen. 1. Fix pieces of tissue or smears in one of the following mixtures:

Absolute (or 95%) alcohol, 9 parts; formalin, 1 part.

Absolute (or 95%) alcohol saturated with picric acid, 9 parts; formalin, 1 part (Rossman's fluid).

Absolute (or 95%) alcohol saturated with picric acid, 85 ml.; formalin, 10 ml.; acetic acid, 5 ml. (Gendre's fluid).

In certain tissues, especially striated muscle of warm-blooded vertebrates, the glycogen disappears very rapidly after death of the animal. It is therefore important to fix the tissue immediately. If the fixative is chilled to a temperature of 5° C. or colder by packing with ice or by refrigeration, this will partially inactivate some of the enzymes involved in hydrolysis of glycogen, although it will also slow the penetration of the tissue by the fixative. In general, fixation at lower temperatures is advised if the glycogen in a particular tissue or organism is not known to be well preserved at room temperature.

The time required for fixation will vary according to the size of the pieces of tissue. However, it is advisable to use the smallest pieces practicable, because if the glycogen is not fixed quickly, much of it may be hydrolyzed by enzymes present in the cells. Twelve hours should be long enough for fixation of pieces of tissue about 3 mm. thick. About an hour should be sufficient for thin smears on coverglasses or slides.

2. Pieces of tissue fixed in a mixture containing picric acid should be washed in 95% alcohol until most of it is removed. An hour or two in 95% alcohol will be long enough for removal of the formalin if formalin-alcohol has been used. The tissue may then be dehydrated in absolute alcohol,

embedded in paraffin, and sectioned. Sections 8 or 10 μ in thickness will be suitable for most purposes. After the sections have been affixed to slides and allowed to dry, the paraffin is removed in toluene and they are passed down the graded series of alcohols to water.

Smears on coverglasses or slides may be transferred from formalin-alcohol to 70% alcohol, and then through the lower grades of alcohol to water. If the fixative contained picric acid, it is advisable to leave them in 70% alcohol until most of the yellow color has been removed.

3. Treat the smears or sections with 4% chromic acid for at least 1 hour.
4. Wash them in running tap water for 5 minutes, then rinse them in distilled water.
5. Stain the preparations for about 15 minutes in the Feulgen reagent (p. 433).
6. Rinse them thoroughly in 3 changes of dilute sulfurous acid (p. 434).
7. Wash them for several minutes in running tap water.
8. If desired, the nuclei may be stained progressively with dilute alum hematoxylin (p. 353); counterstaining with orange G or fast green FCF (see step 5, p. 434, in the method for the Feulgen nucleal reaction) may be helpful in bringing out the general form of the cells or small organisms.
9. Dehydrate the sections or smears, clear them in toluene, and mount them in a resinous medium.

Glycogen will be bright red. The large granules found in gregarines and some other protozoa are especially conspicuous after this method.

Two techniques, among others, may be used to confirm that the material stained by Bauer's method is glycogen or a closely related substance. If representative sections or smears are mounted in a drop of Lugol's iodine solution (p. 369), the glycogen should stain mahogany brown. If comparable preparations are exposed to saliva (full strength, or diluted with 1 to 5 parts of distilled water) at about 37° C. for 30 minutes and then stained by the Bauer method or Lugol's iodine solution, no glycogen should be demonstrated at the usual sites.

Periodic Acid-Schiff Reaction (PAS)

This method is important in histochemistry for demonstrating a variety of carbohydrates, including glycogen, starch, and cellulose. Moreover, it generally stains mucins, matrix of cartilage, collagen, and a number of other secretions or tissue constituents because they contain appreciable amounts of polysaccharides. When used in connection with a compatible nuclear stain and a delicate counterstain for cytoplasm, the PAS technique produces very useful and attractive histological preparations.

If the PAS method is to be used for staining glycogen or other carbohydrates that are dissolved by most fixatives, especially aqueous fixatives, it is necessary to fix the tissue in one of the strong alcoholic solutions recommended for this purpose (see p. 446 for fixatives suitable for preserving glycogen). However, interesting preparations of sections that show mucous glands, cartilage, and a variety of other structures, may be obtained following any one of a number of standard histological fixatives, such as Bouin's fluid, Zenker's fluid, and Heidenhain's "Susa" mixture.

1. Following fixation and appropriate post-fixation treatment, embedding, and sectioning, bring the sections to water.
2. Treat the sections with a 1% aqueous solution of periodic acid for 5 minutes. This will oxidize substances that give a PAS-positive reaction.
3. Wash sections in running water for about 5 minutes.
4. Stain them in the Feulgen reagent (modified Schiff reagent, p. 433) for 10 to 15 minutes.
5. Rinse sections briefly in distilled water, then wash them in running water for about 10 minutes.
6. If a general histological preparation is desired, stain nuclei with alum hematoxylin, preferably by the progressive method (p. 353), and counterstain the cytoplasm lightly with fast green FCF in 70% alcohol (p. 390) (or with orange G in 90% alcohol, p. 386).
7. Dehydrate, clear, and mount the sections.

Formerly, it was customary, after staining in the Feulgen reagent, to wash the preparations in three changes of dilute sulfurous acid, as in step 4 of the Feulgen method (p. 434), or in a freshly prepared 0.5% solution of sodium metabisulfite. This seems to be unnecessary if the sections are thoroughly washed in water after staining.

Strongly PAS-positive substances, such as glycogen and mucopolysaccharides containing hexose sugars, are stained deep reddish-purple. Matrix of cartilage, collagen, and other histological components containing appreciable amounts of PAS-positive compounds are generally rose. Care should be taken in counterstaining, especially with fast green, so as not to spoil an excellent differentiation of such components.

REFERENCES

Bancroft, J. D., 1967. *An Introduction to Histochemical Technique.* New York: Appleton-Century-Crofts.

Chayen, J., Bitensky, L., Butcher, R., and Poulter, L., 1969. *A Guide to Practical Histochemistry.* Philadelphia: J. B. Lippincott Co.

Pearse, A. G. E., 1968. *Histochemistry—Theoretical and Applied,* Vol. 1, 3rd ed. Baltimore: The Williams & Wilkins Co.

CHAPTER 27

Permanent Mounting Media

Among the desirable characteristics that a medium used for preparing permanent mounts should possess, several are of special importance.

Preservative Qualities. The substance must preserve the microscopical structures unaltered for an indefinite time, protecting them from desiccation, oxidation, and microbial decay. It must neither dissolve nor otherwise destroy any of the tissue elements which are to be studied. For example, fatty substances are soluble in the resins commonly used for mounting, as well as in reagents from which tissues are transferred to the resins, and should therefore be either mounted in water-soluble media or treated in such a way that they can be preserved in resinous media.

Stability. The mounting medium should be physically and chemically stable. Evaporation, crystallization, or extensive chemical change in the medium may injure or obscure the tissue elements.

Consistency. A substance which solidifies is preferable, because it will protect the preparation from damage during storage, handling, or examination.

Color. Obviously, the mounting medium should not be deeply colored, as it may mask the color differentiation of certain structures or of the dyes with which they are stained.

Refractive Index. A function as important as preservation is that of increasing the visibility of the structures. The mounting medium should render the specimens transparent enough so that light will readily pass through them, but not to such an extent that important structures become difficult to see.

Transparency and visibility depend upon the physical law that a ray of light (called the "incident ray") passing obliquely from one medium to another, in which its velocity is different, becomes divided into reflected and transmitted rays. The transmitted ray, at the point it enters the new medium, becomes bent, or *refracted*. The extent to which it is bent depends upon two factors: (1) the relative velocity of light in the two media, and (2) the obliquity of the incident ray. These same factors determine also the proportions of the incident ray which are reflected and refracted. The greater the obliquity of the incident ray and the reduction in velocity, the greater will be the proportion of light reflected and the angle at which the transmitted ray is bent. The greater the bending, the smaller will be the angle of the refracted ray (r) to the perpendicular, as compared with the angle of the incident ray (i). The true relations of these angles being those of their sines, the index of refraction is expressed by the following formula:

$$\frac{\text{Sin } i}{\text{Sin } r} = \frac{\text{Index of refraction of refracting medium}}{\text{Index of refraction of incident medium (air = 1)}}$$

Refractive indexes are calculated with reference to passage of light from air into other media, and the refractive index of air is taken as 1. For example, if a ray of light passing through air strikes water at an angle of 60° to the perpendicular, it is bent to an angle of 40° 38′:

$$\frac{\text{Sin } i \text{ (60°), or .86603}}{\text{Sin } r \text{ (40° 38′), or .65115}} = \frac{\text{Refractive index of water}}{1} = 1.33$$

When a ray of light passes into a medium in which its velocity is greater, reflection and refraction occur according to the same laws, but the refracted ray is bent *away* from the perpendicular.

If we apply these laws to a tissue which is permeated and surrounded by a mounting medium, we find that when the refractive index of the medium is either much lower or much higher than that of the tissue elements, a large proportion of the light falling upon the specimen will be reflected at its surface, and a relatively small amount of light—much refracted—will pass through the tissue. However, the difference in refractivity between the tissue elements and the medium permeating them will bring about refraction and reflection at the surfaces of the various structures within the tissues, causing these to be visible as long as the amount of transmitted light is adequate. As the refractive index of the mounting medium increases, refraction and reflection of light between tissue elements diminishes proportionately. More light will pass through the tissues, but the visibility of a particular structure is decreased as its refractive index is approached by that of the medium. When the refractivity of the medium equals that of any colorless structure, all the light passes straight through the latter, without reflection or refraction taking place, and that structure becomes invisible. Should the refractivity of the medium exceed that of the structure, transparency will diminish and the visibility of the structure will be increased.

The practical rule supported by these facts is this: Specimens should be mounted in a medium which has an index of refraction close enough to that of the tissues that these become reasonably transparent, but which is far enough below or above the refractive index of the more important tissue elements to render them visible. *This rule is especially applicable to unstained or lightly stained structures. The absorption of color by strongly stained elements makes them visible (although their boundaries may not appear sharp) in media of equal refractivity.* Unstained specimens, unless they possess considerable natural color, are commonly mounted in glycerol or some other water-soluble medium of low refractivity, or in a highly refractive resin. Stained materials are generally mounted in Canada balsam or in some synthetic resin whose refractive index approximates that of the tissue elements. To improve the visibility of mitotic spindles and other fine structures which stain lightly, material is sometimes mounted in less refractive resins, such as euparal.

Effect upon Stains. Specimens should be mounted in a medium which has the least possible effect upon the dyes with which they have been stained. This subject will be considered in connection with the various mounting media and stains.

Classification of Mounting Media. In early days, specimens to be examined microscopically were mounted either dry or in water. For obvious reasons, neither method was suitable for preservation of soft tissues. Hard structures such as scales, hairs, bones, teeth, and shells keep well when dried, but in this condition they can be studied to advantage only if they are small and not too opaque. Materials and methods for dry mounts are described on page 166. The media which have since been introduced for the preservation and clearing of microscopical material belong to two distinct classes: (1) resins produced by coniferous trees, and synthetic resins having similar properties; (2) water-soluble media, including glycerol, gelatin, various gums, and sugars.

RESINOUS MEDIA

The early English microscopists sought a means to overcome the opacity of dried specimens, and Prichard's mixture of isinglass and a gum (1832) was a first step in the exploration of various substances which might be suitable for the purpose. In 1835, Cooper introduced the resin of the balsam fir, which, because of its high refractive index, rendered dry structures translucent. A few years after this, it became customary to mount soft tissues in glycerol. In 1851, Clarke made the important discovery that tissues of many kinds could be mounted in balsam if the water in the specimens was replaced by alcohol, and the alcohol was then replaced by oil of turpentine.

The value of Clarke's method was quickly recognized by German zoologists, who were beginning to use stains and who felt the need for more transparent mounts. The increasing use of balsam led Stieda to publish, in 1866, a comparison of numerous clearing oils and to propose some improvements in the technique for mounting specimens in balsam. Resinous media, including balsam, are the ones most used for microtechnique.

They preserve tissue structures well, render specimens sufficiently transparent, and harden to form durable preparations.

Instructive summaries of the properties of resinous media, especially synthetic resins, can be found in the contributions of Lillie, Windle, and Zirkle (1950) and Lillie, Zirkle, Dempsey, and Greco (1953).

Synthetic Resins

These products offer many advantages over balsam and other natural resins. Their chemical composition is consistently the same, and in preparations made with them certain dyes which fade rather quickly in balsam are preserved very well. The synthetic resins are readily soluble in volatile hydrocarbons such as toluene and xylene, and harden more quickly than balsam, which contains a variety of slow drying components. Good synthetic resins are nearly colorless from the beginning and do not become yellow or brown with age.

As long as they have an acceptable refractive index, approximating that of fixed proteins (about 1.52 to 1.54) and that of glass (about 1.52 or slightly higher), synthetic resins will probably be the choice of most laboratory workers. For certain purposes, however, balsam and some other natural resins or mixtures may be preferred. One definite advantage of balsam (especially that which has not been dried and re-dissolved in toluene or xylene) is the way it lends itself to being temporarily liquefied by heat without losing its volatile oils too quickly.

A number of suitable synthetic mounting media have appeared on the market. Most of them are β-pinene polymers. The following list may not include some types which are completely acceptable, but it does give most of those which are regularly found in the trade:

Bioloid Synthetic Resin (Will Corporation)
Harleco Synthetic Resin (HRS) (Hartman-Leddon Co., Inc.)
Histoclad (Clay-Adams, Inc.)
Kleermount (Carolina Biological Supply Co., Inc.)
Permount (Fisher Scientific Co.)
Piccolyte (CCM: General Biologicals, Inc.)
Technicon (Technicon Co.)

Most of these resins are sold as solutions in toluene or xylene, although certain of them are available also as lumps which may be dissolved as desired. Those which come packaged in small bottles with applicator rods may be the choice in some laboratories, but the narrow mouths of these bottles make them less convenient to use than standard balsam bottles. Most of the ready-made solutions have a consistency which is appropriate for mounting sections, and become suitable for thicker whole mounts after some of the solvent has evaporated. The evaporation may, of course, be hastened by allowing the dispensing bottle to stand with its lid ajar or by placing the bottle, without its lid, under a larger container inverted over it.

Some of the synthetic resins—or at least certain samples—tend to crystallize after a few years, and therefore to become opaque. To prevent this, add about 5 ml. of terpineol to each 100 ml. of the stock of the medium dissolved in toluene or xylene, or add a few drops to the balsam bottle used for dispensing the medium. Mounts that have already begun to crystallize badly can only be salvaged by carefully soaking off the coverglasses in toluene and making new mounts.

Canada Balsam

This is a mixture of resins, and certain essential oils from which the resins are derived, obtained from the balsam fir (*Abies balsamea*), a tree which ranges from Newfoundland to Lake Superior. Balsam is collected from blisters on the bark, or accumulated in containers placed below cuts made in the tree.

The refractive index of balsam (1.52 to 1.54) makes it excellent for mounting structures which are naturally colored, or which have been stained or impregnated with metals. However, it is similar to the synthetic resins in common use in conferring low visibility upon delicate structures which possess little or no color.

Of the important substances used in microtechnique, balsam is one of the more variable. Because the appearance and durability of slide preparations mounted in balsam depend to a large extent upon the properties of the balsam, selection of a suitable product must be done with care. The properties which should be considered are (1) color, (2) consistency, (3) clarity and freedom from foreign matter, (4) drying qualities, and (5) neutrality.

Natural balsam, containing its essential oils, has a thick, syrupy consistency, and is pale yellow in color. Because xylene, toluene, and other solvents are much cheaper than balsam, this product is sometimes thinned excessively. The best quality, sold as "paper-filtered" Canada balsam, is perfectly clear and free of foreign matter. In thin mounts, a good natural balsam will dry and hold the coverglass firmly in place within 2 weeks at ordinary room temperature. Thicker mounts dry with proportionate slowness. The preparations will keep for many years without showing crystallization of balsam near the edge of the coverglass. Occasionally a natural balsam which fails to harden even after months or years is encountered. This is one of the chief arguments against the use of natural balsam. However, the vigilance required in this matter and the small loss sustained occasionally by discarding a bottle of balsam, are more than counterbalanced by the advantages of its light color and good keeping quality. Balsam is slightly acid in reaction, and should be neutralized by the method to be described.

Balsam which has been dried and re-dissolved in toluene or some other volatile liquid has been recommended by various writers. The chief claim made in its favor is that the natural oils, eliminated by the drying process, are harmful to stains. However, *neutralized* natural balsam preserves stains quite as well as dry balsam which has been re-dissolved and neutralized. It is certainly true that balsam dissolved in xylene or toluene dries more quickly than the natural product, but this advantage is more than offset by the fact that eventually it crystallizes and cracks away from the glass at the margin of the mount. Dry or re-dissolved balsam obtainable in the market generally has a disagreeable deep yellow or brown color, owing to the application of too much heat in the process of drying it.

NEUTRALIZATION OF BALSAM. Balsam which is acid is certain to fade most stains, and it is therefore necessary to neutralize every lot of balsam before putting it to use. The fact that "neutral" appears on the label is not conclusive evidence that the sample is really neutral. Place the natural or re-dissolved balsam in a relatively large bottle or flask and dilute it with toluene or xylene until it reaches a watery consistency. Add sufficient powdered calcium carbonate to form a layer about 5 mm. deep on the bottom of the container. Keep the container stoppered and shake it briskly every day for a week or longer. Then decant the liquid into a glass funnel containing fine filter paper and catch the filtrate in a wide-mouthed bottle. Allow the filtrate to concentrate by evaporation to the consistency of a thick syrup; the process may be accelerated, if necessary, by placing the liquid in an electric oven (not over a flame or in a water bath) at a temperature not exceeding 45° C. (Warning! Do not use an oven in which a spark from the thermostat may lead to an explosion.) Higher temperatures will darken the balsam. Stopper the bottle and store it in a cool, dark place. It is a wise precaution to keep a lump of marble in the stock bottle of balsam.

Gum Damar

The resins dispensed under this name (or as damar balsam) are produced by several species of damar pine (*Agathis*) which grow in the East Indies, Australia, New Zealand, and the Philippines. East Indian damar, the variety ordinarily obtained, is found in commerce as a mixture of lumps and powder, containing considerable trash. The process of clarifying this crude product will be discussed presently.

Damar, dissolved in xylene, toluene, or benzene, is used in the same manner and for the same purposes as Canada balsam, and offers some advantages over the latter. It has a slightly lower refractive index (1.52) than some samples of Canada balsam and some synthetic resins. It therefore gives better definition to unstained or lightly stained structures. Damar is lighter in color than most re-dissolved balsam. In thick mounts it shows the colors of dyes, particularly blue dyes, to better advantage. Another characteristic which makes it desirable for thick mounts is that it dries well to the center of the preparations, so that the specimens are less likely to drift. For these reasons, some devotees of natural resins have advocated substitution of damar for Canada balsam. Others have advised against the use of damar, asserting that it always becomes cloudy or forms a precipitate sooner or later. No doubt this disparity has resulted from the use of different types of damar or different methods of preparing it. If damar is prepared according to the following method, no traces of discoloration or deterioration are likely to be noted.

Pick out lumps of damar, avoiding those which are very dirty, and heat them cautiously until they melt. Pour the melted mass into 6 or 8 times its volume of toluene or xylene, meanwhile stirring the liquid. After a day or more, decant the solution from the sediment into a funnel containing coarse filter paper. Allow the filtrate to stand in a covered vessel for 10 days or longer, during which time a fine, flocculent precipitate will form. Again decant the liquid into a filter, this time using a fine paper. Place the filtrate in an open vessel, in a warm place, until evaporation of the solvent brings it to a syrupy consistency. Store the product in a stoppered bottle. It is unlikely that more precipitate will form. If it does appear, however, separate the clear liquid by decanting it, or dilute it, filter it, and again concentrate it.

Sandarac Media (Gilson's Camsal Balsam; Euparal)

Gum sandarac, also called "cedar gum," is obtained as pale yellow tears from a large pinaceous tree (*Callitris quadrivalvis*) growing in the mountains of North Africa. By dissolving this resin in different organic solvents, Gilson compounded mounting media which have lower refractive indexes than Canada balsam, damar, and most synthetic resins. For this reason, they confer greater visibility upon delicate and lightly colored structures such as mitotic spindle fibers. Another advantage claimed for Gilson's media is that material can be mounted in them directly from 95% alcohol, thus sparing them exposure to the shrinking effects of absolute alcohol and clearing agents such as toluene. However, it is advisable to carry specimens at least as far as absolute alcohol before mounting them in any of these media. *Camsal balsam*, consisting of "camsal" (a mixture of equal quantities of camphor and phenyl salicylate), sandarac, and propyl or isobutyl alcohol, has the lowest refractive index (1.478) and does not clear sufficiently for most purposes. *Euparal* (*diaphane*) contains eucalyptol and paraldehyde in addition to camsal and sandarac. As ordinarily used for mounting, its refractive index is about 1.483, but this rises as it hardens. Euparal clears adequately and gives excellent visibility to most structures. It is best to rinse specimens, before mounting them, in the special solvent called "essence of euparal." Euparal causes hematoxylin stains to fade, and this greatly limits its usefulness. It seems to have no harmful effects upon coal tar dyes, and has proved quite valuable for mounting blood films stained by Giemsa's method. *Green euparal* (euparal vert) contains a salt of copper, and this is supposed to intensify and preserve hematoxylin stains. Its virtues are questioned, and many workers have found it less suitable than plain euparal.

Euparal and essence of euparal are manufactured in England and are available from various dealers in the United States. Camsal balsam seems to have almost disappeared from the trade.

Venetian Turpentine

The oleoresin of the larch, *Larix decidua*, has a refractive index of slightly over 1.54, and in this respect is similar to Canada balsam. It differs from the latter in being completely soluble in absolute alcohol. Therefore, material may be transferred from absolute alcohol to a 10% alcoholic solution of Venetian turpentine. The medium is then allowed to thicken, by evaporation in a desiccator containing soda lime, to the proper consistency for mounting. This eliminates the laborious and time-consuming processes of gradually clearing and infiltrating whole specimens with balsam, and it reduces the possibility that collapse or some other serious distortion may occur. The method is useful for delicate, vesicular organisms such as *Volvox*, and has found favor among botanists for mounting algae such as *Spirogyra*. However, it has several serious disadvantages. A solution of Venetian turpentine in alcohol is so sensitive to water that in very humid weather it may be difficult to transfer material to this solution and get it into the desiccator quickly enough to avoid clouding because of absorption of moisture from the air. Venetian turpentine is a most variable product. Some samples of it will crystallize eventually, and spoil the preparations. This medium also has an undesirable dark color, and commonly remains soft and gummy long after mounting. It preserves hematoxylin stains well, but is not suitable following alcohol-soluble stains. All things considered, it is inferior to Canada balsam and synthetic resins.

Resins of High Refractive Index

Attention has been called to the fact that the visibility of a colorless structure increases as the refractive index of the surrounding medium becomes either higher or lower than its own. Many delicate skeletal structures (diatoms, skeletons of radiolarians, sponge spicules, and thin chitinous structures) have about the same refractive index as most synthetic resins and Canada balsam, and so should not be mounted in it. Aqueous media are much less refractive, but also less permanent. For such materials, it is best to use a very highly refractive resin, because it will give both high visibility and permanence.

For a long time, natural resins called styrax, or storax (refractive index about 1.58) and tolu balsam (refractive index 1.64) were thinned by solvents and used for mounting structures which are difficult to distinguish in other resinous media. One of the synthetic resins, Hyrax (refractive index 1.71) (Fisher Scientific Co.) is excellent for bringing out sculpturings on diatoms, fine setae on appendages of crustacea, and comparable details.

WATER-SOLUBLE MEDIA

The first medium to be used successfully for preservation and clearing of soft tissues was glycerol. It was introduced by Warrington in 1847. Soon afterward, a glycerol jelly which would harden and make more durable mounts than liquid glycerol came into use. During the latter part of the nineteenth century, various kinds of liquid media were devised. Some of these contained copper, mercury, or other chemicals which functioned as fixing agents. Others were solutions of potassium acetate, chloral hydrate, or sugar, and owed their usefulness to high refractive indexes. Liquid or semi-liquid mounts were rendered more or less permanent by applying several coats of cement to the edge of the cover and adjacent part of the slide.

The need for permanent preparations led to development of gum and sugar media which would harden. Owing to the obvious advantages of resins for permanent mounting, the development of methods by which these could be employed routinely caused aqueous media to fall into disuse except for special purposes.

Glycerol

This substance has several uses in microscopy. It will serve as a preservative for unmounted material, and it is a good medium in which to tease apart tissues, because it makes them soft and pliable. Glycerol is also used for making temporary mounts. Having a moderate index of refraction (1.46), it clears unstained specimens sufficiently to show their internal structures, but not enough to render their parts invisible. Material may be brought into glycerol from water or alcohol by placing it in a 10% solution in either liquid and allowing this to concentrate by evaporation, or by adding more glycerol at intervals until the mixture consists almost entirely of glycerol.

Glycerol is used occasionally as a permanent mounting medium for tissues containing lipids, and for nematodes, eggs of certain types, and some specimens possessing a cuticle which is not permeable to resins. It is also useful for some extremely delicate objects. It does not preserve stains well. A serious disadvantage of glycerol in permanent mounts is that the preparations, no matter how they are sealed, are vulnerable. The coverglasses are easily broken and the cement is likely to crack or peel off eventually, allowing the liquid medium to leak out. For permanent mounts of lipids or other materials which cannot be mounted in resins, it is advisable to use glycerol jelly or some other mixture containing a substance which hardens at least slightly.

Gum Arabic (Gum Acacia, Gum Mucilage)

The market product of gum arabic is generally the gum of *Acacia senegal*, cultivated extensively in the Sudan and in Senegal. A useful mounting medium, which sets hard and keeps well, is made by dissolving selected clean gum in distilled water, to make a solution having a consistency about equal to that of glycerol. This is filtered and 2 drops of formalin are added to each 10 ml. The mixture is suitable for mounting crystals, radiolarian skeletons, sponge spicules, diatoms, nematodes, and pressed preparations of tissues impregnated with gold or silver. Material can be mounted from water, but it is well to clear it first in glycerol (gradually, if necessary). Place the specimen upon a slide, drain it and cover it with a drop of the gum solution. Then quickly put in place a circular coverglass, which has been moistened by breathing upon it. Principal disadvantages of this and other gum media lie in the difficulty of avoiding the inclusion of air bubbles in the mounts, and in the tendency of the media to shrink badly. A day or more after making the mounts, it is well to seal them with at least three coats of asphalt cement, gold-size, or Duco cement.

Media Containing Glycerol, Gum Arabic, and Other Water-Soluble Ingredients

Glycerol Jelly

Gelatin	8 gm.
Distilled water	52 ml.
Glycerol	50 ml.
Phenol (crystals)	0.1 gm.

Soak the gelatin in the water for an hour or longer, and then dissolve it by heating in a water bath at a temperature of 65° to 75° C., but no higher than this. Add the glycerol and phenol, stir the mixture, and again heat it in the water bath for 30 minutes. A wide-mouthed bottle with a tight-fitting glass stopper or screw cap is perhaps most convenient for storing the medium. A short time before the medium is to be used, the container should be immersed in a water bath maintained at about 50° C.

Specimens are transferred from glycerol or water (preferably the former) to a drop of glycerol jelly on a slide. This should be done while the slide is on a warming table held at about 40° C., so that the gelatin will remain melted. Cover the preparation with a circular coverglass. When the mount has cooled and become firm, it is cleaned carefully and sealed with asphalt cement or some other suitable sealant (p. 461). Glycerol jelly has about the same optical properties as glycerol, and will harden to some extent. However, it is not as durable as gum media.

Farrant's Medium

Gum arabic	30 gm.
Distilled water	30 ml.
Arsenious oxide	1 gm.

Allow the mixture to stand for several days, and add

Glycerol	30 ml.

After adding the glycerol, dilute the solution with an equal amount of distilled water and filter it. Allow it to evaporate down to about its original volume, add 0.1 gm. of phenol or a crystal of thymol, and bottle it.

Filtering of this and other media containing gum arabic is recommended when mounts of extreme clarity are desired. The type of unsightly, insoluble material with which so many samples of gum arabic are contaminated may settle or come to the surface after the medium has been allowed to stand for a few days, whereupon the conspicuous impurities can be skimmed off or separated by decanting carefully, or both. Purification to this extent may be sufficient for some purposes.

Farrant's medium has the same advantages and disadvantages as the preceding one, and is used in the same way and for the same types of material. It clears the structures more completely than gum without glycerol. This may prove advantageous or not, depending upon the type of material to be mounted in it.

Berlese's Medium

Gum arabic	15 gm.
Distilled water	20 ml.
Dextrose syrup	10 gm.
Chloral hydrate	160 gm.
Acetic acid, glacial	5 ml.

Dissolve the ingredients in a water bath. Filter the solution with the aid of a suction pump; or dilute it, filter it, and allow it to concentrate by evaporation.

This medium, which Berlese used for mounting mites, is excellent also for ticks, aphids, and other small insects and their larvae. Owing to its high content of chloral hydrate, the solution is highly refractive and clears specimens much more than any of the other aqueous media herein described. Living specimens may be placed directly in the medium on a slide, or may first be killed in a 10% aqueous solution of acetic acid. Material which has been preserved in alcohol should be washed thoroughly in 10% acetic acid before it is mounted. Put a circular coverglass upon the preparation, warm the slide, and lay it in a dust-free place to dry for a week or two. Then seal the mount with three coats of a suitable slide-ringing cement.

Hoyer's Medium

Gum arabic	25 gm.
Chloral hydrate	1 gm.
Distilled water	25 ml.

Dissolve the chloral hydrate first, then add the gum arabic. Hold the final mixture in a water bath at about 50° C. until the gum arabic has dissolved completely. Filter the mixture with a suction pump; or dilute it, filter it, and evaporate it down to its original volume.

Hoyer's medium is widely used for arthropods, especially small insects and arachnids. Living specimens, or preserved specimens which have been washed in water, may be transferred directly to it. After applying a coverglass, allow the preparation to lie flat in a dust-free place for at least a few days before sealing the mount with a suitable cement.

Aman's Lacto-Phenol Medium

Phenol (crystals)	20 gm.
Lactic acid	20 gm.
Glycerol	16 ml.
Distilled water	20 ml.

This is a favorite medium (refractive index about 1.44) in which to preserve and clear nematodes for study. The solution may be added, drop by drop, to a small volume of water containing living nematodes. When the resulting mixture contains 10 to 20% lacto-phenol, allow it to concentrate by evaporation of water. Or the nematodes may be fixed in hot 70 or 80% alcohol, then cleared gradually with glycerol, and transferred to the medium.

It is possible, though not completely practicable, to make permanent mounts in this medium, by sealing the preparations with a mixture compounded by melting 2 parts of anhydrous tallow with 5 parts of rosin (colophonium). In most cases, it is probably more satisfactory to mount the worms in glycerol jelly.

DOUBLE MOUNTING

This technique has found favor with commercial preparators of whole mounts of nematodes and some other materials which are mounted most appropriately in glycerol jelly and similar aqueous media. Coverglasses of two sizes are used. The aqueous medium containing the specimen is placed on the larger coverglass and the smaller coverglass is applied over it. After the excess medium has been carefully wiped or trimmed away, the pair of coverglasses is left in a dust-free place to allow the edges of the medium to harden a little more. Then the coverglasses—the smaller one under the larger one—are mounted on a slide over a drop of a resinous medium. A mount of this type is usually just as durable as an ordinary whole mount in a resinous medium. However, if the resinous medium is eroded by glycerol or some other ingredient of the aqueous medium, the preparation will eventually become unsightly, and droplets of the resin may even collect around the specimen.

CEMENTS AND VARNISHES

These substances are used principally to coat the edges of dry mounts, or those made in aqueous media. Such a coating is necessary to prevent the preparation from drying out or absorbing moisture and, in the case of dry or liquid mounts, to hold the coverglass in place. Preparations which must be sealed are generally mounted under circular coverglasses. Care should be taken to avoid using an excess of the medium, but if some does exude it should be carefully trimmed or wiped away, so that the cement will make good contact with absolutely dry surfaces of both the slide and coverglass. Then each slide is placed upon a turntable (Fig. 15, p. 165), centered, and rotated while the cement is being applied with a small camel's hair brush to the edge of the coverglass and adjacent surface of the slide. Each coat is allowed to dry before another is added. Ordinarily, three coats are sufficient. If square or rectangular coverglasses are used, the cement or varnish may be applied free-hand or with the aid of a guide of some sort, but the resulting preparations are seldom so neat as those made with the help of a turntable.

Cements or varnishes are also employed to make shallow cells in which dry specimens or those in aqueous media may be mounted. By applications of successive coats of cement or varnish, a cell may be built up to the required depth. The ring should be of such a width and overall diameter that when the coverglass of chosen diameter is laid in place, at least 2 mm. of the ring will be external to its edges. After the coverglass has been applied, the slide should be warmed a little so that the coverglass may be pressed into the softened cement. Ideally, no mounting medium should exude, and there should be no air pockets due to using insufficient medium. Good judgment and experience will be necessary to prepare mounts which are perfect in this respect.

After the slide has cooled and any excess of medium removed carefully and completely, it is placed on a turntable and several coats of cement applied to make contact with the margins of both the coverglass and the ring.

A wide variety of cements, varnishes, and waxes have been recommended for sealing mounts. Some of them are completely unsatisfactory. The following should prove adequate for most purposes.

ASPHALT CEMENT. Quick-drying asphalt cement (asphaltum), available from dealers in scientific supplies, is one of the most tenacious of cements. It serves well for all of the purposes just described.

GOLD-SIZE. This clear, amber varnish makes a neat finish for preparations. It is satisfactory for sealing dry mounts or those made in any gum media which harden. It also makes good cells for such preparations. It is not suitable for mounts in glycerol or glycerol jelly. These media cause gold-size to become cloudy and to separate from the glass surfaces, and after this happens the mounts leak or dry up.

CELLULOSE VARNISHES OR ENAMELS. "Duco" cement and similar products sold at paint stores are used rather commonly for making cells and for sealing mounts. These give a more elegant finish than does asphalt cement, but are more likely to crack and peel away from the glass. They may be obtained in colorless or colored forms. Although black seems to be most popular with commercial preparators, some workers may wish to take advantage of other colors in working out a rough coding system.

NAIL POLISHES. Many of these, which also come in a wide variety of colors, are quite suitable for sealing mounts.

REFERENCES

LILLIE, R. D., WINDLE, W. F., and ZIRKLE, C., 1950. Interim report of the Committee on Histologic Mounting Media: resinous media. Stain Technology, *25*, 1–9.

LILLIE, R. D., ZIRKLE, C., DEMPSEY, E., and GRECO, J. F., 1953. Final report of the Committee on Histological Mounting Media. Stain Technology, *28*, 57–80.

CHAPTER 28

Drying, Labeling, and Storing Microscopic Preparations

A preparation cannot be considered finished until the coverglass is firmly fixed in place and a label bearing the essential facts about it (or a code number through which accurate information may be found) has been applied to the slide. It should then be put away in a suitable container, where it can be located whenever it is needed. These final steps deserve careful attention, because the permanence and ultimate usefulness of slide preparations may depend to a large extent upon them.

DRYING

The following remarks apply only to mounts made in balsam or other natural or synthetic resinous media. Preparations mounted in glycerol, glycerol jelly, gum acacia, or other non-resinous media are simply sealed with asphalt cement, gold-size, or some other material which will hold the coverglasses in place and prevent escape or evaporation of the mounting medium.

Correct mounting is prerequisite to satisfactory drying. The balsam or other resin must be of the proper consistency for each particular type of mount, and the amount used should be just sufficient to form a narrow rim around the edge of the coverglass after this is applied. For thin sections, use a medium having a rather light, syrupy consistency. For thick sections or entire objects, use a medium which has no more solvent than is necessary to permit handling it conveniently. To facilitate the handling of extremely thick media, it may be advisable to keep them warm. If the medium is too thin, it will gradually recede from the edge of the coverglass as the preparation dries, or the coverglass will be drawn down so tightly as to injure delicate material. On the other hand, if thin sections are mounted in a heavy medium, the coverglass may be supported so high above the specimens as to interfere with the use of objectives having short working distances, and if heat is employed for drying, rivers of the medium may exude. Use enough medium so that it will spread about a millimeter beyond the edge of the coverglass, because some shrinkage will take place during drying. If the amount of medium applied is not sufficient, some parts of the edge of the coverglass may soon be left unsupported and vulnerable to breakage. An extreme shortage of mounting medium will allow the formation of deep air pockets or bubbles under the coverglass, thus encouraging stains to fade. The sticky and unsightly results of using an excess of mounting medium are too familiar to require comment.

In preparing mounts of fragile entire objects, it is important to support the coverglasses. Otherwise, the weight of the cover, and the shrinkage of medium which occurs in drying, may crush such specimens. Devices for supporting coverglasses, together with methods of placing specimens and partially drying the preparations previous to application of coverglasses, are discussed in Chapter 11.

Position and Temperatures for Drying

Slides must remain in a horizontal position until the medium is well-hardened, in order to avoid movement of the coverglasses or specimens. Permanent storage in this position is always best, and is essential in the case of thick whole mounts.

At room temperature, preparations mounted in balsam or certain other natural resins dry slowly, becoming sufficiently hardened for use only after weeks or months. During this slow process the volatile constituents of the medium in central parts of the mount diffuse through the outer layers, as evaporation proceeds at the exposed edges. Thus the medium thickens more uniformly throughout the mount than is the case if the resin at the edges of the cover is rapidly thickened by means of heat. For this and other reasons to be discussed, drying at room temperature is very desirable, although sometimes impractical because of its slowness.

Heat is commonly employed to accelerate drying of balsam and other resins, and, if judiciously applied, does little or no harm. Preparations of most types can safely be dried at temperatures of about 35 to 40° C. Thick nitrocellulose sections, mounted without removal of the embedding medium, should not be exposed to higher temperatures, or the nitrocellulose may shrink and pucker. Temperatures above 40° C. are likely to be injurious to hematoxylin stains in thick objects, and to crystal violet, light green, acid fuchsin, and indigo-carmine in any type of preparations. Temperatures of 54 to 58° C. are often employed for drying whole mounts of unstained or carmine-stained objects, and for paraffin sections stained with hematoxylin and the more permanent aniline dyes such as eosin, orange G, fast green, safranin, and basic fuchsin. Such preparations, if needed for use as soon as possible, are kept at this temperature until they are thoroughly dry, which in the case of balsam mounts requires from 4 or 5 days for thin sections to 2 weeks for thick ones. Paraffin sections thus treated show no ill effects. The effects upon whole mounts are not so favorable. The preparations do not dry well toward their centers, and the medium is likely to shrink and to draw in bubbles of air. Thick objects such as flatworms sometimes tend to become more opaque if dried at temperatures as high as 54 to 58° C. If balsam has been used for mounting, it is likely to become discolored at these higher temperatures.

Synthetic resins dissolved in toluene or xylene dry much more rapidly than balsam. The use of heat is therefore less likely to be necessary in the case of synthetic resins, and if it is applied it will not have to be continued as long as for balsam.

An electric incubator or oven provided with thermostatic control is generally used for drying slides. The slides are laid upon trays, which are separated sufficiently to allow circulation of air, or are placed in racks made by removing the tops and bottoms of old wooden slide boxes. These racks are cheap and conserve space in the oven. In removing the bottom of a box for this purpose, leave a narrow strip at each side in order to retain the slides. When the rack has been filled with slides, tie a string around it to hold them in place.

Very rapid drying is imperative in diagnostic work where paraffin sections, frozen sections, or wet-fixed smears must immediately be prepared for examination with the oil-immersion objective. If the medium used for mounting dries slowly, as does balsam, place the preparations for a half-hour or more on a warming stage or in an oven set at 60 to 70° C. This procedure, however, is suggested only if necessary, and should not be used for preparations of entire objects, which it will cause to shrink and become more opaque.

Cleaning

After the slides have been dried, remove any unsightly excess of mounting medium by the cautious use of a soft cloth moistened with xylene or toluene. If a great deal of the medium has exuded, first scrape off most of it with a scalpel or razor blade.

LABELING

Lay the slide lengthwise before you with its ends so placed that the specimen is oriented properly for study, and fasten the label at the left side. *Be sure to put the label on the same side of the slide as the coverglass.* Square labels of gummed paper, obtainable from dealers in scientific supplies, are ordinarily used, and preferably should be written upon with a waterproof ink such as Higgins Eternal Ink.

Instead of using paper labels, which may come off, some workers coat the end of the slide with very thin balsam or varnish and, when this is dry, write upon it with India ink. The writing, in turn, is covered with more balsam or varnish, to which may be added a coverglass of suitable size.

Various types of rather permanent inks which may be applied with a pen to clean glass surfaces are good for labeling. Several colors are available, so that a color code may be worked out to facilitate cataloguing of slides used in research. These inks are rather viscous, however, and it may be difficult to write enough details directly on the slide if only a little space is available.

If there is a record card or notebook entry describing the material and the way in which it was processed, the code number should be written at the top of the label. If the slide bears serial sections, the number of the series, and the number of the slide within the series to which it belongs, should also appear near the top of the label so that both can be seen without removing the slide from the box. Abbreviated notations of these important facts can be scratched upon the slide, in the course of its preparation, with a diamond-point pencil.

Preparations used in research generally must be accompanied by much more information than can be properly recorded on a slide label. It may be convenient to write some information other than a code number to speed up finding the slides when need for them arises, but the full history of the specimen and technical procedures used to prepare it had best be recorded in full in a card file or notebook. In the case of labels which are supposed to be more or less complete in themselves, and which do not merely refer, by code numbers, to detailed information recorded in a card file or notebook, it is generally advisable to give at least most of the following data:

(1) The name of the organism.
(2) The organ or tissue represented.
(3) The type of preparation (whole mount, paraffin section, plane of sectioning, thickness of sections).
(4) The fixative used.
(5) The method of staining.
(6) Other important facts about procedures used in making the preparation.
(7) The date on which the preparation was completed, unless some other date is more pertinent.

The order in which these facts are written upon the label depends on the way in which the preparations are to be classified and studied. In a general zoological collection or in a collection for embryology or parasitology, the name of the organism (and perhaps also the names of higher taxa to which it belongs) should come first, and then information concerning the organ or tissue, type of preparation, and facts concerning fixation and staining. In histological or cytological collections, the tissue or organ is likely to be written first.

The small size of a standard slide label makes it necessary to abbreviate most words. The following abbreviations, in general use, are suggested:

whole mount—w.m.
transverse section—t.s.
longitudinal section—l.s.
frontal section—fr. s.
sagittal section—sag. s.
serial sections—s.s. (combined with t., l., fr., or sag. to indicate the plane of sectioning [sag. s.s. = sagittal serial sections])
alum hematoxylin and eosin—h. & e.
iron hematoxylin—i.h.
microns—μ
male—♂
female—♀

The following examples of labels illustrate some of the points just outlined:

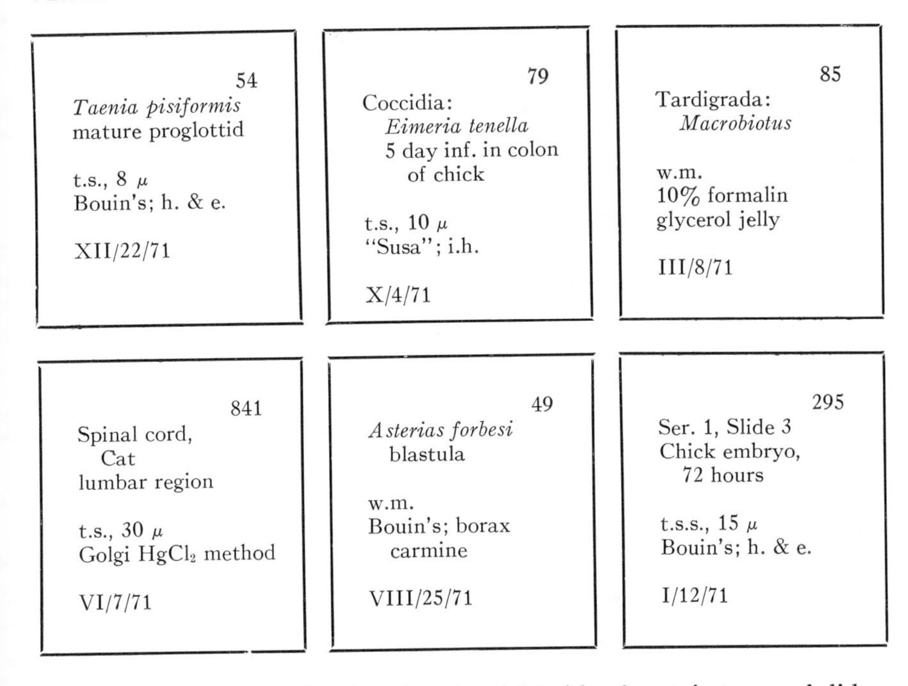

Special warning labels, placed at the right side of certain types of slides, will serve to aid inexperienced students and to prevent wholesale destruction of costly or unusual preparations by classes of beginners. The following are suggested:

Thick mount—do not use high-power!
Unstained—reduce light for examination.
Liquid mount—handle with care!

These warnings can be emphasized by writing them with red ink.

STORAGE

The choice of containers involves several considerations. (1) The preparations must be protected from breakage, light, moisture, and dust. (2) It is generally desirable—and in case of whole mounts essential—that the slides be stored horizontally. (3) It must be possible to remove slides from the containers, and to replace them, with ease and rapidity. (4) Convenience of distribution is important in case the slides are to be used in classwork. (5) Economy may also be an important factor.

Slide boxes with slotted sides are the most economical containers, and meet the needs of the majority of biologists very satisfactorily. Small boxes, for 25 or fewer slides, are made of wood and plastic. The slots in wooden boxes are a little farther apart than those in plastic boxes, and this makes them a little more convenient. Wooden boxes are more serviceable if they have slip-over covers (Fig. 73) rather than fit-in covers, because slides stored in the former type project enough to facilitate their removal.

Boxes holding 100 slides (Fig. 74) are available in wood, wood and cardboard, and plastic; they are generally more useful than smaller boxes for large sets of slides to be assigned to individual students, or for research or clinical preparations which are numbered continuously. They take up less space than 4 boxes holding only 25 slides, and eliminate the inconvenience of having to disarrange one row of small boxes to get at another row behind these on the same shelf.

Regardless of which type of box is to be used, it is always true economy to purchase boxes which have been carefully made. The warping and

Fig. 73. Box with overlapping cover, for 25 slides. (Courtesy of Clay-Adams, Inc., New York.)

collapse of cheap boxes are very likely to lead to disorganization of slide collections and to damage of the preparations.

Boxes of slides should be stored in a dry and moderately cool place, never near a radiator. The boxes should stand on end, so that the slides are horizontal. Shelves for storage of slide boxes are safer if the distance between them is only a little more than the height of the boxes.

For certain purposes, it may be convenient or desirable to store slides flat in trays which fit into a wood or metal cabinet. This type of arrangement is good for whole mounts, because it obviates the danger of leaving a box of slides on its side long enough to permit the specimens to drift in the mounting medium. Having the slides in open trays makes it very easy to read the labels, and therefore to locate quickly a particular slide. However, cabinets with removable trays will prove to be expensive when their storage capacity and cost are considered together. Individual trays, with or without covers, which are laid one on top of another are not particularly useful for permanent storage, because of the inconvenience of having to lift or rearrange the stack to get at a particular series of slides. Even the cheaper trays of this type cost more than slide boxes, when one takes into account their limited capacity.

Metal slide holders, which fit into drawers made for filing 3 × 5 inch cards, may be useful for some purposes. They accommodate slides in ranks of four, but as the slides stand on end, this system is also limited to smear and section preparations.

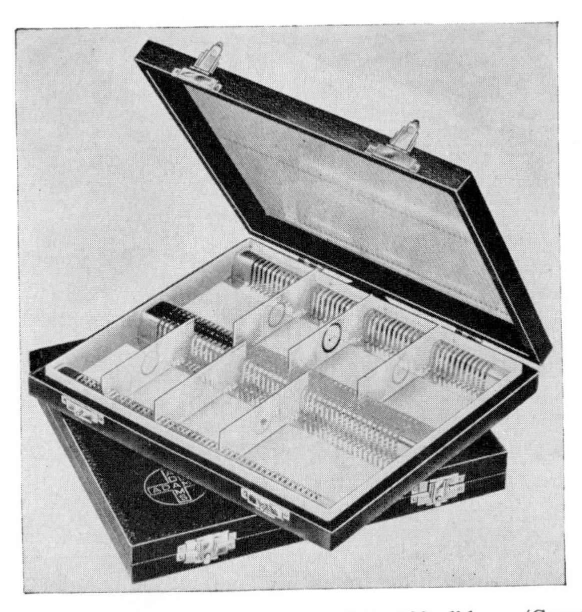

FIG. 74. Box, with hinged cover, to accommodate 100 slides. (Courtesy of Clay Adams, Inc., New York.)

ARRANGEMENT OF SLIDE COLLECTIONS

The importance of arranging a collection of slides according to some rational system will be perfectly evident, because only by this means can a preparation be located when it is needed. Systems of numbering and cataloguing are invented and modified as needed to serve the purposes of those who work with biological material. It is perhaps not desirable to prescribe with exactness for others, because the needs of various teaching and research laboratories are likely to be so different. Accordingly, only a few suggestions will be made to assist those who are responsible for managing a collection of one kind or another.

In the case of slides used for teaching purposes, the arrangement is generally made according to taxonomic groupings or the type of material (histological, cytological, embryological) involved, or a mixture of both. The following labels, such as might be applied to the exteriors of slide boxes or trays, will serve as illustrations.

CILIOPHORA:
HOLOTRICHA

COCCIDIA:
Eimeria tenella (chicken)
Eimeria perforans (rabbit)

COELENTERATA: HYDROZOA:
CALYPTOBLASTEA:
Obelia
Gonothyraea
Aglaophenia

CYTOLOGY
MEIOSIS
Parascaris ♀
Anasa ♂
Dissosteira ♂

EMBRYOLOGY
CHICK, 48 HR.
whole mounts
transverse serial sections
sagittal serial sections

HISTOLOGY
DIGESTIVE SYSTEM
LIVER
cat
cat, injected
human

In teaching and reference collections, it is a good idea to allow for some growth and revision by not filling the boxes or trays to capacity at the beginning. Additional slides may later be added to each series and new items interspersed among the old. This practice eliminates the necessity for frequently redistributing the contents of some boxes or trays because no space has been left for additions. However, it does require more boxes or trays, and therefore also more shelf or cabinet space.

Slides used in research had best be classified according to general groupings of some sort, and then arranged serially by numbers, or perhaps by a decimal system similar to the type used by libraries. If the numerical system is based on the order of accession, it may be inconvenient to scan the pages or cards of the catalogue each time a particular slide series must be located. For this reason, a cross-index to genus and species, or other type of subject involved, is recommended.

The practical value of a system for labeling and arranging finished preparations depends upon the care with which it is maintained. Any system will be worthless unless new slides are promptly and accurately accessioned and put where they belong, and unless slides which have been used are returned to their places as soon as possible.

CHAPTER 29

Summary of Methods Recommended for Various Types of Material

This chapter will serve as a guide for choosing methods to be used in making preparations of materials which are widely used for instruction in various branches of zoology, including histology, and for research purposes.

In briefly outlining the methods recommended for each type of material, procedures for fixation, embedding, staining, and mounting are simply referred to by name. Detailed instructions for carrying out these techniques are not given, except when certain precautions must be observed, or when a modification of the usual method gives superior results. The reader familiar with the basic procedures of microtechnique should have no difficulty in locating and applying the instructions which are pertinent. As a rule, when reference is made to a method which is not widely used, or which requires some special explanation, the number of the page to be consulted is given.

Workers who are prepared to embed in Epon and to cut sections about 1 μ thick with an ultramicrotome will be able to demonstrate many details of morphology, histology, and cytology with remarkable fidelity. The restricted perspective offered by thin sections and the relatively limited choice of good staining methods for Epon-infiltrated tissue are offset by the excellent preservation of the form of cells and of many organelles. The entire process, from fixation to staining and mounting sections, is given in Chapter 18.

The techniques recommended for various materials are, of course, not the only ones which may be used. For certain purposes, entirely different methods may be more appropriate. For example, although the combination of alum hematoxylin and eosin is almost always used for studies on the general structure of a particular organ or tissue, the distribution of connective tissue and mucous glands is likely to be brought out more effectively by the aniline blue in Mallory's triple stain, even though this method of staining may be less suitable in other respects.

It should be noted that when a certain method is recommended, a modification of the conventional procedure may be superior. For instance, when the use of Mallory's triple stain is suggested, the substitution of azocarmine G (as in Heidenhain's "Azan" modification) for acid fuchsin will almost always produce a sharper staining of nuclei. If there is reason to suspect that the entirely different combination of dyes in Masson's trichrome stain might bring out the tissue layers of a particular specimen more favorably than Mallory's method, then it should be tried. Experimentation with other fixatives is also in order when it appears that results obtained by the use of a particular procedure of staining could be better. Only experience enables prediction of the techniques of fixation and staining that are likely to succeed with unfamiliar material.

The beginner must bear in mind that the fixation and subsequent treatment of small invertebrates intended for whole mounts will involve some precautions and procedures which are unnecessary in routine histological work. For example, many small organisms are extremely contractile, fragile, easily ruptured, or likely to become distorted when fixed. Some are covered by a cuticle or exoskeleton which is nearly impermeable to fixatives and to mounting media. Their resistance to infiltration by a mounting medium may lead to their collapse or to the formation of air bubbles within them.

The directions given for processing difficult or delicate organisms, if followed carefully, should result in the preparation of mounts which can be considered successful. But occasionally the student or preparator will encounter animals which are so dense, opaque, or otherwise refractory that there seems to be no practical way of utilizing them for whole mounts.

ZOOLOGICAL MATERIALS

Protozoa

LARGE FREE-LIVING AMOEBAE. Pour the fixative on active specimens in very shallow water. For preparations which show general features of the amoebae, fix in Worcester's fluid (modified, see p. 95) or Bouin's fluid, and stain with dilute acidulated borax carmine; counterstain lightly, if desired, with fast green or some other suitable dye. For precise staining of nuclei, use the iron hematoxylin method following fixation in Bouin's fluid.

SYMBIOTIC AMOEBAE (ENTAMOEBA, ENDOLIMAX, AND OTHER GENERA). Fix smears of fecal material or intestinal contents, preferably on coverglasses, in Schaudinn's fluid or Bouin's fluid; stain with iron hematoxylin.

TESTS OF FORAMINIFERANS. If surface details are to be studied with reflected light, mount dry, against a black background (p. 166); if the tests are to be studied with transmitted light, clear them in toluene and mount them in a resinous medium, supporting the coverglass.

SKELETONS OF RADIOLARIANS. Clean fossil radiolarians with nitric acid; clean radiolarians collected in plankton tows in sodium hydroxide (about 10%); wash thoroughly in water; dehydrate, clear, and mount in a resinous medium.

FLAGELLATES. The flagellates include so many different kinds of organisms that it is impossible to prescribe a few methods which will serve all of them equally well. As a rule, the morphology and cytology of most flagellates can be studied to advantage in preparations stained with iron hematoxylin and by the Feulgen nucleal reaction. Certain types (especially the more complex symbiotic forms, such as trichomonadids) respond very effectively to impregnation by the Protargol method (p. 419).

Generally, the easiest way to handle symbiotic flagellates from the digestive tract or other organs is to prepare smears on coverglasses. If the smears have the appropriate consistency and enough proteinaceous material which will coagulate when they are dropped face down on the surface of the fixative, they should adhere quite well. Judicious addition of some comminuted tissue, egg albumen, blood serum, or some other compatible coagulable substance may be helpful.

Free-living or symbiotic flagellates which are easily cultured and concentrated may be fixed in tubes or dishes, then passed through various reagents by removing each fluid carefully with a pipette after the organisms have settled to the bottom. However, much of the material will probably be lost before the preparations are finished, and certain techniques of staining simply cannot be applied with success when this method of handling is employed. In general, procedures of staining and impregnation, as well as passage through various other reagents involved in dehydration and clearing, can be carried out much more easily if the flagellates are caused to adhere to coverglasses at the time they are fixed. Some culture media contain material which will coagulate upon fixation, thus trapping the flagellates. As a rule, it is necessary to add some foreign material, such as blood serum, egg albumen, or some other proteinaceous substance. Spread the drop of fluid containing the flagellates, together with the coagulable material, in a very thin film on a coverglass, and drop this face down on the surface of the fixative.

The iron hematoxylin method is perhaps the most useful one for demonstrating the general morphology. It brings out the nucleus, kinetoplast (if present), flagella, and some other organelles. The fixatives of Bouin, Hollande, Duboscq and Brasil, Champy, and Schaudinn, as well as a saturated aqueous solution of mercuric chloride with 5% acetic acid, should be tried before staining with iron hematoxylin. The Feulgen reaction for nuclear chromatin and the kinetoplast is likely to give good results after the last two of these fixatives, but material fixed in Bouin's fluid, Hollande's fluid, and Duboscq and Brasil's fluid may also prove suitable for staining by this method.

Impregnation with activated Protargol, following the method outlined for protozoa (p. 419), gives good results with many flagellates, especially trichomonadids and hypermastigotes. It brings out flagella, kinetosomes, and the parabasal body, as well as certain other extranuclear organelles found in some flagellates. Protargol ordinarily does not demonstrate the kinetoplast found in certain apparently lower flagellates, as bodonids, *Cryptobia*, and trypanosomatids. Fixation in Hollande's fluid or Bouin's fluid is recommended before impregnation with Protargol.

Colonies of *Volvox* are large enough to be handled as specimens to be mounted entire. Fixation in 10% formalin and staining in picro-carmine are recommended for preparations intended to show the general organization of the colonies. In any case, alcoholic fixatives are to be avoided. It is advisable to dehydrate the material by the glycerol method (p. 162). Mounting in Venetian turpentine should be considered if the colonies collapse during infiltration with one of the resinous media in common use.

TRYPANOSOMES, MALARIAL PARASITES, AND OTHER BLOOD PROTOZOA. It is routine to make dry smears of infected blood and to stain these by a Romanovski-type stain. Giemsa's method is recommended. Blood films may also be fixed on coverglasses before they have begun to dry out, and the parasites treated as cytological material. Among the fixatives which should be tried before iron hematoxylin are Bouin's fluid and Hollande's fluid. A saturated aqueous solution of mercuric chloride with 5% acetic acid, or Schaudinn's fluid, will probably give good results on material to be stained by the Feulgen method. However, the acetic acid in these fixatives destroys red corpuscles. Trypanosomes are often differentiated very effectively by impregnation with Protargol, following fixation in Hollande's fluid or Bouin's fluid.

GREGARINES. These are found in a variety of invertebrates, especially arthropods and annelids. The following animals are particularly dependable sources: beetles (including *Tenebrio* and *Dermestes*); grasshoppers; marine polychaete annelids; earthworms. The gregarines of earthworms are primarily parasites of the seminal vesicles. Those in arthropods and polychaetes are generally found in the digestive tract; the midgut is the usual site of infection in insects.

For staining with iron hematoxylin, fix smears of the infected organ in Bouin's fluid, Duboscq and Brasil's fluid, or Hollande's fluid. The Protargol method will demonstrate the pellicular fibrils and the elaborations of the epimerite, if this is present; Bouin's fluid and Hollande's fluid are good fixatives for material to be prepared by this technique.

COCCIDIANS IN TISSUE. Fix infected tissue of intestinal mucosa, liver, or other organs in Bouin's fluid, Duboscq and Brasil's fluid, or Gilson's fluid; cut paraffin sections at about 8 μ; stain with iron hematoxylin.

MYXOSPORIDIANS AND MICROSPORIDIANS IN TISSUE. Fix infected tissue in Bouin's fluid or Duboscq and Brasil's fluid; embed in paraffin, and section at about 8 μ; stain with iron hematoxylin.

CILIATES. For routine preparations of ciliates, to show their size and shape, the macronucleus and micronucleus, and the behavior of nuclei during division or conjugation, excellent results may be achieved by fixation in Power's fixative (p. 101), Bouin's fluid, or Kleinenberg's picro-sulfuric acid mixture, staining with dilute acidulated borax carmine, and counterstaining *very lightly* with fast green or indigo-carmine.

To obtain a composite picture of the structure of a ciliate, it will be desirable to use at least three or four different methods. Iron hematoxylin usually brings out the nuclei quite well, and also stains food inclusions; the pattern of ciliary rows will probably not be suitably differentiated. The Feulgen nucleal reaction will stain only the chromatin of the nuclei and of food inclusions, making it possible to establish which of several bodies of similar size and distribution are really micronuclei. The Protargol method, as recommended for protozoa (p. 419), brings out the kinetosomes, enabling the arrangement of ciliary rows on the surface of the body and the disposition of various buccal and adoral membranelles to be worked out. Protargol also usually impregnates the nuclei, although it does not do this in a way which would make the material suitable for cytological study. The Chatton and Lwoff method of impregnation with silver nitrate does not bring out nuclei, but it demonstrates the kinetosomes with remarkable clarity. Klein's technique (p. 405) for ciliates in dried smears is easy and rapid, and often yields very useful preparations, even though the organisms are distorted and the impregnation is limited to those portions of the body surface which are not in contact with the slide.

Bouin's fluid and Hollande's fluid are good fixatives for material to be stained with iron hematoxylin or impregnated with Protargol. Before the Feulgen method, it is generally best to use a saturated aqueous solution of mercuric chloride with 5% acetic acid. For fixation of material to be impregnated with silver nitrate by the method of Chatton and Lwoff, see page 406.

Many of the larger species of ciliates grown in culture or collected from natural sources may be concentrated by centrifugation, discharged into a fixing fluid, then passed through staining solutions and other reagents in centrifuge tubes, short test tubes, or dishes. If the organisms are centrifuged gently (a hand centrifuge is safe) or allowed to settle, and the liquids are removed very carefully with a pipette, the loss of material may be reduced to a minimum. However, it is generally easier to carry ciliates through microtechnical procedures by affixing them to coverglasses at the time of fixation. The procedure recommended for flagellates will work with ciliates. A drop of the fluid containing the ciliates is mixed on a coverglass with a small amount of a coagulable substance such as blood serum, fresh egg albumen (use this sparingly!), or some sticky tissue from an animal. After the smear is spread out in a thin and uniform layer on the coverglass, it is dropped face down on the surface of the fixative in a Petri dish.

Symbiotic ciliates, like symbiotic flagellates, are generally quite easy to handle. The organ or glandular surface of the body in or on which they live will probably provide some coagulable material when a smear is prepared on a coverglass, so that the organisms will adhere quite well unless too much water or isotonic saline solution has been added.

Sponges

FOR GENERAL ORGANIZATION. Fix sponges in a saturated aqueous solution of mercuric chloride, preferably in the field at the time of collection; handle the living specimens very carefully, and if possible do not manipulate them in air before fixation, as air may be trapped in some of the cavities; decalcify or desilicify; stain progressively with alum hematoxylin before embedding in paraffin; cut sections about 8 μ thick and use a minimum of heat in affixing sections to slides; counterstain with eosin after removing paraffin; dehydrate, clear, and mount in resinous medium.

FOR STUDIES OF CELL TYPES, INCLUDING CHOANOCYTES. Prepare according to the Epon method, using buffered osmium tetroxide for fixation. The sections will show many imperfections owing to the presence of spicules, but this is the only good technique for preserving the cells adequately for light microscopy.

CALCAREOUS SPICULES. Clean by soaking pieces of sponge in 10% sodium hydroxide; the process may be accelerated by use of heat; wash in water; dehydrate, clear, and mount.

SILICEOUS SPICULES. Clean by soaking pieces of sponge in 10% sodium hydroxide or in a commercial sodium hypochlorite solution (such as Clorox); wash in water; dehydrate, clear, and mount.

SPONGIN. Use pieces of commercial sponges, or clean pieces of fresh sponges by soaking them in water and then beating out the soft tissues; stain with safranin if desired; wash thoroughly and dry, then pull apart the fibers into pieces which will be thin enough to mount; dip into toluene and mount in a resinous medium.

Coelenterates

HYDRAS. Pour Bouin's fluid heated to about 50° C. over well-separated, expanded (and preferably anesthetized) specimens in shallow water. For whole mounts, stain with precipitated borax carmine and fast green, or progressively with alum hematoxylin; infiltrate slowly with resinous medium before mounting; support coverglass.

For section preparations, embed in paraffin and cut at 6 to 8 μ; stain with iron hematoxylin; counterstain with eosin if desired.

Calyptoblastic (Thecate) Hydroids. Anesthetize healthy and clean specimens by menthol-chloral hydrate method; fix in Bouin's fluid; stain with precipitated borax carmine; counterstain lightly with fast green; infiltrate with resinous medium before mounting; support coverglass.

Gymnoblastic (Athecate) Hydroids. Anesthetize with magnesium chloride or magnesium sulfate; fix in Bouin's fluid; stain with precipitated borax carmine; counterstain lightly with fast green; infiltrate with resinous medium before mounting; support coverglass.

Small Medusae. Anesthetize with urethane or magnesium chloride; stain with precipitated borax carmine; counterstain lightly with fast green; infiltrate with resinous medium before mounting; support coverglass.

Marginal Tentaculocysts of Small Scyphozoans. Quiet medusae with magnesium chloride or magnesium sulfate; cut out pieces containing tentaculocysts from margin of umbrella; fix in 10% formalin (neutral or buffered); stain with alcoholic cochineal; infiltrate with medium before mounting.

Larval Scyphozoans. Anesthetize ephyrulae with urethane, and scyphistomae and strobilae with magnesium chloride or magnesium sulfate; stain with precipitated borax carmine; counterstain lightly with fast green; infiltrate with resinous medium before mounting; support coverglass.

Sea Anemones. Anesthetize with magnesium chloride or magnesium sulfate (isotonic with sea water); when they have become sluggish—this may take a half day or longer—attempt to introduce the anesthetizing solution into the stomodaeum with a pipette of appropriate size, so as to inflate the coelenteron. After it appears that the anemones have become completely non-reactive, inject the fixing fluid into the coelenteron, then bathe the whole specimen with the fixative.

Fix anemones in Bouin's fluid or 10% formalin. For transverse or longitudinal sections of entire animals, select small individuals. Stain progressively with alum hematoxylin before embedding in paraffin; cut sections about 10 μ thick; counterstain with eosin.

For study of gonads, dissect these out of larger anemones after fixation; embed in paraffin; stain ovaries with iron hematoxylin or alum hematoxylin and eosin, and testes with iron hematoxylin.

To demonstrate the discharge of nematocysts and to stain these at the same time, in a temporary mount, prepare a thin smear of tissue from a tentacle or acontium. After the smear has dried, pass it through 95% alcohol (2 minutes) and 100% alcohol (2 minutes), then let it dry again. Apply a coverglass, locate some nematocysts with the microscope, and apply a 1% aqueous solution of methylene blue at the edge of the coverglass. Soon after the nematocysts become hydrated and stained, they will discharge. Dehydrated smears may be stored indefinitely without the nematocysts losing their reactivity to the methylene blue solution.

Platyhelminths

TURBELLARIANS. The morphology of acoels and small rhabdocoels can be studied more effectively in living specimens and serial sections (6 to 8 μ) than in stained whole mounts. Magnesium chloride is usually a good anesthetic. After the specimens have been quieted, they can be picked up with a pipette and dropped into the fixative. Bouin's fluid is recommended for fixation, and iron hematoxylin is perhaps the best stain for bringing out the general morphology. See suggestions given on page 226 for simultaneously embedding several properly oriented small organisms.

Some larger rhabdocoels, such as the fresh water *Mesostoma* and umagillids which live as symbionts in sea urchins and some other marine invertebrates, make excellent whole mounts if fixed in Bouin's fluid and stained by the precipitated borax carmine method; however, they generally need to be flattened slightly during fixation. For material to be sectioned, fix in Bouin's fluid without flattening; stain the sections with iron hematoxylin.

For preparing whole mounts of planarians to show the digestive tract, feed the worms on liver ground with carmine. (Some species will also eat cooked egg yolk mixed with carmine.) Do not fix the worms for several hours, so that the gut will not be overstuffed and intestinal epithelium will have a chance to absorb some of the carmine. Kill the planarians on a glass plate, by letting a drop of 2% nitric acid fall on the specimens. Fix them in 10% formalin, and clear and mount them unstained, or stain them lightly with alum hematoxylin.

Specimens which have not been fed carmine may be stained in borax carmine or alum carmine.

For sections, kill by the method recommended above, and fix in a saturated solution of mercuric chloride or Bouin's fluid; cut paraffin sections 8 or 10 μ thick; unless serial sections are desired, mount selected sections from various regions (anterior to pharynx, through pharynx, posterior to pharynx); stain with alum hematoxylin and eosin, or with iron hematoxylin.

DIGENETIC TREMATODES. Most digenetic flukes, when alive, are extremely active, constantly extending and contracting. If dropped into a fixative at room temperature, they generally contract severely, and may curl and twist as well. Many species, if dropped into an isotonic saline solution containing 10% ethyl alcohol, will quickly become immobile, and may remain moderately well extended. If they are stretched and straightened by manipulation with a camel's hair brush, they may then be fixed, either unflattened or after they have been compressed gently between pieces of glass. For some small species, the weight of a thick coverglass will flatten them to just the right extent. For some larger specimens, another slide or piece of a slide will provide the appropriate weight. If necessary, tie the pieces of glass together.

If the uterus is extensive and so full of eggs that other structures are obscured, put the worms into distilled water until the eggs are shed, then fix them.

As a rule, it is a good idea to fix at least some material unflattened, because even slight pressure may alter the spatial relationships and shapes of certain organs.

Fix the worms, after appropriate anesthetization and flattening, in Gilson's fluid or Bouin's fluid. Separate the pieces of glass between which they are flattened as soon as it is evident that they have been killed, and completely immerse the worms in the fixative.

Small flukes are generally stained very effectively by the precipitated borax carmine method. Borax carmine is better for larger specimens. Alum hematoxylin often gives good results with flukes, especially those of small size.

Infiltrate the specimens slowly with the resinous medium before mounting. Support the coverglass if the specimen is very small or if the coverglass is not likely to remain level otherwise.

Sporocysts, Rediae, and Cercariae. Sporocysts and rediae are rather sluggish and present no particular problems.

Most cercariae are extremely contractile, and many of them will shed the tail if they are not treated with exceptional care. Some species can be gradually warmed, while they are in a dish of isotonic saline solution, until they relax and die; they may then be fixed. Others can be fixed successfully by discharging them with a pipette into a dish of hot fixing fluid.

Fixation in Bouin's fluid, followed by staining by the precipitated borax carmine method and light counterstaining in fast green, often gives excellent results with sporocysts, rediae, and cercariae. Infiltrate them with the resinous medium before mounting, and support the coverglass.

Monogenetic Trematodes. Most of these are external parasites of fishes. Some are very contractile, others are relatively non-contractile. The contractile species respond quite well to anesthetics. Remove them from the fish or other host, place them in a dish of the water in which they have been living, and narcotize them by the menthol-chloral hydrate method, or with chlorobutanol or 10% alcohol; marine species can also be anesthetized by placing them in fresh water, although this may not be a desirable method.

It will be a good idea to fix some specimens unflattened, and others gently flattened between pieces of glass. Any of the fixatives recommended for digenetic trematodes will serve well, and 10% formalin in sea water will give good results with marine species. Staining by the precipitated borax carmine method will produce excellent preparations.

TAPEWORMS. Remove tapeworms from the intestine with care, in order not to lose the scolex or fragment the worms unnecessarily. It is generally safer to open the intestine by tearing it lengthwise, if this is possible, rather than by cutting it with scissors. Some tapeworms will eventually detach themselves if the opened intestine is left in a dish of isotonic saline solution; others can be separated from the tissue by tearing this in such a way that the scolex is freed.

Very small tapeworms are generally fixed, stained, and mounted entire. In order to prevent curling or twisting, straighten each worm on a slide or glass plate, and cover it with another slide which is supported in such a way that it does not crush the worm. Flood the space between the two pieces of glass with the fixative. As soon as the worm appears to have been killed, separate the pieces of glass and transfer the worm to a dish of the fixative.

In the case of tapeworms of moderate to large size, it is generally best to remove the scolex, together with a few of the youngest proglottids, and to dip this (attachment end first) into the fixative. If the scolex is provided with retractable or contractile structures, it should be narcotized, preferably with novocaine or urethane, before fixation.

For fixation of proglottids, the following procedure is recommended. Lay the strobila, or a portion consisting of perhaps 20 or 30 proglottids, on a smooth piece of soft wood, and pin down one end of it with a splinter of hardwood or a glass needle. Gently pull the other end until the body is in a normal state of extension, but no further, and pin it down. Moisten the proglottids with the fixative, adding more as needed, until they become opaque and somewhat hardened. Then remove the pins and transfer the proglottids to a dish of the fixative.

Gilson's fluid is a good fixative for many kinds of tapeworms. After fixation, pass the material through 50% alcohol (two changes; large proglottids should be left in each change for at least 2 hours), and complete the removal of mercury salts in iodized 70% alcohol.

The precipitated borax carmine method is excellent for scoleces and very small tapeworms to be mounted entire. Picro-carmine is recommended for proglottids of moderate to large size, but the conventional borax carmine method sometimes yields fine preparations. Alum hematoxylin is definitely a superior stain for many tapeworms, especially cyclophyllids.

In the case of worms of moderate to large size, it is customary to mount the scolex and short strips of young, mature, and gravid proglottids under one coverglass of appropriate size. The separation into portions of suitable length can be done most conveniently while the material is being washed in alcohol prior to staining. Use a razor blade for this purpose, cutting the pieces while they are on glass plates well moistened with alcohol.

Tapeworms from warm-blooded vertebrates can sometimes be killed without excessive contraction by leaving them in very cold water until they die. A method useful for certain tapeworms is to pick up the strobila with forceps, pull it through a dish of water heated to 65 or 75° C., and then place it into the fixative.

The gravid proglottids of tetrarhynchs and tetraphyllids of elasmobranchs rupture upon being released into sea water, and may collapse into a practically structureless mass. To prevent this, open the spiral valve of the digestive tract and transfer the worms to a mixture of 40% sea water and 60% distilled water (or aged tap water), to which 1 to 3% of urea has been added.

Nemerteans

For histological preparations of large nemerteans, narcotize the worms with propylene phenoxetal (p. 117). After they have been in a fixative a few minutes, cut them with a sharp razor blade into pieces about 1 cm. long, so that the fixative will penetrate more quickly. (If they are not killed before being cut, the short segments of the inverted proboscis may be forced out of the rhynchocoel.) Bouin's fluid is a good fixative for material to be stained with iron hematoxylin or with alum hematoxylin and eosin; Heidenhain's "Susa" fixative generally gives good results before Mallory's triple stain or the "Azan" modification of this method.

Some small nemerteans, including *Malacobdella* (commensal in the mantle cavity of marine pelecypods), make instructive whole mounts if they are fixed, while being flattened to some extent, in Bouin's fluid, and stained by the precipitated borax carmine method. Small nemerteans which are to be sectioned may be fixed entire but should only be kept straight, not flattened, during fixation.

Acanthocephalans

WHOLE MOUNTS. Adult acanthocephalans occur exclusively in the intestines of vertebrate animals. When alive, they generally present a collapsed and slightly flattened appearance, and usually are securely fastened to the wall of the intestine by a proboscis provided with many spines or hooks. If they do not spontaneously detach themselves, it will be necessary to cut out pieces of the intestine and to release each worm individually by tearing the tissue around the proboscis with needles, while observing the process through a wide-field binocular microscope.

After the worms have been removed from the intestine and separated from adhering mucus and debris, place them in 10% ethyl alcohol in an isotonic saline solution. (Simply mix 1 part of absolute alcohol with 9 parts of the saline solution.) This should cause the worms to become more plump (a desirable effect, because it sets apart the body wall from the internal organs), and many will protrude the proboscis. If the proboscis is not protruded, it can probably be forced out by holding the worm down with one camel's hair brush and stroking it with another, from the posterior end to the anterior end, so as to exert pressure on the body wall and thus displace the fluid of the pseudocoelom toward the anterior end. The pressure of the fluid will often force the proboscis out as far as it will go, and if the worm is then quickly dropped into the fixative it will be fixed with the proboscis protruded. Worms which are curled or twisted can be straightened with camel's hair brushes before fixation.

In collecting acanthocephalans, note that the males are usually smaller than the females. Females with irregular brown or blackish masses adhering to the posterior end bear "spermatophores." Very small worms, immature or barely mature, often make especially clear preparations for certain purposes.

The best fixative for acanthocephalans is Carnoy and Lebrun's fluid. Fix small specimens at least an hour or two, then place them in 95% alcohol, and iodize them to remove salts of mercury. The worms may be stored in 70 or 80% alcohol. Other fixatives which give good results are Carnoy's acetic acid-alcohol mixture, Heidenhain's "Susa" fluid, and a saturated aqueous solution of mercuric chloride with 5% acetic acid. Formalin (10%) may also be used.

For staining small worms, less than about 15 mm. long, use borax carmine; alum cochineal gives better results with larger specimens.

After staining and differentiating the specimens, transfer them to water (gradually, if staining or differentiation was carried out in an alcoholic solution), and place them in 10% glycerol covered with a sheet of lens paper. Add a crystal of thymol to discourage the growth of molds. Leave the dish in a warm place. When the solution has become concentrated to approximately the thickness of pure glycerol, examine the specimens. Some may have collapsed. Place the specimens in 95% alcohol; the ones which have collapsed will often become plump again. Stir or shake the alcohol from time to time, to encourage the glycerol to diffuse out of the worms.

At this point, it is desirable to hold large or thick-walled specimens down with a camel's hair brush and to make one or several punctures in the body wall, in places where there are no underlying organs, with "Minuten Nadeln" (see p. 51). With care, inconspicuous punctures no more than 30 μ in diameter can be made. The body wall of small or thin-walled specimens is usually permeable enough that puncturing is unnecessary. Put the worms into clean 95% alcohol for a time, then into absolute alcohol (two changes). Gradually add toluene, or pass the specimens through several successively more concentrated solutions of toluene in absolute alcohol. Leave them for at least an hour in pure toluene.

Before mounting acanthocephalans in balsam or a synthetic resin, put the cleared worms into a thin mixture of the medium in toluene, and allow this to concentrate by evaporation. Protect it from dust during this time. It is also possible to infiltrate the specimens by the method described on page 162.

Some parasitologists put living acanthocephalans into distilled water, and leave them until the osmotic imbibition of water produces enough pressure to force out the proboscis. This method certainly works well in many cases, but it may possibly result in deterioration of internal organs.

Nematodes

SMALL NEMATODES FOR WHOLE MOUNTS. Some species are so delicate that they will rupture in nearly any fixing fluid. Others have a rather impermeable cuticle, and will twist and coil if they are placed into a fixative at room temperature. For this reason, they are routinely fixed in hot (80° C.) 70% alcohol. The addition of 20% glycerol will be helpful in the case of very small worms; 5% is enough for those which are larger. They may be stored in the fixative.

Nematodes are usually mounted unstained in glycerol jelly. However, they may be stained with alum cochineal, cleared *very gradually* (about 10 steps from absolute alcohol to toluene or terpineol-toluene), and infiltrated slowly with a resinous medium. Another good way to prepare nematodes with a very impermeable cuticle for mounting in a resinous medium is to clear them gradually with beechwood creosote, and then to introduce the mounting medium slowly into this.

LARGE NEMATODES. For general morphology, fix pieces in Gilson's fluid; cut paraffin sections about 12 μ thick; stain with alum hematoxylin and eosin.

Rotifers

Anesthetize with 1% cocaine hydrochloride, if available (see p. 44 for suggestions concerning other anesthetics); fix in 10% formalin; stain with picro-carmine; counterstain very lightly with fast green; infiltrate slowly with resinous medium; support coverglass. Consider the possibility of dehydrating in glycerol (p. 162) if the specimens collapse when dehydrated in the usual series of alcohols.

Ectoprocts

Anesthetize fresh-water species, such as *Plumatella*, by the menthol-chloral hydrate method. Some marine forms, such as *Membranipora*, may also be anesthetized in this way, but use cocaine (if available) for *Bugula*. Eucaine may also be tried. Stain with precipitated borax carmine; counterstain lightly with fast green; infiltrate with medium before mounting.

Annelids

LARGE POLYCHAETES. Worms to be fixed for histological preparations should be placed in clean sea water for at least several hours if there is any chance that the digestive tracts contain sand. Narcotize them with magnesium chloride or propylene phenoxetal, then drop them in the fixative, or suspend them in it in the manner suggested for large oligochaetes (see below). After the worms have been in the fixative for a few minutes, cut them with a razor blade into pieces about 1 cm. long; this will enable the fixative to penetrate more rapidly. Cut paraffin sections about 10 or 12 μ thick (or a little thicker, if this seems to improve the perspective). Material fixed in Bouin's fluid generally stains well with alum hematoxylin and eosin; Heidenhain's "Susa" fixative may be recommended for material to be stained by Mallory's triple stain or the "Azan" modification, although it is also a good fixative to use before staining with alum hematoxylin and eosin.

SMALL OLIGOCHAETES AND POLYCHAETES TO BE PREPARED AS WHOLE MOUNTS. Anesthetize with chlorobutanol, magnesium sulfate, or magnesium chloride; lay on a glass plate or in a dish with very little water, and flood with Bouin's fixative; stain with precipitated borax carmine; counterstain with fast green, if desired; infiltrate with medium before mounting.

LARGE OLIGOCHAETES. Earthworms should be washed and placed in a box with moist, clean rags or burlap. In crawling around among the rags, they will rid themselves of sand and other gritty material which may be adhering to them. Next place them between several layers of moist filter paper or paper toweling in a large dish. Do not crowd the worms, as they may die from the accumulation of wastes and products of bacterial decomposition. Adjust the cover so that some air can enter the dish, then put it in a cool place. Renew the moist paper every day, and remove any dead worms. In 2 or 3 days their castings will consist of clean paper, indicating that their intestines are probably free of grit. Then keep them for another day or two between layers of clean, moist cloth, in order that they may empty themselves of paper. Do not attempt to keep them longer, as they are likely to die.

Now place the worms in a covered shallow dish, with only enough water to cover them. Set a bottle of 50% alcohol, the volume of which equals about one-fifth of the amount of water in the dish, on a higher shelf, and allow the alcohol to drip down slowly, by way of a siphon with a finely-drawn glass tip. Regulate the flow with a screw-clamp so that about 2 hours will be required to empty the bottle. Let the worms lie in the weak alcohol until they no longer contract when they are pinched. If the siphon arrangement is not convenient, add a comparable amount of 50% alcohol to the dish by putting in a few drops at frequent intervals.

Fill a tall glass cylinder with Bouin's fixative. Pick up one worm at a time, and dip it repeatedly into this. Then stick a pin through one of its posterior segments and hang it from some convenient support over a dish. Moisten and hang all of the worms in this manner, keeping them well bathed with Bouin's fixative by means of a pipette. The action of gravity will keep them well extended and perfectly straight, even if they still have some tendency to contract. After about half an hour, lay the worms in a long, shallow dish of the fixative, and leave them in this for about an hour. Then subdivide them with a sharp blade. Ordinarily, it is desirable to keep the reproductive organs together in one piece. If cuts are made as suggested in Figure 75, the anterior end will be properly trimmed for sagittal sections, and the remainder of the body for transverse or sagittal sections. Leave the pieces in Bouin's fixative overnight, then transfer them to 70% alcohol for washing.

Embed the pieces in paraffin and cut sections at 10 μ. Make sagittal sections of the anterior 15 or 20 segments; lateral sections of this region will show the reproductive organs, and median sections will show the

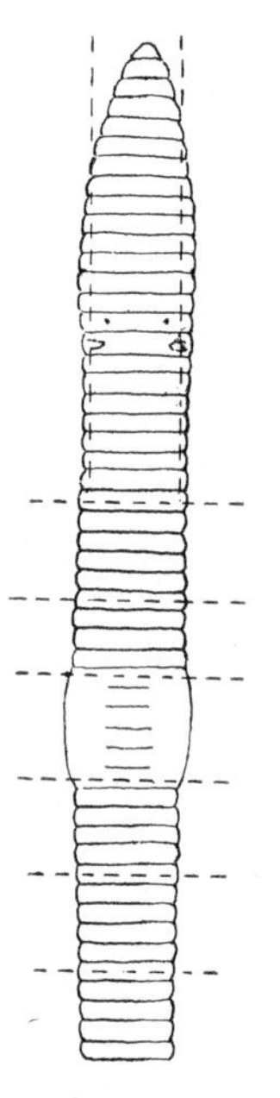

FIG. 75. Scheme for subdividing and trimming an earthworm in preparation for cutting sagittal sections of anterior portion and transverse sections of other regions.

nervous system and digestive system. Make transverse sections through the clitellum and the intestinal region, and also sagittal sections if desired.

Stain the sections with alum hematoxylin and eosin, or with Mallory's triple stain.

Some mud-dwelling aquatic oligochaetes of moderate size are often too delicate to be handled in the manner recommended for large earthworms. Place them in filtered pond water, in a cool place, and change the water on each of 2 or 3 successive days. By the end of this time, it is likely that any gritty material in the gut will have been expelled. Narcotize the animals by addition of 50% alcohol as suggested for earthworms; if this induces rupture of the body wall, try chlorobutanol, magnesium sulfate, or magnesium chloride.

When the worms have been quieted, place them in glass tubes of appropriate length; the inside diameter of these tubes should be sufficient to permit the fixative to bathe the specimens, but not large enough to allow the worms to become extensively kinked during fixation. Bouin's fixative will probably work well. Embed pieces of worms in paraffin and section as necessary to demonstrate morphology. Stain the sections with alum hematoxylin and eosin, or Mallory's triple stain.

LEECHES. Anesthetize with eucaine (p. 116), magnesium chloride, or magnesium sulfate. If whole mounts of small species are to be prepared, compress them slightly between two slides, and fix in Gilson's fluid; separate the slides as soon as it is apparent that the leeches will not contract or become too thick once the upper slide is removed; stain with precipitated borax carmine; counterstain very lightly with fast green if desired.

If sections are to be prepared, fix extended and anesthetized specimens in Bouin's fluid; it may be necessary to place the leeches between two slides or glass plates so that they do not curl or contract excessively when they are placed in the fixative, but they should not be flattened to the extent that internal structure is distorted. Embed whole small leeches, or portions cut from very large specimens, in paraffin; section transversely and sagittally; stain with alum hematoxylin and eosin, or Mallory's triple stain.

Arthropods

Crustacea. SMALL CRUSTACEA FOR WHOLE MOUNTS. Fix in 95% alcohol; stain with precipitated borax carmine; if desired, counterstain very lightly with fast green; infiltrate with resinous medium before mounting; support coverglass.

TISSUES OF LARGE CRUSTACEA. Dissect out organs and tissues, and fix, section, and stain these as directed under Histological Materials.

Insects and Arachnids: Whole Mounts. SMALL INSECTS AND ARACH-

NIDS. Fix in 95% alcohol, keeping appendages spread out; bleach if heavy pigmentation will interfere with study; clear and mount unstained. Oil of wintergreen (synthetic) is very good for clearing.

SMALL AND TRANSPARENT LARVAE. Fix larvae (such as those of mosquitoes) in formalin-alcohol; stain with alum cochineal; counterstain very lightly with fast green; infiltrate with resinous medium before mounting.

LARGE SPECIES. To prepare mounts of appendages, mouth parts, wings, and comparable structures, remove them from animals fixed in 95% alcohol; bleach if necessary; clear and mount without staining.

To prepare mounts of entire specimens, fix in 95% alcohol; soak in 10% sodium hydroxide to destroy soft parts; slit open body, if possible, and remove remnants of tissues; then flatten between slides, dehydrate, clear (preferably with oil of wintergreen), and mount.

WING SCALES OF LEPIDOPTERA. Scrape off and prepare dry mounts.

Insects and Arachnids: Sections. EGGS AND EMBRYOS. Puncture shells, fix in Kleinenberg's picro-sulfuric mixture or Duboscq and Brasil's fluid; double embed in nitrocellulose and paraffin; cut sections at thickness desired; stain with iron hematoxylin, or alum hematoxylin and eosin.

SECTIONS THROUGH ENTIRE ORGANISMS OR APPENDAGES. Make openings in exoskeleton to permit fixative to penetrate rapidly, or inject the fixative; fix in Duboscq and Brasil's fluid; when internal tissues are hardened, cut into pieces about 3 or 4 mm. thick; double embed in nitrocellulose and paraffin, and section; stain with alum hematoxylin and eosin or by the best method for structures to be demonstrated. With this technique, it is not necessary to soften the exoskeleton with acids or other strong chemicals.

SECTIONS OF VARIOUS ORGANS. Dissect out organs, subdivide them if necessary, and fix in an appropriate fixative; section by either paraffin or nitrocellulose method; choose staining method according to nature of tissues to be stained.

Molluscs

RADULAS OF SNAILS AND CHITONS. Dissect out the radular mass from a fresh animal and soak it in weak (about 10%) sodium hydroxide overnight, or as long as is necessary to remove the fleshy parts. Wash the radula repeatedly, brushing off adhering debris with camel's hair brushes if necessary. Most radulas are easily stained. Picro-carmine (or alcoholic cochineal) followed by a counterstain of fast green or indigo-carmine produces very fine preparations.

Radulas of some gastropods and chitons tend to twist, or to roll into a tube, with the teeth facing inward. In such situations, straighten the radula and keep it flattened gently between two pieces of glass until it is finally cleared and ready for mounting.

It is difficult to make good preparations of radulas from animals which have been preserved in formalin. However, when the radula is straightened, cleared, and mounted, it may be satisfactory for analysis of the arrangement of teeth.

WHOLE MOUNTS OF ENTIRE CLAMS. Very small clams, such as the young within the brood chamber of the common fresh-water *Pisidium*, make useful whole mounts. Place a number of clams on the bottom of a dish in a minimum amount of water. As soon as the valves gape and the foot is protruded in efforts to crawl, quickly pour hot Bouin's fixative over them. Many will die with the foot extended. Stain them with borax carmine or alum cochineal.

SECTIONS OF ENTIRE CLAMS. Use small specimens, no larger than about 1 cm. in length; prop valves apart with toothpick; fix in Bouin's fluid; harden in 70% alcohol; decalcify with hydrochloric acid, unless decalcification was completed by fixative; embed in paraffin, and cut transverse sections at about 12 or 14 μ; stain with alum hematoxylin and eosin, or with Mallory's triple stain.

CTENIDIUM OF A CLAM. Pry open the shell of the clam, cut the adductor muscles and hinge ligament, and remove one valve of the shell. Turn back the mantle in order to expose the ctenidium on that side of the body. Take hold of the free margin of the ctenidium, being sure to include both lamellae, and hold it in such a way that its line of attachment is exposed. Then cut it off as close as possible to the rest of the body, so that the juncture of the two lamellae is preserved. Lay the ctenidium on a smooth piece of soft wood and pour over it enough Bouin's fixative to cover it but not enough to float it. In about 15 minutes take a sharp razor blade and divide the gill, parallel to its filaments, into strips about 1 cm. wide. Place the strips in Bouin's fluid for several hours, then transfer them to 70% alcohol for washing.

Embed the pieces of ctenidium in paraffin and cut them transversely (*i.e.*, in a plane parallel to the individual filaments) at 8 or 10 μ. For general structure of the ctenidium, stain some sections with iron hematoxylin and orange G, others with Mallory's triple stain.

For bringing out cilia and basal bodies sharply, impregnation of ctenidial tissue with activated Protargol often gives excellent results. The modification described for protozoan material (p. 419) is recommended; it may be applied to paraffin sections of ctenidia fixed in Bouin's fluid or Hollande's fluid.

OTHER INDIVIDUAL ORGANS OF MOLLUSCS. Follow methods recommended for histology and cytology in general.

Echinoderms

Almost all echinoderms can be quickly and easily narcotized by the use of a solution of magnesium chloride which is isotonic with sea water.

SEA STARS. For general morphology, fix small specimens (about 3 cm. in diameter) in a saturated solution of mercuric chloride; cut out pieces from which sections are to be made; decalcify with hydrochloric acid; stain pieces progressively with dilute alum hematoxylin; section by nitrocellulose method, or by double embedding in nitrocellulose and paraffin; counterstain with eosin, and finish preparations.

PEDICELLARIAE. Dissect off from surface; fix in 10% formalin; stain lightly with alcoholic cochineal or eosin.

SKELETAL PLATES OF HOLOTHURIANS. In the case of thick-walled species, such as *Cucumaria*, place pieces of body wall in 10% sodium hydroxide until soft parts are dissolved; wash plates in water; dehydrate, clear, and mount.

In the case of thin-walled species, such as *Leptosynapta*, fix small pieces of body wall in 95% alcohol; dehydrate, clear, and mount unstained.

Hemichordates

DOLICHOGLOSSUS, SACCOGLOSSUS, AND RELATED GENERA. Anesthetize with magnesium chloride or magnesium sulfate; fix in Gilson's fluid; cut into small pieces, but keep the proboscis, collar region, and part of the trunk of at least one specimen together; embed in paraffin; section continuous anterior portion sagittally (median sections will be most instructive); also cut transverse sections of various regions of proboscis, collar, and trunk; stain sections with alum hematoxylin and eosin.

Tunicates

SMALL TUNICATES, SUCH AS PEROPHORA AND CLAVELINA. Anesthetize with magnesium chloride or magnesium sulfate; fix in 10% formalin; stain with precipitated borax carmine; counterstain lightly with fast green, if desired; infiltrate with mounting medium; support coverglass to keep it level.

LARVAE OF TUNICATES. Fix in Bouin's fluid; stain with dilute acidulated borax carmine, or precipitated borax carmine; counterstain lightly with fast green if desired; infiltrate slowly with mounting medium; support coverglass.

Cephalochordates

BRANCHIOSTOMA ("AMPHIOXUS"). For whole mounts of young specimens (preferably about 8 to 15 mm.), anesthetize with magnesium chloride or magnesium sulfate; fix in Bouin's fluid; stain with precipitated borax carmine; counterstain very lightly with fast green; infiltrate with mounting medium; support coverglass to keep it level.

For sections of mature animals, anesthetize and fix as directed for small specimens; cut into 6 or 8 mm. lengths; stain progressively with dilute alum hematoxylin; embed in paraffin; use *sliding microtome* to cut sections about 12 μ thick; mount representative sections from various regions (as pharynx, mid-body, and tail) on each slide; remove paraffin and counterstain with eosin before mounting.

Cyclostomes

Ammocoetes Larvae. Anesthetize with magnesium chloride or magnesium sulfate, and fix in Bouin's fluid. Stain small specimens (about 8 to 12 mm.) with precipitated borax carmine; counterstain lightly with fast green; infiltrate with mounting medium; support coverglasses to keep them level.

For sections, select larvae about 30 to 40 mm. long. Make preparations as recommended for *Branchiostoma*; however, a rotary microtome may be used for sectioning ammocoetes larvae.

Elasmobranch and Teleost Fishes

Placoid Scales. Clean these by treatment with 10% sodium hydroxide; wash thoroughly; dehydrate, clear, and mount in a resinous medium.

Developing Placoid Scales. Fix pieces of skin of very young individuals (as dogfish "pups") in Bouin's fluid; embed in paraffin; stain sections with alum hematoxylin and eosin.

Ganoid Scales. Soak pieces of skin in 10% sodium hydroxide until the scales can be separated and cleaned; wash them thoroughly; dehydrate, clear, and mount in a resinous medium, supporting the coverglass to keep it level.

Cycloid and Ctenoid Scales. Gently scrape off the scales and preserve them in alcohol; select small scales, and stain these lightly with fast green; dehydrate, clear, and mount in a resinous medium.

Tissues in General. See Histological Materials (below).

Amphibians, Reptiles, Birds, Mammals

Tissues in General. See Histological Materials (below).

Embryos. See pages 507–511, under Embryological Materials.

HISTOLOGICAL MATERIALS

Epithelial Tissues

Simple Squamous Epithelium. Try making silver nitrate preparations of mesentery (p. 404).

Simple Columnar Epithelium. Follow directions for stomach, small intestine, gallbladder.

Ciliated Columnar Epithelium. Follow directions for trachea.

Transitional Epithelium. Follow directions for ureter, bladder.

Stratified Squamous Epithelium. Follow directions for esophagus, skin, vagina, or tongue.

Dissociation Methods for Epithelial Tissues. See page 180.

Connective Tissues

Areolar Tissue. Stretch pieces of subcutaneous tissue over rings or necks of vials; fix in 10% formalin; stain with Verhoeff's elastic tissue stain and eosin; cut up into smaller pieces before mounting.

White Fibrous (Collagenous) Tissue. Make teased preparations (see p. 177).

Adipose Tissue. Fix tissue in 10% formalin and make sections by the freezing method; stain with Sudan IV; mount in glycerol jelly.

Cartilage

Hyaline Cartilage. Fix small pieces of tracheal or costal cartilage in Bouin's fluid; embed in paraffin and cut sections about 12 μ thick; stain with alum hematoxylin and eosin; or fix in Zenker's fluid and stain sections with Mallory's triple stain.

Elastic Cartilage. For typical structure, fix pieces of the epiglottis of a dog or other small mammal in Zenker's fluid; stain by Verhoeff's elastic tissue stain and eosin. For cartilage of external ear, obtain material from rabbit and prepare in the same way; the cells are very large.

White Fibro-Cartilage. Obtain pieces of intervertebral discs of some large mammal, such as a cow; fix in 10% formalin; prepare sections by freezing method and stain with Mallory's triple stain or alum hematoxylin and picro-fuchsin.

Bone

Compact Bone. For decalcified sections, fix 6 to 8 mm. lengths of a small bone (such as the ulna or radius of a cat) in 10% formalin or Zenker's fluid; decalcify it; cut longitudinal and transverse sections by the nitrocellulose method; stain with alum hematoxylin and eosin.

For ground sections, see page 320.

Cancellous Bone. For decalcified sections, follow the procedure recommended for compact bone. For ground sections, see page 320.

Developing Bone. Use long bones of the appendages of fetal or new-born individuals; fix the bone in Zenker's fluid and stain paraffin sections (10 to 12 μ thick) in Mallory's triple stain; or fix in Bouin's fluid and stain with alum hematoxylin and eosin.

Muscle

Smooth Muscle. For sections, cut transverse slices of small intestine; slit them open and pin them out; strip off the mucosa; pin out the remaining tissue, which is largely muscle, and fix it in Bouin's fluid; embed in paraffin and cut sections, 6 to 8 μ thick, parallel to fibers of the circular muscle layer; stain with alum hematoxylin and eosin.

For dissociation preparations, see page 181.

Striated Muscle. Pin out very thin slices of skeletal muscle and fix them in Bouin's fluid; embed in paraffin. For general histological purposes, cut longitudinal and transverse sections at 10 μ; stain with alum hematoxylin and eosin. For details of striations, make sections 4 or 5 μ thick (the fibers are likely to splinter to some extent) and stain with iron hematoxylin.

Cardiac Muscle. To demonstrate intercalated discs, fix thin slices in nitric acid-alcohol (p. 75); make paraffin sections 8 μ thick; stain with iron hematoxylin. Heart muscle from a dog or sheep is better than material from a cat or rat.

Blood and Blood-Forming Organs

Blood Films. Prepare smears as directed on page 170; stain by Giemsa's method, at pH of 6.75 (see p. 394); leave unmounted, or mount under No. 1 coverglass.

Bone Marrow. Stain smear preparations in the same way as blood films. For sections, fix small pieces in Zenker's fluid; dehydrate and infiltrate with paraffin very gradually; cut sections at 3 or 4 μ; stain with alum hematoxylin and eosin.

Lymphatic System

Lymph. Obtain lymph from the thoracic duct; prepare smears; stain by Giemsa's method.

Lymph Nodes. Cut into slices about 4 mm. thick and fix in Zenker's fluid; cut paraffin sections at 5 to 7μ; use minimum of heat to spread sections on slides; stain with alum hematoxylin and eosin; dehydrate, coat with 1% Parlodion, and harden in chloroform; clear in terpineol-toluene; transfer sections to mounting medium diluted with 9 parts of terpineol-toluene; allow to dry; clean mounting medium from back and outlying areas of slide; apply mounting medium and coverglass over sections. This special treatment prevents the sections from cracking apart as they are likely to do otherwise.

SPLEEN. For sections, use the method recommended for lymph nodes. Smear preparations may be stained by Giemsa's method.

THORACIC DUCT. Follow directions given for arteries and veins.

LYMPH PLEXUSES. An excellent demonstration may be made by stripping off the surface layer of the central tendon of the diaphragm, and impregnating this with silver nitrate (p. 403). However, the method often fails.

Circulatory System

HEART. For general morphology, fix the entire heart of a mouse in Bouin's fluid and make longitudinal sections by the paraffin method; stain with alum hematoxylin and eosin.

For details of structure of cardiac muscle, fix thin slices in nitric acid-alcohol (p. 75); cut paraffin sections at 8 μ; stain with iron hematoxylin. Cardiac muscle from a dog or sheep is more favorable than tissue from a cat or rat.

For demonstrating Purkinje fibers, cut out pieces of the atrioventricular bundle of His from a sheep heart. Prepare and stain sections as recommended for cardiac muscle.

ARTERIES and VEINS. Pin out veins or associated arteries and veins as directed on page 123; fix in Bouin's fluid or Zenker's fluid; embed in paraffin and cut transverse sections at about 10 μ; stain by Verhoeff's method for elastic tissue, and counterstain with eosin.

Skin

THIN SKIN; SKIN WITHOUT HAIR. Pin out and fix in Bouin's fluid; embed in paraffin using toluene or terpineol-toluene (never xylene) as an intermedium between alcohol and paraffin; cut sections 10 μ thick; stain with alum hematoxylin and eosin.

THICK SKIN. Pin out and fix in Bouin's fluid; embed in paraffin, using toluene or terpineol-toluene as an intermedium between alcohol and paraffin; cut sections 10 μ thick; for general histology, stain with alum hematoxylin, and counterstain with either eosin or picro-fuchsin.

SCALP OR OTHER SKIN WITH HAIR. Pin out and fix in Bouin's or Zenker's fluid; embed in nitrocellulose, carefully orienting the tissue so that perfect longitudinal sections of some hair follicles will be obtained; section at 15 μ; stain with alum hematoxylin and eosin.

Respiratory System

NASAL PASSAGES. For general morphology, fix the snout of a new-born cat or rabbit in Bouin's fluid or Zenker's fluid; embed in paraffin and cut transverse sections about 12 or 14 μ thick; stain with alum hematoxylin and eosin or with Mallory's triple stain (the latter gives especially good results after fixation in Zenker's fluid).

To demonstrate olfactory cells, strip off the olfactory membrane and fix this in Flemming's "weak" fixative; embed in paraffin and cut sections at about 5 μ; stain with iron hematoxylin. The gold chloride method for peripheral nerve endings (p. 425) should also be tried.

EPIGLOTTIS. See method for Elastic Cartilage.

TRACHEA. Rinse out with isotonic saline solution; cut into short pieces; fix in Bouin's fluid and stain sections with alum hematoxylin and eosin, or fix in Zenker's fluid and stain with Mallory's triple stain.

LUNG. Inflate collapsed lung with Bouin's fixative; cut out smaller pieces after the lung has hardened (do this while the lung is completely immersed); stain with alum hematoxylin and eosin.

Digestive System

TEETH. For general histology, cut out portions of jaw, each with a tooth; fix in 10% formalin; decalcify; prepare sections by nitrocellulose method; select median sections; stain with alum hematoxylin and eosin.

For details of hard parts, make transverse sections by grinding (p. 320).

For development, fix entire snout and jaw of a small fetal pig in Bouin's fluid; embed in paraffin and cut transverse sections of jaw at 10 or 12 μ; stain with alum hematoxylin and eosin. The preparations may also show developing hairs.

TONGUE. For general histology, cut out slices from various parts of tongue; fix in Bouin's fluid; embed in paraffin; stain with alum hematoxylin and eosin.

For demonstrating taste buds, fix foliate papillae in Bouin's fluid; cut thin (about 6 μ) paraffin sections; stain with iron hematoxylin.

For nerve endings, try methods of Golgi and Ramón y Cajal.

ESOPHAGUS. Place 2 cm. lengths in Bouin's fluid; after they have hardened somewhat, cut out 5 mm. lengths; embed in paraffin and make transverse sections; stain with alum hematoxylin and eosin.

STOMACH, SMALL INTESTINE, LARGE INTESTINE. Slit open these organs (unless they are less than 5 mm. in diameter); rinse with isotonic saline solution; cut out pieces about 1 by 2 cm. from various histological regions (cardiac, fundic, and pyloric portions of stomach; duodenum, ileum, and colon) and transitions (junctions of esophagus and stomach, and stomach and duodenum; ileocecal valve); pin out pieces and fix in Bouin's fluid or Zenker's fluid; trim tissue when hardened, being careful to cut in definite relation to axis of organ, so as to facilitate orientation later on; cut paraffin sections at about 10 μ; stain material with alum hematoxylin and eosin, or Mallory's triple stain (the latter is generally more effective after fixation of tissue in Zenker's fluid).

SALIVARY GLANDS. Fix in saturated solution of mercuric chloride in isotonic saline solution; cut paraffin sections at 10 μ; stain with alum hematoxylin and eosin, and with Mallory's triple stain.

PANCREAS. Fix thin slices in Zenker's fluid; cut paraffin sections at 8 or 10 μ; stain with alum hematoxylin and eosin.

LIVER. For general histology, fix thin slices of tissue in Bouin's fluid; cut paraffin sections 8 or 10 μ thick; stain with alum hematoxylin and eosin.

For details of bile canalicules, impregnate by Ramón y Cajal's method, and cut thin paraffin sections; or impregnate by Golgi's method and cut nitrocellulose sections.

To demonstrate glycogen, use Bauer's method (p. 446) or the periodic acid-Schiff technique (p. 447).

GALLBLADDER. Wash out with isotonic saline solution; inflate with Bouin's fixative and immerse until hardened; cut out small pieces and embed these in paraffin; cut sections about 8 μ thick; stain with alum hematoxylin and eosin.

Excretory System

KIDNEY. For general histology, fixation in Bouin's fluid or Allen's B 15 modification is recommended. In the case of the rat and large animals, it will be desirable to inject the fixative (warmed to body temperature) into the aorta proximal to the renal arteries (p. 129), and then to remove the kidneys and immerse them in the fixative. Subdivide larger kidneys, after they have been hardened somewhat by the fixative, in such a way that it will be easy to orient the pieces for sagittal and transverse sections. Embed the tissue in paraffin and cut sections at 10 μ; stain with alum hematoxylin and eosin.

BLADDER. Inflate the bladder by injection with Bouin's fluid, then immerse it in the fixative; cut it into smaller pieces and embed these in paraffin; section at 10 μ; stain with alum hematoxylin and eosin.

It will also be desirable to prepare sections of pieces of a contracted bladder, not inflated with fixative before immersion.

URETER. Fix pieces in Bouin's fluid, embed in paraffin, and section at 10 μ; stain with alum hematoxylin and eosin. The ureter of the horse is especially good for showing transitional epithelium.

Reproductive System

TESTIS AND OVARY. See suggestions under Cytological Materials.

SPERMATOZOA. Stain dry-fixed smears by Giemsa's method. Smears fixed in a saturated aqueous solution of mercuric chloride with 5% acetic acid should be stained with iron hematoxylin. Alum hematoxylin and eosin (or erythrosin) should also be tried with wet-fixed smears. In the case of spermatozoa of mammals, this combination often brings out the tail and middle piece in slightly different shades which contrast with the nuclear stain taken up by the head.

PROSTATE GLAND. Fix small pieces or slices of prostate gland in Orth's fluid; cut paraffin sections 8 μ thick; stain with alum hematoxylin and eosin.

UTERUS. Fix small pieces of the wall, or segments of the entire uterus of smaller animals, in Bouin's fixative; cut paraffin sections at 10 μ; stain with alum hematoxylin and eosin.

Central Nervous System

GENERAL MORPHOLOGY OF BRAIN AND SPINAL CORD. Harden for 1 or 2 hours in 10% formalin, aiding penetration by injection if the piece of tissue is large; cut transverse slices of spinal cord, brain stem, and various regions of cerebrum, and sagittal slices of cerebellum; continue fixation in 10% formalin; cut paraffin sections at about 10 μ; stain with alum hematoxylin and eosin.

MORPHOLOGY, DISTRIBUTION, AND RELATIONS OF NERVE CELLS, AXONES, AND DENDRITES. Cut out pieces as directed above, and use Golgi's mercuric chloride or chrome-silver method. Make thick sections by the nitrocellulose method.

FIBER TRACTS. Use the Weigert-Pal method (p. 371).

NEUROFIBRILS. Use Ramón y Cajal's methods (p. 410).

NEUROGLIA. Use Golgi's chrome-silver method (p. 408).

Peripheral Nerves

MYELINATED NERVES. Fix pieces in 10% formalin; impregnate with osmium tetroxide; embed some pieces in nitrocellulose, then embed the nitrocellulose blocks in paraffin (p. 269); section transversely, remove paraffin with toluene, and mount. Stain some pieces progressively with alum hematoxylin and tease apart before mounting, to show nuclei of the neurilemmas.

UNMYELINATED NERVES. Fix pieces of the vagus and cervical sympathetic nerves of a cat (these are together in one sheath) in 10% formalin; impregnate them with osmium tetroxide (p. 444); embed in nitrocellulose, then in paraffin (p. 269); cut transverse sections, and stain these with acid fuchsin (0.5% aqueous solution) to bring out the axis cylinders.

Special Sense Organs

EYE. Open the eye and fix it in Perenyi's fluid, Zenker's fluid, or Helly's fluid. For general histology, embed the eye in nitrocellulose; make complete horizontal sections; stain with alum hematoxylin and eosin or Heidenhain's "Azan" modification of Mallory's triple stain (this method is likely to give better results if Zenker's fluid or Helly's fluid has been used for fixation).

For detailed study of the retina, remove this from an eye fixed in Perenyi's fluid; embed it in nitrocellulose, then in paraffin; section at about 4 or 5 μ; stain with iron hematoxylin. Also try methods of Golgi and Ramón y Cajal.

EAR. In a guinea pig, study the orientation of various parts of the ear in the petrous portion of the temporal bone. Remove this complex; cautiously shave away the bone to open the membranous labyrinth in 2 or 3 places; fix in Bouin's fluid; wash and harden in 70% alcohol; decalcify; embed in nitrocellulose, orienting carefully so that true axial sections of the cochlea may be made; cut sections at about 10 μ if possible; stain with alum hematoxylin and eosin. The sections should show structure of both the cochlea and semicircular canals.

OLFACTORY MUCOUS MEMBRANE. Dissociate in Ranvier's one-third alcohol; make temporary mounts. Also make sections as suggested under Respiratory System.

Nerve Endings

SENSORY AND MOTOR NERVE ENDINGS IN GENERAL. Try the Bodian technique (p. 418). Two other methods are recommended, but neither is completely predictable. See supravital methylene blue technique (p. 53) and gold chloride method (p. 425).

PACINIAN CORPUSCLES. These may be identified in stained sections of pancreas, skin, and loose connective tissue. It is possible also to make whole mount preparations by cutting out a piece of cat mesentery which contains one or more Pacinian corpuscles, fixing this in Zenker's fluid, and staining it with 0.5% acid fuchsin.

Endocrine Glands

THYMUS AND THYROID GLANDS. Fix thin slices in Bouin's fluid; embed in paraffin and section at 7 or 8 μ; stain with alum hematoxylin and eosin.

PITUITARY BODY. Remove the entire complex carefully and fix it in Bouin's fluid; embed in paraffin and cut sagittal sections at about 7 or 8 μ; stain with alum hematoxylin and eosin.

ADRENAL GLAND. Fix thin slices in Helly's fixative; cut paraffin sections at about 10 μ; stain with alum hematoxylin and eosin. For demonstration of chromaffin granules, cut sections at about 20 μ; leave unstained, or tint lightly with safranin.

PINEAL BODY. Remove carefully and fix in Zenker's fluid. (If the gland is large, cut it into thin slices before fixation.) Embed in paraffin; section at 4 or 5 μ; stain with alum hematoxylin and eosin.

CYTOLOGICAL MATERIALS

Mitosis

TESTIS OF CRAYFISH. Crayfishes of the genus *Pacifastacus*, native to the western United States, provide excellent material. The eastern crayfishes (*Cambarus* and other genera), some of which have been introduced in the west, seem not to be as good. The chromosomes of crayfishes are small and numerous, but the spindles of the mitotic figures are very clear. In coastal regions of California and southern Oregon, late July or August is the most favorable time to make collections; farther north, or at higher elevations, September is likely to be better. Fix the testis in Bouin's fluid for 6 to 24 hours, and embed the tissue in paraffin after very gradual dehydration and infiltration. Cut longitudinal sections at 8 μ, and use a minimum of heat in spreading the sections on slides. Stain the preparations with iron hematoxylin, allowing the spindles to retain some of the dye. Coat the sections with Parlodion before clearing them.

ROOT-TIPS OF PLANTS. Root-tips of certain species of monocotyledonous plants provide very useful cytological material, and preparations of them are routinely made in courses in animal microtechnique. The onion, *Allium cepa*, is most commonly used, but representatives of some other genera show larger and clearer mitotic figures. *Tradescantia virginica*, the spiderwort, is one of the best species for the purpose, but is less convenient to handle because it must be grown in soil, and the root-tips must therefore be carefully cleaned before fixation. *Hyacinthus* and *Narcissus* (especially *N. tazetta*, forms of which include the so-called Chinese lily and paperwhite narcissus) are also excellent.

To obtain root-tips of onions and other bulbous plants of the same general type, place the bulbs on the mouths of jars or bottles with their lower ends just touching water. The roots will sprout in a few days and will begin to grow down into the water. To obtain material showing an abundance of mitotic figures, fix the root-tips at about noon or 1 P.M., or at about 11 P.M. or midnight. Simply remove the bulbs, and with sharp scissors cut off the terminal 1 cm. of the roots. This will include the meristematic region. Allow the tips to fall into Karpechenko's fixative. Leave them in the fixative for 6 or 8 hours, or overnight, then wash them in running water for about 6 hours.

Embed the root-tips in paraffin, carrying out dehydration with alcohol very gradually. Replacement of absolute alcohol with toluene and infiltration with paraffin should also be as gradual as possible.

Cut longitudinal sections (and transverse sections, if desired) at 8 μ. Only the more nearly median longitudinal sections are likely to provide views of a considerable portion of the meristematic region. Use as little heat as possible in affixing the sections to slides. Stain with iron hematoxylin. Destaining with picric acid often brings out the spindles better than if ferric ammonium sulfate is used.

Epidermis of Amphibian Larvae. Tangential sections of the epidermis of the tail region of salamander larvae often show abundant mitotic figures. Fix the tadpoles in Bouin's fluid, cut off the caudal portions, and embed these in paraffin. Cut sections 8 to 10 μ thick and stain them with iron hematoxylin.

It is also possible to show mitotic stages (although most of these will be seen in polar views) in flat mounts of the epidermis of the caudal region. After fixation of the larvae in Bouin's fluid, strip off pieces of the epidermis from the tail, wash them in 70% alcohol to remove the picric acid, and pass them down the graded series of alcohols to water. Stain them with iron hematoxylin or dilute alum hematoxylin (used as a progressive stain), then dehydrate, clear, and mount them.

Salivary Gland Chromosomes

The giant chromosomes of the salivary glands of *Drosophila* larvae in the fourth (last) instar have been very thoroughly investigated by geneticists and cytologists. These chromosomes are in a state of synapsis nearly comparable to that involving homologous chromosomes before the first meiotic divisions in oögenesis and spermatogenesis. By correlating the results of genetic breeding experiments with deletions, inversions, and other anomalies observed in the chromosomes of the salivary glands, and by studying the percentages of crossing-over taking place between genes in the same linkage group, it has been possible to establish the approximate position of hundreds of alleles on the four pairs of chromosomes of *Drosophila*.

Drosophila melanogaster and certain other species are routinely cultured in laboratories of genetics. Best results are obtained with plump and well-fed larvae which are about to pupate. Male larvae are distinguishable due to the fact that their gonads can be seen through the body wall as large, clear spaces posterior to the middle of the body. The smaller gonads of females are not externally visible.

Removal of Salivary Glands. Place the larva in a small dish containing enough Ringer's solution to cover it. Set the dish upon a black background under a binocular dissecting microscope. Hold the larva in place with a blunt needle manipulated by the left hand. With a very sharp needle held in the right hand, cut the dorsal body wall for most of its length, beginning at a point just posterior to the jaws. Locate the two salivary glands, one of which lies at each side of the pharynx, just back of the jaws and under a branch of the large fat body. Each gland is a transparent, cylindrical body tapering anteriorly to its duct. The ducts from the two salivary glands join to form a short common duct leading to the pharynx. Remove the glands and free them from adhering fat.

TEMPORARY ACETO-CARMINE AND ACETO-ORCEIN PREPARATIONS. With a pipette, transfer the glands to a clean slide and remove excess fluid from them. Place a drop or two of iron aceto-carmine (p. 341) or aceto-orcein (p. 342) upon them, and then a No. 1 coverglass. Crush the glands by pressing gently upon the coverglass with a needle. Pry up the coverglass a little and then let it down again, so that the staining solution bathes the tissue thoroughly. When inspection under the oil-immersion objective or high-power dry objective shows that the chromosomes are sharply stained, draw off most of the liquid by touching a piece of filter paper or bibulous paper to the margins of the coverglass. This will cause the coverglass to be drawn closer to the slide, and will help to spread out the chromosomes. The preparation should be sealed with melted vaspar (a mixture of equal parts of paraffin and petroleum jelly).

PERMANENT PREPARATIONS STAINED BY THE FEULGEN NUCLEAL REACTION. Remove the glands in the same way as directed above. Fix them for an hour in a saturated aqueous solution of mercuric chloride with 5% acetic acid. During fixation and subsequent steps in the procedure, it will be convenient to place the glands in a little basket made by tying nylon bolting cloth over one end of a glass tube about 1 cm. long and 1 cm. in diameter; pieces of string tied to the other end of the tube will serve as handles.

After fixation, transfer the salivary glands to 50% alcohol for about 15 minutes, then put them, for an equal period of time, into 70% alcohol containing enough tincture of iodine to give it a light straw color. Leave them in clean 70% alcohol until any discoloration due to iodine has been washed out. Following this, transfer the glands through 50, 30, and 15% alcohol (about 10 minutes in each grade) to distilled water. Stain them by the Feulgen method (steps 2 to 4, p. 433). Dehydrate the specimens gradually and clear them in terpineol or beechwood creosote.

Place one or two glands in a small drop of the clearing agent in the center of a slide. Prepare a small circular coverglass by placing upon it a small drop of moderately thick mounting medium. With a piece of filter paper, absorb most of the clearing agent surrounding the tissue. Invert the coverglass and place it upon the glands, applying gentle pressure to facilitate spread of the mounting medium. Immediately place the slide under a binocular dissecting microscope. Hold a pair of curved forceps with the convex sides of their tips upon the coverglass and spanning the material. Press the forceps upon the coverglass to flatten the glands, moving the forceps about to insure that the tissue will be compressed nearly uniformly. Then let the preparation stand until the mounting medium has hardened sufficiently to permit examination with the oil-immersion objective.

In good preparations, the chromatin is stained deep magenta, and the cytoplasm is colorless.

Spermatogenesis

Certain types of organisms have been routinely used for study of chromosomal changes involved in the development of male germ cells. Methods for making preparations of some of these will be explained, and applications of the same methods to other organisms will suggest themselves.

GRASSHOPPERS (ORTHOPTERA). If possible, obtain testes from a large species. The best season for collecting is generally late spring or early summer, but varies with species and locality.

Fix the testes in Allen's PFA 3 modification of Bouin's fluid for 12 to 24 hours. Dehydrate the tissue very gradually, and embed it in paraffin. Cut sagittal sections, about 8 μ thick. During affixation of sections to slides, use as little heat as possible to spread them. Stain with iron hematoxylin and counterstain lightly with orange G. Coat the sections with 1% Parlodion and clear and mount them.

ANASA TRISTIS (HEMIPTERA). This bug is widely used to show the behavior of the unpaired X chromosome during meiosis. Material is generally collected in August. Make preparations as suggested for testes of grasshoppers.

AMPHIBIA. Urodeles are generally preferable to anurans, because of their very large cells and linear arrangement of testicular follicles. The best time for taking material will have to be established by trial, unless information concerning a particular species is recorded. Species of *Ambystoma* and *Necturus* in the eastern United States are in suitable condition during late July and early August, and the small *Batrachoseps attenuatus* found in central California will show all stages in spermatogenesis if collected in late May or June.

Bouin's fluid is an excellent fixing agent; leave the testes in it for 12 to 24 hours. Embed in paraffin after very gradual dehydration. Make longitudinal sections of the testes, and use as little heat as possible in affixing the sections to slides. Stain with iron hematoxylin, and counterstain if desired with orange G. Coat sections with Parlodion before clearing and mounting them.

BIRDS AND MAMMALS. It is best to obtain material from young mature specimens, rather than from older individuals. The chicken is a commonly used avian species, and the rat and mouse are favorite mammals for this purpose.

Immediately after killing the animal, insert a cannula into the posterior part of the aorta or branch artery supplying the testis; wash out the blood with a warm isotonic saline solution, and then inject the fixative warmed to body temperature.

Bouin's fluid and Allen's B 15 modification are excellent fixatives for testes of birds and mammals. If the testes can be cut with a razor blade into slices about 2 or 3 mm. in thickness as soon as they are removed from the animal, this should be done, and the slices placed in an ample amount of the fixative. If they are too soft to cut immediately, allow the entire testes to harden for a time in the fixative and then slice them.

If Bouin's fluid or Allen's modification is used, fixation at about 37° C. is recommended; at this temperature, 3 hours will be long enough for thin slices. At room temperature, at least 12 hours should be allowed. After fixation, place them for about the same length of time (at room temperature) in equal parts of the fixative and 70% alcohol; leave them in 70% alcohol until most of the picric acid is washed out of them.

Embed the slices of testis in paraffin, after gradual dehydration. Cut sections 5 to 8 μ in thickness, affix them to slides with a minimum of heat, and stain them with iron hematoxylin. It is a good idea to coat the sections with Parlodion before clearing and mounting them.

If it is not practicable to perfuse the testis with fixative before removal from the animal, it is important that it be cut into small pieces soon after it shows signs of being hardened by the fixative.

Flemming's stronger fixative is also excellent for cytological work with the testis. If possible, cut it into slices immediately after removal. If the tubules are not cohesive enough for slicing, place the entire organ into the fixative and cut it as soon as it has hardened a little. Fixation in Flemming's fluid for 18 to 24 hours should be followed by washing in running water for about 12 hours. Iron hematoxylin and Flemming's triple stain bring out mitotic figures and chromosomes very sharply after this mixture.

The early development of the testis is well illustrated by material taken from chick embryos. Embryos of about 5 days after the start of incubation show indifferent gonadal tissue. At the hatching stage (21 days) the testis consists of hollow cords partially lined by primordial germ cells. Fixation in Bouin's fluid followed by staining of sections with iron hematoxylin gives good results.

SMEARS OF TESTIS TISSUE. To supplement the study of sections, smears are of great value in counting chromosomes, because the entire chromosome complex may often be seen with its members spread apart. Methods for making temporary and permanent smear preparations are outlined on page 176.

Oögenesis

PARASCARIS EQUORUM (SYNONYM: ASCARIS MEGALOCEPHALA). This common nematode parasite of the horse provides classical material for the study of oögenesis, fertilization, and early development. One cytological variety ("univalens") possesses only two chromosomes; it is prevalent in horses in the western United States. The other variety ("bivalens") has four chromosomes; it is the more common of the two in horses of the eastern United States. The latter is the favorite type. Its eggs are somewhat larger, and most published accounts deal with it.

To obtain material suitable for cytological studies, it is best to go to a slaughterhouse with thermos bottles filled with an isotonic solution warmed to about 37° C. The worms are obtained by stripping the small intestine of a freshly killed horse between the fingers. Place them in the warm saline solution and transport them back to the laboratory as quickly as possible. Pin out the adult females in a dissecting pan. Slit them open and remove the two ovaries and the uterus which is continuous with them.

The ovary proper consists of a central rachis surrounded by primary oöcytes. These are more or less cylindrical in the distal portion, but become more nearly spherical as the ovary approaches the uterus. The cytological stages in the uterus, passing from its distal to its proximal portion, are: primary oöcytes before entrance of sperm, primary oöcytes showing the entrance of the sperm, then the two meiotic divisions, and finally the pronuclear stages. To obtain the stages of cleavage, it will be necessary to place the uteri in dishes of saline solution and to incubate them at 37° C. for 8 to 12 hours. The more advanced stages will then be found at the proximal ends of the uteri.

The pronuclear and cleavage stages in the proximal part of the uterus are enclosed by rather thick shells, and Carnoy and Lebrun's fluid is one of the few mixtures which will penetrate these rapidly enough to insure good fixation. This fixative may be used for the entire uterus, but Duboscq and Brasil's fluid is superior for cytological purposes. It is therefore advisable to fix the distal portion, containing earlier stages, in Duboscq and Brasil's mixture, and to use Carnoy and Lebrun's fluid for the proximal portion only. By examining a few eggs teased out of the uterus at intervals, it may be possible to establish the point at which it should be divided.

Rinse the pieces of uterus before dropping them into the fixative. Material fixed in Carnoy and Lebrun's fluid should remain in this for about 2 hours; fixation in Duboscq and Brasil's fluid should be continued for at least 4 hours, but may be prolonged for 24 hours. After appropriate post-fixation washing (which, for material fixed in Carnoy and Lebrun's fixative, will include treatment with iodine-alcohol), subdivide the pieces of uterus into lengths of about 2 cm. These should be placed in separate vials and arranged in the order in which they were taken from the intact uterus.

Embed the pieces in paraffin, following gradual dehydration, de-alcoholization, and infiltration. Section them longitudinally, at 8 to 10 μ. If desired, sections from several regions may be mounted, one below the other, on the same slide, so that all of the stages found in one uterus can be seen on a single preparation.

For most purposes, staining with iron hematoxylin gives superior results. Mordant the sections in ferric ammonium sulfate for 2 to 3 hours, then stain them for 2 to 3 hours. Destaining by the picric acid method is recommended.

If sections from several regions of the uterus are mounted on the same slide, mordanting with 5% ferric chloride and destaining in picric acid yield preparations in which the staining of different stages is rather uniform. The phosphotungstic acid hematoxylin method is also good when several stages are mounted on the same slide.

It is advisable to coat sections with 1% Parlodion after dehydrating them, and to mount them under coverglasses of No. 1 thickness.

BIRDS. Fix portions of ovaries of young mature animals in Heidenhain's "Susa" fluid. After washing and removing salts of mercury in iodine-alcohol, dehydrate the tissue gradually. Replacement of alcohol with toluene or terpineol-toluene, and infiltration with paraffin, should also be as gradual as possible. Cut sections 10 to 12 μ thick and stain these with alum hematoxylin and eosin, or with Heidenhain's "Azan" modification of Mallory's triple stain.

The early development of the avian ovary is demonstrated very well by the chick. Sections of embryos 5 days after the start of incubation show the early indifferent stage. At hatching (21 days), the ovigerous strands, with oögonia and young follicles, project into the stroma. Bouin's fluid is a good fixative, and alum hematoxylin and eosin or iron hematoxylin are suitable stains.

MAMMALS. The ovaries of young mature female cats, rabbits, rats, and mice furnish excellent material for the study of the growth of oögonia into primary oöcytes. Ovaries of these small animals may be fixed entire; those of larger species should be cut into slices before fixation. Bouin's fluid is a good fixative, and should be allowed to act for 6 to 24 hours, depending upon the size of the pieces of tissue.

After fixation and appropriate washing, dehydrate the material gradually, and also carry out the processes of replacement of alcohol and infiltration with paraffin slowly. Cut sections about 10 μ thick. It is a good idea to affix at least two rows of sections to each slide; this will insure that median sections of Graafian follicles will be in each preparation. Alum hematoxylin and eosin, iron hematoxylin, and Heidenhain's "Azan" modification of Mallory's triple stain give good results. The latter brings out the general organization of ovarian tissue especially well.

Sections of the ovary of a new-born kitten, prepared by the techniques described above, are very instructive. They show Pfluger's egg tubes, containing very young oögonia, and primordial ovarian follicles deeper in the stroma. From about 2 days before birth to about 3 days after birth, the ovary of a kitten will show stages in the prophase of the first reduction division. The nuclei then go into a germinal vesicle stage and do not complete the first reduction division until just before the prospective ovum escapes from the Graafian follicle. The first day after the birth of the kitten is said to be the best time for fixing ovaries to show these prophase stages.

EMBRYOLOGICAL MATERIALS

Annelids

CLEAVAGE STAGES OF POLYCHAETES. Fix in Kleinenberg's picro-sulfuric acid mixture, or in a saturated aqueous solution of mercuric chloride with 5% acetic acid; stain regressively with alum hematoxylin; clear, infiltrate with resinous medium, and mount; support coverglass.

TROCHOPHORE LARVAE AND STAGES OF METAMORPHOSIS. Fix in Bouin's fluid or Kleinenberg's fluid; stain by the precipitated borax carmine method or in alum cochineal; counterstain lightly with fast green or indigo-carmine; clear, infiltrate with resinous medium, and mount; support coverglass.

Arthropods and Crustacea

NAUPLIUS LARVAE. Fix in 95% alcohol; stain with alum cochineal; counterstain lightly with fast green; support coverglass.

ZOAEA AND MEGALOPS LARVAE. Prepare as directed for small crustacea (p. 487).

Molluscs

CLEAVAGE STAGES. Remove eggs from capsules; fix in Kleinenberg's fluid; stain regressively with alum hematoxylin; infiltrate slowly with resinous medium; support coverglass.

TROCHOPHORE AND VELIGER LARVAE. Prepare as directed for echinoderm larvae.

GLOCHIDIA OF FRESH WATER CLAMS. Prepare as directed for nauplius larvae of crustacea.

Echinoderms

CLEAVAGE STAGES OF SEA STARS, SEA URCHINS, AND SAND DOLLARS. Fix in Bouin's fluid; stain by the dilute acidulated borax carmine method; clear, and infiltrate very slowly with resinous medium, in order to prevent wrinkling of the fertilization membrane; support coverglass. It is a good idea to stain early (up to 16-cell) stages and later stages separately, as they take up the stain at different rates; they may be mixed before mounting.

BLASTULAS AND GASTRULAS. Fix in Bouin's fluid; stain by the acidulated borax-carmine method; counterstain lightly with fast green or indigo-carmine if desired; clear and infiltrate very slowly with resinous medium; support coverglass.

BIPINNARIA, BRACHIOLARIA, AND AURICULARIA LARVAE. Prepare as directed for blastulas and gastrulas.

PLUTEUS LARVAE. Fix in neutral 10% formalin; stain with safranin or alcoholic cochineal; infiltrate very slowly with resinous medium; support coverglass. Avoid use of acid fluids at any stage in preparation; these will dissolve the spicules.

Tunicates

EARLY STAGES. Prepare as directed for cleavage stages of echinoderm larvae.

LARVAE ("TADPOLES"). See Tunicates under Zoological Materials.

Cephalochordates

EARLY STAGES OF BRANCHIOSTOMA ("AMPHIOXUS"). Prepare as directed for cleavage stages and larvae of echinoderms. Also make paraffin sections 10 μ thick and stain them with iron hematoxylin.

Teleost Fishes

Very favorable material is provided by species in which the underside of the blastodisc is not deeply concave; this makes it possible to free the blastodisc easily from practically all traces of yolk. *Gadus* (cod), *Fundulus* (killifish), and *Coregonus* (whitefish) are among the fishes providing excellent material. The eggs of these, and other species in which the egg is not provided with a heavy capsule, may be fixed by simply placing them in Bouin's fluid or some other fixative having considerable penetrating power. The heavy capsules of the eggs of Salmonidae should be opened with fine needles before they are placed in the fixing fluid, and subsequently removed completely. In all cases, eggs of teleost fishes should be hardened in alcohol after fixation, the chorion removed, and the blastodisc separated from the yolk mass. The blastodisc may then be sectioned or prepared as a whole mount.

Amphibians

The coats of albuminous jelly which surround the eggs of amphibians are a considerable obstacle to fixation. Frog eggs in appropriate stages of development should be cut out of the mass in which they are laid, freed of as much jelly as possible, and placed in a large volume of Smith's fixative. After 24 hours, shake them gently; the remainder of the jelly will probably fall off.

The eggs of some salamanders are separated from the albuminous matrix by a rather large fluid-filled space, and can be liberated by shaving off layers of the firm jelly with a razor blade. When the cavities containing several eggs have been opened, invert the mass in water and the eggs will drop out. With a large pipette, transfer them to Smith's fixative for 24 hours. After fixing amphibian eggs in Smith's fluid, wash them and store them in 10% formalin until they are needed.

Small aquatic larvae should simply be dropped into Bouin's fixative or Zenker's fixative. The latter is especially good if Mallory's triple stain is to be used to bring out hyaline cartilage boldly. In the case of larger tadpoles, the body cavity should be opened before they are placed in the fixative.

Birds

CHICK EMBRYOS. The chick is routinely used in studies of vertebrate development, and the preparation of chick embryos is therefore of special interest to teachers, students, and technicians. The following method may be applied to embryos of any bird, but those which are very small will, of course, be difficult to handle.

The age of the embryo is measured by the number of hours it has developed, at the normal temperature of incubation (38° C.) after the end of the latent period. The duration of the latent period varies according to the length of time which elapses between laying and the beginning of incubation. If the eggs have stood for 3 or 4 days, which is usually the case, one should allow 10 or 11 hours for them to warm up and begin to develop. Accordingly, the eggs should remain in the incubator for 34 or 35 hours to reach the 24-hour stage, and 58 or 59 hours to reach the 48-hour stage.

The following materials should be ready for use: a shallow dish (such as a culture dish) filled to a depth of about 5 cm. with an isotonic saline solution warmed to about 38° C. Syracuse dishes; a medicine dropper with curved tip; heavy scissors; fine scissors; fine forceps with curved tips.

With the sharp point of the heavy scissors, punch a tiny hole in the shell of the egg, about half-way between its large end and its middle; take care that the point penetrates only about 2 mm. Place the egg in the dish of saline solution, and turn it so that the little hole is directed toward you and at the same time somewhat downward. Insert the point of the scissors into the hole, holding the scissors as nearly parallel to the shell as possible, and cautiously cut through the shell. Cut toward the bottom of the dish, and keep the egg turning toward the scissors. In this way, cut about three-quarters of the way around the shell, then break it apart and remove it from the dish. When the yolk is liberated, the side having the blastoderm will come to lie upward. With very sharp and fine scissors, carefully cut around the edge of the blastoderm, just outside of the vascular area. Then fill the medicine dropper with some of the saline solution and insert its curved tip under the blastoderm. Gently waft it away from the yolk by forcing out a gentle stream of saline solution. Immerse a clean Syracuse dish in the liquid, and with the aid of forceps float the blastoderm into it. Lift the Syracuse dish from the liquid and carefully pour most of the liquid from it. Make sure that the blastoderm is right side up, and then use fine forceps to remove the vitelline membrane which covers it. Withdraw the rest of the saline solution, being careful to keep the blastoderm well spread and to avoid sucking any part of it into the pipette.

If the embryo is to be mounted entire, almost all of the liquid may be removed by running the pipette around the blastoderm, so that the blastoderm will be spread smoothly on the bottom of the dish. If the embryo is to be sectioned, however, a very thin layer of saline solution should remain on the bottom of the dish, and the blastoderm should not be stretched. The tension which results when the blastoderm is flattened will cause minute tears in the delicate tissues of the embryo, chiefly in the lateral folds. These are not likely to show in whole mounts, but render material unfit for sectioning.

Hold a pipette containing Kleinenberg's picric acid-sulfuric acid fixative very close to the embryo, and allow a very small amount of this to flow over and moisten the embryo, but do not immerse it. After a minute or two, add a little more. Then allow the well-moistened embryo to stand about 4 or 5 minutes; at the end of this time it will be somewhat hardened. Next add enough of the fixative to immerse the embryo completely. If the process has been carried out properly, the blastoderm will remain quite flat and it will not adhere to the dish. When the embryo is well hardened (about 3 to 8 hours, depending on its size), transfer it to 70% alcohol, and change this as often as necessary to remove the picric acid.

For whole mounts, stain the embryos with alum cochineal, borax carmine, or dilute alum hematoxylin (progressive method). Infiltrate them slowly with neutral balsam or a synthetic medium, and mount them in cells of metal, glass, or cellulose acetate, or support the coverglass with glass rods or strips.

For sections, stain the embryos progressively with dilute alum hematoxylin before embedding them in paraffin. Make transverse and sagittal serial sections about 12 or 15 μ thick. After the sections are dry and the paraffin has been removed, transfer them to two changes of absolute alcohol, and coat them with Parlodion. This coating will help keep the sections intact if the coverglass is accidentally contacted by one of the objectives of short working distance. Then clear the preparations in toluene and mount them in a resinous medium.

If desired, the sections may be counterstained lightly with eosin in 90 or 95% alcohol after removal of paraffin and replacement of toluene with absolute alcohol. Following counterstaining, rinse the sections in clean 95% alcohol; dehydrate them in absolute alcohol, coat them with Parlodion, and clear and mount them.

Mammals

Pig, rat, and rabbit embryos are easy to obtain, and provide excellent material for teaching and research purposes. All traces of blood should be washed from the embryos with an isotonic saline solution warmed to body temperature. For routine embryological work, Bouin's fluid and Zenker's fluid are good fixatives.

BLASTODERMIC VESICLES. These should be opened with fine needles before being placed in the fixative. Prepare serial transverse sections by the paraffin method, stain them with alum hematoxylin and eosin, or with iron hematoxylin.

EMBRYOS OF 3 TO 12 MM. Carefully remove the embryos from the uterus, open the amnion with fine scissors, and place the embryos in the fixative. The length of time which they should remain in the fixative will vary according to their size; 18 to 24 hours will be sufficient for embryos about 10 mm. long. Trim away the amnion before proceeding with staining and embedding.

For general structure, stain entire embryos progressively with dilute alum hematoxylin. Embed them in paraffin and cut serial sections (transverse, sagittal, and frontal) at about 15 μ. Remove the paraffin in toluene, transfer the sections through two changes of absolute alcohol, and counterstain them with eosin in 90% alcohol. Then dehydrate the sections, coat them with Parlodion, and clear and mount them.

Alum cochineal is also good for staining entire embryos before embedding. After sectioning and affixing sections to slides, bring the sections gradually to water, and counterstain them with Mallory's aniline blue-orange G mixture.

For cellular details, it is advisable to cut sections at about 8 or 10 μ, and to stain them with iron hematoxylin.

EMBRYOS LARGER THAN 12 MM. Remove the amnion before placing the embryos into the fixative. After they have hardened for an hour or two, cut them with a very sharp razor blade into two or more pieces. This must be done with foresight. If representative sections are to be taken from various portions of the embryo, or if nearly continuous serial sections are to be prepared of all or part of the embryo, the cuts should be made through regions which are not likely to be important.

Larger embryos should be sectioned before staining. The nitrocellulose method is recommended for specimens which are over 25 mm. long. For general structure, sections 15 or 20 μ thick should prove suitable. Alum hematoxylin and eosin and Mallory's triple stain are good.

If cellular details in certain organs of very large embryos are to be studied, it will probably be best to dissect out these organs and fix them separately. Cut sections at 8 or 10 μ, and stain them with iron hematoxylin, or by other suitable histological methods.

Table of Weights and Measures

WEIGHTS

Conversion of Metric Units to Avoirdupois Units

1 gram (gm.; g.) = weight of 1 milliliter (ml.) of water at maximum density = 15.432 grains = 0.0352 ounce
1 kilogram (kg.) = 1,000 grams = 35.2 ounces = approximately 2.2 pounds

Conversion of Avoirdupois Units to Metric Units

1 grain (gr.) = 0.065 gram
1 dram = 27.34 grains = 1.77 grams
1 ounce = 16 drams = 437.5 grains = 28.35 grams
1 pound (lb.) = 16 ounces = 7,000 grains = 453.6 grams

LINEAR MEASURES

Conversion of Metric Units to English Units

1 micron (mu or μ) = 0.001 millimeter = approximately $\frac{1}{25{,}000}$ inch
1 millimeter (mm.) = 0.001 meter = approximately $\frac{1}{25}$ inch
1 centimeter (cm.) = 10 millimeters = 0.01 meter = approximately $\frac{2}{5}$ inch
1 meter (m.) = 100 centimeters = 39.37 inches = 3.28 feet = approximately 1.1 yards

Conversion of English Units to Metric Units

1 inch (in.) = 25.4 millimeters = 2.54 centimeters
1 foot (ft.) = 30.48 centimeters = 0.3048 meter
1 yard (yd.) = 0.914 meter

LIQUID MEASURES

Conversion of Metric Units to United States (U.S.) Units

1 milliliter (ml.) or cubic centimeter (cc.) = 0.001 liter = 0.034 fluid ounce
1 liter (l.) = 1,000 milliliters = approximately 34 fluid ounces = 2.11 pints = 1.057 quarts

Conversion of United States Units to Metric Units

1 fluid ounce = 29.57 milliliters
1 pint = 16 fluid ounces = 473.2 milliliters
1 quart = 32 fluid ounces = 946.4 milliliters = approximately 0.95 liter
1 gallon (gal.) = 128 fluid ounces = approximately 3.79 liters

CENTIGRADE AND FAHRENHEIT THERMOMETER SCALES

To convert from degrees Fahrenheit (°F.) to degrees Centigrade (°C.), subtract 32 from the Fahrenheit reading and then multiply the remainder by $\frac{5}{9}$. Example: 104° F. = $(104 - 32) \times \frac{5}{9}$, or 40° C.

To convert from degrees Centigrade to degrees Fahrenheit, multiply the Centigrade reading by $\frac{9}{5}$ and add 32 to the product. Example: 35° C. = $(35 \times \frac{9}{5}) + 32$, or 95° F.

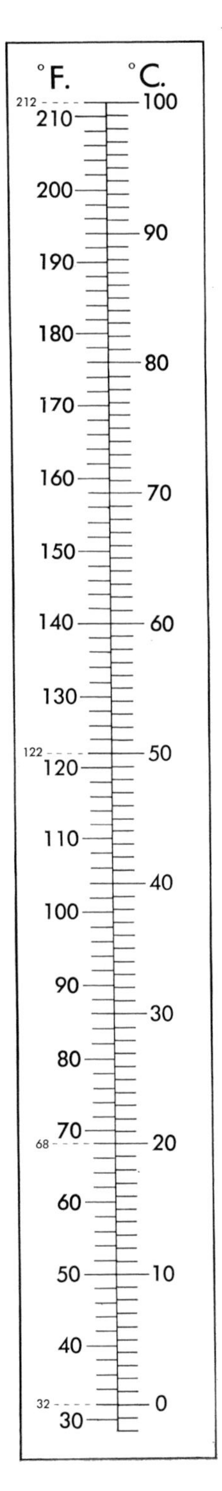
°F.
°C.
212
210
200
190
180
170
160
150
140
130
122
120
110
100
90
80
70
68
60
50
40
32
30
100
90
80
70
60
50
40
30
20
10
0

Index

G

H

I

Q

R

S

T

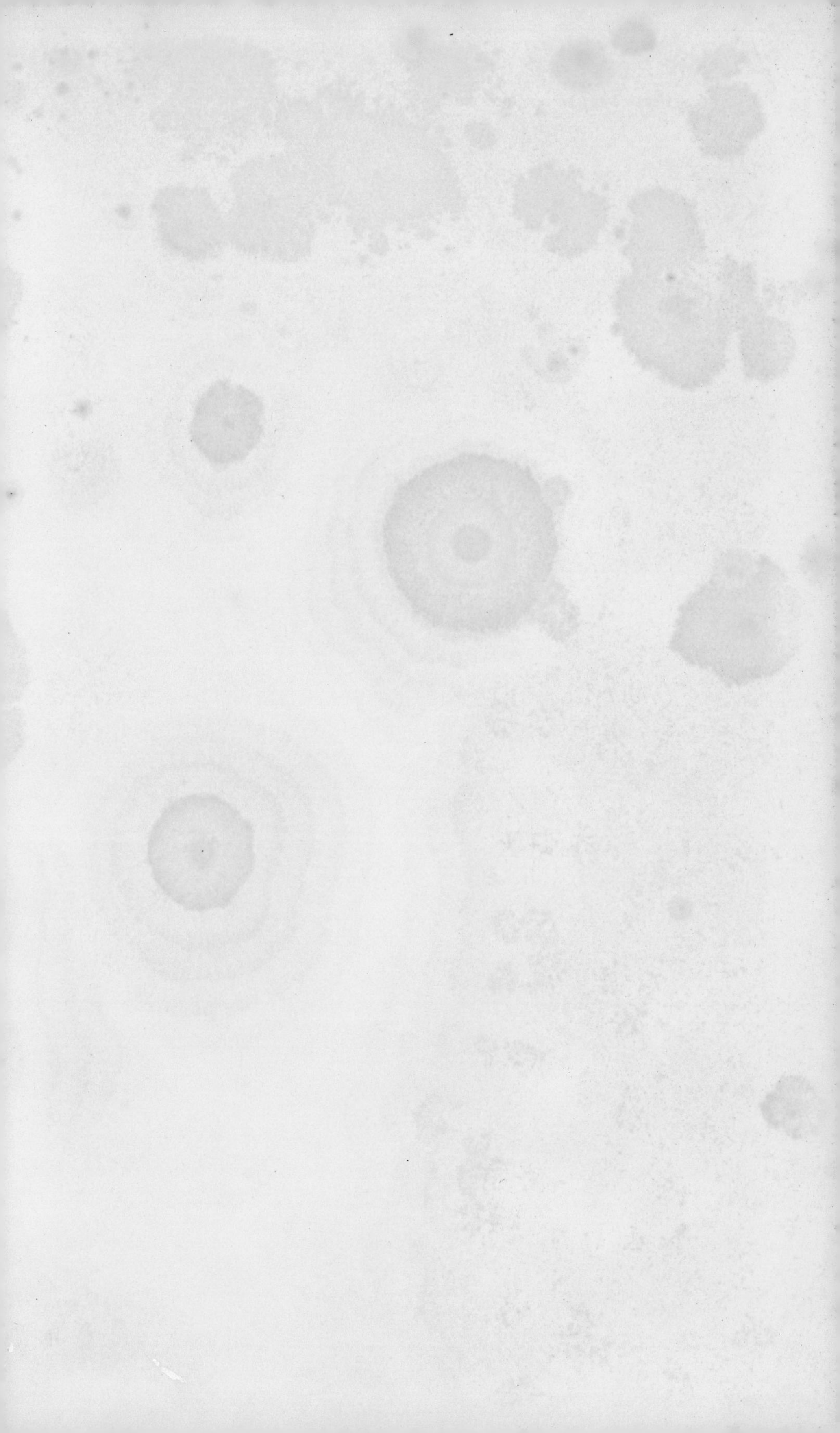